COMMON BEAN IMPROVEMENT IN THE TWENTY-FIRST CENTURY

Developments in Plant Breeding

VOLUME 7

Common Bean Improvement in the Twenty-First Century

Edited by

SHREE P. SINGH
University of Idaho,
Kimberly, U.S.A.

Springer-Science+Business Media, B.V.

A C.I.P. Catalogue record for this book is available from the Library of Congress.

ISBN 978-90-481-5293-3 ISBN 978-94-015-9211-6 (eBook)
DOI 10.1007/978-94-015-9211-6

Printed on acid-free paper

Contents

Contributors

James R. Baggett
Department of Horticulture
Oregon State University
Corvallis, OR 97331-7304, U.S.A.
E-mail: baggettj@bcc.orst.edu

James S. Beaver
Agronomy and Soils
University of Puerto Rico
Mayaguez, PR 00681-9030
E-mail: j_beaver@rumac.upr.clu.edu

Mark A. Brick
Soil and Crop Science
Colorado State University
Fort Collins, CO 80523-1170, U.S.A.
E-mail: mbrick@lamar.colostate.edu

Daniel G. Debouck
Centro Internacional de Agricultura Tropical (CIAT)
A.A. 6713, Cali, Colombia
E-mail: D.DEBOUCK@CGIAR.ORG

Paul Gepts
Agronomy and Range Science
University of California
1 Shields Avenue
Davis, CA 95616-8515, U.S.A.
E-mail: plgepts@ucdavis.edu

Kenneth F. Grafton
Plant Sciences
North Dakota State University
Fargo, ND 58105, U.S.A.
E-mail: grafton@plains.Nodak.edu

Hans-Jörg Jacobsen
Universität Hannover
Herrenhaeuserst 2, 30419
Hannover, Germany
E-mail:
jacobsen@mbox.lgm.uni-hannover.de

James D. Kelly
Crop and Soil Sciences
Michigan State University
East Lansing, MI 48824, U.S.A.
E-mail: kellyj@pilot.msu.edu

Judith M. Kolkman
Crop and Soil Sciences
Michigan State University
East Lansing, MI 48824, U.S.A.
E-mail: Kolkmanj@pilot.msu.edu

Phillip N. Miklas
USDA-ARS-IAREC
24106 N. Bunn Rd.
Prosser, WA 99350, U.S.A.
E-mail: pmiklas@tricity.wsu.edu

James R. Myers
Department of Horticulture
Oregon State University
Corvallis, OR 97331-7304, U.S.A.
E-mail: myersja@bcc.orst.edu

Thandiwe Nleya
Crop Development Centre
University of Saskatchewan
51 Campus Drive, Saskatoon
Saskatchewan S7N 5A8, Canada
E-mail: nleya@sask.usask.ca

Frank B. Peairs
Bioagricultural Sciences & Pest Management
Colorado State University
Fort Collins, CO 80523-1177, U.S.A.
E-mail: fbpeairs@lamar.colostate.edu

Idupulapati M. Rao
Centro Internacional de Agricultura Tropical (CIAT)
A. A. 6713, Cali, Colombia
E-mail: I.RAO@CGIAR.ORG

Kristin A. Schneider
Crop and Soil Sciences
Michigan State University
East Lansing, MI 48824, U.S.A.
E-mail: Schneid1@pilot.msu.edu

Howard F. Schwartz
Bioagricultural Sciences & Pest Management
Colorado State University
Fort Collins, CO 80523-1177, U.S.A.
E-mail: hfspp@lamar.colostate.edu

Shree P. Singh
Department of Plant, Soil, and Entomological Sciences
University of Idaho
3793 North 3600 East
Kimberly, ID 83341-5076, U.S.A.
E-mail: singh@kimberly.uidaho.edu

Michael Thung
(formerly affiliated with CIAT)
Rua 15, No. 87. Apto. 801
Condominio Cote D'Or
74140-030 Goiânia/GO, BRAZIL
E-mail: mthung@ultra.internetional.com.br

Albert Vandenberg
Crop Development Centre
University Of Saskatchewan
51 Campus Drive, Saskatoon,
Saskatchewan S7N 5A8, Canada
E-mail: vandenberg@sask.usask.ca

Acknowledgments

I am profoundly grateful to each contributing author and co-author of the chapters in this book who took keen interest in and prepared the initial draft and returned the revised manuscript so promptly. My heartfelt thanks are also due to the following reviewers who painstakingly offered their constructive and thoughtful critiques:

George Abawi
Wayne Adams
Steve Antonius
Mark Bassett
Edwin Bingham
Mark Brick
Fernando Correa
Dermot Coyne
Michael Dickson
Paul Gepts
Kenneth Grafton
Earl Gritton
Pierre Hucl
Gale Kleinkopf
John Kraft
Andries Libenberg
Rojelio Lépiz
Steve Magnuson
Henry Mayland
Phillip McClean
Robert McMillan Jr.
Phillip Miklas
Krishna Mohan
David Mok
David Nuland
Fred Muehlbauer
James Myers
Soon Park
Juan Carlos Rosas
Howard Schwartz
Mathew Silbernagle
Albert Slinkard
Robert Sojka
James Steadman
Albert Vandenberg
Greg Varner
Clibas Vieira
Oswaldo Voysest
Giles Waines
David Webster
Norman Weeden
David Willmot
James Wright

The generous help of Patricia Zamorano de Montoya during the initial stages of the preparation of several manuscripts and their distribution to authors and reviewers is gratefully acknowledged. Similarly, I extend my sincere thanks to Cindy Strausbaugh for English editing of the manuscripts. I am grateful to Paul Kalkman of Kluwer Academic Publishers for his advice on using the Kluwer style files. I am deeply indebted to Anita Fahrenwald for her keen interest in this project and for meticulously reformatting and preparing the final camera-ready version of all chapters and subject index. Last, but not least, I am very grateful to my wife, Nirmala Singh, for diligently proof reading all manuscripts and for her patience while I was preoccupied with many aspects of preparation of this book.

Preface

The common bean (*Phaseolus vulgaris* L.) is the most important pulse crop in the world. It is an important source of calories, proteins, dietary fibers, minerals, and vitamins for millions of people in both developing and developed countries worldwide. It complements cereals and other carbohydrate-rich foods in providing near-perfect nutrition to people of all ages. Moreover, a regular intake of beans helps lower cholesterol and cancer risks. Despite the fact that per capita consumption of common bean in some developed countries (e.g., the U.S.A.) has been increasing over the last several years, in general, the average global per capita consumption is declining because production is unable to keep up with the population growth. Moreover, increasing demand for pesticide-free food products, concern for natural resources conservation, and the need to reduce production costs offer daunting challenges to the twenty-first century policy makers, bean growers, and researchers alike. High yielding, high quality bean cultivars that require less water, fertilizers, pesticides, and manual labor combined with integrated management of abiotic and biotic stresses will have to be developed. Eminent bean researchers were invited to contemplate these issues, prepare a state-of-the-art account on most relevant topics, and offer their insight into research directions into the twenty-first century.

Four excellent books have been published covering various aspects of the common bean since 1980. These books are: 1) *Bean Production Problems in the Tropics* (1st ed. 1980, 2nd ed. 1989), H.F. Schwartz & M.A. Pastor-Corrales (eds.), CIAT, Cali, Colombia; 2) *Genetic Resources of Phaseolus Beans* (1988), P. Gepts (ed.), Kluwer, Dordrecht, Netherlands; 3) *Common Beans: Research for Crop Improvement* (1991), A. van Schoonhoven & O. Voysest (eds.), C.A.B. Int., Wallingford, U.K. & CIAT, Cali, Colombia; and

4) Phaseolus spp.: *Bean Science* (1997), R.K. Maiti, Science Publishers, Enfield, U.S.A. Significant advances have since been made in several important areas. For example, there is clearer understanding of the organization of the genetic diversity at the genus and species levels as well as within *P. vulgaris*. Several linkage maps using molecular markers have been developed, and the use of molecular marker-assisted selection and the pyramiding of major genes are becoming routine now. The concept of integrated genetic improvement and pest and crop management is evolving rapidly. Thus, this book is supplementary to its predecessors. It emphasizes recent advances in maximizing the use of genetic diversity within common bean and its related species for integrated improvement of cultivars of major market classes, using the best of conventional and modern genetics and breeding technologies. It also deals with the integrated management of abiotic and biotic stresses affecting global common bean production and the availability of this nutritious food product.

The book encompasses 14 chapters. Chapter 1 describes the global importance of the dry and snap beans, and different market classes of dry beans including popular and specialty beans, their usage, major production regions, and constraints. Genetic diversity among *Phaseolus* species; grouping of species within the section *Phaseolus* sensu stricto; primary, secondary, and tertiary gene pools in relation to the common bean; and useful traits available within them for common bean improvement are reviewed in Chapter 2. Molecular linkage maps and their integration, tagging, and mapping of phenotypic traits including those involved in domestication, and their use in introgression and germplasm conversion form the main theme of Chapter 3. Morphological, biochemical, and DNA-based marker-assisted selection for qualitative and quantitative traits of agronomic importance and their limitations are emphasized in Chapter 4. Chapter 5 addresses in vitro regeneration, present difficulties and successes with direct and indirect transformation, and production of transgenic beans.

Traits deficient in dry bean cultivars and potential sources of those traits in related species are discussed in Chapter 6. Integrated genetic improvement, including breeding strategies and selection methods for introgression and accumulation of favorable alleles into cultivars from across races and gene pools within *P. vulgaris*, from its wild relatives, and from wild and cultivated members of its primary, secondary, and tertiary gene pools, is also discussed. Chapters 7 reviews development of new plant types in other food legumes and describes breeding strategies for improvement of plant types for major bean growing environments. Limitations to bean yield improvement, different strategies used in the past, and the three-tiered approach for maximizing bean yield through plant breeding are discussed in Chapter 8. Genetic improvement of cultivars of medium, small, and large-seeded races

Durango, Mesoamerica, and Nueva Granada (which together occupy more than 85% of total dry bean production worldwide) is covered in Chapters 9, 10, and 11, respectively. Chapter 12 describes different snap bean types, important production regions and problems, and reviews genetic improvement of important traits. It also discusses production of stock seeds.

Major soil, water, and temperature related stresses (abiotic) affecting global production of common bean and their integrated managements are addressed in Chapter 13. Similarly, the final Chapter 14 describes integrated management of major biotic production constraints including diseases, insects, and weeds.

It is my sincere hope that aggregates of the most relevant topics covered in this book and its timeliness will increase our awareness of common bean, its production problems, and genetic diversity available within the species and its relatives. It should also provide alternative strategies and breeding methods for maximizing use of genetic diversity for integrated genetic improvement of common bean cultivars in the major market classes of global importance. Improved cultivars combined with the integrated management of the most pertinent biotic and abiotic stresses will contribute to the improvement of production and consumption of common bean, and help conserve natural resources well into the twenty-first century despite its accompanying challenges.

Shree Singh
Technical Editor
University of Idaho
May 1999

Chapter 1

Production and Utilization[1]

Shree P. Singh
Plant, Soil and Entomological Sciences, University of Idaho, 3793 North 3600 East, Kimberly, Idaho 83341-5076, U.S.A.

Key words: common bean races, common bean usage, dry bean, primary gene pools, growth habits, market classes, popping bean, production problems, production regions, snap bean

1. INTRODUCTION

Several books, book chapters, symposia proceedings, bulletins, and review articles covering different aspects of this general topic have been published during the last few decades (Adams et al., 1985; Allavena, 1984; Araya & Beck, 1995; Beebe, 1989; CIAT, 1981, 1985, 1989; Gepts, 1988; Graham, 1978; Graham & Ranalli, 1997; Laing et al., 1984; Maiti, 1997; Park & Buzzell, 1995; Robertson & Frazier, 1978; Schoonhoven & Voysest, 1991; Schwartz & Pastor-Corrales, 1989; Singh, 1989, 1992; Singh & Voysest, 1997; Thung & Oliveira, 1998; Vieira, 1967; Vieira et al., 1998; Voysest, 1983, 1998; Wall, 1973; White et al., 1988; Wortmann et al., 1998; Zaumeyer & Meiners, 1975; Zaumeyer & Thomas, 1957; Zimmermann et al., 1988). Readers interested in details may refer to these and other publications. This chapter will provide an overview and supplementary information.

[1] Published as Idaho Agricultural Experiment Station Journal Article no. 99708. University of Idaho, College of Agriculture, Moscow, ID.

2. THE COMMON BEAN

Among major food legumes the common bean (*Phaseolus vulgaris* L.) is the third most important worldwide, superceded only by soybean [*Glycine max* (L.) Merr.] and peanut (*Arachis hypogea* L.). However, it is not clear if the statistics for peanut refer to the pod or seed yield, and in many countries soybean is used primarily as oil and animal feed. Among the pulses (i.e., annual leguminous food crops that are harvested for dry seeds) the common bean is by far the most important (Table 1). Moreover, of over 30 different *Phaseolus* species of American origin (Debouck, 1991; see also Chapter 2 by Debouck; Delgado Salinas, 1985) although four other species, namely, tepary (*P. acutifolius* A. Gray), scarlet runner (*P. coccineus* L.), Lima (*P. lunatus* L.), and year-long bean (*P. polyanthus* Greenman) were also domesticated (Gepts & Debouck, 1991), none is as important and popular worldwide as *P. vulgaris*. The common bean is the most widely distributed, grown on all continents except Antarctica, and occupies more than 90% of production areas sown to *Phaseolus* species in the world. Present day cultivation of *P. polyanthus* is largely restricted to the highlands of Guatemala and Mexico, and *P. acutifolius* is grown from Nicaragua to the southern U.S.A. While the cultivation of *P. coccineus* and *P. lunatus* has expanded to other regions of the world including Europe, Africa, and Asia, none of the four species alone annually occupies 100,000 ha worldwide. It is estimated that the total area annually sown in the world under these four groups of cultigens may not exceed 200,000 ha.

Those who rely entirely on the FAO production statistics for dry bean must be aware that despite the separation of the American *Phaseolus* species from their African and Asiatic counterparts nearly a quarter of a century ago (Maréchal et al., 1978; Westphal, 1974) the FAO data collectors and collaborators continue to disregard taxonomic facts and combine under "dry beans" data on at least six species, some belonging to entirely different genera and of origins in different continents. For example, according to the FAO statistics India is the largest producer of dry bean (4.9 million MT in 1997) in the world. However, to the best of my knowledge, the production of the common bean of American origin (*P. vulgaris*) is not even 5% of the total for India. The same can be said for many other countries in Asia, Africa, Europe, and the Americas. On the American continent, for example, in most northern and northeastern states of Brazil (Vieira, 1988) and in Paraguay, the cowpea [*Vigna unguiculata* L. (Walp.)] predominates. Thus, the Americas are the largest common bean (*P. vulgaris*) producing region and Brazil is the largest producer and consumer in the world today. Using the FAO and USDA data, the approximate values for area harvested and production of dry bean for various regions of the world in 1997 are given in

Table 2. According to these estimates, Africa, Asia, and Europe, where the crop was introduced after the discovery of the Americas in the fifteenth century, follow the lead of the Americas in dry bean production.

Table 1. Area harvested and production of major pulse crops in the world in 1997[1]

Identification	Area harvested (10^6 ha)	Production (10^6 MT)	Yield/ha (kg)
Common bean (*Phaseolus vulgaris*)[2]	14.3	11.6	810
Pea (*Pisum sativum*)	6.2	11.4	1,840
Chickpea (*Cicer arietinum*)	11.3	8.8	780
Cowpea (*Vigna unguiculata*)	6.8	2.5	370
Pigeonpea (*Cajanus cajan*)	5.2	2.5	480
Lentil (*Lens culinaris*)	3.3	2.9	880
Broad bean (*Vicia faba*)	2.3	3.3	1,430

[1] Source: FAO statistical data base of July 1998.
[2] After adjusting values for Asia and Brazil for different *Vigna* spp.

Table 2. Major dry bean producing regions in the world in 1997[1]

Region	Area harvested (10^3 ha)	Production (10^3 MT)	Yield /ha (kg)
Africa	3,260	2,120	650
Asia[2]	2,076	2,211	1,065
Europe	568	589	1,040
The Americas	8,398	6,687	796
Canada and U.S.A.	784	1,473	1,880
Caribbean	158	95	600
Mexico and Central America	2,296	1,432	620
South America[2]	5,160	3,687	715
World	14,302	11,607	810

[1] Source: FAO statistical data base of July 1998.
[2] After adjusting values for Asia and Brazil for different *Vigna* spp.

In the Americas, common bean production in Brazil (2.5 million MT) is followed by the U.S.A. (1.3 million MT), Mexico (0.98 million MT), Argentina, Canada, Colombia, Nicaragua, Honduras, Guatemala, El Salvador, Peru, Haiti, Ecuador, Chile, Venezuela, and the Dominican Republic. Of these, the production has increased substantially in the last 50 years in Argentina, Bolivia, Brazil, Canada, and the U.S.A., largely due to the increase in area planted in these countries. Among Asian countries, China, Iran, Japan, and Turkey are the major producers of the common bean. In Africa, Burundi, Ethiopia, Malawi, Republic of South Africa, Rwanda, Tanzania, Uganda, and Zimbabwe, among others, form the list of important common bean producing nations. From the statistics in recent years it is becoming evident that in Europe (and Asia), Albania, Belarus, Bulgaria, Croatia, Greece, Italy, Moldova Republic, Poland, Romania, Spain, Ukraine, and Yugoslavia are major producers of common bean.

3. COMMON BEAN USAGE

3.1 Introduction

The common bean is cultivated primarily for its green pods, green shelled seeds, and dry seeds. The cultivars for green-pod harvests are also called French, garden, green, snap, or stringless bean, but here they will be referred to as snap bean. Fully developed green pods of these cultivars have reduced fibers or no fibers in the pod walls and sutures. In some Central and Eastern African and Latin American countries, young tender leaves or flowers are also harvested as fresh vegetables. However, the largest production (>14 million hectares) and consumption are of dry bean (Table 2), followed by a much lower level of production for snap bean cultivars. In addition, green leaves, stems, and shelled pods are fed to cattle; and dry plant stubbles are used as feed for cattle, ploughed under in order to increase soil organic matter, or used as fuel for cooking.

3.2 Popping Bean

It is believed that before pottery and cooking in boiling water were invented, the natives in the highlands of the Andes (certain regions of Peru and Bolivia) used to roast dry bean. These special landraces or heirloom cultivars, popularly known as ñuñas, are still grown in both Peru and Bolivia for this type of consumption. Recently, there has been some interest in breeding ñuñas for North America and other parts of the world.

3.3 Green Shelled Bean

The cultivars harvested for green shelled bean are often large-seeded cream mottled (cranberry), red mottled (Pompadour, Diacol Calima), pink mottled (Miss Kelly), or white mottled (Coscorrón). The distinguishing characteristic of such cultivars is that the pods change color (turn red or purple, with or without stripes) when fresh seed is ready to be harvested for consumption. Some small and medium-seeded cultivars of Middle American origin also have attractive pod colors when maturing. The pods not harvested for green shelled seeds are allowed to mature normally on the plant to be harvested as dry bean. Thus, these are dual-purpose cultivars.

3.4 Snap Bean

Since the last quarter of the nineteenth century, and especially within the last 50 years, there has been a major effort to develop snap bean cultivars. Most of the work has been done in Europe and the U.S.A. Dramatic changes and large variations in plant type, maturity, fruiting pattern, and the length, shape, color, fleshiness, and other pod characteristics of snap bean cultivars have resulted. Both determinate bush and indeterminate climbing snap bean cultivars exist. The latter permit multiple harvests (pickings) over a longer period of time and much higher yields per unit area of cropped land than their bush type counterparts. However, they are labor-intensive and require substantial initial investments for installing trellises or stakes required for climbing and full plant growth and development. The cultivation of climbing snap bean cultivars is more popular in China, home gardens in Europe, winter sowings in Florida (U.S.A.), and near larger cosmopolitan cities in Latin America and other developing nations.

Snap bean cultivars with flat or cylindrical pods, yellow (waxy types) or green colors, and long or short pods are all used for fresh, frozen, and canning purposes. As far as cylindrical types are concerned, there is a growing demand for smaller, thinner, and darker green pods, especially in France. In North America, Europe, and other temperate regions, consumers often have direct access to fresh snap bean during the summer months. During winter months, snap bean, like other fresh vegetables, is produced in relatively warmer regions, in greenhouses, or is imported from other countries. In tropical and subtropical regions of Latin America, fresh harvests of snap bean are available throughout the year. For details regarding snap bean readers should refer to Chapter 12 by Myers & Baggett. The remainder of the discussion in this chapter will refer to dry bean.

3.5 Dry Bean

Consumer preferences for seed size, color, shape, and brilliance or seedcoat luster of dry bean vary a great deal. It is not uncommon to find these differences in taste among regions within a country. For example, in northeastern Brazil light colored cream and cream striped beans are preferred, whereas in the southern region black beans predominate. Similarly, medium-seeded light colored beans of different types (e.g., Bayos, pintos, Flor de Mayo, etc.) are preferred in the central highlands of Mexico, whereas in the coastal-southern part of the country, small blacks are favored.

In Canada and the U.S.A., pinto, red kidney, and small white (navy or pea) beans are preferred. In Africa, Asia, and Europe large-seeded beans of various colors, though not black, are popular.

3.5.1 Consumption

The per capita consumption of dry bean may vary from country to country and region to region within a country. The social and economic status of the family may also play an important role. There is often a higher consumption of bean among low-income families both on farms and in urban areas. In Latin America the highest per capita consumption of dry bean is in Brazil and Mexico with more than 13 kg per year. In central and eastern African countries (e.g., Rwanda and Burundi), per capita consumption is over 40 kg per year.

3.5.2 Food Preparation

Many types of food are prepared from dry bean. Dishes range from simply beans boiled in water to more sophisticated preparations of creams, soups, stews, pastes, cakes, chips, salads, baked beans, and other products. Also, different dry beans are used in the preparation of very prestigious dishes, some of which are national favorites. Small black beans are used for "Feijoãda" in Brazil; "Gallo Pinto" is popular in Costa Rica and Nicaragua. Other popular national dishes are "Moros y Cristianos" or "Casados" in Cuba, large cream mottled Cargamanto and round red colored Radicals and Sangretoro for "Plato Montañero" in Colombia, and the use of extremely large white Faba Granja or Fabes for "Fabada" in Spain. In the Orient (from Thailand to Japan) and in Peru, large-seeded Andean beans are used in sweet dishes for dessert.

Dry beans have high nutritional value, especially in conjunction with cereals and other carbohydrate-rich foods; beans play a role in the reduction of cholesterol and cancer risk (Andersen & Gustafson, 1989; Andersen et al., 1984; Tietyen-Clark, 1986). For those reasons, dry bean consumption is increasing in higher income families in countries such as the U.S.A. and Canada. Research is also underway to use dry bean in baby food.

4. GROWTH HABITS

Variation in growth habit appears to be continuous from determinate bush to indeterminate, extreme climbing types. However, for simplicity, agronomic value, and because of their adaptation to different cropping systems, Singh (1982) used the type of terminal bud (vegetative versus reproductive), stem strength (weak versus strong), climbing ability (nonclimber versus strong climber), and fruiting patterns (mostly basal versus along entire stem length or only in the upper part) to classify growth habits into four major classes. These are: type I = determinate upright or bush, type II = indeterminate upright bush, type III = indeterminate, prostrate, nonclimbing or viny semiclimbing, and type IV = indeterminate, strong climbers.

5. GENE POOLS AND RACES

Debouck in Chapter 2 of this book covers the organization of diversity among *Phaseolus* species; the primary, secondary, and tertiary gene pools in relation to the common bean; and details regarding origin and domestication of common bean. The common bean primary gene pool comprises both the cultigens and wild populations. Only gene pools and races found among common bean cultigens will be briefly mentioned here. Evans (1973) was the first to recognize the two major groups of dry bean germplasm: large-seeded Andean and small-seeded Middle American, and she further divided them into five races. Unequivocal support for the existence of the two gene pools was provided subsequently by: 1) establishing the relationship between the seed size (small versus large) and the *Dl* genes (*Dl-1* versus *Dl-2*) (Shii et al., 1980) involved in the F_1 hybrid incompatibility (Gepts & Bliss, 1985; Singh & Gutiérrez, 1984), 2) phaseolin seed proteins (Gepts & Bliss, 1985; Gepts et al., 1986), 3) allozymes (Singh et al., 1991c), 4) morphological traits (Singh et al., 1991b), and 5) DNA markers (Khairallah et al., 1990).

Singh (1989) described in considerable detail the patterns of variation and organization of genetic diversity among common bean cultigens. Moreover, based on morphological, biochemical, adaptation, and agronomic traits and geographical distribution in their primary centers of origin and domestication, Singh et al. (1991a) further divided the Andean and Middle American gene pools into six races: Andean (all large-seeded) = Chile, Nueva Granada, and Peru; Middle American = Durango (medium-seeded semi-climber), Jalisco (medium-seeded climber), and Mesoamerica (all small-seeded). For details regarding characteristics of these races, readers should refer to the literature cited. Of these races, cultivars of Durango, Mesoamerica, and Nueva Granada are planted in more than 80% of the area under dry bean worldwide. Today, cultivation of the Jalisco race is restricted to the highlands of Mexico, and the Peru race is limited to the Andean highlands.

6. MARKET CLASSES

Readers interested in details about the different market classes of dry beans in the world should refer to Voysest & Dessert (1991). Only a brief account and supplementary information about beans of major economic importance belonging to each of the six races will be presented here.

Table 3. Major market classes of dry bean, their production region, and estimates of cultivated area

Races and bean type	Production region	Area (10^3ha)
Small-seeded race Mesoamerica		
Black	Argentina, Brazil, Venezuela, the Caribbean, Middle and North America	3,500
Carioca (cream striped)	Brazil, Bolivia	2,000
Jalinho (beige)	Brazil	500
Mulatinho (cream)	Brazil	500
Red	Central America, China	250
White	Africa, China, North America	250
Medium-seeded race Durango		
Bayo (beige)	Mexican highlands	800
Flor de Mayo (pink speckled)[1]	Mexican highlands	250
Great northern (white)	Europe, North America, West Asia	700
Ojo de Cabra (cream striped)	Mexican highlands	150
Pink	North America	20
Pinto (cream with greenish brown spots)	North America	800
Red	North America, West Asia	30
Large-seeded race Nueva Granada		
Alubia (white elongated)	Argentina, Europe, North Africa, West Asia	250
Azufrado (sulfur colored)	Pacific Coast of Mexico and Peru	150
Calima (red mottled)	Africa, Andes, Caribbean	1,500
Manteca (beige)	Andes	100
Cranberry (cream with red spots)[2]	Africa, Asia, Europe, North and South America	800
Dark red kidney	Africa, Andes, North America	500
Light red kidney	Africa, North America	300
Radical (oval/round red)[2]	Andes	50

[1] Belongs to race Jalisco.
[2] Also includes races Chile and Peru.

6.1 Popular Beans

6.1.1 Small-Seeded Beans of Race Mesoamerica

Black, beige (Jalinho), cream (Mulatinho), cream striped (Carioca), red, and white colored beans belonging to race Mesoamerica (Singh et al., 1991a) are each grown on at least one quarter million ha worldwide (Table 3). With the exception of small white canning beans, these are mostly grown and consumed on the American continent, chiefly in Mexico, Central America, Brazil, Venezuela, Cuba, and other Caribbean countries. In Brazil, black, beige, cream, and cream striped beans are the most popular. The small black bean has by far the largest production, is the most widely grown on the American continent, and is produced from Canada to Chile and Argentina.

6.1.2 Medium-Seeded Beans of Race Durango

Among the medium-seeded beans of race Durango, Bayos (beige), great northern (white), Ojo de Cabra (cream striped), pinks, pintos (cream spotted), and reds are the most popular types. Pintos are chiefly produced and consumed in North America. Chile and the Dominican Republic also grow and consume some pintos. The great northerns are mostly produced in the U.S.A. and Canada and exported to Europe and the Middle East. The reds and pinks are also grown in Iran and some European countries on a very small scale. Ojo de Cabra and Bayos are grown and consumed only in the central and northern highlands of Mexico.

6.1.3 Large-Seeded Beans of Race Nueva Granada

Among all large-seeded cultivated dry bean of Andean origin, the red kidney (light and dark), red mottled, and cream mottled (popularly known as cranberry) types are the most popular worldwide. Of these popular types, large elongated kidney and cylindrical shaped red and red mottled beans belong to race Nueva Granada. The cream mottled types are found in all three races originating in the Andes. However, the most internationally known market class of true cranberry beans (large oval or round shaped) belongs to the race Chile. The largest-seeded oval shaped cream mottled Cargamanto beans (aggressive climbing growth habit IV) belong to race Peru and are still more appreciated in Antioquia and other coffee-growing regions of Colombia. The large white kidney (Riñón) or cylindrical shaped (Alubia) beans are produced on a small scale in the U.S.A. and Chile. These kidney or cylindrical beans are grown annually on over 50,000 ha in

Argentina for export to Europe (especially Spain and Portugal), the Middle East, and northern Africa.

6.2 Specialty Beans

In addition to the dry beans of major economic importance, there are some dry beans that are produced and consumed on a smaller scale within a specific country or region. These dry beans are highly valued and appreciated, at least regionally or locally, and they merit mention. Since their breeding may not be covered in this book, some general information regarding these will be included here.

6.2.1 Medium-Seeded Flor de Mayo of Race Jalisco

These pink speckled beans are produced under both irrigation and rainfed farming systems on approximately 250,000 ha in the central highlands of Mexico. The landraces maintain their seed color long after harvest and it is believed that they also cook easily. Both climbing type IV or pole beans (e.g., Rosa de Castilla) and semiclimbing or large viny type III (Flor de Mayo) growth habit cultivars are grown. Variants or closely related types include Flor de Abril, Flor de Junio, and Manzano.

High yield potential and moderate levels of tolerance to low soil fertility and water stress are common characteristics of these cultivars. Early maturity, resistance to rust [caused by *Uromyces appendiculatus* (Pers.) Ung.] and bean common mosaic (BCM, caused by a potyvirus), as well as upright growth habit are essential for beans grown under irrigation as spring crops (March-May). Other desirable pest resistances for a successful cultivar especially as a rainfed summer crop in the humid and semiarid highlands of Mexico may include anthracnose [caused by *Colletotrichum lindemuthianum* (Sacc. & Magn.) Bri. & Cav.], angular leaf spot [caused by *Phaeoisariopsis griseola* (Sacc.) Ferr.], common bacterial blight [caused by *Xanthomonas campestris* pv. *phaseoli* (Smith) Dye], Mexican bean beetle (*Epilachna varivestis* Mulsant), and bean pod weevil (*Apion godmani* Wagner and *A. aurichalceum* Wagner). Moderate to high levels of tolerance to water and low soil fertility stresses would also be useful to stabilize and increase production in these environments.

6.2.2 Large-Seeded Manteca of Races Nueva Granada and Peru

This general class of bean possesses wide variation in seed size (100 seed weight range from 40 to 80 g) and shape, and is known under different local names from Mexico to Brazil and Chile. As a group they may occupy over

150,000 ha. The seeds are of different tones or shades of beige with or without a marked light or dark brown or golden colored hilum ring or stripes, and of round, oval, or elongated shape. This class includes cultivars from races Nueva Granada (e.g., Bayo Chimu, Bola 60 Dias, Canario 107, Jalo EEP 558), Chile (Burros Argentinos), and Peru (Bolón Bayo, Bola Canario). These cultivars can range from determinate type I to extreme climbing type IV, grown from sea level in coastal Peru to over 2,500 m elevation in the highlands of Ecuador. Medium-seeded Amarillos, Garbancillo Zarco, and Frijola of race Jalisco from Mexico, Amarillos Gigantes (large-seeded) from Peru, and Jalinho (small-seeded) from Brazil could also be included in this group to broaden the genetic base of these groups of cultivars.

As there is a great diversity in seed type, plant type, and climatic region where these beans are grown, different combinations of production constraints exist which affect this group. In general, most biotic and abiotic production constraints occurring in tropical and subtropical Latin America are also problematic for these beans.

6.2.3 Large-Seeded Azufrados, Canarios or Peruanos of Race Nueva Granada

The sulfur colored large oval or elongated seeds of this group are highly appreciated in the central and northern Pacific coastal regions and in the highlands of Mexico and coastal Peru (where they are locally known as Canario). The total area given to these beans is estimated at more than 100,000 ha in Mexico alone. The old Mexican cultivars (Canarios) were initially introduced from Peru, hence the name Peruano is still popular in Mexico. However, at present the most popular improved cultivar in Mexico is Mayocoba, a determinate type I.

Among several production constraints, resistance to bean golden mosaic (BGM, caused by a geminivirus) and BCM are essential for the successful cultivation of beans in the Pacific coastal regions of Mexico. For cultivation in coastal Peru resistance to rust and nematodes are also required. These cultivars are poorly adapted in the Mexican highlands, the U.S.A., and Canada where there is growing interest in their production.

6.2.4 Large-Seeded Radical, Sangretoro, Bola Roja or Bolón Rojo of Race Peru

These large round or oval red-seeded cultivars are highly valued in Colombia and Ecuador. Although the total area under these beans is approximately 50,000 ha, they are among the most highly priced beans in

Colombia. Most of the popular cultivars are of climbing type IV, but some determinate type I (e.g., Radical San Gil, Chocho) and indeterminate type III (e.g., Chaucha Colorada and Don Timoteo) cultivars belonging to races Chile and Nueva Granada also are found. The latter often possess relatively smaller seeds and fetch lower prices.

In addition to other desirable traits, resistance to anthracnose, angular leaf spot, rust, and halo blight (caused by *Pseudomonas syringae* pv. *phaseolicola* Burkh.) are essential in these cultivars for their successful and sustained cultivation in the highlands of Colombia, Ecuador, and Peru.

6.2.5 Large-Seeded Ñuñas of Race Peru

These popping beans are currently grown in the highlands of Peru and Bolivia. Although their total hectarage may be well under 20,000 ha, they potentially offer an interesting and highly valuable nutritive snack food for people of both the developing and the developed world. All known landraces are of climbing type IV, and possess medium to large seeds. Yields are low when ñuñas are intercropped with maize (*Zea mays* L.), but farmers using trellises harvest up to 4,000 kg per ha.

Resistance to anthracnose, BCM, halo blight, and rust are essential for improved cultivars for the Andean highlands. For cultivation in North America and Europe, cultivars with earliness, insensitivity to long days, and dwarf bush growth habits are required. Resistance to BCM, common bacterial blight, halo blight, white mold, and rust are also essential.

6.2.6 Large-Seeded Faba or Fabes of Race Nueva Granada

These are among the largest-seeded cultivated common bean. The 100-seed weight may easily exceed 100 g. All present-day cultivars are climbing type IV. Although the elongated cylindrical shaped seeds are white, the maturing pods possess light purple or pink colored stripes like cultivar Coscorrón of race Chile (see below). Currently, these cultivars are grown in the Principado de Asturias (northern Spain) and adjoining provinces. The total under this group of cultivars is less than 5,000 ha. The much smaller seeded and hence less appreciated cylindrical shaped Alubias and kidney shaped Riñon are grown in Galicia, Castilla y León, and other provinces in Spain, Portugal, North Africa, West Asia, coastal Peru, and Argentina. Small-scale production either already occurs or attempts are being made to grow these types in other Latin American countries, the U.S.A., and Canada.

Resistance to anthracnose, BCM, common bacterial blight, rust, and halo blight are required in these types for cultivation in Spain, Portugal, and other European countries. For Argentina, resistance to BGM, white mold, root rot, and angular leaf spot are essential in addition to resistance to the above diseases. For North American environments, earliness and insensitivity to long days are required to be successful cultivars.

6.2.7 Large-Seeded Tórtolas and Coscorrón of Race Chile

Large-seeded gray Tórtolas and white spotted Coscorrón beans are grown and consumed in Chile. About a decade ago they were cultivated on almost 50,000 ha. However, there has been a gradual reduction in their hectarage in Chile due to greater emphasis on export of fruit and vegetable crops in recent years. Chilean landraces are type III and are late maturing, taking more than 120 days to harvest.

Of the above mentioned specialty dry beans, currently there is no breeding of any consequence for Bolón Bayo, Cargamanto, Garbancillos, Radical, and Ñuñas in their natural habitat, and they remain unimproved to date.

7. DRY BEAN PRODUCTION

As noted earlier, dry bean is presently grown annually on more than 14 million hectares worldwide (Table 1). From its origin and initial domestication regions in Andean South and Middle America, dry bean production and consumption have expanded into other parts of the Americas (from about 35°S to >50°N latitude and from sea level to >3,000 m altitude), Africa, Asia, Europe, and other parts of the world within the last five centuries (Table 2).

7.1 Day Length and Temperature

The common bean is a short-day crop (White & Laing, 1989) and its growth and development are favored by mildly cool environments. Thus, in environments with 16 to 18°C mean growing temperatures with about 12 h day-length, and free from abiotic and biotic stresses, most cultivars complete their growing cycle from germination to seed maturity in 100 to 130 days. However, photoperiod insensitive cultivars that are successfully grown at higher latitudes (>14 hr daylength) in Canada, the U.S.A., Europe, Japan, and other parts of the world have either evolved or been developed by breeding. Most cultivars grown in the highlands of Mexico, Central America, and the Andes are often highly sensitive to long photoperiod and high temperatures and will not complete their growing cycle under these conditions.

7.2 Growing Season

Below 2,000 m elevation in tropical and subtropical Latin America and Africa, dry bean is grown twice during the year, often coinciding with the prevalent bimodal rainfall patterns of these regions. In temperate regions at higher latitudes, dry bean is usually grown as either a spring or summer crop. In many regions (e.g., the Middle East, Iran, Europe, Canada, the U.S.A., Chile, and hilly regions of India, Nepal, and Pakistan) the dry bean is cultivated in the spring or summer together with other cool season food legumes such as chickpea, pea, lentil, and broad bean. Often the latter cool season food legumes are sown several weeks earlier; bean plantings are done later when soil and air temperatures are relatively higher. Thus, while rainfed cultivation (e.g., in Argentina, Mexican highlands, southwestern Canada, and northeastern and midwestern U.S.A.) may be practiced, especially in areas with more than 400 mm annual rainfall, the dry bean may require supplemental irrigation for secured harvests and higher yields. It may also be grown entirely as an irrigated crop (e.g., in Chile; Idaho and Washington, U.S.A.; western Asia; and April to July plantings in central Brazil). In regions with warm or hot summers, the common bean is grown in the autumn (e.g., northeastern Argentina), spring (Indo-gangetic plains of India), or winter (in many countries in Africa, Brazil, and the Caribbean).

Whereas in warm, tropical regions, the crop is harvested within approximately 75 days, in the highlands (above 2,000 m elevation) of the Andes (Colombia, Ecuador, Bolivia, and Peru) the climbing bean crop may often take more than 250 days to mature for harvest. In the humid highlands of Guatemala and Mexico, and in Asturias, Spain, the climbing cultivars usually take approximately 150 days to maturity. At higher latitudes in temperate climates, dry beans of growth habits I, II, and III are harvested within 100 to 120 days from planting.

7.3 Cropping Systems

Cultivars of growth habits I, II, and III are grown in monoculture as well as under different relay, strip, and intercropping systems throughout the world (Singh, 1992; Woolley et al., 1991). The type IV climbing cultivars always require support. Thus, these are grown either in association with maize, cassava (*Manihot esculenta* Crantz), and other crops, or they are grown on trellises or stakes. The type IV climbing cultivars are popular in regions such as Asturias, Spain; Antioquia, Colombia; or Florida, U.S.A. where highly priced snap (McClasan or Dade in Florida, U.S.A.) or dry bean cultivars (Cargamanto in Colombia or Faba Granja in Spain) are grown for higher yields and multiple harvests (pickings). Although intercropping is often more profitable (Francis & Sanders, 1978) and may be favored for sustainable farming, bean yield reductions occur with intercropping for cultivars of all growth habits (Clark & Francis, 1985).

7.4 Agronomic Practices

Although dry bean is grown in a wide range of soil types, light loamy soils with pH between 5.5 to 7.0 and rich in organic matter are more suitable for good crop production. A 100 to 120 day crop with a seed yield of 2,500 kg per ha will usually remove 60 to 80 kg of soil nitrogen and 40 kg of phosphorus. In acidic soils deficient in nitrogen and phosphorus and containing toxic levels of aluminum and/or manganese, it is essential to use appropriate corrective measures. These measures include the adequate use of lime, and fertilizers rich in nitrogen and phosphorus, as well as other major and minor elements (Howeler, 1980; Thung, 1990; Thung & Oliveira, 1998; Thung & Rao in Chapter 13 of this book). Similarly, in somewhat alkaline soils, deficiency of micro elements (iron, boron) is common and use of gypsum and sulfur rich fertilizers becomes necessary.

As farmers move toward good conservation practices, they are increasingly using minimum or no tillage and organic and green manures, crop rotation, and integrated pest management practices. These practices enhance and conserve soil fertility and moisture and control soil erosion, weeds, diseases, and insects. Moreover, since in most traditional bean growing regions nodulation and nitrogen fixation are common, use of inoculants with the most effective and competitive *Rhizobium* strains may be promoted while minimizing use of nitrogenous fertilizers at the time of sowing or restricting nitrogen to foliar use only. For high yields (>2,000 kg per ha) chemical fertilizers are used in most bean production areas. In nutrient deficient soils, cultivars tolerant of and efficient in use of low soil phosphorus and nitrogen are required.

7.5 Important Production Regions

7.5.1 Brazil

In Brazil, there are three distinct growing seasons; the wet season from September to December (Safras das Aguas), dry season from January to April (Safra da Seca), and winter planting from April to July (Tercera Epoca) under the central pivot irrigation systems. The yields are highest during winter planting and lowest during the dry season, especially if the crop is grown in monoculture (C. Vieira, personal communication, 1998). Similarly, in the semiarid regions of northeastern Brazil (e.g., Bahia,

Pernambuco, Sergipe, Alagoas), bean yields are much lower than in the central and southern states. Parana, Bahia, and Minas Gerais are the largest producers of dry bean. Lighter colored cream and cream striped beans are preferred in the northern and central regions, whereas black bean predominates in the southern states. A more diverse group of beans is popular in the central region (e.g., Minas Gerais and São Paulo).

7.5.2 North America

In Mexico, much of the area under dry bean is situated in the highlands and yield is considerably lower than that in the coastal regions of Nayarit and Sinaloa. Zacatecas, Durango, and Chihuahua have the largest dry bean production areas, totaling about one million ha. Light colored Bayo, Garbancillo, Flor de Mayo, and pinto beans are preferred in the highlands. Yellow or sulfur colored Azufrado is popular in the North Pacific coastal regions, and small black in the south (Veracruz, Chiapas) is popular.

In the U.S.A., North Dakota (pinto) and Michigan (navy and black), and in Canada, Manitoba (pinto, black, and navy) and Ontario (navy) are the largest bean growing states and provinces.

7.5.3 Central America

In Central America, Nicaragua, Guatemala, Costa Rica, Honduras, and El Salvador grow a total of approximately 400,000 ha of dry bean, mostly small-seeded black and red. Production of dry bean in Panama is less than 10,000 ha. Moreover, large-seeded Andean beans of red and red mottled color are preferred.

7.5.4 The Caribbean

In the Caribbean, Cuba, the Dominican Republic, and Haiti are the major producers of dry bean. In Cuba, small black is more popular. Although small black is also grown in smaller areas in the Dominican Republic and Haiti, large-seeded Andean beans of red, pink, and cream of both solid and spotted types predominate.

7.5.5 Andean Region

In Colombia, large red, red mottled, and cream mottled (in Antioquia) beans predominate. Although most of these types are also grown in Ecuador, the most preferred types for local consumption are large-seeded beige beans (Bayo and Canario). Among the Andean countries, Peru has more diverse bean types. Large white beans (Caballeros) are preferred in the highlands, and small white (Panamito or navy), beige, cream, red, and yellow colored Bayo and Canarios are consumed in the coastal region. Argentina grows large white (Alubia), small black, large red kidney, and other miscellaneous colored beans. In Chile, gray, white spotted, and cranberry beans are more popular for domestic consumption although some pintos and other types of beans are grown and consumed. Small black bean is grown for export to Latin America and navy and medium-seeded great northern beans are grown for Europe and the Middle East.

7.5.6 Europe, Africa, and Asia

In Europe, Africa, and Asia, large-seeded Andean beans of white, red, red mottled, cream mottled, and beige color usually predominate. Some small and medium-seeded dry beans of various colors and Middle American origin are also grown for export markets, in marginal environments because of their higher levels of tolerance to abiotic and biotic stresses, and in areas where high yields are sought such as in central and east African countries.

8. PRODUCTION PROBLEMS

The common bean suffers from both abiotic and biotic production constraints (Graham, 1978; Graham & Ranalli, 1997; Schwartz & Pastor-Corrales, 1989; Singh, 1992; Thung & Oliveira, 1998; Wall, 1973; White et al., 1988; Wortmann et al., 1998; Zaumeyer & Meiners, 1975; Zaumeyer & Thomas, 1957; Zimmermann et al., 1988; see also Chapter 13 by Thung & Rao and Chapter 14 by Schwartz & Peairs). Some important abiotic and biotic constraints of regional and global importance are listed in Table 4.

8.1 Soil Fertility and Water Stress

Among the abiotic constraints low soil fertility in general but in particular deficiency of nitrogen and phosphorus as well as aluminum and manganese toxicity are probably the most widely distributed (Araya & Beck, 1995; Howeler, 1980; Thung, 1990; Thung & Oliveira, 1998; Wortmann et al., 1998; see also Chapter 13 by Thung & Rao). Similarly, some form of water stress or drought is a widespread phenomenon throughout most bean production regions (White et al., 1988). In Latin America, drought is common and can be frequent in northeastern Brazil, coastal Peru, and in the central and northern highlands of Mexico. Complete crop failures under dryland conditions are not uncommon in these areas. In regions where the crop is planted towards the end of the rainy season (e.g., September-December in Central America) moderate water stress frequently occurs.

8.2 Temperature Stress

High temperatures (>30°C day and/or >20°C night) in tropical lowlands (below 650 m elevation) and during summer at higher latitudes (e.g., California, Colorado, Idaho, Nebraska, Washington, and Wyoming in the U.S.A.) can severely limit bean production. Recurring low temperatures (below 10°C) as well as frost during the beginning and end of the growing season in the highlands (above 2,000 m elevation) of Latin America and at higher latitudes (e.g., in the U.S.A. and Canada) can also reduce bean yields. High yielding, early maturing, and cold and heat tolerant cultivars could stabilize production and expand cultivation in these areas.

Table 4. Some common bean production constraints of regional and global importance

Production constraint	Occurrence	Yield loss (%)
Abiotic		
Drought	Global	10-100
Low soil fertility[1]	Global	30-100
Biotic		
Diseases		
Bean common mosaic	Global	50-70
Bean golden mosaic	Argentina, Brazil, Caribbean, Middle America	50-90
Common bacterial blight	Global	20-50
Halo blight	African and Latin American highlands, Europe, North America	20-50
Angular leaf spot	Africa, Latin America	40-80
Anthracnose	Global	10-100
Root rots[2]	Global	10-90
Rust	Global	40-80
White mold	Brazil, Argentina, North America	30-90
Insects		
Bean fly	Africa	30-100
Bean pod weevil	Mexico, Central America	10-80
Bruchids	Global	10-40
Leafhoppers	Global	10-100

[1] May include deficiency of phosphorus, nitrogen, and other elements and toxicity of aluminum and manganese.

[2] May include complex of two or more fungi (*Aphanomyces, Fusarium, Macrophomina, Pythium, Rhizoctonia, Sclerotium, Thielaviopsis,* and others).

8.3 Bacterial Diseases

Common bacterial blight is a widespread problem from tropical to temperate bean growing environments (Saettler, 1989; Yoshii et al., 1976). In relatively cooler and wetter areas, halo blight and bacterial brown spot (caused by *Pseudomonas syringae* pv. *syringae* van Hall) cause severe yield losses (Saettler & Potter, 1970).

8.4 Fungal Diseases

Angular leaf spot, anthracnose, and rust are considered among the most widely distributed foliar fungal diseases that cause severe yield losses of common bean in Latin America, Africa, and other parts of the world. Various root rots (Abawi, 1989; Keenan et al., 1974) in most bean growing environments, web blight [caused by *Thanatephorus cucumeris* (Frank) Donk.] in the warm humid tropics, and white mold [caused by *Sclerotinia sclerotiorum* (Lib) de Bary] (Kerr et al., 1978) or ascochyta blight [caused by *Phoma exigua* var. *diversispora* (Bub.) Boerma] in cool wet regions, occasionally become severe on the common bean.

8.5 Viral Diseases

BCM in most bean production regions and BGM in tropical and subtropical Central America, coastal Mexico, the Caribbean, Brazil, and Argentina (Bird et al., 1973; Costa & Cupertino, 1976; Gámez, 1971; Morales & Niessen, 1988) cause severe yield losses in common bean. Sugarbeet curly top in the northwestern U.S.A. and bean yellow mosaic in some European countries and the Middle East, North Africa, and Asia can also cause severe yield losses in susceptible cultivars.

The literature cited at the beginning of this chapter provides more information about diseases. Schwartz & Pastor-Corrales (1989), Zaumeyer & Meiners (1975), and Zaumeyer & Thomas (1957) are particularly useful references.

8.6 Insects

Leafhoppers *Empoasca kraemeri* Ross & Moore (in the tropics and subtropics) and *E. fabae* Harris (in the temperate and cooler environments) are the most widely distributed insect problem in bean fields, especially in relatively drier areas. As noted earlier, bean pod weevil causes severe damage in the highlands of Mexico, in Guatemala, El Salvador, Honduras, and Nicaragua (Beebe et al., 1993; Garza et al., 1996; Guevara-Calderón, 1961). In the highlands of Mexico and in the U.S.A., Mexican bean beetles also cause severe leaf damage, especially in late maturing cultivars. Bean fly (*Ophiomyia phaseoli* Tryon) is by far the most damaging insect of beans in Africa (Karel & Autrique, 1989; Karel & Matee, 1986; Wortmann et al., 1998). The bean weevils, *Zabrotes subfasciatus* Boheman (in warm tropical and subtropical environments) and *Acanthoscelides obtectus* (Say) (in cool and temperate environments), cause severe problems when dry beans are not properly stored. Additional information on bean insects can be found elsewhere (Altieri et al., 1978; Cardona, 1989; Karel & Autrique, 1989; Schoonhoven & Cardona, 1980).

8.7 Weeds0

Many broadleaf and grassy weeds invade bean fields. The composition of the weed population and the most dominant weeds in bean fields vary from region to region and depend upon several factors. These factors include the growing environments (dry versus wet; warm versus cool), agronomic management of not only the standing bean crop but also other crops grown on the farm, and the history of the fields being used for bean cultivation. Other important factors are the cropping systems, tillage systems (e.g., minimum or no tillage versus conventional methods), growth habit and competitive ability of bean cultivars, planting density (i.e., row spacing and spacing between plants within rows), moisture availability, and the control measures.

9. THE FUTURE OF THE COMMON BEAN

The diversity in common bean types, their multipurpose usage in a variety of food preparations, their wide adaptation, ability to be grown in different cropping systems, and ability to fix biological nitrogen make them a valuable food crop worldwide. These properties, in addition to the highly nutritive protein content of dry bean when combined with cereal grains and their ability to help prevent cancer and lower cholesterol, should help increase the demand for them on a global basis. In developed countries such as the U.S.A., the per capita consumption has been increasing steadily. As awareness of the above properties of common bean increases, consumption is likely to rise. Similarly, diversification and publicity for processed bean products (e.g., chips for snacks, instant beans for cream or soups), availability of pre-cooked beans in different forms, and internationalization of food habits could increase global bean production and consumption. Other important factors are the use of common bean in baby foods, the reduction of flatulence, antinutritional factors, and cooking time.

Availability of broadly adapted high yielding, high quality, early maturing cultivars that can be directly harvested mechanically and that are less dependent on water, fertilizers, pesticides, and manual labor should lead to a global expansion of common bean cultivation.

REFERENCES

Abawi, G.S. 1989. Root rots. pp. 105-157. *In* H.F. Schwartz & M.A. Pastor-Corrales (eds.), Bean Production Problems in the Tropics. 2nd ed. CIAT, Cali, Colombia.

Adams, M.W., D.P. Coyne, J.H.C. Davis, P.H. Graham & C.A. Francis. 1985. Common bean (*Phaseolus vulgaris* L.). pp. 433-476. *In* R.J. Summerfield & E.H. Roberts (eds.), Grain Legume Crops. Collins, London, U.K.

Allavena, A. 1984. Beans (*Phaseolus).* pp. 137-168. *In* W.R. Sharp, D.A. Evans & P.V. Amirato (eds.), Handbook of Plant Cell Culture. Vol. 2. MacMillan, New York, U.S.A.

Altieri, M.A., C.A. Francis, A.van Schoonhoven & J.D. Doll. 1978. A review of insect prevalence in maize (*Zea mays* L.) and bean (*Phaseolus vulgaris* L.) polyculture systems. Field Crops Res. 1:33-49.

Andersen, J.W. & N.J. Gustafson. 1989. Hypocholesterolemic effects of oat and bean products. Michigan Dry Bean Digest 13:2-5.

Andersen, J.W., L. Story, B. Sieling, W-J.L. Chen, M.S. Petro & J. Story. 1984. Hypocholesterolemic effects of oat-bran or bean intake for hypercholesterolemic men. Amer. J. Clinical Nutri. 40:1146-1155.

Araya, R. & D. Beck (eds.). 1995. Memoria del Taller Internacional sobre Bajo Fosforo en el Cultivo de Frijol. Univ. de Costa Rica, San Jose, Costa Rica.

Beebe, S. (ed.). 1989. Current Topics in Breeding of Common Bean. CIAT, Cali, Colombia.

Beebe, S., C. Cardona, O. Díaz, F. Rodríguez, E. Mancia & S. Ajquejay. 1993. Development of common bean (*Phaseolus vulgaris* L.) lines resistant to the pod weevil, *Apion godmani* Wagner, in Central America. Euphytica 69:83-88.

Bird, J., J. Sánchez & N.G. Vakili. 1973. Golden-yellow mosaic virus of beans (*Phaseolus vulgaris*) in Puerto Rico. Phytopathology 63:1435 (astract).

Cardona, C. 1989. Insects and other invertebrate bean pests in Latin America. pp.505-570. *In* H.F. Schwartz & M.A. Pastor-Corrales (eds.), Bean Production Problems in the Tropics. 2nd ed. CIAT, Cali, Colombia.

CIAT (Centro Internacional de Agricultura Tropical). 1981. Potential for Field Beans in Eastern Africa. CIAT, Cali, Colombia.

CIAT (Centro Internacional de Agricultura Tropical). 1985. Potentials for Field Beans (*Phaseolus vulgaris* L.) in West Asia and North Africa. CIAT, Cali, Colombia.

CIAT (Centro Internacional de Agricultura Tropical). 1989. Advances in Bean (*Phaseolus vulgaris* L.) Research and Production. CIAT, Cali, Colombia.

Clark, E.A. & C.A. Francis. 1985. Bean-maize intercrops: a comparison of bush and climbing bean growth habits. Field Crops Res. 10:151-166.

Costa, C.L. & F.P. Cupertino. 1976. Avaliacão das perdas na producão do feijoeiro causadas pelo virus do mosaico dourado. Fitopat. Bras. 1:18-25.

Debouck, D.G. 1991. Systematics and morphology. pp. 55-118. *In* A. van Schoonhoven & O. Voysest (eds.), Common Beans: Research for Crop Improvement. C.A.B. Int., Wallingford, U.K. & CIAT, Cali, Colombia.

Delgado Salinas, A. 1985. Systematics of the genus *Phaseolus* (Leguminosae) in North and Central America. Ph.D. dissertation. Univ. of Texas, Austin, U.S.A.

Evans, A.M. 1973. Commentary upon plant architecture and physiological efficiency in the field bean. pp.279-286. *In* D. Wall (ed.), Potential of Field Beans and Other Food Legumes in Latin America. CIAT, Cali, Colombia.

Francis, C.A. & J.H. Sanders. 1978. Economic analysis of bean and maize systems: monoculture versus associated cropping. Field Crops Res. 1:319-335.

Gámez, R. 1971. Los virus del frijol en Centroamérica. I. Transmisión por moscas blancas (*Bemisia tabaci* Genn.) y plantas hospedantes del virus del mosaico dorado. Turrialba 21:22-27.

Garza, R., C. Cardona & S.P. Singh. 1996. Inheritance of resistance to the bean-pod weevil (*Apion godmani* Wagner) in common beans from Mexico. Theor. Appl. Genet. 92:357-362.

Gepts, P. (ed.). 1988. Genetic Resources of *Phaseolus* Beans. Kluwer, Dordrecht, Netherlands.

Gepts, P. & F.A. Bliss. 1985. F_1 hybrid weakness in the common bean: differential geographic origin suggests two gene pools in cultivated bean germplasm. J. Hered. 76:447-450.

Gepts, P. & D. Debouck. 1991. Origin, domestication, and evolution of the common bean (*Phaseolus vulgaris* L.). pp. 7-53. *In* A. van Schoonhoven & O. Voysest (eds.), Common Beans: Research for Crop Improvement. C.A.B. Int., Wallingford, U.K. & CIAT, Cali, Colombia.

Gepts, P., T.C. Osborn, K. Rashka & F.A. Bliss. 1986. Phaseolin protein variability in wild forms and landraces of the common bean (*Phaseolus vulgaris*): evidence for multiple centers of domestication. Econ. Bot. 40:451-468.

Graham, P.H. 1978. Problems and potentials of field beans (*Phaseolus vulgaris* L.) in Latin America. Field Crops Res. 1:295-317.

Graham, P.H. & P. Ranalli. 1997. Common bean (*Phaseolus vulgaris* L.). Field Crops Res. 53:131-146.

Guevara-Calderón, J. 1961. El combate del picudo del ejote mediante la combinaicón de variedades resistentes e insecticidas. Agric. Tec. Mex. 1(12):17-19.

Howeler, R.H. 1980. Nutritional disorders. pp. 341-362. *In* H.F. Schwartz & G.E. Gálvez (eds.), Bean Production Problems: Disease, Insect, Soil and Climatic Constraints of *Phaseolus vulgaris*. CIAT, Cali, Colombia.

Karel, A.K. & A. Autrique. 1989. Insects and other pests in Africa. pp.455-504. *In* H.F. Schwartz & M.A. Pastor-Corrales (eds.), Bean Production Problems in the Tropics. 2nd ed. CIAT, Cali, Colombia.

Karel, A.K. & J.J. Matee. 1986. Yield losses in common beans following damage by bean fly *Ophiomyia phaseoli* Tryon Diptera: agronyzidae. Annu. Rpt. Bean Improv. Coop. 29:115-116.

Keenan, J.G., H.D. Moore, N. Oshima & L.E. Jenkins. 1974. Effect of bean root rot on dry and pinto bean production in southwestern Colorado. Plant Dis. Rptr. 58:890-892.

Kerr, E.D., J.R. Steadman & L.A. Nelson. 1978. Estimation of white mold disease reduction of yield and yield components of dry edible beans. Crop Sci. 18:275-279.

Khairallah, M.M., M.W. Adams & B.B. Sears. 1990. Mitochondrial DNA polymorphisms of Malawian bean lines: further evidence for two major gene pools. Theor. Appl. Genet. 80:753-761.

Laing, D.R., P.G. Jones & J.H.C. Davis. 1984. Common bean (*Phaseolus vulgaris* L.). pp. 305-352. *In* P.R. Goldsworthy & N. Fisher (eds.), The Physiology of Tropical Field Crops. Wiley, New York, U.S.A.

Maiti, R.K. 1997. *Phaseolus* spp.: Bean Science. Science Publishers, Enfield, U.S.A.

Maréchal, R., J.M. Mascherpa & F. Stainier. 1978. Etude taxonomique d'un groupe complexe d'especes des generes *Phaseolus* et *Vigna* (Papilionaceae) sur la base de donées morphologiques et polliniques, traitées par l'analyse informatique. Boissiera 28:1-273.

Morales, F.J. & A.I. Niessen. 1988. Comparative responses of selected *Phaseolus vulgaris* germplasm inoculated artificially and naturally with bean golden mosaic virus. Plant Dis. 72: 1020-1023.

Park, S.J. & R.I. Buzzell. 1995. Common bean. pp. 1-16. *In* A.E. Slinkard & D.R. Knott (eds.), Harvest of Gold: The History of Field Crop Breeding in Canada. Univ. of Saskatchewan, Saskatoon, Canada.

Robertson, L.S. & R.D. Frazier (eds.). 1978. Dry Bean Production-Principles and Practices. Michigan State Univ. Ext. Bull. E-1251, East Lansing, U.S.A.

Saettler, A.W. 1989. Common bacterial blight. pp. 261-283. *In* H.F. Schwartz & M.A.Pastor-Corrales (eds.), Bean Production Problems in the Tropics. 2nd ed. CIAT, Cali, Colombia.

Saettler, A.W. & H.S. Potter. 1970. Chemical control of halo bacterial blight in field beans. Michigan Agr. Exp. Sta. Res. Rept. 98, East Lansing, U.S.A.

Schoonhoven, A. van & C. Cardona. 1980. Insects and other bean pests in Latin America. pp. 363-412. *In* H.F. Schwartz & G.E. Gálvez (eds.), Bean Production Problems: Disease, Insect, Soil and Climatic Constraints of *Phaseolus vulgaris*. CIAT, Cali, Colombia.

Schoonhoven, A. van & O. Voysest (eds.). 1991. Common Beans: Research for Crop Improvement. C.A.B. Int., Wallingford, U.K. & CIAT, Cali, Colombia.

Schwartz, H.F. & M.A. Pastor-Corrales (eds.). 1989. Bean Production Problems in the Tropics. 2nd ed. CIAT, Cali, Colombia.

Shii, C.T., M.C. Mok, S.R. Temple & D.W.S. Mok. 1980. Expression of developmental abnormalities in hybrids of *Phaseolus vulgaris* L.: interaction between temperature and allelic dosage. J. Hered. 71:218-222.

Singh, S.P. 1982. A key for identification of different growth habits of *Phaseolus vulgaris* L. Annu. Rpt. Bean Improv. Coop. 25:92-95.

Singh, S.P. 1989. Patterns of variation in cultivated common bean (*Phaseolus vulgaris*, Fabaceae). Econ. Bot. 43:39-57.

Singh, S.P. 1992. Common bean improvement in the tropics. Plant Breed. Rev. 10:199-269.

Singh, S.P., P. Gepts & D.G. Debouck. 1991a. Races of common bean (*Phaseolus vulgaris*, Fabaceae). Econ. Bot. 45:379-396.

Singh, S.P. & J.A. Gutiérrez. 1984. Geographical distribution of DL_1 and DL_2 genes causing hybrid dwarfism in *Phaseolus vulgaris* L., their association with seed size, and their significance to breeding. Euphytica 33:337-345.

Singh, S.P., J.A. Gutiérrez, A. Molina, C. Urrea & P.Gepts. 1991b. Genetic diversity in cultivated common bean. II. Marker-based analysis of morphological and agronomic traits. Crop Sci. 31:23-29.

Singh, S.P., R. Nodari & P. Gepts. 1991c. Genetic diversity in cultivated common bean. I. Allozymes. Crop Sci. 31:19-23.

Singh, S.P. & O. Voysest (eds.). 1997. Taller de Mejoramiento de Frijol para el Siglo XXI: Bases para Una Estrategia para América Latina. CIAT, Cali, Colombia.

Thung, M. 1990. Phosphorus: a limiting nutrient in bean (*Phaseolus vulgaris* L.) production in Latin America and field screening for efficiency and response. pp. 501-521. *In* N. El Bassam, M. Dambroth & B.G. Loughman (eds.), Genetic Aspects of Plant Mineral Nutrition. Kluwer, Dordrecht, Netherlands.

Thung, M.D.T. & I.P. de Oliveira. 1998. Problemas Abióticos que Afectam a Produção do Feijoeiro é Seus Métodos de Controle. EMBRAPA- CNPAF, Santo Antônio de Goias, Brazil.

Tietyen-Clark, J. 1986. The benefits of bean. pp. 1. *In* J.W. Andersen (ed.), Diabetes and Nutr. News. HCF Diabetes Foundation, Lexington, U.S.A.

Vieira, C. 1967. O Feijoeiro-Comum: Cultura, Doenças e Molhoramiento. Univ. Rural do Estado de Minas Gerais, Viçosa, Brazil.

Vieira, C. 1988. *Phaseolus* genetic resources and breeding in Brazil. pp. 467-483. *In* P. Gepts (ed.), Genetic Resources of *Phaseolus* Beans. Kluwer, Dordrecht, Netherlands.

Vieira, C., T.J. de Paula Jr. & A. Borém (eds.). 1998. Feijão: Aspectos Gerais e Cultura no Estado de Minas. UFV, Viçosa, Brazil.

Yoshii, K., G.E. Gálvez & G. Alvarez. 1976. Estimation of yield losses in beans caused by common blight. Proc. Amer. Phytopath. Soc. 3:298-299.

Voysest, O. 1983. Variedades de Frijol en América Latina y su Origen. CIAT, Cali, Colombia.

Voysest, O. (ed.). 1998. An Ecoregional Framework for Bean Germplasm Development and Natural Resources Research. CIAT, Cali, Colombia.

Voysest, O. & M. Dessert. 1991. Bean cultivars: classes and commercial seed types. pp.119-162. *In* A. van Schoonhoven & O. Voysest (eds.), Common Beans: Research for Crop Improvement. C.A.B. Int., Wallingford, U.K. & CIAT, Cali, Colombia.

Wall, D. (ed.). 1973. Potentials of Field Beans and Other Legumes in Latin America. CIAT, Cali, Colombia.

Westphal, E. 1974. Pulses in Ethiopia, their Taxonomy and Agricultural Significance. Agric. Res. Rpt. No.815. Center for Agricultural Publishing and Documentation, Wageningen, Netherlands.

White, J., G. Hoogenboom, F.J. Ibarra-Pérez & S.P. Singh (eds.). 1988. Research on Drought Tolerance in Common Bean. CIAT, Cali, Colombia.

White, J. & D.R. Laing. 1989. Photoperiod response of flowering in diverse genotypes of common bean (*Phaseolus vulgaris*). Field Crops Res. 22:113-128.

Woolley, J., R. Lépiz I., T. De A. Portes e Castro & J. Voss. 1991. Bean cropping systems in the tropics and subtropics and their determinants. pp. 679-706. *In* A. van Schoonhoven & O. Voysest (eds.), Common Beans: Research for Crop Improvement. C.A.B. Int., Wallingford, U.K. & CIAT, Cali, Colombia.

Wortmann, C.S., R.A. Kirkby, C.A. Eledu & D.J. Allen. 1998. Atlas of Common Bean (*Phaseolus vulgaris* L.) Production in Africa. CIAT, Cali, Colombia.

Zaumeyer, W.J. & J.P. Meiners. 1975. Disease resistance in beans. Annu. Rev. Phytopath. 13: 313-334.

Zaumeyer, W.J. & H.R. Thomas. 1957. A Monographic Study of Bean Diseases and Methods for their Control. U.S.D.A. Agric. Tech. Bull. No.868. US Govt. Print. Off., Washington DC, U.S.A.

Zimmermann, M.J.O., M. Rocha & T. Yamada (eds.). 1988. Cultura do Feijoeiro: Fatores que Afetam a Produtividade. Associacão Brasileira para Pesquisa da Potassa e do Fosfato, Piracicaba, Brazil.

Chapter 2

Diversity in *Phaseolus* Species in Relation to the Common Bean

Daniel G. Debouck
Centro Internacional de Agricultura Tropical (CIAT), A.A. 6713, Cali, Colombia

Key words: cultivated and wild *Phaseolus* species, evolution under domestication, founder effect, gene pools, grouping of species, useful traits

1. INTRODUCTION

The genus was originally defined by Linnaeus (see Delgado Salinas, 1985; Westphal, 1974). The poor initial definition of the genus, together with the biological wealth of tropical forms in this group of legumes, resulted in the naming of hundreds of species (over 400), especially in the period 1810-1910. Early reviews by Bentham (1840), Hassler (1923), and Piper (1926), however, contributed to the clarification of natural groups at a higher level, and lead to the definition of several sections. Consolidation of these sections, mostly after 1950, thanks to the contributions of Urban (1928), Verdcourt (1970), Maréchal et al. (1978), and Lewis & Delgado Salinas (1994), resulted in several new genera, including *Vigna, Phaseolus* sensu stricto, *Macroptilium, Ramirezella*, and recently *Misanthus*. At the International Legume Conference of 1978, a definition of the genus was narrowly defined for a natural group of American legumes within the Phaseolinae with the following main attributes: stipules not extending below insertion, presence of uncinate hairs, floral bracts persistent up to or past flowering, absence of extra floral nectaries, and style not extending beyond the stigma.

The grouping of species within the *Phaseolus* sensu stricto, gene pools in relation to the common bean (*Phaseolus vulgaris* L.), the founder effect in the five cultigens, evolution under domestication, and some useful traits available in different species for genetic improvement of the common bean will be covered in this chapter. For additional details on these topics readers should refer to the literature cited.

2. THE GROUPING OF SPECIES WITHIN *PHASEOLUS*

If a definition of the genus is now agreed upon by most legume taxonomists, the exact number of species within the genus is still unknown. The reviews by Delgado Salinas (1985), Lackey (1983), and Debouck (1991) indicate 36, 31, and 52 species, respectively. The list presented in Table 1 amends my previous work, thanks to several communications with Dr. G. Freytag during 1992-98. Despite the fact that many, if not all, species of this list are well accepted at present, it still seems premature to set a definitive number and nomenclature of *Phaseolus* species, as field work continues to reveal new taxa (e.g., *P. costaricensis*: Debouck et al., 1989a; Freytag & Debouck, 1996). Other studies, particularly with the help of biochemical (e.g., Jaaska, 1996) and molecular (e.g., Delgado Salinas et al., 1998; Hamann et al., 1995; Jacob et al., 1995; Schumann & Nagl, 1995) markers, have contributed to confirm taxa already known (many of them presented in Table 1). In some cases the relationships among them and their phylogeny have also been clarified. Field work and molecular taxonomy data are complementary and can contribute substantially to improve the grouping of species in the near future.

Based on floral morphology, Maréchal & Baudoin (1978) divided the genus into three sections: *Phaseolus, Alepidocalyx,* and *Minkelersia*, with 24, 3, and 4 species, respectively. Lackey (1983) merged the last two sections, and recognized two sections, *Phaseolus* and *Minkelersia*, with 24 and 7 species, respectively. Delgado Salinas (1985) suggested four sections: *Chiapasana* (with 1 species), *Phaseolus* (25 species), *Minkelersia* (8 species), and *Xanthotricha* (2 species). Information from chloroplast DNA polymorphisms and DNA sequences might confirm such groupings. Data available so far (e.g., Delgado Salinas et al., 1993) indicate that there is some reason to recognize four such sections without multiplying the intrageneric classifications. The largest section, *Phaseolus*, however, seems heterogeneous on the basis of cpDNA polymorphism (Delgado Salinas et al., 1993). Recently, Delgado Salinas et al. (1998) indicated the monophyly of the genus with seven major lineages. Results presented elsewhere (Fofana et al., 1999) also show the presence of at least two lineages, one including the common bean and another one including the Lima bean. The natural groups of species (on the basis of morphology, hybridology, palynology, and molecular genetics) which reflect different lines of evolution and speciation are indicated below. Apart from the groups with few obviously related species, there are several isolated species that could represent unique phylum or taxa for which companion species are yet to be discovered.

Table 1. List of *Phaseolus* sensu stricto species (revised after Debouck, 1991)

Section Chiapasana
(sensu Delgado Salinas, 1985)

chiapasanus Piper

Section Phaseolus

acutifolius Asa Gray [1]
altimontanus Freytag & Debouck
angustissimus Asa Gray
carteri Freytag & Debouck
coccineus L. [1]
costaricensis Freytag & Debouck
esperanzae Seaton
filiformis Bentham
glabellus Piper
grayanus Woot. & Standl.
jaliscanus Piper
laxiflorus Piper
leptophyllus G. Don
leptostachyus Bentham
lunatus L. [1]
macrolepis Piper
maculatus Scheele
marechalii Delgado
micranthus Hook. & Arn.
Microcarpus Mart.
Minimiflorus Norvell ex Freytag & Debouck
mollis Hook. & Arn.
neglectus Hermann
oaxacanus Rose
oligospermus Piper
pachyrrhizoides Harms
parvifolius Freytag
pedicellatus Bentham
Polyanthus Greenman [1]
Polymorphus S. Watson
Polystachyus Britt., Stern & Pogg
purpusii Brandegee
ritensis Jones
salicifolius Piper
sonorensis Standley
talamancensis Debouck
tuerckheimii Donnell-Smith
vulgaris L.[1]
xolocotzii Delgado

Section Minkelersia

(sensu Lackey, 1983; including *Alepidocalyx* M.M.S.)

amblyosepalus (Piper) Morton
nelsonii Maréchal, Mascherpa & Stainier
parvulus Greene
pauciflorus Sessé & Mociño
perplexus Delgado
plagiocylix Harms
pluriflorus Maréchal, Mascherpa & Stainier
tenellus Piper

Section Xanthotricha
(sensu Delgado Salinas, 1985)

esquicensis Freytag
gladiolatus Freytag & Debouck
hintonii Delgado
magnilobatus Freytag & Debouck
xanthotrichus Piper
zimapanensis (Delgado) Jaaska

[1]Species with domesticated forms.

The isolation of *P. chiapasanus* from Oaxaca and Chiapas, Mexico, is not fully confirmed by cpDNA polymorphisms (Delgado Salinas et al., 1993). However, *P. chiapasanus* is so distinct, because of its gigantism of flowers and pods and because all its parts turning black on drying, that it could deserve a section of its own. No other related species has been found. Two other isolated species, *P. leptophyllus* (from central Guerrero, Mexico) and *P. sonorensis* (from south-eastern Sonora, Mexico) look morphologically distinct from the other species of section *Phaseolus*, the first because of its flap-like foliar margins and the second because of its larger flowers. They are not at all related nor related to *P. chiapasanus* nor to any other species of the genus on the basis of current evidence. I am, however, reluctant to create a novel section for each of them.

P. vulgaris could be chosen as representative of the *Phaseolus* section, with long pedicels in comparison to calyx, short calyx teeth in comparison to tube, a 2-revolutions keel, and a 2-10 ovules pod. This section includes the five cultivated species, all coming from wild ancestors still existing in parts of the Neotropics through changes induced by domestication (Debouck & Smartt, 1995). Each cultivated species forms a primary gene pool with its wild ancestral form (Table 2 and Figure 1) (Smartt, 1985). Secondary and tertiary gene pools may exist for all of them, depending on phylogenetic events that lead to the formation of the biological species, long before domestication took place (Debouck & Smartt, 1995). As detailed below, while three cultivated species, *P. vulgaris, P. polyanthus*, and *P. coccineus* have been part of the same evolutionary phylum, the Lima bean (*P. lunatus*), and, to a lesser degree, the tepary (*P. acutifolius*), are not evolutionarily related to this phylum nor to each other. The phylum of *P. lunatus* seems to be the largest, while *P. acutifolius* would have the smallest phylum. Another large phylum in the section *Phaseolus* is the one of *P. pedicellatus*.

In contrast with the synonymy proposed by Delgado Salinas (1985), it seems preferable, pending further information, not to merge the species related to *P. pedicellatus*. This species with lobed leaves is known from the Eje Volcánico to the Huasteca and southern Nuevo León, Mexico. Its range to the West and the South needs to be documented. So, the following taxa are still maintained apart: *P. esperanzae* (from the northern slope of Eje Volcánico, Mexico), *P. grayanus* (from Arizona, U.S.A. to Durango, Mexico), *P. laxiflorus* (from southern Huasteca, Mexico), *P. oaxacanus* (from the central mountains of Oaxaca, Mexico), *P. polymorphus* (from western San Luis Potosi and Guanajuato, Mexico), and *P. purpusii* (only from the type locality in San Luis Potosi, Mexico). There is recent evidence (Jaaska, 1996) to separate *P. grayanus* from *P. pedicellatus*.

Table 2. Gene pools of the two main *Phaseolus* bean crops and their number of accessions available from CIAT

Primary	Secondary			Tertiary		
	Close	Intermediate	Distant	Close	Intermediate	Distant
Common bean: cultivated forms (over 29,000)		*P. polyanthus*: cultivated forms (338)	*P. coccineus*: cultivated forms (574)		*P. acutifolius*: cultivated forms (163)	
Common bean: Wild forms (over 1,315)	*P. costaricensis*: wild only (20)	*P. polyanthus*: wild forms (7)	*P. coccineus*: wild forms (110)		*P. acutifolius*: wild forms (156) *P. parvifolius*: wild only (6)	*P. filiformis*: wild only (10)
Lima bean: cultivated forms (1,867)						
Lima bean: wild forms (178)	*P. pachyrrhizoides*: wild only (16)			*P. maculatus*: wild only (25) *P. ritensis*: wild only (16)	*P. jaliscanus*: wild only (2)	*P. salicifolius*: wild only (1)

Source: adapted from Singh et al. (1997b).

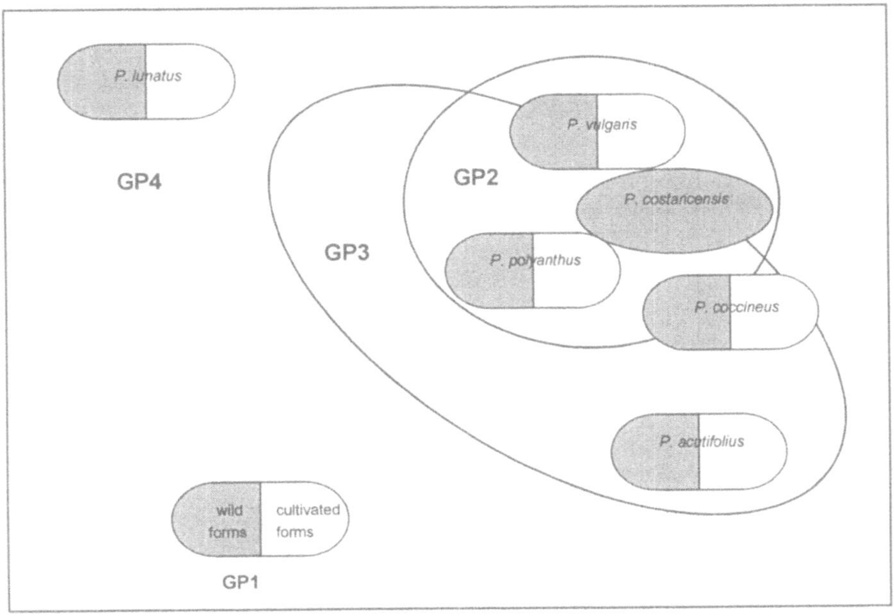

Figure 1. Schematic representation of the different gene pools of the five cultivated species of *Phaseolus* beans, in relation to the common bean. The primary gene pool of each species includes cultivated and wild forms (after Debouck & Smartt, 1995).

The northwestern flat highlands of Mexico, with pine-oak grasslands, harbor *P. maculatus*, which is replaced by *P. ritensis* in the mountainous ranges to the west and the north (Nabhan et al., 1980). Two poorly known species exist in pine forests of the Sierra Madre Occidental of Mexico, *P. salicifolius* in Sinaloa and *P. jaliscanus* in Jalisco (Nabhan, 1990). Recently identified by Delgado Salinas (1985), *P. marechalii* is distributed in the pine forests of both slopes of the Eje Volcánico of central Mexico and *P. xolocotzii* on the northern slope of Sierra Madre del Sur. All these species might be related, with varying levels of compatibility to the tertiary gene pool of *P. lunatus* (Fofana et al., 1999; Katanga & Baudoin, 1987a, 1990; Zink & Nagl, 1998).

The eastern mountainous range of Mexico has been less explored than the species-rich Sierra Madre Occidental, yet it harbors species such as *P. altimontanus* and *P. neglectus*, with probably a few more variants of the latter. Around the Gulf of Mexico, in cloud forests from Tamaulipas to Chiapas, one can find *P. glabellus*, once thought to be close to *P. coccineus*, but now known to be completely distant (Llaca et al., 1994; Schmit et al., 1993, 1996).

A small group of species (i.e., *P. angustissimus, P. carteri, P. filiformis*), from very dry habitats of northwestern Mexico and the southwestern U.S.A., has noteworthy rugose seed testa. This trait is probably related to germination ability under low soil moisture. Available evidence (see below) points to their phylogenetic distance from the common bean.

P. microcarpus, distributed from Durango to Chiapas, Mexico, and eastern Guatemala (Delgado Salinas, 1985), is easily identifiable because of its rhomboid 1-2 seed pods. A poorly collected and perhaps related species with a prow-ended pod is *P. minimiflorus* from the hillsides of Sinaloa to Colima, Mexico.

Perhaps only one species, *P. polystachyus*, is endemic to the U.S.A., and its presence in Nuevo León, Mexico, is still disputed. Two variants of *P. polystachyus* are usually recognized: *P. sinuatus* Nutt. from Florida and *P. smilacifolius* Pollard from the Carolinas. If these variants still exist, they need further documentation. *P. polystachyus* is remotely related to the Lima bean (Lorz, 1952). Other endemic species, which are not widely distributed, include *P. macrolepis* from Central Guatemala and *P. talamancensis* from eastern Costa Rica. So far I don't know whether these species are related, though they are not related to any other group of known species.

As discussed elsewhere (Caicedo et al., 1999; Debouck, 1996; Fofana et al., 1999), the Lima bean probably has an Andean origin, with its secondary gene pool made of South American species such as *P. pachyrrhizoides* and *P. mollis* (restricted to the Galapagos). Its large-seeded wild form extends from Cajamarca, Peru to Imbabura, Ecuador, and possibly into Colombia (Toro Ch. et al., 1993), while its small-seeded form extends at lower altitudes from Sinaloa, Mexico to Salta, Argentina (Gutiérrez Salgado et al., 1995). As shown elsewhere (Caicedo et al., 1999), the group of Andean species, *P. augusti* Harms, *P. bolivianus* Piper, and *P. pachyrrhizoides* is related to the Lima bean, and probably constitutes a single species with geographic variants.

P. oligospermus extends from Chiapas, Mexico (although it might have a related group of populations in western Jalisco, Mexico) to Costa Rica, and *P. tuerckheimii* is distributed from Guatemala to western Panama. Although these species are often confused, they are quite distinct and perhaps remotely related to the Lima bean (Jacob et al., 1995).

As indicated by several authors, *P. leptostachyus* and *P. micranthus* could form a group of its own, the only group in the genus with $2n = 2x = 20$ (while all other species are of $2n = 2x = 22$). Given the extent of its range (from Nuevo León, Mexico to Costa Rica; Debouck, 1991), *P. leptostachyus* may harbor several variants (previously named *P. intonsus* or *P. opacus*).

The species of the other sections *Minkelersia* and *Xanthotricha* are probably far too distant in relation to the common bean (see below), and shall thus not be considered here. For details about these two sections, readers may refer to other works (Delgado Salinas, 1985; Lackey, 1983; Maréchal et al., 1978).

3. THE GROUPING OF SPECIES IN RELATION TO THE COMMON BEAN

I shall review here the relationships between different species and the common bean (considering only the wild relatives), starting with *P. vulgaris* as the study of its primary gene pool might give insight into the origin of the phylum to which it belongs.

3.1 The Primary Gene Pool

Two major gene pools were first recognized in the wild form, matching with its geographic distribution (Gepts, 1988; Gepts et al., 1986; Koenig & Gepts, 1989). A third gene pool in the northern Andes was described later (Debouck et al., 1993). It contains unique genetic diversity, probably not the result of crosses between wild forms of the Middle American and Andean gene pools occurring there in sympatry (Becerra Velásquez & Gepts, 1994; Khairallah et al., 1992; Tohme et al., 1996). This third gene pool could be ancestral in the evolution of wild common bean (Kami et al., 1995; Tohme et al., 1996). The present location of this gene pool in the subhumid montane forests of southern Ecuador and northern Peru (Debouck et al., 1989b, 1993) might be a relict one. Due to climatic oscillations during the early Pleistocene it might have occupied a broader habitat that included Colombia and Central America (Debouck, 1996). The finding of a third gene pool of wild common bean in Pacific northern South America was interesting for two reasons. First, it brought light to the "direct migration" scenario; none of the major wild gene pools would derive straight from the other, but rather they would all come from an ancestral stock. Second, in comparison to species of the *P. coccineus* complex, the similarities between the phaseolin of this third gene pool, the 'I' phaseolin (Koenig et al., 1990), and the globulin of *P. polyanthus*, particularly the wild *P. polyanthus* from central Guatemala (Schmit & Debouck, 1991), are perhaps not fortuitous, and suggest affinities with this species. Finally, there might be a fourth gene pool in Colombia (Tohme et al., 1996) that is still poorly understood. The relationships of this wild gene pool with the wild form of southern Ecuador and northern Peru, and with the wild forms of eastern Central America need to be clarified.

3.2 The Secondary Gene Pool

Affinities between the common bean and other *Phaseolus* species, particularly *P. coccineus*, have been suspected by taxonomists and geneticists. Linnaeus established the two species in 1753. In 1763 he considered *P. coccineus* as a variety of *P. vulgaris* (Westphal, 1974), acknowledging the small differences between these two taxa. This morphological similarity may have encouraged scholars to cross them early in the history of plant breeding (Fermond, 1855). Mendel provided information about the cross between *Phaseolus nanus* L. (= *P. vulgaris*) and *P. multiflorus* Lam. (= *P. coccineus*) in 1866 (Blanc, 1984; Grant, 1975). In the period 1940-1985, the two species were crossed frequently (Alvarez et al., 1981; Baggett, 1956; Bannerot, 1979; Baudoin et al., 1985; Cheng et al., 1981; Ibrahim & Coyne, 1975; Lamprecht, 1945; Le Marchand, 1971; Manshardt & Bassett, 1984; Park & Dhanvantari, 1984; Thomas, 1964; Wall, 1970; Wall & York, 1957) for a wide range of purposes. The goals of these projects ranged from the development of new ornamentals (Lamprecht, 1945) to the transfer of disease resistance (Baggett, 1956; Park & Dhanvantari, 1984; Steadman, 1993) and yield (Wilkinson, 1983). In spite of this work, practical achievements, in terms of release of commercial cultivars, have been few (Hucl & Scoles, 1985; Singh, 1992).

Authors also realized that the reciprocal cross using *P. coccineus* as female was difficult (Bannerot, 1979; Freytag, 1975; Ockendon et al., 1982; Smartt et al., 1974), and observed that in a few generations segregants naturally reverted back to the cytoplasm donor parent. Major genes have already established a barrier between these two species, and chromosome pairing is not perfect (Cheng et al., 1981; Guo et al., 1991, 1994; Maréchal, 1971; Shii et al., 1982) (Table 3). *P. coccineus* and *P. vulgaris* may have ancestral genetic stock in common (Hervieu et al., 1993, 1994), as demonstrated by the similarity of many nuclear genes (Jaaska, 1996; Kloz & Klozová, 1974; Sullivan & Freytag, 1986). They may have evolved for a long time as distinct species, as differences between the two can be detected at the cpDNA level (Llaca et al., 1994; Schmit et al., 1993). Geneticists attempted to cross *P. vulgaris* with wild variants of *P. coccineus*, since reproductive isolation appears to be partly the result of domestication (Miranda Colín & Evans, 1973). The wild form of *P. coccineus*, previously known as *P. obvallatus* Schlecht. or *P. formosus* Kunth, was used for a few crosses (Le Marchand, 1971; Maréchal, 1971), but it was insufficiently exploited. Two variants of wild *P. coccineus*, *P. griseus* Piper (with scarlet flowers and silvery pubescence, from Jalisco, Mexico) and *P. striatus* Brandegee (with purple flowers, from Puebla, Mexico; reported as *purpurascens*: Jacob et al., 1995; Schmit et al., 1991), could perhaps be distinguished at the infraspecific level. Their use in wide crossing has been limited (Schmit et al., 1991).

Another related species is *P. polyanthus* (synonyms: *P. dumosus* McFady., and *P. coccineus* subsp. *darwinianus* Hernández & Miranda) (Table 2 and Figure 1). Based on morphology and cytology, *P. polyanthus* was thought to be the result of a natural cross between cultivated *P. coccineus* and *P. vulgaris* (Hernández Xolocotzi et al., 1959). Later, on the basis of allozymic evidence, this species was considered to be a particular variant of *P. coccineus* (Piñero & Eguiarte, 1988). The discovery of wild *P. polyanthus* in central Guatemala suggested that there might instead be a fifth case of domestication in the genus from a different wild ancestor (Schmit & Debouck, 1991). *P. polyanthus* cpDNA was shown to be close to that of *P. vulgaris*, and particularly the wild *P. vulgaris* from Ecuador with 'I' phaseolin (Schmit et al., 1993). Later, Llaca et al. (1994) demonstrated that the nuclear genome of *P. polyanthus* is close to that of *P. coccineus*, and confirmed that the chloroplast genome is close to that of *P. vulgaris*. In wide crosses, *P. polyanthus* crossed more easily with *P. coccineus* and related forms than with *P. vulgaris* (Maréchal, 1971), particularly if the later is the pollen parent (Camarena & Baudoin, 1987). These facts supported the hypothesis of a distinct species (Smartt, 1973), and could be explained in the following way: *P. polyanthus* belongs to the phylum of *P. vulgaris*, but its nuclear genome has been introgressed with *coccineus* genes, as these three taxa share some ancestral genetic stock (Hervieu et al., 1993, 1994). *P. polyanthus* would thus have a "reticulate" origin (Llaca et al., 1994), and such introgression makes the use of *P. polyanthus* in widecrossing with *P. vulgaris* to be of limited potential (Table 3). The question of whether all *P. polyanthus* accessions, particularly the wild ones, are equally contaminated, still needs to be addressed.

The search for additional wild forms of *P. coccineus* in Central America lead to the discovery in Costa Rica of a novel species with epigeal germination (like *P. vulgaris*), bright cerise flowers and cross pollination reproductive biology (like *P. coccineus*) (Debouck et al., 1989a). Morphological differences were strong enough to recognize it as a separate species, *P. costaricensis* Freytag & Debouck (Freytag & Debouck, 1996). Its cpDNA was shown to be close to that of *P. vulgaris* (Schmit et al., 1993). So far, it is not certain that its nuclear genome has been contaminated by *P. coccineus* genes. If not, possibilities may exist for easier wide-crossing and genetic recombination with *P. vulgaris*. *P. costaricensis* can be crossed naturally with *P. polyanthus* (Debouck et al., 1989a), and *P. costaricensis* crosses with *P. vulgaris* (Singh et al., 1997a). Finding a new species suggests there may be even more new species in the phylum of *P. vulgaris*. According to available cpDNA evidence (Schmit et al., 1993), the two species, namely, *P. polyanthus* and *P. costaricensis* share the same polymorphisms with *P. vulgaris*. From herbarium and field surveys, it seems that there might be more species resembling *P. polyanthus* in Central America and western Mexico (McVaugh, 1987), and a search for them should be a priority for future germplasm exploration.

Table 3. Chromosome pairing at metaphase I of interspecific hybrids between different *Phaseolus* species

Cross[1]	Caryotype	References
coccineus x *polyanthus*	10.98 II + 0.04 I	Maréchal, 1971
vulgaris x *coccineus*	10.84 II + 0.32 I	Maréchal, 1971
coccineus x *vulgaris*	obtained; not computed	Shii et al., 1982
vulgaris x *polyanthus*	10.56 II + 0.88 I	Maréchal, 1971
polyanthus x *vulgaris*	obtained; not computed	Camarena & Baudoin, 1987
vulgaris x *acutifolius*	6.0 II + 8.0 I	Katanga & Baudoin, 1990
acutifolius x *vulgaris*	8.1 II + 5.5 I	Prendota et al., 1982
vulgaris x *filiformis*	8.1 II + 5.7 I	Maréchal & Baudoin, 1978
vulgaris x *angustissimus*	obtained; sterile	Belivanis & Doré, 1986
vulgaris x *lunatus*	obtained; sterile	Kuboyama et al., 1991
vulgaris x *maculatus*	3.0 II + 15.0 I	Katanga & Baudoin, 1990
lunatus x *jaliscanus*	9.97 II + 2.06 I	Katanga & Baudoin, 1990
lunatus x *polystachyus*	6.0 II + 10.0 I	Dhaliwal et al., 1962
lunatus x *ritensis*	10.02 II + 1.95 I	Katanga & Baudoin, 1990
lunatus x *salicifolius*	9.19 II + 3.61 I	Katanga & Baudoin, 1990

[1]The female parent of each cross is shown on the left.

Linnaeus and Mendel probably did not foresee that their work would generate so many attempts to cross *P. coccineus* with *P. vulgaris*, and, indirectly, numerous efforts to assemble and study the germplasm of "*P. coccineus* and its allies" as described by Piper (1926). The interest in the 'coccinoids' generated one unexpected byproduct: some coccinoids were demonstrated not to belong to that group; so, *P. glabellus* was shown to be a distinct taxon (Hamann et al., 1995; Llaca et al., 1994; Schmit et al., 1993, 1996). Other coccinoids were shown to belong to the *P. vulgaris* phylum. Hucl & Scoles (1985) and Maréchal et al. (1978) suggest that the coccinoids shall be the reservoir of diversity with greatest potential once the primary gene pool and the *P. vulgaris* phylum have been fully explored.

3.3 The Tertiary Gene Pool

Tepary bean, *P. acutifolius*, has some useful attributes. These include tolerance to drought (Freeman, 1918; Hendry, 1919; Parsons & Howe, 1984), high temperatures (Garver, 1934; Petersen & Davis, 1982), and common bacterial blight (Coyne et al., 1963; Singh & Muñoz, 1999;

Valladares et al., 1977; Zapata et al., 1985). These advantages have inspired bean breeders to explore the possibilities of wide crosses with common bean. Several attempts were made (Alvarez et al., 1981; Al-Yasiri & Coyne, 1964; Andrade-Aguilar & Jackson, 1988; Honma, 1956; Jung et al., 1992; Mok et al., 1978; Smartt, 1970; Thomas & Waines, 1980, 1984), but success was limited, and embryo rescue was required to recover hybrid progenies. More recently, with the careful choice of donor parents (Federici & Waines, 1989; Nelson & Ascher, 1984; Parker & Michaels, 1986; Thomas & Waines, 1984) and the application of the congruity backcross scheme (Brown et al., 1996; Haghighi & Ascher, 1988; Mejía-Jiménez et al., 1994), a second series of crosses was attempted with greater success. Given the need for embryo rescue techniques and the complete sterility of F_1 hybrids, it seems reasonable to consider tepary as belonging to the tertiary gene pool of common bean, a conclusion reached by many authors (Andrade-Aguilar & Jackson, 1988; Prendota et al., 1982; Pueyo & Delgado Salinas, 1997; Smartt, 1970).

One species distributed around the Gulf of California, *P. filiformis*, has been crossed with common bean (Federici & Waines, 1988; Hucl & Scoles, 1985; Maréchal & Baudoin, 1978; Petzoldt & Dickson, 1987), the latter being used as female, but hybrid plants showed complete sterility. *P. angustissimus* has also been crossed successfully as male parent with the help of *in vitro* techniques (Belivanis & Doré, 1986; Petzoldt & Dickson, 1987). While these two species were thought to be closely related (Maréchal et al., 1978), their hybrid was almost completely sterile (Katanga & Baudoin, 1987b). Chromosome doubling has been attempted (Belivanis & Doré, 1986; Weilenmann de Tau et al., 1986), but as in previous cases (with *P. coccineus*: Smartt, 1970; Smartt & Haq, 1972; with *P. acutifolius*: Prendota et al., 1982; Thomas & Waines, 1984), this technique may not be useful, given the problems with the exploitation of amphidiploids. *P. filiformis* and *P. angustissimus* should be considered as belonging at best to the tertiary gene pool of the common bean, but at a different level in relation to *P. acutifolius*, as the tepary could not be used as a bridge species (Hucl & Scoles, 1985).

3.4 The Quaternary Gene Pool and Beyond

Researchers (Evans, 1980; Le Marchand et al., 1976) have dreamed about combining the productivity of Lima bean and its adaptation to tropical conditions with the wide acceptance of the common bean. Sadly, all attempts to cross *P. lunatus* with *P. vulgaris* have failed to produce a viable and fertile hybrid (Al-Yasiri & Coyne, 1966; Leonard et al., 1987; Mok et al., 1978). The early success claimed by Honma & Heeckt (1959) with the use of heterozygous parents is doubtful, as mentioned elsewhere (Hucl & Scoles, 1985; Smartt, 1979). Recently, this cross was re-attempted (Cabral & Crocomo, 1989; Kuboyama et al., 1991), but apparently without success beyond the early generations. The reciprocal cross, *P. lunatus* x *P. vulgaris* was even more difficult to obtain (Rabakoarihanta et al., 1979). The difficulty in getting fertile viable hybrids between common and Lima beans confirms their respective taxonomic positions in the genus. On the basis of palynological and hybridological evidence, these two species are believed to be at the extremes of an orthogenetic sequence (Le Marchand & Maréchal, 1977; Maréchal et al., 1978). Many species would fall between these species extremes, but most species have not been studied in great detail so far. Recently, Delgado Salinas et al. (1998) and Fofana et al. (1999) have indicated that the common bean and the Lima bean belong to different evolutionary pathways, thus explaining the phylogenetic distance.

The cross with *P. ritensis* used as male parent (Petzoldt & Dickson, 1987) seems difficult. The cross attempted earlier by Braak & Kooistra (1975) was realized with *P. maculatus*, and made possible using embryo rescue. The slight improvement in comparison to the cross made with *P. ritensis* (a hybrid could be obtained with *P. maculatus* but with 14 univalents) indicates, however, that such species are quite distant from the common bean. In contrast, they seem to be close to the Lima bean with which they have been successfully crossed (Katanga & Baudoin, 1990; Le Marchand et al., 1976). Curiously, more species, up to six so far (Katanga & Baudoin, 1987a; Le Marchand et al., 1976; Lorz, 1952), seem to be crossable with Lima bean than with common bean.

As expected because of the taxonomic distance (Maréchal et al., 1978), crosses between *Phaseolus* and *Macroptilium*, another genus distributed mainly in Brazil (Barbosa Fevereiro, 1986-87), have failed (Evans, 1980). Attempts to cross the common bean with the Asiatic species of *Vigna* subgenus *Ceratotropis* (Maréchal et al., 1978) such as *Vigna umbellata* (Thunb.) Ohwi & Ohashi, *Vigna mungo* (L.) Hepper, and *Vigna angularis* (Willd.) Ohwi & Ohashi [previously named *P. calcaratus* Roxb., *P. mungo* L. and *P. angularis* (Willd.) Wight, respectively, Debouck, 1991] have all failed (Al-Yasiri & Coyne, 1966). Possibilities of widecrossing with other members of the subtribe Phaseolinae still seem remote at this stage, but once again, many possibilities within the genus *Phaseolus* are unexplored so far (for instance to the author's knowledge no crosses were ever attempted with species of the sections *Minkelersia* or *Xanthotricha*).

4. THE FOUNDER EFFECT IN FIVE CULTIGENS

Genetic changes under domestication indicate an overall reduction of diversity, contrasting with an increased diversity in morphological traits (mostly the ones under direct, conscious selection). The reduction of genetic diversity called "founder effect" has been best demonstrated with molecular markers that are thought to be neutral towards the selection processes (Kimura, 1991). This reduction has been shown in *P. vulgaris* (Gepts et al., 1986), *P. lunatus* (Gutiérrez Salgado et al., 1995), *P. acutifolius* (Schinkel & Gepts, 1988), and *P. polyanthus* (Schmit & Debouck, 1991). The founder effect can be explained by the fact that just a few wild populations have been involved in the domestication process; why to go back to the wild once you have got success in the domestication? Thus only one part of the wild gene pool has been included into that of the cultigens. The founder effect is neither uniform across species nor across gene pools within species. For instance, the founder effect would be less pronounced in the Andean region as compared to Middle America, for both the common and Lima beans. This would be due to the number and location of wild populations involved in domestication. An extreme example is that of tepary. With very few wild populations domesticated probably in western Mexico (Garvin & Weeden, 1994), almost all future breeding potential would depend on the exploitation of wild germplasm (Debouck & Smartt, 1995). Crosses between wild populations and cultigens during incipient domestication through the "wild-weed-crop complex" (Beebe et al., 1997) might lessen the founder effect, although wild traits, usually being dominant, are likely to be eliminated soon (see below). The diversity of cytoplasms in the gene pools of the cultigens might also be affected.

Additional studies of the founder effect in the five cultigens are needed because of its practical implications for evolutionary studies and crop improvement. For example, resistance to bruchids (*Zabrotes subfasciatus* Boheman) was found only in wild common bean (Schoonhoven & Cardona, 1982; Schoonhoven et al., 1983). The resistance is due to a storage protein arcelin (Osborn et al., 1988), which exists predominantly in western Mexico (Acosta Gallegos et al., 1998) at low frequencies in the original wild populations (Osborn et al., 1986) where bean domestication probably did not occur.

5. EVOLUTION UNDER DOMESTICATION

The comparative evolution of the cultivated species under domestication has often been reviewed (Smartt, 1976, 1988, 1990) with little attention paid, however, to *P. polyanthus* (Table 4). As in most crops, changes in the harvested parts, pod and seed, as measured by size and 100-seed weight, are the most striking differences between the cultivated forms and their wild ancestors. Gigantism is, not surprisingly, also found in leaves, and, to a lesser degree, in flowers. Use of flowers and leaves for food (in southeastern Mexico and in the Great Lakes region of Africa, respectively) is probably recent for indirect selective pressure to be significant. Another recent selection in common bean, but so far not in the other species, is for very large pods, as in the yard-long bean, *Vigna unguiculata* (L.) Walp. cv-gr. *sesquipedalis* (Westphal, 1974). Apart from increase in seed size, reduction in hard seededness and dormancy has been obtained in the five cultigens, probably from unconscious selection first (the plants with even germination being eventually harvested) and later conscious selection. There has been selection for reduced antinutritional factors (i.e., antitrypsin factors, cyanide glycoside) in all five cultigens. Some landraces of *P. lunatus* still contain high levels of linamarin (Baudoin et al., 1991), indicating that they may have been domesticated for purposes other than food (Debouck, 1989).

Table 4. Evolutionary changes under domestication in the five cultivated *Phaseolus* species[1]

Traits	*Pv*	*Pp*	*Pc*	*Pa*	*Pl*
1. Seed					
Size	C	C	C	C	C
Dormancy	C	C	C	C	C
Hard seedcoat	C	C	C	C	C
Colors & patterns	C	C	C	C	C
2. Pod					
Lack of dehiscence	C	C	C	c	c[1]
Gigantism	C	C	c	c	c
3. Growth habit					
Few nodes	C	C	C	c	C
Terminal racemes	C	C	C	c[1]	C
Annualism	C	C	C	C	C
4. Photoperiod					
Insensitiveness	C	c	C	C	C
5. Biochemical changes	C	C[1]	C	C	C[1]

[1] *Pv* = *P. vulgaris*; *Pp* = *P. polyanthus*; *Pc* = *P. coccineus*; *Pa* = *P. acutifolius*; *Pl* = *P. lunatus;* C = change occurred; c = little or no change; C[1], c[1] = see text for additional explanations.

Interestingly, high linamarin seems independent of seed color. As expected, white-seeded cultivars have been selected in all five cultigens (with associated white flowers and non-pigmented vegetative parts). This mutation appears from time to time in wild populations (observed in wild *P. vulgaris* in Apurimac, Peru and in *P. lunatus* in Yucatan, Mexico); it was noticed and picked up by pre-Columbian agriculturists.

The reduction of seed dispersal mechanisms is also a preferred trait in most crops. Selection might well be unconscious, that is, traits were selected without deliberate efforts, but as the result of incipient domestication practices (Heiser, 1988; Pernès, 1983). In wild populations, such practices favored individuals with less pod shatter. In *Phaseolus* beans, the lack of pod dehiscence is due to the combination of two characters: reduction of parchment layer and absence of fibers along the two pod sutures. The complete "stringless" character is unknown in *P. acutifolius* and in *P. polyanthus*. The control of both characteristics by few major genes (Bassett, 1996) would allow breeders to alter the character in few generations. A reduced parchment layer has been observed in prehistoric beans from Ocampo, Mexico (Kaplan, 1965), indicating that this selection pressure was applied early in the domestication process. Common bean may have been domesticated first as snap bean because before ceramic times little was known about detoxifying antinutritional factors (Debouck, 1990).

Major evolutionary changes have occurred in growth habit. The original viny habit with profuse branching, seen in most wild species, was selected against to obtain a quick harvest. Terminal inflorescence causing reduced growth combined with few nodes on the main stem (i.e., bush determinate growth habit) has not yet been reported in *P. polyanthus* (Basurto Peña et al., 1996). But accessions with terminal raceme have been found in wild tepary (Debouck, 1992), indicating that the potential for changing growth habit is there. The dwarf growth habit already existed in the common bean in pre-Columbian times (Kaplan & Kaplan, 1988), suggesting mutations occurred early in crop history and there was keen interest by agriculturists to select and keep them.

In contrast, the selection of daylength-insensitive cultivars could be a modern post-Columbian activity, although the Araucanians in central Chile may have already started some selection (Wilhelm de Mosbach, 1992). This trait, combined with the dwarf bush growth habit, is believed to be the driving force in the ever-increasing expansion of the common bean to higher latitudes, in temperate summer environments of the northern hemisphere.

As pointed out elsewhere (Gepts & Debouck, 1991; Koinange et al., 1996), most of the evolutionary changes in beans are due to mutations of just a few genes, the increase in seed size being one noteworthy exception. With the possible evolution towards cleistogamy in common, Lima, and tepary bean, such evolutionary changes could have been relatively rapid. From Table 4, it seems that the species with the least realized potential is *P. polyanthus*, or, in other words, the cultigen with the highest potential for novel crop developments. Since this crop shows the least progress in increase of seed size in comparison to its wild ancestral form, one could conclude that this species was domesticated more recently than the other cultigens, although archaeological records are lacking. Though both species are largely allogamous, the exceptional increase in seed size in *P. coccineus* in comparison to its wild ancestor is impressive (its outbreeding behavior would thus not be the sole explanation).

Because of the extent of distribution of their respective wild ancestors and the large number of domestication events, common and Lima bean cultigens may have a wider genetic base than *P. acutifolius*, *P. coccineus*, and *P. polyanthus*. This can perhaps partly explain the vast ecological expansion of these two *Phaseolus* beans, in comparison to the limited success of the other three species. Evolutionary progress is diverse among the five bean cultigens, in terms of yield, seed size, growth habit, and ecological adaptation, and no single factor seems to account for that diversity (Debouck, 1992). Following Vavilov's "law of homologous series" (Vavilov, 1922), substantial genetic progress through breeding would thus be possible.

6. USEFUL TRAITS IN DIFFERENT SPECIES

The ecology of the original habitats of several species is likely to have selected for traits against abiotic stresses. Not surprisingly, *P. angustissimus, P. maculatus,* and *P. ritensis* are tolerant to frost (Buhrow, 1980). The 'coccinoids' are generally tolerant to low temperatures as well (Bannerot, 1979). Sources of tolerance against salinity are likely to be found in wild teparies, *P. filiformis,* and *P. minimiflorus*. Many *Phaseolus* species thrive in dry habitats, but as desert ephemerals they escape most of the drought stress. Some drought tolerance has been reported in *P. angustissimus* and *P. filiformis* (Buhrow, 1981).

Limited evaluation has been systematically done against pests and diseases. In spite of that, alien germplasm seems promising for common bean improvement. Good sources of tolerance to leafhopper (*Empoasca kraemeri* Ross & Moore) exist in *P. acutifolius* (C. Cardona, unpublished). Resistance to bruchids, due to seed protein arcelin, has been found in wild *P. vulgaris* (Acosta Gallegos et al., 1998; Cardona & Kornegay, 1989; Schoonhoven et al., 1983). It seems that arcelin-like proteins exist in other wild species as well (Mirkov et al., 1994; Pueyo & Delgado Salinas, 1997).

P. polyanthus is particularly known for its resistance to ascochyta blight (Schmit & Baudoin, 1992) as well as to white mold (Hunter et al., 1982). *P. coccineus* has been known as a source of resistance to anthracnose (Hubbeling, 1957), root rots (Wilkinson, 1983), white mold (Abawi et al., 1978), and bean yellow mosaic (Baggett, 1956). *P. acutifolius* and *P. costaricensis* might be useful for resistance to bean golden mosaic (Miklas & Santiago, 1996; Singh et al., 1997b). Sources of resistance to ascochyta blight have also been reported in *P. oligospermus*, *P. pedicellatus*, and *P. salicifolius* (Lepoivre & Baudoin, 1994).

Some specific attributes of wild species *per se* might be useful. For instance, hypogeal germination from many wild species, especially the *P. coccineus* group, might be used against bean fly (*Ophiomyia phaseoli* Tryon) in eastern Africa. The same group might be used as source of extrorse stigma and nectaries in hybrid bean programs. Unexpected variation exists for a few photosynthesis traits in wild common bean (González et al., 1995; Lynch et al., 1992). These traits might be useful to increase bean productivity.

7. CONCLUSIONS

Interspecific hybridization has been practiced in *Phaseolus* beans perhaps as long as for any other crop, since attempts were made even before genetics as a science came into being; yet practical success has been relatively limited (Hucl & Scoles, 1985; Singh, 1992). In the beginning, the species crossed with the common bean were probably not the ones with which some success could have been expected.

P. vulgaris is not an isolated species; a few other taxa belong to its phylum. In addition, interspecific hybridization was mostly restricted to the crossing of cultigens, where reproductive barriers might be stronger (Evans, 1980). Moreover, several workers (Andrade-Aguilar & Jackson, 1988; Rabakoarihanta et al., 1979; Thomas & Waines, 1984) have shown the importance of cytoplasmic effects for successful wide-crossing, starting with common bean itself (Singh, 1992), and predictive tools in this regard have been lacking. It should be now possible, however, to follow useful traits in segregating populations with DNA-based markers such as RFLPs, RAPDs, or SCARs (Adam-Blondon et al., 1994; Guo et al., 1994; Nodari et al., 1993; Schneider et al., 1997; see also Chapter 4 by Kelly & Miklas). If such tracking is done, it may be possible to use wild species for difficult traits such as photosynthesis (González et al., 1995; Lynch et al., 1992), seed yield (Singh et al., 1995) or for unsuspected traits (Rick & Smith, 1953). Progress is being made in understanding *Phaseolus* phylogeny (Delgado Salinas et al.,

1998). With increasing availability of molecular markers (Kelly & Miklas, 1998; see also Chapter 4) and development of linkage maps (Freyre et al., 1998; see also Chapter 3 by Gepts) wide-crossing should have greater potential.

ACKNOWLEDGMENTS

This work has been made possible thanks to several grants to the author from the Centro Internacional de Agricultura Tropical (Cali, Colombia), the International Board for Plant Genetic Resources (Rome, Italy), the United States Department of Agriculture (Beltsville, USA), and the World Conservation Union (Gland, Switzerland). The author acknowledges the support from national programs of plant genetic resources of Mexico, Guatemala, Costa Rica, Colombia, Ecuador, Peru, Bolivia, Chile, and Argentina. The author also expresses his deepest appreciation to Dr. G.F. Freytag for novel information about *Phaseolus* species. Comments on former versions of the manuscript by Prof. J.P. Baudoin and Drs. S. Beebe, S.P. Singh, J. Smartt, and J. Tohme have been particularly useful and appreciated. Thanks are due to Sandra Albarracín for the preparation of this manuscript.

REFERENCES

Abawi, G.S., R. Provvidenti, D.C. Crosier & J.E. Hunter. 1978. Inheritance of resistance to white mold disease in *Phaseolus coccineus*. J. Hered. 69:200-202.

Acosta Gallegos, J., C. Quintero, J. Vargas, O. Toro, J. Tohme & C. Cardona. 1998. A new variant of arcelin in wild common bean, *Phaseolus vulgaris* L., from southern Mexico. Genet. Resources & Crop Evol. 45:235-242.

Adam-Blondon, A.F., M. Sévignac, H. Bannerot & M. Dron. 1994. SCAR, RAPD and RFLP markers linked to a dominant gene (*Are*) conferring resistance to anthracnose in common bean. Theor. Appl. Genet. 88:865-870.

Alvarez, M.N., P.D. Ascher & D.W. Davis. 1981. Interspecific hybridization in *Euphaseolus* through embryo rescue. HortScience 16:541-543.

Al-Yasiri, S.A. & D.P. Coyne. 1964. Effect of growth regulators in delaying pod abscission and embryo abortion in the interspecific cross *Phaseolus vulgaris* x *P. acutifolius*. Crop Sci. 4:433-435.

Al-Yasiri, S.A. & D.P. Coyne. 1966. Interspecific hybridization in the genus *Phaseolus*. Crop Sci. 6:59-60.

Andrade-Aguilar, J.A. & M.T. Jackson. 1988. Attempts at interspecific hybridization between *Phaseolus vulgaris* L. and *P. acutifolius* A.Gray using embryo rescue. Plant Breed. 101:173-180.

Baggett, J.R. 1956. The inheritance of resistance to strains of bean yellow mosaic virus in the interspecific cross *Phaseolus vulgaris* x *P. coccineus*. Plant Dis. Rptr. 40:702-707.

Bannerot, H. 1979. Cold tolerance in beans. Annu. Rpt. Bean Improv. Coop. 22:81-84.
Barbosa Fevereiro, V.P. 1986-87. *Macroptilium* (Bentham) Urban do Brasil (Leguminosae - Faboideae - Phaseoleae - Phaseolinae). Arq. Jard. Bot. Rio de Janeiro 28:109-180.
Bassett, M.J. 1996. List of genes - *Phaseolus vulgaris* L. Annu. Rpt. Bean Improv. Coop. 39:1-19.
Basurto Peña, F.A., D. Martínez Moreno, M.A. Martínez Alfaro & A. Castellanos Villegas. 1996. Frijol gordo abreviador, una forma precoz de *Phaseolus coccineus* L. spp. *darwinianus* Hernández X. & Miranda C., en Nauzontla, Pue. Geografía Agrícola (México) 22-23:95-102.
Baudoin, J.P., J.P. Barthélémy & V. Ndungo. 1991. Variability of cyanide contents in the primary and secondary genepools of the lima bean, *Phaseolus lunatus* L. FAO/IBPGR Plant Genet. Resources Newsl. 85:5-9.
Baudoin, J.P., R. Maréchal, E. Otoul & F. Camarena. 1985. Interspecific hybridizations within the *Phaseolus vulgaris* - *Phaseolus coccineus* L. complex. Annu. Rpt. Bean Improv. Coop. 28:64-65.
Becerra Velásquez, V.L. & P. Gepts. 1994. RFLP diversity of common bean (*Phaseolus vulgaris*) in its centres of origin. Genome 37:256-263.
Beebe, S., O. Toro Ch., A.V. González, M.I. Chacón & D.G. Debouck. 1997. Wild-weed-crop complexes of common bean (*Phaseolus vulgaris* L., Fabaceae) in the Andes of Peru and Colombia, and their implications for conservation and breeding. Genet. Resources & Crop Evol. 44:73-91.
Belivanis, T. & C. Doré. 1986. Interspecific hybridization of *Phaseolus vulgaris* L. and *P. angustissimus* A. Gray using *in vitro* embryo culture. Plant Cell Rep. 5:329-331.
Bentham, G. 1840. De leguminosarum generibus commentationes. Annln. Wien. Mus. Naturg. 2:61-142.
Blanc, M. 1984. Gregor Mendel: la légende du génie méconnu. La Recherche 151:46-59.
Braak, J.P. & E. Kooistra. 1975. A successful cross between *Phaseolus vulgaris* L. and *P. ritensis* Jones with the aid of embryo culture. Euphytica 24:669-679.
Brown, R., N. Anderson & P. Ascher. 1996. Congruity backcross (CBC) hybrids used to create three-species hybrids with *Phaseolus acutifolius* cytoplasm. Annu. Rpt. Bean Improv. Coop. 39:98-99.
Buhrow, R. 1981. Stress tolerance of Southwestern Phaseolinae. Annu. Rpt. Bean Improv. Coop. 24:94-95.
Buhrow, R.A. 1980. Frost tolerance of the Phaseolinae of the Southwestern United States. Annu. Rpt. Bean Improv. Coop. 23:62-64.
Cabral, J.B. & O.J. Crocomo. 1989. Interspecific hybridization of *Phaseolus vulgaris, P. acutifolius* and *P. lunatus* using *in vitro* technique. Turrialba 39:243-246.
Caicedo, A.L., E. Gaitán, M.C. Duque, O. Toro Chica, D.G. Debouck & J. Tohme. 1999. AFLPs fingerprinting of Lima bean (*Phaseolus lunatus* L.) and related wild species of South America. Crop Sci. 39:(in press).
Camarena, F. & J.P. Baudoin. 1987. Obtention des premiers hybrides interspécifiques entre *Phaseolus vulgaris* et *Phaseolus polyanthus* avec le cytoplasme de cette dernière forme. Bull. Rech. Agron. Gembloux 22:43-55.
Cardona, C. & J. Kornegay. 1989. Use of wild *Phaseolus vulgaris* to improve beans for resistance to bruchids. pp. 90-98. *In* S. Beebe (ed), Current Topics in Breeding of Common beans. CIAT, Cali, Colombia.
Cheng, S.S., M.J. Bassett & K.H. Quesenberry. 1981. Cytogenetic analysis of interspecific hybrids between common bean and scarlet runner bean. Crop Sci. 21:75-79.

Coyne, D.P., M.L. Schuster & S. Al-Yasiri. 1963. Reaction studies of bean species and varieties to common blight and bacterial wilt. Plant Dis. Rptr. 47:534-537.

Debouck, D.G. 1989. Early beans (*Phaseolus vulgaris* L. and *P. lunatus* L.) domesticated for their aesthetic value? Annu. Rpt. Bean Improv. Coop. 32:62-63.

Debouck, D.G. 1990. Wild beans as a food resource in the Andes. Annu. Rpt. Bean Improv. Coop. 33:102-103.

Debouck, D.G. 1991. Systematics and morphology. pp. 55-118. *In* A. van Schoonhoven & O. Voysest (eds.), Common Beans: Research for Crop Improvement. C.A.B. Int., Wallingford, U.K. & CIAT, Cali, Colombia.

Debouck, D.G. 1992. Frijoles, *Phaseolus spp*. pp. 45-60. *In* E. Hernández Bermejo & J. León (eds.), Cultivos Marginados: Otra Perspectiva de 1492. FAO, Rome, Italy.

Debouck, D.G. 1996. Colombian common and Lima beans: views on their origin and evolutionary significance. Revista CORPOICA 1:7-15.

Debouck, D.G., R. Araya Villalobos, R.A. Ocampo Sánchez & W.G. González Ugalde. 1989a. Collecting *Phaseolus* in Costa Rica. FAO/IBPGR Plant Genet. Resources Newsl. 78/79:44-46.

Debouck, D.G., R. Castillo & J. Tohme. 1989b. Observations on little-known *Phaseolus* germplasm of Ecuador. FAO/IBPGR Plant Genet. Resources Newsl. 80:15-21.

Debouck, D.G. & J. Smartt. 1995. Beans, *Phaseolus* spp. (Leguminosae-Papilionoideae). pp. 287-294. *In* J. Smartt & N.W. Simmonds (eds.), Evolution of Crop Plants. 2nd ed. Longman, London, U.K.

Debouck, D.G., O. Toro, O.M. Paredes, W.C. Johnson & P. Gepts. 1993. Genetic diversity and ecological distribution of *Phaseolus vulgaris* (Fabaceae) in northwestern South America. Econ. Bot. 47:408-423.

Delgado Salinas, A. 1985. Systematics of the genus *Phaseolus* (Leguminosae) in North and Central America. Ph.D. dissertation. Univ. of Texas, Austin, U.S.A.

Delgado Salinas, A., A. Bruneau & J.J. Doyle. 1993. Chloroplast DNA phylogenetic studies in New World Phaseolinae (Leguminosae: Papilionoideae: Phaseoleae). Syst. Bot. 18:6-17.

Delgado Salinas, A., T. Turley & M. Lavin. 1998. Cultivated and wild species of *Phaseolus* in phylogenetic context. Amer. J. Bot. (supplem) 85:123. (abstract).

Dhaliwal, A.S., L.H. Pollard & A.P. Lorz. 1962. Cytological behavior of an F_1 species cross (*Phaseolus lunatus* L. var Fordhook x *Phaseolus polystachyus* L.). Cytologia 27:369-374.

Evans, A.M. 1980. Structure, variation, evolution, and classification in *Phaseolus*. pp. 337-347. *In* R.J. Summerfield & A.H. Bunting (eds.), Advances in Legume Science. Royal Botanic Gardens, Kew, U.K.

Federici, C.T. & J.G. Waines. 1988. Interspecific hybrid compatibility of selected *Phaseolus vulgaris* L. lines with *P. acutifolius* A. Gray, *P. lunatus* L., and *P. filiformis* Bentham. Annu. Rpt. Bean Improv. Coop. 31:201-202.

Federici, C.T. & J.G. Waines. 1989. Interspecific hybridization of common beans with tepary beans. Annu. Rpt. Bean Improv. Coop. 32:70-71.

Fermond, C. 1855. Recherches sur les fécondations réciproques de quelques végétaux. Bull. Soc. Bot. Fr. 2:748-754.

Fofana, B., J.P. Baudoin, X. Vekemans, D.G. Debouck & P. du Jardin. 1999. Molecular evidence for an Andean origin and a secondary gene pool for the Lima bean (*Phaseolus lunatus* L.) using chloroplast DNA. Theor. Appl. Genet. 98:202-212.

Freeman, G.F. 1918. Southwestern beans and teparies. Bulletin No. 68. Univ. Arizona Agric. Exp. Stn, Tucson, U.S.A.

Freyre, R., P.W. Skroch, V. Geffroy, A.F. Adam-Blondon, A. Shirmohamadali, W.C. Johnson, V. Llaca, R.O. Nodari, P.A. Pereira, S.M. Tsai, J. Tohme, M. Dron, J. Nienhuis, C.E. Vallejos & P. Gepts. 1998. Towards an integrated linkage map of common bean. 4. Development of a core linkage map and alignment of RFLP maps. Theor. Appl. Genet. 97:847-856.

Freytag, G.F. 1975. Research related to the origin and improvement of the common bean (*Phaseolus vulgaris* L.). pp. 159-163. *In* J. Bird & K. Maramorosch (eds), Tropical Diseases of Legumes. Academic Press, New York, U.S.A.

Freytag, G.F. & D.G. Debouck. 1996. *Phaseolus costaricensis*, a new wild bean species (Phaseolinae, Leguminosae) from Costa Rica and Panama, Central America. NOVON 6:157-163.

Garver, S. 1934. The Redfield tepary bean, an early maturing variety. J. Amer. Soc. Agron. 3:397-403.

Garvin, D.F. & N.F. Weeden. 1994. Isozyme evidence supporting a single geographic origin for domesticated tepary bean. Crop Sci. 34:1390-1395.

Gepts, P. 1988. A middle American and an Andean common bean gene pool. pp. 375-390. *In* P. Gepts (ed.), Genetic Resources of *Phaseolus* Beans. Kluwer, Dordrecht, Netherlands.

Gepts, P. & D.G. Debouck. 1991. Origin, domestication, and evolution of the common bean (*Phaseolus vulgaris* L.). pp. 7-53. *In* A. van Schoonhoven & O. Voysest (eds.), Common Beans: Research for Crop Improvement. C.A.B. Int., Wallingford, U.K. & CIAT, Cali, Colombia.

Gepts, P., T.C. Osborn, K. Rashka & F.A. Bliss. 1986. Phaseolin protein variability in wild forms and landraces of the common bean (*Phaseolus vulgaris* L.): evidence for multiple centers of domestication. Econ. Bot. 40:451-468.

González, A., J. Lynch, J.M. Tohme, S.E. Beebe & R.E. Macchiaveli. 1995. Characters related to leaf photosynthesis in wild populations and landraces of common bean. Crop Sci. 35:1468-1476.

Grant, V. 1975. Genetics of Flowering Plants. Columbia Univ. Press, New York, U.S.A.

Guo, M., D.A. Lightfoot, M.C. Mok & D.W.S. Mok. 1991. Analyses of *Phaseolus vulgaris* L. and *P. coccineus* Lam. hybrids by RFLP: preferential transmission of *P. vulgaris* alleles. Theor. Appl. Genet. 81:703-709.

Guo, M., M.C. Mok & D.W.S. Mok. 1994. RFLP analysis of preferential transmission in interspecific hybrids of *Phaseolus vulgaris* and *P. coccineus*. J. Hered. 85:174-178.

Gutiérrez Salgado, A., P. Gepts & D.G. Debouck. 1995. Evidence for two gene pools of the Lima bean, *Phaseolus lunatus* L., in the Americas. Genet. Resources & Crop Evol. 42:15-28.

Haghighi, K.R. & P.D. Ascher. 1988. Fertile, intermediate hybrids between *Phaseolus vulgaris* and *P. acutifolius* from congruity backcrossing. Sex. Plant Reprod. 1:51-58.

Hamann, A., D. Zink & W. Nagl. 1995. Microsatellite fingerprinting in the genus *Phaseolus*. Genome 38:507-515.

Hassler, E. 1923. Revisio specierum austro-americanarum generis *Phaseoli* L. Candollea 1:417-472.

Heiser, C.H. 1988. Aspects of unconscious selection and the evolution of domesticated plants. Euphytica 37:77-81.

Hendry, G.W. 1919. Climatic adaptations of the white tepary bean. J. Amer. Soc. Agron. 11:247-252.

Hernández Xolocotzi, E., S. Miranda Colín & C. Prywer. 1959. El origen de *Phaseolus coccineus* L. *darwinianus* Hdz. X. & Miranda C., *subspecies nova*. Rev. Soc. Mex. Hist. Nat. 20:99-121.

Hervieu, F., H. Bannerot & G. Pelletier. 1994. A unique cytoplasmic male sterility (CMS) determinant is present in three *Phaseolus* species characterized by different mitochondrial genomes. Theor. Appl. Genet. 88:314-320.

Hervieu, F., L. Charbonnier, H. Bannerot & G. Pelletier. 1993. The cytoplasmic male-sterility (CMS) determinant of common bean is widespread in *Phaseolus coccineus* L. and *Phaseolus vulgaris* L. Curr. Genet. 24:149-155.

Honma, S. 1956. A bean interspecific hybrid. J. Hered. 47:217-220.

Honma, S. & O. Heeckt. 1959. Interspecific hybrid between *Phaseolus vulgaris* and *P. lunatus*. J. Hered. 50:233-237.

Hubbeling, N. 1957. New aspects of breeding for disease resistance in beans (*Phaseolus vulgaris* L.). Euphytica 6:111-141.

Hucl, P. & G.J. Scoles. 1985. Interspecific hybridization in the common bean: a review. HortScience 20:352-357.

Hunter, J.E., M.H. Dickson, M.A. Boettger & J.A. Cigna. 1982. Evaluation of plant introductions of *Phaseolus* spp. for resistance to white mold. Plant Dis. 66:320-322.

Ibrahim, A.M. & D.P. Coyne. 1975. Genetics of stigma shape, cotyledon position, and flower color in reciprocal crosses between *Phaseolus vulgaris* L. and *Phaseolus coccineus* (Lam.) and implications in breeding. J. Amer. Soc. Hort. Sci. 100:622-626.

Jaaska, V. 1996. Isoenzyme diversity and phylogenetic affinities among the *Phaseolus* beans (Fabaceae). Pl. Syst. Evol. 200:233-252.

Jacob, M., D. Zink & W. Nagl. 1995. RFLPs of the rRNA genes in the genus *Phaseolus*. Genet. Resources & Crop Evol. 42:97-106.

Jung, G., D.P. Coyne & P. Read. 1992. Interspecific hybridization of *Phaseolus vulgaris* x *Phaseolus acutifolius*. Annu. Rpt. Bean Improv. Coop. 35:206.

Kami, J., V. Becerra Velásquez, D.G. Debouck & P. Gepts. 1995. Identification of presumed ancestral DNA sequences of phaseolin in *Phaseolus vulgaris*. Proc. Natl. Acad. Sci. U.S.A. 92:1101-1104.

Kaplan, L. 1965. Archeology and domestication in American *Phaseolus* (beans). Econ. Bot. 19:358-368.

Kaplan, L. & L.N. Kaplan. 1988. *Phaseolus* in archaeology. pp. 125-142. *In* P. Gepts (ed), Genetic Resources of *Phaseolus* Beans. Kluwer, Dordrecht, Netherlands.

Katanga, K. & J.P. Baudoin. 1987a. Obtention de nouveaux hybrides interspécifiques avec *Phaseolus lunatus* L. Bull. Rech. Agron. Gembloux 22:33-42.

Katanga, K. & J.P. Baudoin. 1987b. Obtention et observations d'un nouvel hybride interspécifique entre deux espèces sauvages: *Phaseolus filiformis* Benth. x *Phaseolus angustissimus* A. Gray. Bull. Rech. Agron. Gembloux 22:153-160.

Katanga, K. & J.P. Baudoin. 1990. Analyses méiotiques des hybrides F_1 et étude des descendances F_2 chez quatre combinaisons interspécifiques avec *Phaseolus lunatus* L. Bull. Rech. Agron. Gembloux 25:237-250.

Kelly, J.D. & P.N. Miklas. 1998. The role of RAPD markers in breeding for disease resistance in common bean. Mol. Breed. 4:1-11.

Khairallah, M.M., B.B. Sears & M.W. Adams. 1992. Mitochondrial restriction fragment length polymorphisms in wild *Phaseolus vulgaris* L.: insights on the domestication of the common bean. Theor. Appl. Genet. 84:915-922.

Kimura, M. 1991. The neutral theory of molecular evolution: a review of recent evidence. Jap. J. Genet. 66:367-386.

Kloz, J. & E. Klozová. 1974. The protein euphaseolin in Phaseolinae - a chemotaxonomical study. Biol. Plant. 16:290-299.

Koenig, R. & P. Gepts. 1989. Allozyme diversity in wild *Phaseolus vulgaris*: further evidence for two major centers of genetic diversity. Theor. Appl. Genet. 78:809-817.
Koenig, R.L., S.P. Singh & P. Gepts. 1990. Novel phaseolin types in wild and cultivated common bean (*Phaseolus vulgaris*, Fabaceae). Econ. Bot. 44:50-60.
Koinange, E.M.K., S.P. Singh & P. Gepts. 1996. Genetic control of the domestication syndrome in common bean. Crop Sci. 36:1037-1045.
Kuboyama, T., Y. Shintaku & G. Takeda. 1991. Hybrid plant of *Phaseolus vulgaris* L. and *P. lunatus* L. obtained by means of embryo rescue and confirmed by restriction endonuclease analysis of rDNA. Euphytica 54:177-182.
Lackey, J.A. 1983. A review of generic concepts in American Phaseolinae (Fabaceae, Faboideae). Iselya 2:21-64.
Lamprecht, H. 1945. A new ornamental plant from a cross between the garden bean and the scarlet runner bean. Agric. Hortique Genetica 3:14-32.
Le Marchand, G. 1971. Observations sur quelques hybrides dans le genre *Phaseolus*. 1. Le problème des incompatibilités interspécifiques. Bull. Rech. Agron. Gembloux 6:441-452.
Le Marchand, G. & R. Maréchal. 1977. Chromosome pairing in interspecific hybrids reveal the value of pollen morphology for deducting phylogenetic affinities in the genus *Phaseolus*. pp. 335-337. *In* Proceedings 8th Congr. Eucarpia, Interspecific hybridization in plant breeding, Madrid, Spain.
Le Marchand, G., R. Maréchal & J.C. Baudet. 1976. Observations sur quelques hybrides dans le genre *Phaseolus*. 3. *P. lunatus*: nouveaux hybrides et considérations sur les affinités interspécifiques. Bull. Rech. Agron. Gembloux 11:183-200.
Leonard, M.F., L.C. Stephens & W.L. Summers. 1987. Effect of maternal genotype on development of *Phaseolus vulgaris* L. x *P. lunatus* L. interspecific hybrid embryos. Euphytica 36:327-332.
Lepoivre, P. & J.P. Baudoin. 1994. Caractérisation de l'agent responsable de l'ascochytose de *Phaseolus vulgaris* en Afrique et en Amérique du Sud. Bull. Séances Acad. Roy. Sci. Outre-Mer (Belgique) 39:557-566.
Lewis, G.P. & A. Delgado Salinas. 1994. *Mysanthus*, a new genus in tribe Phaseoleae (Leguminosae: Papilionoideae) from Brazil. Kew Bull. 49:343-351.
Llaca, V., A. Delgado Salinas & P. Gepts. 1994. Chloroplast DNA as an evolutionary marker in the *Phaseolus vulgaris* complex. Theor. Appl. Genet. 88:646-652.
Lorz, A.P. 1952. An interspecific cross involving the Lima bean *Phaseolus lunatus* L. Science 115:702-703.
Lynch, J., A. González, J.M. Tohme & J. García. 1992. Variation in characters related to leaf photosynthesis in wild bean populations. Crop Sci. 32:633-640.
Manshardt, R.M. & M.J. Bassett. 1984. Inheritance of stigma position in *Phaseolus vulgaris* x *P. coccineus* hybrid populations. J. Hered. 75:45-50.
Maréchal, R. 1971. Observations sur quelques hybrides dans le genre *Phaseolus*. 2. Les phénomènes méiotiques. Bull. Rech. Agron. Gembloux 6:461-489.
Maréchal, R. & J.P. Baudoin. 1978. Observations sur quelques hybrides dans le genre *Phaseolus*. 4. L'hybride *Phaseolus vulgaris* x *P. filiformis*. Bull. Rech. Agron. Gembloux 13:233-240.
Maréchal, R., J.M. Mascherpa & F. Stainier. 1978. Etude taxonomique d'un groupe complexe d'espèces des genres *Phaseolus* et *Vigna* (Papilionaceae) sur la base de données morphologiques et polliniques, traitées par l'analyse informatique. Boissiera 28:1-273.
McVaugh, R. 1987. Flora Novo-Galiciana - A descriptive account of the vascular plants of western Mexico - Volume 5 Leguminosae. The Univ. of Michigan Press, Ann Arbor, U.S.A.

Mejía-Jiménez, A., C. Muñoz, H.J. Jacobsen, W.M. Roca & S.P. Singh. 1994. Interspecific hybridization between common and tepary beans: increased hybrid embryo growth, fertility, and efficiency of hybridization through recurrent and congruity backcrossing. Theor. Appl. Genet. 88:324-331.

Miklas, P.N. & J. Santiago. 1996. Reaction of selected tepary bean to bean golden mosaic virus. HortScience 31:430-432.

Miranda Colín, S. & A.M. Evans. 1973. Exploring the genetic isolating mechanisms between *Phaseolus vulgaris* and *P. coccineus* Lam. Annu. Rpt. Bean Improv. Coop. 16:39-41.

Mirkov, T.E., J.M. Wahlstrom, K. Hagiwara, F. Finardi-Filho, S. Kjemtrup & M.J. Chrispeels. 1994. Evolutionary relationships among proteins in the phytohemagglutinin-arcelin-α-amylase inhibitor family of the common bean and its relatives. Plant Mol. Biol. 26:1103-1113.

Mok, D.W.S., M.C. Mok & A. Rabakoarihanta. 1978. Interspecific hybridization of *Phaseolus vulgaris* with *P. lunatus* and *P. acutifolius*. Theor. Appl. Genet. 52:209-215.

Nabhan, G.P. 1990. Wild *Phaseolus* ecogeography in the Sierra Madre Occidental, Mexico:areographic techniques for targeting and conserving species diversity. IBPGR, Rome, Italy.

Nabhan, G.P., J.W. Berry & C.W. Weber. 1980. Wild beans of the Greater Southwest: *Phaseolus metcalfei* and *P. ritensis*. Econ. Bot. 34:68-85.

Nelson, B.M. & P.D. Ascher. 1984. Influence of parental lines on making *Phaseolus vulgaris* - *P. acutifolius* hybrids. Annu. Rpt. Bean Improv. Coop. 27:40-41.

Nodari, R.O., S.M. Tsai, R.L. Gilbertson & P. Gepts. 1993. Towards an integrated linkage map of common bean. 2. Development of an RFLP-based linkage map. Theor. Appl. Genet. 85:513-520.

Ockendon, D.J., L. Currah & J.D. Taylor. 1982. Transfer of resistance to halo-blight (*Pseudomonas phaseolicola*) from *Phaseolus vulgaris* to *P. coccineus*. Annu. Rpt. Bean Improv. Coop. 25:84-85.

Osborn, T.C, D.C. Alexander, S.S.M. Sun, C. Cardona & F.A. Bliss. 1988. Insecticidal activity and lectin homology of arcelin seed protein. Science 240:207-210.

Osborn, T.C, T. Blake, P. Gepts & F.A. Bliss. 1986. Bean arcelin. 2. Genetic variation, inheritance and linkage relationships of a novel seed protein of *Phaseolus vulgaris* L. Theor. Appl. Genet. 71:847-855.

Park, S.J. & B.N. Dhanvantari. 1984. Breeding for resistance to common bacterial blight in interspecific crosses between *Phaseolus vulgaris* and *P. coccineus*. Annu. Rpt. Bean Improv. Coop. 27:179-180.

Parker, J.P. & T.E. Michaels. 1986. Genetic control over hybrid plant development in interspecific crosses between *Phaseolus vulgaris* L. and *P. acutifolius* A. Gray. Annu. Rpt. Bean Improv. Coop. 29:22-23.

Parsons, L.R. & T.K. Howe. 1984. Effects of water stress on the water relations of *Phaseolus vulgaris* and the drought resistant *Phaseolus acutifolius*. Physiol. Plant 60:197-202.

Pernès, J. 1983. La génétique de la domestication des céréales. La Recherche 146:910-919.

Petersen, A.C. & D.D. Davis. 1982. Yield response of *Phaseolus vulgaris* L. and *Phaseolus acutifolius* subjected to water stress. Annu. Rpt. Bean Improv. Coop. 25:53-54.

Petzoldt, R. & M.H. Dickson. 1987. Interspecific hybridization of *Phaseolus vulgaris* with *P. angustissimus*, *P. filiformis* and *P. ritensis*. Annu. Rpt. Bean Improv. Coop. 30:94-95.

Piñero, D. & L. Eguiarte. 1988. The origin and biosystematic status of *Phaseolus coccineus* subsp. *polyanthus*: electrophoretic evidence. Euphytica 37:199-203.

Piper, C.V. 1926. Studies in American Phaseolinae. Contr. US Nat. Herb. 22:663-701.

Prendota, K., J.P. Baudoin & R. Maréchal. 1982. Fertile allopolyploids from the cross *Phaseolus acutifolius* x *P. vulgaris*. Bull. Rech. Agron. Gembloux 17:177-190.

Pueyo, J.J. & A. Delgado Salinas. 1997. Presence of ∝-amylase inhibitor in some members of the subtribe Phaseolinae (Phaseoleae: Fabaceae). Amer. J. Bot. 84:79-84.

Rabakoarihanta, A., D.W.S. Mok & M.C. Mok. 1979. Fertilization and early embryo development in reciprocal interspecific crosses of *Phaseolus*. Theor. Appl. Genet. 54:55-59.

Rick, C.M. & P.G. Smith. 1953. Novel variation in tomato species hybrids. Amer. Nat. 87:359-373.

Schinkel, C. & P. Gepts. 1988. Phaseolin diversity in the tepary bean, *Phaseolus acutifolius* A. Gray. Plant Breed. 101:292-301.

Schmit, V. & J.P. Baudoin. 1992. Screening for resistance to *Ascochyta* blight in populations of *Phaseolus coccineus* L. and *P. polyanthus* Greenman. Field Crops Res. 30:155-165.

Schmit, V., J.P. Baudoin & B. Wathelet. 1991. Considerations about the taxonomical position of various taxa of the *P. coccineus* complex on the basis of seeds protein electrophoresis and experimental hybridizations. Annu. Rpt. Bean Improv. Coop. 34:87-88.

Schmit, V. & D.G. Debouck. 1991. Observations on the origin of *Phaseolus polyanthus* Greenman. Econ. Bot. 45:345-364.

Schmit, V., D.G. Debouck & J.P. Baudoin. 1996. Biogeographical and molecular observations on *Phaseolus glabellus* (Fabaceae, Phaseolinae) and its taxonomic status. Taxon 45:493-501.

Schmit, V., P. du Jardin, J.P. Baudoin & D.G. Debouck. 1993. Use of chloroplast DNA polymorphisms for the phylogenetic study of seven *Phaseolus* taxa including *P. vulgaris* and *P. coccineus*. Theor. Appl. Genet. 87:506-516.

Schneider, K.A., M.E. Brothers & J.D. Kelly. 1997. Marker-assisted selection to improve drought resistance in common bean. Crop Sci. 37:51-60.

Schoonhoven, A. van & C. Cardona. 1982. Low levels of resistance to the Mexican bean weevil in dry beans. J. Econ. Entomol. 75:567-569.

Schoonhoven, A. van, C. Cardona & J. Valor. 1983. Resistance to the bean weevil and the Mexican bean weevil (Coleoptera:Bruchidae) in noncultivated common bean accessions. J. Econ. Entomol. 76:1255-1259.

Schumann, K. & W. Nagl. 1995. PCR amplification of the phaseolin-genes in 18 species of the genus *Phaseolus*. Annu. Rpt. Bean Improv. Coop. 38:107-108.

Shii, C.T., A. Rabakoarihanta, M.C. Mok & D.W.S. Mok. 1982. Embryo development in reciprocal crosses of *Phaseolus vulgaris* L. and *P. coccineus* Lam. Theor. Appl. Genet. 62:59-64.

Singh, S.P. 1992. Common bean improvement in the tropics. Plant Breed. Rev. 10:199-269.

Singh, S.P., D.G. Debouck & W.M. Roca. 1997a. Successful interspecific hybridization between *Phaseolus vulgaris* L. and *P. costaricensis* Freytag & Debouck. Annu. Rpt. Bean Improv. Coop. 40:40-41.

Singh, S.P., A. Molina & P. Gepts. 1995. Potential of wild common bean for seed yield improvement of cultivars in the tropics. Can J. Plant Sci. 75:807-813.

Singh, S.P. & C.G. Muñoz. 1999. Resistance to common bacterial blight among *Phaseolus* species and common bean improvement. Crop Sci. 39:80-89.

Singh, S.P., W.M. Roca & D.G. Debouck. 1997b. Ampliación de la base genética de los cultivares de frijol: hibridación interespecífica en especies de *Phaseolus*. pp. 9-19. *In* S. P. Singh & O. Voysest (eds.), Taller de Mejoramiento de Frijol para el Siglo XXI: Bases para Una Estrategia para América Latina. CIAT, Cali, Colombia.

Smartt, J. 1970. Interspecific hybridization between cultivated American species of the genus *Phaseolus*. Euphytica 19:480-489.
Smartt, J. 1973. The possible status of *Phaseolus coccineus* L. ssp. *darwinianus* Hdz. X. et Miranda C. as a distinct species and cultigen of the genus *Phaseolus*. Euphytica 22:424-426.
Smartt, J. 1976. Comparative evolution of pulse crops. Euphytica 25:139-143.
Smartt, J. 1979. Interspecific hybridization in the grain legumes - A review. Econ. Bot. 33:329-337.
Smartt, J. 1985. Evolution of grain legumes. IV. Pulses in the genus *Phaseolus*. Expl. Agric. 21:193-207.
Smartt, J. 1988. Morphological, physiological and biochemical changes in *Phaseolus* beans under domestication. pp. 143-161. *In* P. Gepts (ed.), Genetic Resources of *Phaseolus* Beans. Kluwer, Dordrecht, Netherlands.
Smartt, J. 1990. Vavilov's Law of Homologous Series and *de novo* crop plant domestication. Biol. J. Linnean Soc. 39:27-38.
Smartt, J. & N. Haq. 1972. Fertility and segregation of the amphidiploid *Phaseolus vulgaris* L. x *P. coccineus* L. and its behaviour in backcrosses. Euphytica 21:496-501.
Smartt, J., N. Haq & M. Nassar. 1974. The production of interspecific hybrids using *Phaseolus coccineus* L. as seed parent. Annu. Rpt. Bean Improv. Coop. 17:80-81.
Steadman, J.R. 1993. White mold (*Sclerotinia*) resistance identified in national bean nursery. Annu. Rpt. Bean Improv. Coop. 36:178-179.
Sullivan, J.G. & G. Freytag. 1986. Predicting interspecific compatibilities in beans *(Phaseolus)* by seed protein electrophoresis. Euphytica 35:201-209.
Thomas, C.V. & J.G. Waines. 1980. Interspecific hybridization between *Phaseolus vulgaris* and *P. acutifolius*. Annu. Rpt. Bean Improv. Coop. 23:75-77.
Thomas, C.V. & J.G. Waines. 1984. Fertile backcross and allotetraploid plants from crosses between tepary beans and common beans. J. Hered. 75:93-98.
Thomas, H. 1964. Investigations into the inter-relationships of *Phaseolus vulgaris* L. and *P. coccineus* L. Genetica 35:59-74.
Tohme, J., D.O. González, S. Beebe & M.C. Duque. 1996. AFLP analysis of gene pools of a wild bean core collection. Crop Sci. 36:1375-1384.
Toro Ch., O., L. Lareo & D.G. Debouck. 1993. Observations on a noteworthy wild lima bean, *Phaseolus lunatus* L., from Colombia. Annu. Rpt. Bean Improv. Coop. 36:53-54.
Urban, I. 1928. Plantae cubenses novae vel rariores a clo. Er. L. Ekman lectae. IV. Symbolae Antillanae 9:433-568.
Valladares, N., D.P. Coyne, M.L. Schuster & B. Hoff. 1977. Reaction of *Phaseolus* germplasm to different strains of *Xanthomonas phaseoli* and *X. phaseoli* var. *fuscans*. Annu. Rpt. Bean Improv. Coop. 20:74-75.
Vavilov, N.I. 1922. The law of homologous series in variation. J. Genet. 12:47-89.
Verdcourt, B. 1970. Studies in the Leguminosae-Papilionoideae for the flora of tropical East Africa. IV. Kew Bull. 24:507-569.
Wall, J.R. 1970. Experimental introgression in the genus *Phaseolus*. I. Effect of mating systems on interspecific gene flow. Evolution 24:356-366.
Wall, J.R. & T.L. York. 1957. Inheritance of seedling cotyledon position in *Phaseolus* species. J. Hered. 48:71-74.
Weilenmann de Tau, E., J.P. Baudoin & R. Maréchal. 1986. Obtention d'allopolyploïdes fertiles chez le croisement entre *Phaseolus vulgaris* et *Phaseolus filiformis*. Bull. Rech. Agron. Gembloux 21:35-46.

Westphal, E. 1974. Pulses in Ethiopia, their taxonomy and agricultural significance. Agr. Res. Rpt. No. 815. Centre for Agricultural Publishing and Documentation, Wageningen, Netherlands.
Wilhelm de Mosbach, E. 1992. Botánica Indígena de Chile. Editorial Andres Bello, Santiago, Chile.
Wilkinson, R.E. 1983. Incorporation of *Phaseolus coccineus* germplasm may facilitate production of high yielding *P. vulgaris* lines. Annu. Rpt. Bean Improv. Coop. 26:28-29.
Zapata, M., G.F. Freytag & R.E. Wilkinson. 1985. Evaluation for bacterial blight resistance in beans. Phytopathology 75:1032-1039.
Zink, D. & W. Nagl. 1998. Interspecific microsatellite-primed PCR analysis in twenty different *Phaseolus* species. Annu. Rpt. Bean Improv. Coop. 41:109-110.

Chapter 3

Development of an Integrated Linkage Map

Paul Gepts
Agronomy and Range Science, University of California, 1 Shields Avenue, Davis, CA 95616-8515, U.S.A.

Key words: core linkage map, disease resistance genes, domestication syndrome, gene introgression, germplasm conversion, RAPD, RFLP

1. INTRODUCTION

Linkage maps have become widespread and essential genetic tools for crop improvement and other biological studies. Indeed, assembling a linkage map is now possible, even for species, cultivated or wild, that do not have a long and intensive history of genetic research. Previously, linkage maps existed only for a few experimental organisms, such as *Drosophila*, maize (*Zea mays* L.), and tomato (*Lycopersicon esculentum* L.). This change is due to the development of molecular marker technologies that have led to a large increase in the number of polymorphic loci within any given segregating population. Before the advent of molecular markers, phenotypic markers were generally used. Accumulating enough of these markers in the same population was time-consuming and difficult (because of the often-deleterious nature of these markers or epistatic interactions). Alternatively, results from different segregating populations had to be combined leading to uncertainties as to allelism and actual map location.

Since Botstein et al. (1980) made the seminal proposal of using RFLPs as genetic markers, many RFLP maps have been produced. Other types of molecular markers have been introduced since then. These markers include RAPDs, DAFs or AP-PCR (Caetano-Anollès et al., 1991; Welsh & McClelland, 1990; Williams et al., 1990), minisatellite markers or VNTRs (variable number of tandem repeats) (Dallas, 1988; Nakamura et al., 1987), microsatellites or SSRs (simple sequence repeats) (Tautz, 1989), ISSRs (inter simple sequence repeats) (Zietkiewicz et al., 1994), and AFLPs (amplified fragment length polymorphisms) (Vos et al., 1995). Overall, the trend has been toward an increased reliance on PCR (polymerase-chain-reaction)-based markers because of their relative speed and reduced DNA amount requirements. There is no perfect marker yet, but each of these markers has its advantages and disadvantages as discussed in Gepts (1993) and Rafalski & Tingey (1993).

The large number of linkage maps developed in recent years (e.g., see abstracts of the Plant and Animal Genome Conference: http://probe.nalusda.gov:8000/otherdocs/ pg/index.html) is testimony to their importance as tools for crop improvement. Uses of maps include the map-based cloning of agronomically important genes (e.g., disease resistance genes, reviewed by Hammond-Kosack & Jones, 1997), comparative mapping (e.g., grass family: Ahn & Tanksley, 1993), elucidation of the inheritance of quantitative traits (e.g., Lee, 1995; Paterson, 1995; Tanksley, 1993), relationship between QTLs (quantitative trait loci) and "candidate" genes (i.e., genes that provide a molecular or biochemical basis for the observed effect of a genome region on a quantitative trait) (e.g., McMullen et al., 1998; Prioul et al., 1997), marker-assisted selection and backcross-conversion in breeding (e.g., Tanksley et al., 1996), and, last but not least, analysis of germplasm diversity and conversion of exotic germplasm (Gepts, 1993).

Common bean, *Phaseolus vulgaris* L. (2n = 2x = 22), has a long history of genetic studies. Mendel (1866), Johannsen (1909), and Sax (1923) used common bean to make important, historic contributions to genetic science. As an experimental organism, common bean has the advantage of having a small genome (0.65 pg/haploid genome or 635 mbp), which is one of the smallest genomes in the legume family (Arumuganathan & Earle, 1991). Its size is only six and 1.6 times that of the *Arabidopsis* and rice genomes, respectively. It has also a predominantly selfing reproductive mode (although this is a disadvantage from other perspectives). An additional advantage is the availability of large germplasm collections (e.g., at CIAT, Colombia, and at the USDA Western Regional Plant Introduction Station in Pullman, Washington, U.S.A.; 30,000 and 13,000 accessions, respectively) and of breeding materials, such as segregating populations (both early and late generations) and nearly-isogenic lines (see Chapter 4 by Kelly & Miklas). In addition, there is a good understanding of the organization of genetic diversity in the primary gene pool of the species (Gepts, 1993, 1998b).

An additional genetic asset of common bean is the availability of several linkage maps. In this chapter, I review these linkage maps and the efforts made to establish an integrated map. I then discuss some important applications of these linkage maps, including: 1) comparative mapping in the Phaseoleae tribe to which *P. vulgaris* belongs; 2) genomic distribution of disease resistance and other phenotypic trait loci in the common bean genome; 3) an analysis of the domestication syndrome and the consequences for introgression and conversion of exotic (particularly wild) germplasm; and 4) a study of the inheritance of a genetically complex disease, common bacterial blight.

2. LINKAGE MAPS IN COMMON BEAN

Several linkage maps have been developed in common bean (Table 1). They can be distinguished by several characteristics, including the types of parents used, the segregating generation in which they were established, the traits segregating in each population, the type of markers, the number of markers (and, therefore, the degree of saturation of the genome), and the total map length. Because these maps have been published or are about to be published, I will not describe them in further detail. The reader is referred to the original publications for further information.

In spite of their many differences, these maps have the same characteristic as to the choice of the parents. Common bean consists of two major gene pools (Gepts, 1993, 1998b; Gepts & Bliss, 1985). Therefore, the parents were chosen to belong to different gene pools as experiments have shown that polymorphism among genotypes is increased markedly in that situation (Haley et al., 1994d; Nodari et al., 1992). This choice was made out of concern for low levels of polymorphism within the primary gene pool of common bean. These concerns have not been borne out (see references just cited). To the contrary, as new marker types have become available such as RAPDs (Caetano-Anollès et al., 1991; Welsh & McClelland, 1990; Williams et al., 1990) and AFLPs (Vos et al., 1995), it has become increasingly easier to identify the requisite number of polymorphic markers. Indeed, whereas, between-gene-pool polymorphism reaches 80% for RFLPs and RAPDs, intra-gene-pool polymorphism reaches 30-40% for RFLPs (Nodari et al., 1992) and 60% for RAPDs (Haley et al., 1994d). Therefore, it is to be expected that in the near future more maps will be based on crosses between parents belonging to the same gene pool, whether they belong to different races or the same race (Singh et al., 1991). This trend is illustrated by the map recently developed by Ewing et al. (Table 1) in the cross between

Table 1. Overview of molecular linkage maps in common bean

Population (generation)	Traits/genes[1]	Markers			Length	References
		RFLPs	RAPDs	AFLPs	(cM)	
BAT93 x Jalo EEP558 (F_2)	ALS, anthracnose, BCM, CBB, rust; *Rhizobium* nodulation; *V*, *C*	194	27	--	1,200	Gepts et al., 1993; Nodari et al., 1993a
BAT93 x Jalo EEP558 (RI)		120	470	--	1,200	Freyre et al., 1998
Midas x G12873 (RI)	*Ppd*, *fin*, *St*, *y*, *P*; number of nodes and pods, seed weight; dormancy	83	--	--	1,600	Koinange et al., 1996
DIACOL Calima x XR235-1-1 (BC)	CBB	224	21	--	1,200	Vallejos et al., 1992; Yu et al., 1998
Corel x EO2 (BC)	Anthracnose; *Ms-8*, *SGou*	51	100	--	1,200	Adam-Blondon et al., 1994b
Sierra x AC1028 (RI)	Drought	--	42	--	155	Schneider et al., 1997
Sierra x LEF-2RB (RI)		--	34	--	125	
BAC 6 x HT 7719 (RI)	CBB, rust, web blight	--	75	--	545	Jung et al., 1996
PC-50 x XAN-159	CBB, *C*, *V*	--	168	--	426	Jung et al., 1997
A55 x G122 (RI)	Yield in Andean x Middle American cross, *C*	5	--	245	1,631	Johnson, 1997;
California Dark Red Kidney x Yolano (RI)	Yield in Andean x Middle American cross, *rk*	8	--	196	1,487	W.C. Johnson & P. Gepts, unpublished results
Sierra x Olathe (RI)	Rust	–	72	11	743	Ewing, McClean, Prischman & Grafton, personal communication

[1] Resistance to ALS= angular leaf spot, BCM= bean common mosaic, CBB= common bacterial blight.

cultivars Sierra and Olathe, both of which belong to race Durango of the Middle American gene pool.

An additional common characteristic, at least among the maps that include enough markers (Adam-Blondon et al., 1994b; Nodari et al., 1993a; Vallejos et al., 1992), is that total map length appears to be close to 1,200 cM. This may increase somewhat once all currently unassigned markers are attached to a linkage group, but not by much (100-200 cM). The average relationship between physical and genetic distance of 450-500 kb/cM corresponds well with measured values by pulsed field gel electrophoresis around the phaseolin seed protein locus (*Phs*) of 400 kb/cM (Llaca & Gepts, 1996). What remains to be determined is the range of values around this average.

3. INTEGRATION OF LINKAGE MAPS

3.1 General Approach

The general approach followed to integrate mapping information in beans has been to establish a stable, immortalized mapping population, i.e., a recombinant inbred (RI) population resulting from single-seed descent of an F_2 population. Shared markers permit correlation between the core map developed in this immortalized population and other maps. In a third step, the general map location of markers and genes in the other maps can be established in the core map by extrapolation.

The core mapping population was established in the cross BAT93 x Jalo EEP558 (BJ) (Figure 1). The choice of the two parents was guided by two main arguments: 1) they belong to different gene pools (BAT93 is Middle American and Jalo EEP558 is Andean), therefore, polymorphism at the molecular level between the two parents is maximized (Nodari et al., 1992), and 2) several host-microorganism interactions segregate in this population including resistance to bean common mosaic (BCM; *I* gene), common bacterial blight (CBB) [caused by *Xanthomonas campestris* pv. *phaseoli* (Smith) Dye], angular leaf spot [caused by *Phaeoisariopsis griseola* (Sacc.) Ferr.], rust [caused by *Uromyces appendiculatus* (Pers.) Ung., and anthracnose [caused by *Colletotrichum lindemuthianum (Sacc. & Magn.)* Bri. & Cav. Bri. & Cav.] (CIAT, 1980). In addition, the two parents also differ for their *Rhizobium* nodulation capacity (Nodari et al., 1993b). The population currently consists of 75 RI lines that have gone beyond the F_{10} generation.

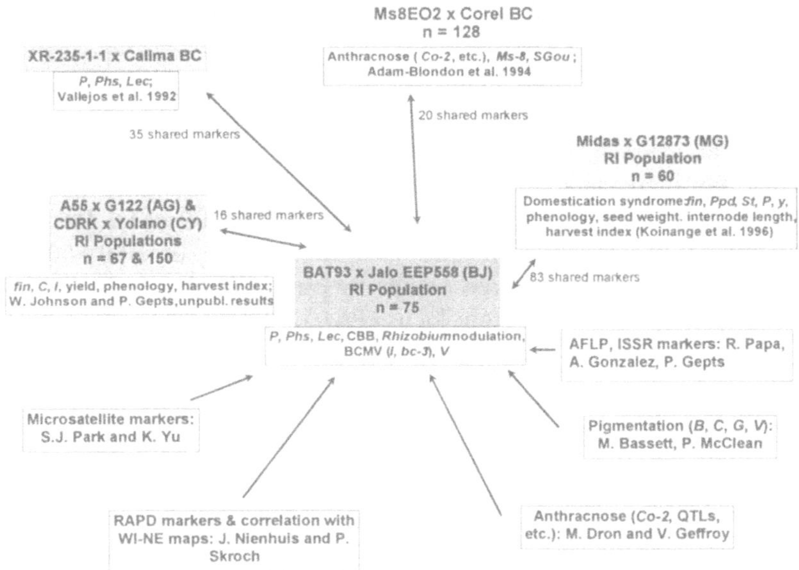

Figure 1. The central role of the BAT93 x Jalo EEP558 recombinant inbred population in the integration of molecular linkage maps of common bean (Gepts, 1998a).

3.2 The Core Linkage Map

This map is more fully described in Freyre et al. (1998). Briefly, some 550 molecular markers, principally RFLP and RAPD markers, have been mapped in the RI population of the BAT93 x Jalo EEP558 cross. These include 45 RFLP markers mapped in the F_2 generation of this cross (Appendix A; Nodari et al., 1993a). These markers allow a correlation to be established between the F_2 and RI maps of this population. In addition, sequences of known molecular and biochemical function have been mapped as RFLPs (Table 2; Freyre et al., 1998; V. Geffroy, personal communication, 1998). Four hundred and thirty RAPD markers have also been added. Some 240 markers are placed unequivocally as framework markers (log-likelihood difference >2.0) and 329 are placed in intervals (log-likelihood ratio difference <2.0). Forty-six markers remain unassigned to a linkage group. Some 20% loci showed a distorted segregation (P = 0.05).

Table 2. Clones of known molecular or biochemical function mapped in common bean (modified from Geffroy, 1997)

Gene (symbol)	Gene bank accession number[1]	Linkage group	References
α-amylase inhibitor (*αAI*) - arcelin (*Arc*) - lectin (*Lec*)	X02409, U10352	B4	Hoffman & Donaldson, 1985; Mirkov et al., 1994; Moreno & Chrispeels, 1989; Osborn et al., 1988
Cell division cycle (*Cdc-2*)	M99497	B6	Miao et al., 1993
Cellulase (*Cel*)	U34754	B2	Tucker & Milligan, 1991
Chalcone isomerase (*ChI*)	Z15046	B7	Mehdy & Lamb, 1987
Chalcone synthase (*ChS*)	X06411	B2	Ryder et al., 1987
Chitinase (*Ch*)	M19052, S43926	B9	Broglie et al., 1986; Hedrick et al., 1988
Chlorophyll a/b-binding protein (*Cab-1*)	X57082	B4	Thompson et al., 1983
Cinnamyl-alcohol dehydrogenase (*Cad*)	J03825	B9	Walter et al., 1988
Elongation factor (*Ef*)	NA	B7	J. Creusot & M. Dron, personal communication
Glucanase (*Gluc*)	X53129	B9	Edington et al., 1991
Glutamine synthetase (*Gs*)	X68598	B1, B8	Hirel et al., 1992
Glycine-rich protein (*Grp1.8-1* and *Grp1.8-2*)	X13596	B2, B3	Keller et al., 1988
Heat-shock protein (*Hsp70*)	X66874	B6	Vidal et al., 1993
Hydroxyproline-rich glycoprotein (*Hrgp3.6*)	A29356	B7	Corbin et al., 1987
Hydroxyproline-rich glycoprotein (*Hrgp213*)	B29356	B6	Corbin et al., 1987

Gene (symbol)	Gene bank accession number[1]	Linkage group	References
Leghemoglobin (*LegH*)	K03152	B7	Lee & Verma, 1984
Lipoxygenase (*Lox*)	X63521	B5	Meier et al., 1993
Pathogenesis-related proteins (*PvPR-1* and *PvPR-2*)	X61365, X61364	B3, B2	Walter et al., 1990
Pectinesterases (*Vpe-2* and *Vpe-3*)	X68029, X68028	B2, B10	Ebbelaar et al., 1996
Phosphoenolpyruvate carboxylase (*PepC*)	M83086	B11	Pathirana et al., 1992
Peroxidase (*Per*)	NA	B7	J. de Oliveira & M. Dron, personal communication
Phaseolin (*Phs*)	X52626, U01131, J01263, M13758	B7	Anthony et al., 1990; Kami & Gepts, 1994; Slightom et al., 1983
Phenylalanine ammonialyase (*Pal-1* and *Pal-2*)	M11939	B1, B7	Edwards et al., 1985
Polygalacturonase-inhibiting protein (*Pgip*)	X64769	B2	Toubart et al., 1992
Ribosomal RNA (*rRNA*)	-	B4, B10	Jorgensen et al., 1987
Ribulose bisphosphate carboxylase, small subunit (*Rbcs*)	AF028707	B4	Knight & Jenkins, 1992
Sucrose synthase (*SS*)	AF030231	B3	Fuller et al., 1983; Thummler & Verma, 1987
Uricase (*Uri-1* and *Uri-2*; *Uri-3*)	D86930, U72663	B9 and B7; B2	Capote-Mainez & Sánchez, 1997; Nguyen et al., 1985

[1]Http://www.ncbi.nlm.nih.gov/Entrez/.

The core map consists of 11 linkage groups, numbered from B(ean)1 to B11 (Appendix A). Linkage group D(avis)1 originally identified by Nodari et al. (1993a) has been split into two linkage groups, B1 and B11, in the interval marked by RFLP loci *D1032* and *D1228*. Linkage group D14, identified in the F_2 generation of the BAT93 x Jalo EEP558 population, has been attached to linkage group B4 through the mapping of RAPD marker, *AM04.950* (Appendix A). Linkage groups D12 and D13 had previously been attached to linkage groups D6 (now B6) and D7 (now B7), respectively (Gepts et al., 1993). Linkage groups D11 and D15 remain unattached to the larger core linkage groups. The length of the core map is around 1226 cM or an average of 2 cM per interval. The current map length corresponds closely to the predicted map length for the BJ population (Gepts et al., 1993).

3.3 Correlation with RFLP Maps

Based on an exchange of RFLP probes (Appendix A), it has been possible to correlate 10 of 11 linkage groups (Table 3) between the Davis (Nodari et al., 1993a) and Paris maps (Adam-Blondon et al., 1994b). With the exception of linkage group pairs D2 and P(aris)2 and B11 and P12, which have only one common marker, and linkage groups D8 and P8, where two Paris probes map on the same locus in the Davis map, it has also been possible to co-orient the linkage groups. The Paris map has one additional linkage group (P10) with few markers for which no correspondence has yet been established with Davis linkage groups.

The mapping results of shared markers between the two maps has allowed a comparison of the portability of RFLP and RAPD markers between different labs. When the same restriction enzyme was used to digest genomic DNA, 11 of 12 RFLP probes (or 92%) mapped to the same linkage group. When a different restriction enzyme was used, seven of 15 RFLP bands (or 47%) mapped to the same linkage group, whereas eight of 15 bands (or 53%) mapped to a different linkage group. The decreased co-localization observed when a different restriction enzyme is used suggests that (small) multigene families can be frequent and that the different members of the family can be distributed over several linkage groups (see also below). When RAPD markers were analyzed only 16 of 26 bands could be scored in both laboratories. Of these 16, six bands (or 38%) mapped to the same linkage group and 10 (62%) to a different linkage group. RAPD markers, therefore, do not seem reliable markers to correlate maps across laboratories, which is consistent with results of Jones et al. (1997).

A combination of 42 markers, consisting of Florida RFLP markers (showed with prefix Bng) and previously shared markers, has led to a

Table 3. Correlation among linkage groups of common bean (Core, Davis, Florida, Paris, and classical maps), mungbean, and soybean

Core (RI) (Freyre et al., 1998)	Davis (F_2) (Nodari et al., 1993a)	Florida (F_2) (Vallejos et al., 1992)	Paris (BC) (Adam-Blondon et al., 1994b)	Classical map (Bassett, 1991)	Mungbean map (Boutin et al., 1995)	Soybean map (Boutin et al., 1995)
B1	D1[a]	H	P1	IV	2, 6	B2,F,G
B2	D2	D	P2	III, VI	1, 3, 7, 13	A1,A2,B1,D,G,K,Q
B3	D3	C	P3		1, 4	A1,A2,C1,C2,F,I,J,K
B4	D4	B	P4	XII		K,L
B5	D5	E	P5		2, 12	A2,C1
B6	D6	G	P6		7	A2,B2,E,G,H,J,L,N
B7	D7	A	P7	X, XIII	4, 10	D1,E,F,H,I,O,Q,
B8	D8	F	P8		2, 6	A2,B2,D,G,K,O,P
B9	D9	K	P9		1, 7, 8, 9, 10	A1,A2,B1,C1,C2,E,G,I, K,O,Y
B10	D10	I				A2,F,N
B11		J	P12		5	B1,D,H,O,W
			P10			
	D11[b]		P11			
	D12-D15[b-d]					

[a] Linkage group D1 has been split into linkage groups B1 and B11.
[b] Linkage groups D11 and D15 remain unlinked.
[c] Linkage groups D12 and D13 have been attached to linkage groups B6 and B7, respectively (Gepts et al., 1993).
[d] Linkage group D14 has been attached to linkage group B4 (Freyre et al., 1998).

correlation of all linkage groups in the Davis and Florida maps (Table 3; Appendix A). The shared markers included *I* (on Davis linkage group D2 and Florida linkage group D), *Skdh* (D3 and C), α*AI* (D4 and B), *Diap-1* and *Aco-2* (D5 and E), and *Phs* and *P* (D7 and A). Except linkage group B9, all linkage groups can also be oriented with respect to each other. On linkage group B9, the two markers assayed, *Bng102* and *Bng228*, were mapped at a LOD threshold > 2.0 in both the core map and the Florida map. However, their close linkage suggests caution in the determination of the respective orientation of the linkage groups.

Mapping of some Bng clones has allowed mapping of traits mapped in other populations. Thus, markers *Bng171*, *Bng 173*, and *Bng122*, which map on linkage group B1 near RFLP marker *D0140*, are linked to a gene conferring partial resistance to soybean cyst nematode (SCN) (Concibido et al., 1996; Danesh et al., 1998). Markers *Bng102* and *Bng228*, on linkage group B9, are linked to the nuclear male fertility restorer gene *Fr* of common bean (He et al., 1995).

Overall, correlation efforts among common bean maps have allowed us to learn the approximate, relative location of more than 1,000 markers. These correspond to approximately 190 RFLP markers on the Davis F_2 map, 430 RAPD markers on the core map, 240 markers on the Florida map, and 160 markers on the Paris map. Appendix A shows map alignments for all linkage groups. These same figures can also be downloaded from the Gepts lab web site (http://agronomy.ucdavis.edu/ gepts/mapdata2.htm). Using these maps, geneticists and breeders can choose a set of markers for their region of interest and determine a more precise order of markers for that region in their own mapping populations.

3.4 Correlation with RAPD Maps

Skroch et al. (1996) described a strategy and preliminary results for using the RAPD map constructed in the BAT93 x Jalo EEP558 RI population to correlate the RAPD maps of Jung et al. (1996, 1997). Therefore, the RFLP maps described here will also be correlated with these RAPD maps. Correlations with other maps could be established using PCR-based markers. SCAR (sequence-characterized amplified regions) or ASAP (allele-specific associated primers) markers are currently distributed on linkage groups 1 (near the *Ppd* and *Hr* genes, Gu et al., 1998), 2 (near *I*, Melotto et al., 1996), 6 (near *bc-3*, Johnson et al., 1997), 8 and 10 (*J1ks* and *J1ds*, respectively; Adam-Blondon et al., 1994a), and 11 (near *Co-2*: *H20s* and *F3s*, Adam-Blondon et al., 1994a). Additional, as yet unmapped, SCAR markers have been developed for *Co-4*^{2} (Young et al., 1998), *Ur-5* (Melotto & Kelly, 1998), and two QTLs for common bacterial blight resistance (Jung et al.,

1998). In addition, clones with known sequences have been mapped on linkage groups 1 (*Pal-1*), 2 (*PvPR-2*, *Pgip*, *Cel*, *ChS* and *ChS-2*, *Vpe-2*, and *Uri-3*), 3 (*PGRP1.8-3*, *PvPR-1*, and *SS*), 4 (*Lec-Arl-αAI*, *Hrgp4-1*, *Rbcs*, and *Cab-1*), 5 (*Lox-1*), 6 (*Hsp70* and *Cdc-2*), 7 (*ChI*, *Ef*, *Lec-2* and *Lec-3*, *LegH*, *Per*, *Phs*, and *Uri-2*), 8 (*GS-c* and *Lox-2*), 9 (*Gluc*, *Ch*, and *Cad*), and 10 (*Vpe-3*) as shown in Figure 2.

Based on published sequences for these genes, PCR primers can be designed so that polymorphisms between parents of the different mapping populations can be detected, either directly or after restriction digestion (so-called sequence-tagged-sites or STSs in mammalian genetics). Primers have been developed for one gene just mentioned, the *LegH* gene, by Uchiumi et al. (1998), although the authors did not investigate the polymorphism revealed by the amplified sequences. Additional *P. vulgaris* genes exist; their sequences are known, but they have yet to be mapped (search nucleotide sequence listing for *P. vulgaris* genes with the Entrez search system: http://www.ncbi.nlm.nih.gov/htbin-post/Entrez/query?db=n_r). These sequences can be then also transformed in STSs.

3.5 Genes for Phenotypic Traits Located on the Core Map

The segregation of the following phenotypic traits has been investigated in this population: BCM resistance controlled by the *I* gene (Freyre et al., 1998; Nodari et al., 1993a), common bacterial blight resistance and *Rhizobium* nodulation (Nodari et al., 1993b), anthracnose resistance (V. Geffroy, M. Sévignac, P. Gepts, T. Langin & M. Dron, unpublished results), seed proteins (*Phs*, *Lec-αAI-Arl*), and plant or seed pigmentation (genes *B*, *C*, *G*, and *V*) (M.J. Bassett, P. McClean & P. Gepts, unpublished results). Through alignment with other linkage maps, other genes controlling phenotypic traits have also been placed on the core map (see below).

The BJ RI population is, therefore, a core mapping population in common bean because of the intense mapping activities in this population. In addition, it is a spill population because of the correlation available between the map obtained in this population and in other populations. The wide range of phenotypic traits analyzed in this population make it a significant resource in genetic studies of common bean.

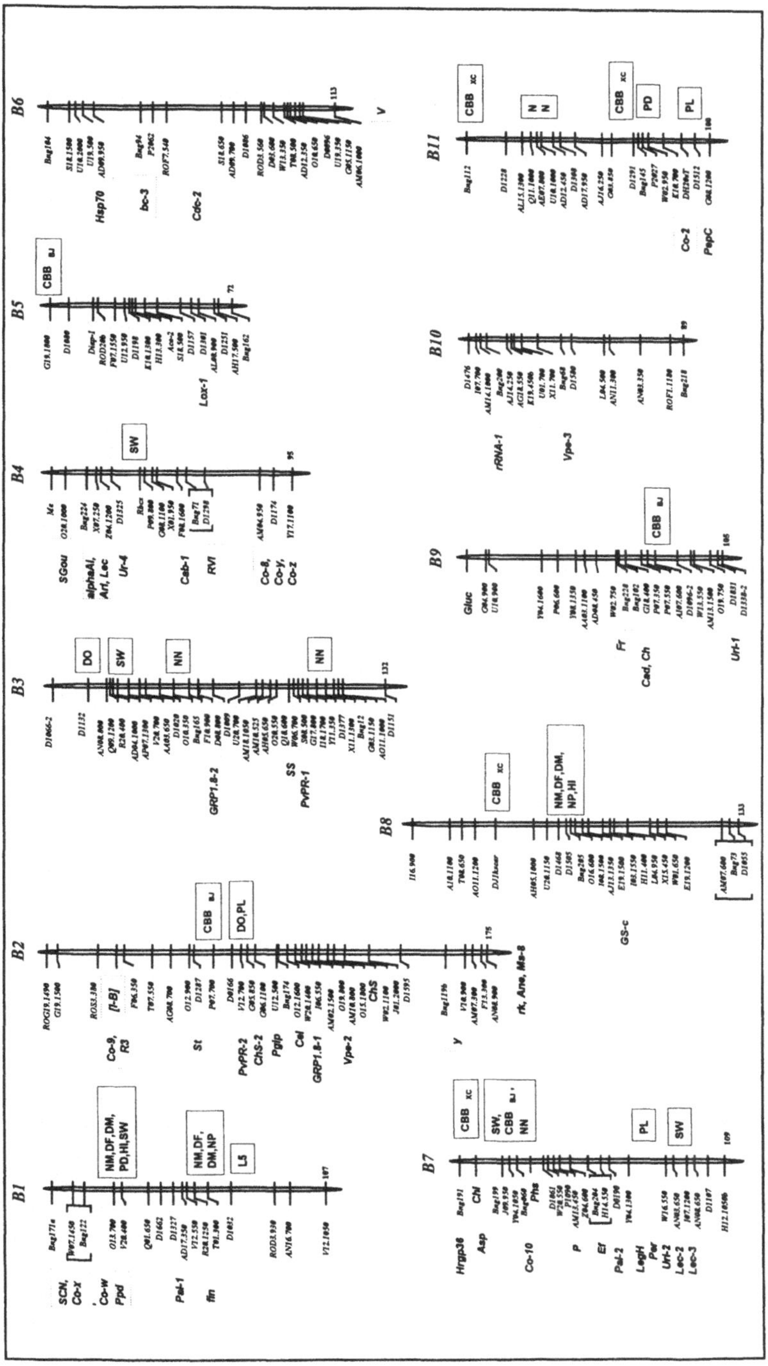

Figure 2. For explanation, see details at the bottom of page 66.

4. INTERSPECIFIC AND INTERGENERIC ALIGNMENT OF LINKAGE MAPS

Boutin et al. (1995) conducted an alignment or comparative mapping study among three species belonging to the Phaseoleae tribe: common bean and mungbean (*Vigna radiata* L.Wilczek), belonging both to the Phaseolinae subtribe, and the more distant soybean [*Glycine max* (L.) Merr.], which belongs to the Glycininae subtribe. Previous studies by Palmer (1987) had shown a 7% nucleotide substitution in cpDNA between common bean and mungbean, but a 13% substitution rate between common or mungbean and soybean. The different degrees of cpDNA divergence were paralleled by the differences in alignment conservation in the nuclear genomes of these three species. Whereas several linkage groups of common bean consisted of only one or two mungbean linkage groups (and vice versa), only short (12-14 cM) linkage segments were conserved and scattered on the soybean linkage groups.

The results of Boutin et al. (1995) suggest that there is no unified Phaseoleae genome as there is a grass genome (Bennetzen & Freeling, 1993). A unified Phaseolinae genome, encompassing the genomes of common bean, cowpea (*Vigna unguiculata* L. Walp.), and mungbean (Fatokun et al., 1992; Menancio-Hautea et al., 1993; Menéndez et al., 1997) genomes, can be constructed. The linkage map of Vallejos et al. (1992) and the present results play an important role in developing such a system. Yu et al. (1998), for example, in establishing their map in the DIACOL Calima x XR235-1-1 RI population used a combination of common bean and mungbean clones.

Although no unified linkage map can probably be established between the Phaseolinae and soybean, identification of micro-alignments can lead to

Previous page: Figure 2. Genomic distribution of genes with a biochemical function, major genes coding for phenotypic traits, and QTLs. Each linkage group is represented by its core map version (see Figure 2). To the left of each linkage group are the framework molecular markers (smaller font), the biochemical genes (larger font), and major genes for phenotypic trait (shaded boxes). For explanations on symbols, see Tables 4 and 5 and the original references (Adam-Blondon et al., 1994b; Nodari et al., 1993a; Vallejos et al., 1992). *Co-8* to *Co-z* loci are anthracnose resistance loci from Geffroy (1997). To the right (boxed two-letter symbols), are the QTLs mapped by Nodari et al. (BJ population; 1993b), Yu et al. (DX population; 1998), and Koinange et al. (1996). CBB: common bacterial blight resistance, DF and DM: number of days to flowering and to maturity, DO: seed dormancy, HI: harvest index, L5: length of the 5th internode, NM: number of nodes on the main stem, NN: *Rhizobium* nodulation, NP: number of pods, PD: photoperiod-induced delay in flowering, PL: pod length, SW: seed weight. Location of genes is approximate as most were not directly mapped in the BJ population.

local increases in marker density as illustrated by Danesh et al. (1998) for a gene for resistance to the soybean cyst nematode (SCN, *Heterodera globodera*) in soybean. Here, clones from common bean and mungbean complemented soybean clones to increase the local marker density and facilitate the development of bacterial artificial library contigs (i.e., an ordered array of overlapping clones) around the major SCN resistance locus (*rhg1*). The common bean homologue of this locus appears to map on linkage group 1 in common bean, although the existence of a nematode resistance gene at that location in the genome of common bean has not yet been verified (Figure 2).

It remains to be determined whether conserved linkages between the Phaseoleae, in general, and the Phaseolinae, in particular, and other sections of the Fabaceae (e.g., pea, chickpea, or lentil: Weeden et al., 1992) can be identified.

5. CYTOGENETICS AND PHYSICAL MAPPING

Most species of the genus *Phaseolus* sensu stricto, including *P. vulgaris*, have a chromosome number of 2n = 2x = 22. Three species show an aneuploid reduction to 2n = 2x = 20, *P. leptostachyus*, *P. micranthus*, and *P. macvaughii* (Mercado-Ruaro & Delgado Salinas, 1998). Although cytological analyses are difficult because the mitotic and meiotic chromosomes are small (approximately 1-3 μm), most chromosomes are metacentric or submetacentric (Maréchal, 1971; Mercado-Ruaro & Delgado Salinas, 1998). Maréchal (1971) observed an average of 1.9 chiasmata per bivalent in *P. vulgaris* meioses, or one crossing-over per chromosome arm. These values are consistent with the overall map length of 1,200 cM for the common bean map (Freyre et al., 1998).

A possible alternative to observations of mitotic and meiotic chromosomes are analyses of polytene chromosomes in embryo suspensor cells (Nagl, 1969, 1974, 1985). The suspensor, a short-lived set of cells at the base of the embryo, is characterized by a high rate of metabolic activity, which may account for the presence of polytene chromosomes. In contrast with dipteran polytene chromosomes, plant polytene chromosomes have poorly developed banding patterns, variable degrees of chromatin condensation, and lack of somatic pairing (Nagl et al., 1983), which make distinguishing individual chromosomes difficult, except for the presence of nucleolar-organizing regions. Alternative identification methods are being sought such as the use of centromeric heterochromatin morphology (Nenno & Nagl, 1998). Using the latter method, eight chromosomes of the haploid complement can be identified, whereas the distinction among the three remaining chromosomes remains difficult.

In situ hybridizations on polytene (and other) chromosomes have been performed for repetitive sequences (rRNA) and single- or low-copy-number sequences, polygalacturonase-inhibiting protein (PGIP), phaseolin (Phs), and leghemoglobin (LegH). In fluorescent *in situ* hybridization (FISH) studies of embryo suspensor polytene chromosomes of *P. coccineus*, the first report of the use of FISH in plant polytene chromosomes, Nenno et al. (1994) identified three chromosomes with 45S (18S, 5.8S, and 25S) rRNA sequences. These results were confirmed by Guerra et al. (1996) using FISH on root tip mitotic and anther tapetal polytene chromosomes of *P. coccineus*, by Ferreira Fernández (1996) using FISH on root mitotic chromosomes, and by Shi et al. (1996) using FISH on root protoplast interphase and metaphase chromosomes. Nevertheless, Uchiumi et al. (1998) observed four rRNA loci following FISH. The cause of these discrepancies has not been identified yet. Shi et al. (1996) also observed at least three locations for 5S rRNA sequences on chromosomes different from those harboring the 45S rRNA genes. Recently, Nenno et al. (1998) performed FISH with a synthetic oligonucleotide [$(TTTAGGG)_5$] corresponding to the *Arabidopsis thaliana* telomere sequence. Distinct signals were obtained at the chromosome termini of *P. vulgaris* and *P. coccineus* and in the centromere regions of *P. acutifolius*.

In an attempt to map a low-copy-number sequence, Frediani et al. (1993) used a radioactively-labeled PGIP DNA probe to determine the genome distribution of this small multigene family by *in situ* hybridization of polytene chromosomes of *P. vulgaris* embryo suspensor cells. They observed one hybridization region suggesting that this gene family is concentrated in a single region of the bean genome. The *Pgip* locus maps on linkage group B2 of the core map (Figure 2) (V. Geffroy, P. Gepts & M. Dron, unpublished results). Nenno et al. (1994) located the phaseolin locus by FISH on *P. coccineus* chromosome 7, which corresponds to linkage group B7 of the core linkage map (Nodari et al., 1993a). Uchiumi et al. (1998) were able to locate LegH sequences on pair of, presumably homologous, mitotic chromosomes using a combination of *in situ* PCR on root mitotic chromosome spreads with LegH-specific primers, followed by FISH. The assignment to a single chromosome pair is consistent with earlier observations by Lee & Verma (1984) and the genetic mapping of this sequence to linkage group B7 (Gepts et al., 1993; Koinange et al., 1996) (Figure 2).

These results, obtained with repetitive and low-copy-number sequences, suggest that a complete molecular karyotype can be established for common bean and related species. In turn, this karyotype can help in completing the coverage of the molecular linkage map, in particular by determining the location on the linkage map of centromeric and telomeric regions (Shippen & McKnight, 1998). Of particular interest in this regard is the development of a bacterial artificial chromosome (BAC) library for common bean (S. Mackenzie, personal communication, 1998). Individual BAC clones can be mapped onto the common bean linkage map and, in turn, their map location can be correlated with their cytological location obtained by FISH. The advantage of using BAC clones in FISH is that they provide a larger labeling target and, therefore, probes with a higher activity. Of course, this BAC library has many other applications, including comparative genome mapping and map-based cloning. The library contains some 33,000 clones with an average insert size of 100 kb, which represents a 4.5 x genome coverage (S. Mackenzie, personal communication, 1998).

6. ISSUES ADDRESSED WITH MOLECULAR LINKAGE MAPS

6.1 Tagging and Mapping Genes for Phenotypic Traits

Disease resistance genes are the most abundant category of genes tagged or mapped in common bean. A recent review by Kelly & Miklas (1998; also see Chapter 4) reported that 23 RAPD and 5 SCAR markers had been identified, tagging 15 major disease resistance genes. This work reflects the importance of disease resistance in bean breeding (Schwartz & Pastor-Corrales, 1989). Only major genes for resistance to viral and fungal diseases have been mapped so far. Two important genes for resistance to BCM are the *I* and *bc-3* (Drijfhout, 1978). These genes have been tagged (Haley et al., 1994a, c; Johnson et al., 1997; Melotto et al., 1996) and mapped on linkage groups 2 and 6, respectively (Johnson et al., 1997; Nodari et al., 1993a) (Table 4; Figure 2). Substantial progress has been made recently in our understanding of the resistance to another viral disease, bean golden mosaic (BGM). Resistance to this disease originated in race Durango and may be governed by major genes, one of which (*bgm-1*) has been tagged (Urrea et al., 1996).

At least seven genes have been identified for resistance to anthracnose (Kelly & Young, 1996). Four of them have been tagged (*Co-1*, *Co-2*, *Co-5*, and *Co-6*) (Table 3), and for one of them (*Co-2*) a map position has been obtained (Figure 2, linkage group 11; Adam-Blondon et al., 1994b; Gepts et al., 1993) and two SCAR markers have been developed (Adam-Blondon et al., 1994a; Geffroy et al., 1998). Recently, Young et al. (1998) described the gene *Co-7*, one of three genes present in landrace Colorado de Teopisca (G 2333), although no tests of allelism were conducted with the other six anthracnose resistance genes.

Six genes for resistance to rust have been described and an additional four have been proposed (Kelly et al., 1996). Four genes have been tagged with RAPD markers (Table 3), and one of them (*Ur-4*) has been mapped (Figure 2) onto linkage group 4 (Gepts et al., 1993; Miklas et al., 1993). In addition, a rust resistance gene associated with small pustules (*Ur-9*) was tagged and mapped on a RAPD map by Jung et al. (1996).

Resistance to the seed weevil *Zabrotes subfasciatus* Boheman is conditioned by the seed protein arcelin. Genes for arcelin are part of small multigene family with lectins and α-amylase inhibitors (Hoffman & Donaldson, 1985; Moreno & Chrispeels, 1989; Osborn et al., 1988). This multigene family has been mapped to linkage group 4 (Figure 2; Gepts et al., 1993).

An analysis of the inheritance of the domestication syndrome (Koinange et al., 1996; see section 6.2) has allowed us to map the genes coding for several important morphological or physiological traits. Several of them had already been identified in the cultivated gene pool (Leakey, 1988) (Table 5; Figure 2). The analysis has also allowed us to align the classical and molecular maps to some extent (Table 2). The domestication syndrome is the suite of traits distinguishing wild and cultivated plants. In common bean, these traits include seed dispersal and dormancy, growth habit, gigantism, and phenology (for further details, see Koinange et al., 1996). The gene for determinacy (*fin*) has been mapped to linkage group 1, where it is linked to the *Ppd* gene for photoperiod sensitivity. This linkage had been observed earlier by Coyne (1967, 1970). Further studies by Bassett (1997) have shown that *fin* and *Z*, a gene responsible for partial seedcoat coloration, are tightly linked (Table 5). The *B* gene, which contributes to seed color, is tightly linked to the *I* gene for BCM resistance on linkage group B2 (Kyle & Dickson, 1988; Temple & Morales, 1986) and is loosely linked to the *St*, conditioning the presence of strings in pod sutures (Bassett, 1991). The *y* gene for yellow ("wax") pod color is also located on linkage group B2 (Koinange et al., 1996), which shows a correspondence with classical linkage map groups III and VIII (Table 2).

Additional seed color genes linked to linkage group B2, but for which a specific location in the gene order has not been determined yet, are *rk* (red kidney seed color) and *Ane* (Anebulosus). Gene *V* (flower color) maps to linkage group B6 (Nodari et al., 1993a). This map location is confirmed by the mapping on the same linkage group of *Cor* (Nodari et al., 1993a), which is synonymous with v^{lae}, an allele of *V* (Bassett, 1995). Gene *P*, responsible

Table 4. Major disease resistance genes, marker tags, and map location in common bean

	Resistance gene		Marker tags or most closely linked markers				
Disease or insect	Symbol (previous symbol)	Gene pool	RFLP	RAPD	SCAR	Linkage group	References
Viruses							
Bean common mosaic	*I*	M	-	$W13^{690}$	Yes	B2	Haley et al., 1994c; Melotto et al., 1996
	bc-3	M	*Bng94, D1050, P2062*	$S13^{660}$, $AD19^{690}$, $C20^{460}$, $C11^{350/420}$	Yes	B6	Haley et al., 1994a; Johnson et al., 1997
Bean golden mosaic	*bgm*	M	-	$R2^{570/530}$	No	-	Urrea et al., 1996
Fungi							
Rust	*Ur-3*	M	-	$K14^{620}$	No	-	Haley et al., 1994b
	Ur-4 *(Up-2)*	A	*D0168, D1011*	$A14^{1100}$	No	B4	Miklas et al., 1993
	Ur-5 (B190)	M	-	$F10^{970}$, $I19^{460}$	Yes	-	Haley et al., 1993; Melotto & Kelly, 1998
	Ur-9 (URPR1)	A	-	$AJ16^{250}$, $H19^{1050}$	No	-	Jung et al., 1996
	Ur-11	-	-	$AC20^{490}$ $AE19^{890}$	No	-	Johnson et al., 1995

Disease or insect	Resistance gene: Symbol (previous symbol)	Resistance gene: Gene pool	Marker tags or most closely linked markers: RFLP	Marker tags or most closely linked markers: RAPD	Marker tags or most closely linked markers: SCAR	Linkage group	References
Ashy stem blight	*Mp-1*	M	-	$B386^{900}$	-	-	Olaya et al., 1996
	Mp-2	M	-	$B459^{1600}$	-	-	
Anthracnose	*Co-1* (*A*)	A	-	$F10^{530}$	No	-	Young & Kelly, 1997
	Co-2 (*Are*)	M	*P3040a*, *D1512*	$H20^{450}$, Q^{1440}, $B355^{1000}$	Yes	B11	Adam-Blondon et al., 1994a; Geffroy et al., 1998; Gepts et al., 1993; Young & Kelly, 1996
	$Co\text{-}4^2$	M	-	$AS13^{950}$, $AL9^{740}$	Yes	-	Young et al., 1998
	Co-5 (*Mexique 3*)	M	-	$AB3^{450}$	No	-	Young & Kelly, 1997
	Co-6	M	-	$AH1^{780}$, $AK20^{890}$	No	-	Young & Kelly, 1997
Insects							
Mexican bean weevil	*Arl*	M	*Arl- Lec-αAI*	-	No	B4	Gepts et al., 1993; Hoffman & Donaldson, 1985; Moreno & Chrispeels, 1989; Osborn et al., 1988

Table 5. Major genes coding for morphological and physiological traits in common bean

Trait	Gene	Marker tags or most closely linked markers			Linkage	References
		RFLP	RAPD	SCAR	group	
Morphological traits						
Determinacy	*fin*	*D1325*	-	No	B1	Coyne, 1967, 1970; Koinange et al., 1996
Partial seedcoat color	*Z*	*D1325*	*AM13*^{1350}, *M19*^{400}	No	B1	Bassett, 1997; Brady et al., 1998
Pod suture fibers ("strings")	*St*	*D1026*	-	No	B2	Bassett, 1991; Koinange et al., 1996
Yellow pod color ("wax")	*y*	*D1619*	-	No	B2	Bassett, 1991; Koinange et al., 1996
Seed color	*B*	-	*W13*^{690}	Yes	B2	Haley et al., 1994c; Kyle & Dickson, 1988; Melotto et al., 1996; Nodari et al., 1993a; Temple & Morales, 1986
Red kidney seed color	*rk*	-	-	No	B2	Bassett, 1991; Koinange et al., 1996
Anebulosus	*Ane*	-	-	No	B2	Bassett, 1991; Koinange et al., 1996
Flower color, seed corona	*V* (= *Cor*)	-	-	No	B6	Nodari et al., 1993a
Primary pigmentation	*P*	*P1090*	-	No	B7	Arndt & Gepts, 1989; Koinange et al., 1996; Vallejos et al., 1992
Seed shininess	*asp*	-	-	No	B7	W.C. Johnson & P. Gepts, unpublished results; M.J. Bassett, personal communication
Physiological traits						
Photoperiod sensitivity	*Ppd*	D1492-3	P5	-	B1	Coyne, 1967, 1970; Gu et al., 1993, 1995; Koinange et al., 1996
	Hr	D1479	BC303	-	-	Gu et al., 1994, 1995, 1998

for the presence or absence of anthocyanin pigments in vegetative and reproductive organs, is linked, although not tightly, to the *Phs* (phaseolin seed protein) locus on linkage group B7 (Arndt & Gepts, 1989; Koinange et al., 1996; Vallejos et al., 1992). The *asp* (asper, M.J. Bassett, personal communication, 1998) gene, responsible for seedcoat dullness, is loosely linked to the *Phs* locus (Arndt & Gepts, 1989; Ferreira Fernández, 1996). Finally, gene *C* for seed color and the tightly linked genes or alleles controlling some seed color patterns, such as the mottled, striped, or rhomboid flecking (Leakey, 1988), are located on linkage group B8 (W.C. Johnson & P. Gepts, unpublished results).

These analyses show that for some traits such as disease resistances, genes often cluster, whereas for others, such as pigmentation, they are generally unlinked or loosely linked. Geffroy et al. (1998) showed that the 450 bp *PvH20* fragment corresponding to the $ROH20^{450}$ RAPD marker tagging the *Co-2* gene contains a sequence coding for a putative polypeptide with six imperfect leucine-rich repeats, similar to many disease resistance genes (Hammond-Kosack & Jones, 1997). Low-stringency hybridizations with *PvH20* after *Dra*I digestion revealed complex hybridization patterns, strongly suggesting that *PvH20* is part of a multigene family. In addition, genotypes carrying *Co-2* showed an identical hybridization pattern that was different from the ones shown by susceptible genotypes (*co-2*). Mapping of the polymorphic bands in the EO2 x Corel backcross and BJ RI populations (Table 1) showed that this family clusters near the *Co-2* locus at one end of linkage group B11 (Figure 2). Additional resistance gene analogs (RGAs) have been identified by Rivkin et al. (1999) and Geffroy (1997) based on conservation of specific sequence motifs in cloned resistance genes from other plant species. Rivkin et al. (1999) identified RGAs on linkage groups B, D, F, and I of the Florida map and ND1 and ND4 of the North Dakota map (Table 1). One RGA mapped at some 17 cM of the *Ur-3* gene on linkage group ND1. Geffroy (1997) identified a cluster of RGAs on linkage group B4 of the BJ RI population. In addition, three major genes as well as QTLs for anthracnose resistance also mapped to this cluster. These genes conferred resistance against either Andean or Middle American strains of the anthracnose pathogen, suggesting that the cluster existed before the divergence into the two major bean gene pools and that different resistance specificities had been recruited from this gene cluster during the co-evolution of this host-pathogen relationship.

Further mapping experiments conducted by Geffroy (1997) in the BJ RI population showed that anthracnose resistance genes also cluster with genes for resistance (or their homologues) to other diseases. These include a homologue to the SCN resistance gene on linkage group B1, the *I* gene for BCM resistance on B2, and a major QTL for CBB resistance and *Rhizobium* nodulation on B7. In the *I* gene region, a gene for resistance against *Pseudomonas syringae* pv. *phaseolicola* Burkh. was also located (Crute & Pink, 1996).

In contrast, seed pigmentation genes may be loosely linked or not linked at all (Table 5 and Figure 2). Among the seed color genes mapped is the *P* gene (B7), which controls the presence or absence of pigmentation whatever the color, the *B* gene (B2), which, with *v*, leads to brownish colors, and the *V* gene (B6), which controls flower color and is generally associated with mauve to black seed colors. Seed color pattern genes include the *Z* gene (B1) for partial seed coloration and the v^{lae} gene (B6) for seed corona color pattern (formerly *Cor*). The absence of strong linkages may be partially responsible for the wide range of colors and color patterns observed in the progeny of some common bean crosses.

QTL analyses have been performed for several traits, including disease resistances, *Rhizobium* nodulation (Nodari et al., 1993b), plant architecture (Jung et al., 1996), canning quality (Walters et al., 1997), and drought tolerance (Schneider et al., 1997). Diseases include common bacterial blight (Jung et al., 1996, 1997; Nodari et al., 1993b; Yu et al., 1998) and web blight [caused by *Thanatephorus cucumeris* (Frank) Donk.] (Jung et al., 1996).

6.2 Gene Introgression and Germplasm Conversion

Analyses with biochemical and molecular markers have shown that the domesticated gene pool in common bean has a strongly reduced level of genetic diversity compared with its wild counterpart (Gepts et al., 1986; Koenig et al., 1990; Sonnante et al., 1994). In particular, commercial bean classes, such as pink and kidney beans, may be particularly lacking genetic diversity. This dearth of diversity is consistent with the pedigree of cultivars in these classes. Both these factors suggest that these classes resulted from crosses between closely related parents (McClean et al., 1993). Lack of genetic diversity may also explain in part why yields over the 1930-1990 period have only doubled. Such modest increases, which are also characteristic of other crops such as soybean, potato (*Solanum tuberosum* L.), and wheat (*Triticum aestivum* L.), can be attributed in part to the limited introduction of genetic diversity in the elite cultivated gene pool (Gepts, 1998b).

Wild beans provide a largely untapped source of genetic diversity (Kornegay et al., 1993; Sonnante et al., 1994). An obvious difficulty in introgressing genetic diversity from this source is the sheer lack of adaptation of wild beans to cultivation. Wild and domesticated beans show marked phenotypic differences (called the domestication syndrome after Hammer, 1984) concerning seed dispersal and dormancy, growth habit, photoperiod sensitivity, and consumer-related traits, such as pod and seed size, pigmentation, and quality (Gepts & Debouck, 1991). Traditional quantitative genetics posited that such large phenotypic differences had to be controlled by many genes, each with small effects. Such genetic control would make it very difficult to recover the domesticated phenotype following a wild x domesticated cross. Conversely, the existence of genes with large effects controlling part or all of the expression of domestication syndrome traits would ease this recovery.

To analyze the inheritance of the domestication syndrome, an RI population was developed from the cross between the wax snap bean cultivar Midas (of Andean origin) and the wild bean G 12873 (of Middle American origin). The recombinant inbred population was evaluated in multiple environments [e.g., field vs. greenhouse; high latitude (Davis, U.S.A., 38° N) vs. low latitude (Popayán, Colombia, 2° N)]. Replicated trials were conducted to measure quantitative traits. A molecular linkage map based on RFLPs was developed while the evaluations were going on. The development of the linkage map was eased by the high level of molecular polymorphism existing between the two parents, which belong to two diverged gene pools (Nodari et al., 1992).

The results of this mapping experiment were summarized. Many traits of the domestication syndrome are controlled at least in part by genes with major effect ($R^2 > 30\%$), such as a quantitative trait locus (QTL) controlling 50% of seed dormancy. Multiple regression analyses showed that genetic causes accounted for more than 50% of phenotypic variation, suggesting a high heritability for most traits. Linkage analyses suggested that genes of the domestication syndrome were non-randomly distributed in the genome, with a concentration of genes controlling growth habit and photoperiod sensitivity on linkage group 1, seed dispersal and dormancy on linkage group 2, seed weight on linkage group 7, and earliness on linkage group 8.

The simple genetic control of the domestication syndrome has several implications. First, the existence of major genes and the high heritability of several components of the domestication syndrome is a possible reflection of the selection process during domestication. Whether this selection took place unconsciously or consciously, it is likely that relatively simple selection schemes such as bulk or single plant selection would have increased chances of recovering useful variants. A corollary of this observation is that selection for a domesticated phenotype would have taken place rapidly.

Second, recovering a domesticated phenotype in the progeny of wild x domesticated crosses should be easy. An example of this type of cross is provided by the successful introduction of the *Arl* gene coding for the insecticidal arcelin protein from wild into cultivated beans (Kornegay et al., 1993; Osborn et al., 1988). The gene was introduced by direct selection of the protein product by immunological tests in the progeny of cultivated recipient x wild donor lines (Kornegay et al., 1993). The *Arl* gene, located on linkage group B4, does not appear to be linked to major genes for the domestication syndrome (Koinange et al., 1996), which explains why Kornegay et al. (1993) were able to recover the *Arl* gene in progenies with a fully domesticated phenotype. Several caveats should be taken into account regarding introgression from wild beans. The presence of linkage may slow the recovery of useful recombinants. This analysis was able to detect only genes with an $R^2 > 10\%$. Genes with a smaller effect remained undetected, but may be important in recovering the full expression of the domestication syndrome. It is likely that appropriate mating systems involving some form of backcrossing need to be identified to transfer quantitative genetic variation efficiently in wild x domesticated crosses. Standard pedigree methods relying primarily on selfing are insufficient based on our experience with this recombinant inbred population and early generation trials (Singh et al., 1995).

Third, our results lead to hypotheses about which regions should be introgressed into exotic germplasm as part of a conversion program to enhance its adaptation and allow its evaluation and use in breeding programs. A condition *sine qua non* for evaluation under temperate conditions is photoperiod insensitivity (conditioned by allele *ppd*). Because *ppd* is linked to *fin*, which conditions determinacy and is linked to a quantitative gene controlling shorter internodes, the entire *ppd-fin* genome segment could be introgressed. In this way, one could transform climbing, indeterminate, photoperiod-sensitive germplasm into bush, determinate, photoperiod-insensitive germplasm that should be more amenable to evaluation under the standard monoculture, mechanized agricultural practices of temperate regions.

An additional region of interest is the region on linkage group 2 spanning the *st* gene (conditioning the absence of pod fibers) and the principal dormancy gene. Reduced dormancy would lead to more uniform, higher germination rates in conversion programs of wild beans. Although the *St* gene would prevent seed dispersal and would be useful in snap bean germplasm improvement, expression of this gene might be too strong for dry bean. A quantitative analysis needs to be conducted to identify major quantitative trait loci that condition a partial reduction in pod fiber content.

Last, many genes distinguishing the wild and domesticated parents in this study also distinguish genotypes in the cultivated gene pool as suggested by the similarity in linkages in this and previous studies. Many linkages observed with classical markers were obtained with European genotypes (see work of Lamprecht & Prakken summarized in Leakey, 1988; also consult the article of Lamprecht, 1940, in English translation in http://beangenes.cws.ndsu.nodak.edu/papers/lamp1940.htm). These European genotypes are primarily of Andean origin, but they also include Middle American genotypes (Gepts & Bliss, 1988). In contrast, the results of Koinange et al. (1996) were obtained using a snap bean of Andean ancestry and a Middle American wild bean. The fact that a wide range of geographic origins show similar linkage relationships provides additional support for the narrow basis of the domesticated gene pool.

6.3 Dissection of Complex Traits: Common Bacterial Blight Resistance

Common bacterial blight (CBB) is a damaging disease with broad distribution in warmer, more humid areas (Saettler, 1989). The most effective control method seems to be genetic resistance. Progress toward obtaining highly resistant germplasm has been slow in part because of the quantitative and organ-specific nature of the resistance. QTL analyses have provided information on the inheritance of this resistance. Nodari et al. (1993b) analyzed resistance to CBB in the $F_{2:3}$ generation of the BJ population. They identified four QTLs, located on linkage groups 2, 5, 7, and 9, for resistance of the first trifoliolate leaf, with R^2 values ranging from 13 to 32% of total phenotypic variation. All resistance alleles originated in the resistant parent, BAT 93. Further analyses in the BJ RI population have confirmed the QTLs on linkage groups 7 and 9 (R. Freyre & P. Gepts, unpublished results). The QTL on linkage group 7 has a major effect ($R^2 >$ 30-40%) and may be conserved across strains and genetic backgrounds. The control of CBB resistance by both major and minor genes observed in this experiment was consistent with classical genetic studies suggesting both qualitative and quantitative inheritance for this trait (reviewed by Singh, 1991).

More recently, Yu et al. (1998) analyzed the inheritance of CBB resistance in an RI population of cross DIACOL Calima x XR-235-1-1. Previous research had established an RFLP linkage map in a BC_1 population of this cross (Vallejos et al., 1992). DIACOL Calima is a Colombian cultivar of Andean background. The resistance in breeding line XR235-1-1 is derived from *P. coccineus* after introgression into a Middle American *P. vulgaris* background (Freytag et al., 1982). Four QTLs were identified for resistance of the third or fourth trifoliolate leaf. Two QTLs originated in the resistant parent [linkage groups A (=7) and F (=8)] and two in the susceptible parent (linkage group J or 11). Their R^2 values, ranging between 11 and 15 to 40% of phenotypic variation, could be attributed to genetic causes based on multivariate analyses. The QTL identified on linkage group

A appears to overlap with the one identified in the BJ population (Nodari et al., 1993b).

Valladares-Sánchez et al. (1979), Schuster et al. (1983), and Singh & Muñoz (1999) observed differential leaf and pod reaction to CBB, suggesting that resistance in these two organs might be differentially controlled. This possibility was investigated by Jung et al. (1996) with a QTL analysis based on a RAPD linkage map (see Table 1) in the RI population BAC 6 x HT 7719, both genotypes of race Mesoamerica of the Middle American gene pool. BAC 6 is a Brazilian breeding line with resistance to CBB and rust and susceptibility to web blight (WB). HT 7719 is a breeding line with resistance to WB but susceptibility to CBB and rust. Inoculations were conducted with two *Xcp* strains, one of which was applied to the first trifoliolate, later-developed trifoliolates, pods, and seeds, allowing a comparison of resistance QTLs in these different organs. Jung et al. (1996) observed a quantitative type of inheritance for CBB resistance in this cross, with both strains, and at all developmental stages. However, in contrast with earlier results (e.g., Valladares-Sánchez et al., 1979), high levels of phenotypic correlations (r= 0.80) were observed between leaf and pod resistance. This observation was consistent with the substantial overlap in QTLs identified for CBB resistance in leaves and pods (see Table 3 in Jung et al., 1996). Conversely, the lower level of phenotypic correlations (r = 0.20-0.25) between seed resistance to CBB, on one hand, and leaf and pod resistances, on the other, was consistent with fewer shared QTLs. Interestingly, all three seed resistance QTLs were contributed by the susceptible parent, whereas for leaf and pod resistance both parents contributed resistance QTLs.

In a further study, Jung et al. (1997) investigated the inheritance of *P. acutifolius*-derived resistance to CBB in the cross PC-50 (susceptible) x XAN 159 (resistant interspecific cross-derived line). *P. acutifolius* displays much higher levels of CBB resistance than *P. vulgaris* (Drijfhout & Blok, 1987; Schuster et al., 1983; Singh & Muñoz, 1999). Hence, introgression of this resistance has been attempted. Part of the reason for the success of this introgression between two more distantly related species may be because *P. acutifolius* contains major genes for CBB resistance (Drijfhout & Blok, 1987) or at least QTLs with major effect. Indeed, in the study of Jung et al. (1997), a major QTL was identified with effects in all plant organs and with the two bacterial strains investigated. Linkage to the *V* locus suggests a location on linkage group 6 of the core map for this major QTL (Figure 2). Further QTLs, with smaller effects were noted, including some originating in the susceptible parent.

In summary, QTL mapping has helped us confirm and gain additional understanding of the genetic control of a trait as complex as CBB resistance.

Particular features of this inheritance are the large number of genes involved (when considering different genetic backgrounds, different strains, and successive developmental stages or organs), the relatively small magnitude of the effect of some of the genes involved (with notable exceptions), and the contribution of resistance genes by both resistant and susceptible parents.

7. FUTURE MAPPING

Mapping is a necessary complement to tagging of genes of agronomic interests for marker-assisted selection (as envisioned by Kelly & Miklas, 1998) and not just an exercise of academic interest. For example, if molecular tags turn out to be useful only in certain backgrounds, then mapping the original tags and searching among the linked markers can identify other potential marker candidates. Mapping of different tags for putative alleles is a test of allelism that replaces the traditional test, which is based on an actual cross. Such crosses may be difficult to obtain because of the lack of adaptation of the parents or hybrid dysgenesis.

Clusters of disease resistance genes provide another example. Genetic and physical mapping experiments have revealed that resistance genes tend to cluster. In plants, including common bean (see above), one can find, within the same cluster, genes for resistance to different strains of the same pathogen or different pathogens, including nematodes and insects, besides viruses, bacteria, and fungi (Hammond-Kosack & Jones, 1997). Likewise, major genes and QTLs for disease resistance coexist within the same cluster (V. Geffroy, P. Gepts & M. Dron, unpublished results). Clearly, these observations could not have been obtained without a comprehensive linkage mapping approach.

The mapping of the domestication syndrome discussed earlier provides a final example. The genomic region encompassing the *fin* and *Ppd* regions on linkage group B1 should be very important in conversion programs to adapt exotic germplasm to temperate regions. Again, identification of this region would not have been possible without linkage studies.

What can be done, therefore, to improve the current mapping work in common bean? There are at least three broad areas in which further work can be achieved:

1. Genetic mapping:
   a) Increase the genome coverage to attach all currently unassigned markers. This may be achieved primarily by using different types of crosses. Currently, most mapping populations have been obtained from Andean x Middle American crosses. Alternative populations include crosses within gene pools or interspecific

crosses between *P. vulgaris* and *P. costaricensis*, *P. polyanthus*, or *P. coccineus*.

b) Pursue the mapping of major genes and QTLs for agronomic traits, especially in multiple environments and in different genetic backgrounds, to analyze trait stability and identify QTLs with more stable expression. A list of named genes in *Phaseolus* can be found at http://beangenes.cws.ndsu.nodak.edu/genes/genlist3.htm. Among these are genes that are agronomically important and deserve to be mapped such as *C* and *T* for seed pigmentation, *bc-u* and *Mel* for disease or pest resistance, *Dl* and *lcr* for hybrid dysgenesis, etc.
c) Correlate RFLP and RAPD maps.
d) Develop a set of PCR-based, universal markers to correlate linkage maps. These markers would rely on full-length (>20 bp) PCR primers (to assure reproducibility), show a high level of polymorphisms between and within gene pools, and be distributed on all 11 linkage groups (to increase the probability of map correlation or alignment).
e) Determine the relationship between genetic and physical maps in different regions of the genome and genetic and environmental effects on recombination.

2. Cytogenetics and physical mapping:
   a) More systematic (i.e., on each linkage group) physical mapping of markers to establish a correspondence between the genetic map, physical map, and karyotypes (especially of polytene chromosomes).
   b) Map BAC clones on chromosomes and develop contigs on selected regions of the genome containing genes of agronomic importance.
   c) Map telomeres and centromeres to establish a correspondence between the genetic and karyotypic map and assess the degree of genome coverage (in the case of telomeres).
   d) Develop a full set of trisomic stocks. Five primary trisomics have been developed already (Ashraf & Bassett, 1987). A complete set would become a significant tool for mapping of new loci as has been done in other crops (Khush et al., 1984; Satovic et al., 1997; Schondelmaier & Jung, 1997).
3. Conversion programs:
   a) Develop PCR-based markers characterizing key regions in the genome (both regions with essential adaptation traits and other regions to be recovered during the backcross program).

b) Conduct a QTL analysis on pod fiber content (seed dispersal mechanism).
c) Attempt actual conversion experiments to evaluate changes in adaptation and other phenotypic traits.

Since the publication of the first molecular maps by Vallejos et al. (1992), Nodari et al. (1993a), Adam-Blondon et al. (1994b), substantial progress has been made in increasing the coverage of the linkage map, developing mapping populations, broadening the types of molecular markers used, correlating these maps, and placing both major genes and QTLs for agronomic traits. Much needs to be done, however, to make the common bean linkage map a fully developed genetic tool for use in basic genetic studies and breeding.

ACKNOWLEDGMENTS

My research has been funded by USDA (Plant Genome, OICD), USAID (PSTC, Bean Cowpea CRSP), IBPGR, the McKnight Foundation, the California Agricultural Experiment Station, and the California Crop Improvement Association. I thank the following individuals from my lab for their enthusiastic participation in the research presented here: V. Becerra, R. Freyre, W. Johnson, J. Kami, R. Koenig, E. Koinange, V. Llaca, R. Papa, M. Paredes, and G. Sonnante. Collaborators are also gratefully acknowledged: D. Debouck and S. Singh (CIAT, Colombia), J. Acosta (INIFAP, Mexico), A. Delgado (UNAM, Mexico), and V. Geffroy and M. Dron (Université de Paris XI, Orsay, France). I thank S. Mackenzie, P. McClean, and T. Uchiumi for providing information before publication.

REFERENCES

Adam-Blondon, A.F., M. Sévignac, H. Bannerot & M. Dron. 1994a. SCAR, RAPD and RFLP markers tightly linked to a dominant gene (*Are*) conferring resistance to anthracnose in common bean. Theor. Appl. Genet. 88: 865-870.

Adam-Blondon, A.F., M. Sévignac & M. Dron. 1994b. A genetic map of common bean to localize specific resistance genes against anthracnose. Genome 37: 915-924.

Ahn, S. & S.D. Tanksley. 1993. Comparative linkage maps of the rice and maize genomes. Proc. Natl. Acad. Sci. U.S.A. 90: 7980-7984.

Anthony, J.L., R.A. Vonder Haar & T.C. Hall. 1990. Nucleotide sequence of an alpha-phaseolin gene from *Phaseolus vulgaris*. Nucl. Ac. Res. 18: 3396.

Arndt, G.C. & P. Gepts. 1989. Segregation and linkage for morphological and biochemical markers in a wide cross in common bean (*Phaseolus vulgaris*). Annu. Rpt. Bean Improv. Coop. 32: 68-69.

Arumuganathan, K. & D.E. Earle. 1991. Nuclear DNA content of some important plant species. Plant Mol. Biol. Rep. 9: 208-218.

Ashraf, M. & M.J. Bassett. 1987. Five primary trisomics from translocation heterozygote progenies in common bean, *Phaseolus vulgaris* L. Theor. Appl. Genet. 74: 346-360.

Bassett, M.J. 1991. A revised linkage map of common bean. HortScience 26: 834-836.

Bassett, M.J. 1995. The dark corona character in seedcoat of common bean cosegregates with the pink flower allele v^{lae}. J. Amer. Soc. Hort. Sci. 120: 520-522.

Bassett, M. J. 1997. Tight linkage between the *fin* locus for plant habit and the *Z* locus for partly colored seedcoat patterns in common bean. J. Amer. Soc. Hort. Sci. 122: 656-658.

Bennetzen, J.L. & M. Freeling. 1993. Grasses as a single genetic system: genome composition, collinearity and compatibility. Trends Genet. 9: 259-260.

Botstein, D., R.L. White, M. Skolnick & R.V. Davis. 1980. Construction of a genetic linkage map in man using restriction fragment length polymorphisms. Amer. J. Hum. Genet. 32: 314-331.

Boutin, S., N. Young, T. Olson, Z. Yu, R. Shoemaker & C. Vallejos. 1995. Genome conservation among three legume genera detected with DNA markers. Genetics 38: 928-937.

Brady L., M. Bassett & P.E. McClean. 1998. Molecular markers associated with *T* and *Z*, two genes controlling partly colored seed coat patterns in common bean. Crop Sci. 38: 1073-1075.

Broglie, K.E., J.J. Gaynor & R.M. Broglie. 1986. Ethylene-regulated gene expression: Molecular cloning of the genes encoding an endochitinase from *Phaseolus vulgaris*. Proc. Natl. Acad. Sci. U.S.A. 83: 6820-6824.

Caetano-Anollès, G., B.J. Bassam & P.M. Gresshoff. 1991. DNA amplification fingerprinting using very short arbitrary oligonucleotide primers. Bio/Technology 9: 553-557.

Capote-Mainez, N. & F. Sánchez. 1997. Characterization of the common bean uricase II and its expression in organs other than nodules. Plant Physiol. 115: 1307-1317.

CIAT (Centro Internacional de Agricultura Tropical). 1980. Bean Program Annual Report. CIAT, Cali, Colombia.

Concibido, V., N. Young, D. Lange, R. Denny, D. Danesh & J. Orf. 1996. Targeted comparative genome analysis and qualitative mapping of a major partial-resistance gene to the soybean cyst nematode. Theor. Appl. Genet. 93: 234-241.

Corbin, D., N. Sauer & C. Lamb. 1987. Differential regulation of a hydroxyproline-rich glycoprotein gene family in wounded and infected plants. Mol. Cell. Biol. 7: 4337-4344.

Coyne, D.P. 1967. Photoperiodism: inheritance and linkage studies in *Phaseolus vulgaris*. J. Hered. 58: 313-314.

Coyne, D.P. 1970. Genetic control of a photoperiod-temperature response for time of flowering in beans (*Phaseolus vulgaris*). Crop Sci. 19: 246-248.

Crute, I. & D. Pink. 1996. Genetics and utilization of pathogen resistance in plants. Plant Cell 8: 1747-1755.

Dallas, J.F. 1988. Detection of DNA "fingerprints" of cultivated rice by hybridization with a human minisatellite DNA probe. Proc. Natl. Acad. Sci. U.S.A. 85: 6831-6835.

Danesh, D., S. Peñuela, J. Mudge, R. Denny, H. Nordstrom, J. Martínez & N. Young. 1998. A bacterial artificial chromosome library for soybean and identification of clones near a major cyst nematode resistance gene. Theor. Appl. Genet. 96: 196-202.

Drijfhout, E. 1978. Genetic interaction between *Phaseolus vulgaris* and bean common mosaic virus with implications for strain identification and breeding for resistance. Agric. Res. Rpt. 872. Centre for Agricultural Publishing and Documentation, Wageningen, Netherlands.

Drijfhout, E. & W.J. Blok. 1987. Inheritance of resistance to *Xanthomonas campestris* pv. *phaseoli* in tepary bean (*Phaseolus acutifolius*). Euphytica 36: 803-808.

Ebbelaar, M., G. Tucker, M. Laats & C. van Dijk. 1996. Characterization of pectinases and pectin methylesterase cDNAs in pods of green beans (*Phaseolus vulgaris* L.). Plant Mol. Biol. 31: 1141-1151.

Edington, B.V., C.J. Lamb & R.A. Dixon. 1991. cDNA cloning and characterization of a putative 1,3-α-D-glucanase transcript induced by fungal elicitor in bean cell suspension cultures. Plant Mol. Biol. 16: 81-94.

Edwards, K.E., C.L. Cramer, G.P. Bolwell, R.A. Dixon, W. Schuch & C.J. Lamb. 1985. Rapid transient induction of phenylalanine ammonia-lyase mRNA in elicitor-treated bean cells. Proc. Natl. Acad. Sci. U.S.A. 82: 6731-6735.

Fatokun, C.A., D.I. Menancio-Hautea, D. Danesh & N.D. Young. 1992. Evidence for orthologous seed weight genes in cowpea and mung bean based on RFLP mapping. Genetics 132: 841-846.

Ferreira Fernández, J. 1996. Caracterización y mapas genéticos en *Phaseolus vulgaris* L. para la mejora genética de la variedad Granja Asturiana. Doctoral thesis, Univ. de Oviedo, Asturias, Spain.

Frediani, M., R. Cremonini, G. Salvi, C. Caprari, A. Desiderio, R. D'Ovidio, F. Cervone & G. De Lorenzo. 1993. Cytological localization of the *PGIP* genes in the embryo suspensor cells of *Phaseolus vulgaris* L. Theor. Appl. Genet. 87: 369-373.

Freyre, R., P. Skroch, V. Geffroy, A.F. Adam-Blondon, A. Shirmohamadali, W. Johnson, V. Llaca, R. Nodari, P. Pereira, S.M. Tsai, J. Tohme, M. Dron, J. Nienhuis, C. Vallejos & P. Gepts. 1998. Towards an integrated linkage map of common bean. 4. Development of a core map and alignment of RFLP maps. Theor. Appl. Genet. 97: 847-856.

Freytag, G.F., M.J. Bassett & M. Zapata. 1982. Registration of XR-235-1-1 bean germplasm. Crop Sci. 22: 1268-1269.

Fuller, F., P.W. Kunstner, T. Nguyen & D.P.S. Verma. 1983. Soybean nodulin genes: analysis of cDNA clones reveals several major tissue-specific sequences in nitrogen-fixing root nodules. Proc. Natl. Acad. Sci. U.S.A. 80: 2594-2598.

Geffroy, V. 1997. Dissection génétique de la résistance à *Colletotrichum lindemuthianum*, agent de l'anthracnose, chez deux génotypes représentatifs des pools géniques de *Phaseolus vulgaris*. Doctoral thesis. Institut National Agronomique, Paris, France.

Geffroy, V., F. Creusot, J. Falquet, M. Sévignac, A.F. Adam-Blondon, H. Bannerot, P. Gepts & M. Dron. 1998. A family of LRR sequences in the vicinity of the *Co-2* locus for anthracnose resistance in *Phaseolus vulgaris* and its potential use in marker-assisted selection. Theor. Appl. Genet. 96: 494-502.

Gepts, P. 1993. The use of molecular and biochemical markers in crop evolution studies. Evol. Biol. 27: 51-94.

Gepts, P. 1998a. Genome mapping resources in common bean: towards a unified linkage map in recombinant inbred population BAT93 x Jalo EEP558. Annu. Rpt. Bean Improv. Coop. 41: 91-92.

Gepts, P. 1998b. Origin and evolution of common bean: past events and recent trends. HortScience 33: 1124-1130.

Gepts, P. & F.A. Bliss. 1985. F_1 hybrid weakness in the common bean: differential geographic origin suggests two gene pools in cultivated bean germplasm. J. Hered. 76: 447-450.
Gepts, P. & F.A. Bliss. 1988. Dissemination pathways of common bean (*Phaseolus vulgaris*, Fabaceae) deduced from phaseolin electrophoretic variability. II. Europe and Africa. Econ. Bot. 42: 86-104.
Gepts, P. & D.G. Debouck. 1991. Origin, domestication & evolution of the common bean, *Phaseolus vulgaris*. pp. 7-53. *In* A. Van Schoonhoven & O. Voysest (eds.), Common Beans: Research for Crop Improvement. C.A.B. Int., Wallingford, U.K. & CIAT, Cali, Colombia.
Gepts, P., R. Nodari, R. Tsai, E.M.K. Koinange, V. Llaca, R. Gilbertson & P. Guzmán. 1993. Linkage mapping in common bean. Annu. Rpt. Bean Improv. Coop. 36: xxiv-xxxviii.
Gepts, P., T.C. Osborn, K. Rashka & F.A. Bliss. 1986. Phaseolin-protein variability in wild forms and landraces of the common bean (*Phaseolus vulgaris*): evidence for multiple centers of domestication. Econ. Bot. 40: 451-468.
Gu, W., N.F. Weeden, J. Yu & D. Wallace. 1995. Large-scale, cost-effective screening of PCR products in marker-assisted selection applications. Theor. Appl. Genet. 91: 465-470.
Gu, W.K., N.F. Weeden, D.H. Wallace & S. Singh. 1993. A DNA marker for *ppd*, a gene conferring insensitivity to photoperiod in common bean. Annu. Rpt. Bean Improv. Coop. 36: 1-2.
Gu, W., N.F. Weeden, J. Zhu & D. Wallace. 1994. Identification of a DNA marker for *Hr*, a gene that interacts with *Ppd* to confer extreme photoperiod sensitivity. Annu. Rpt. Bean Improv. Coop. 37: 125-126.
Gu, W.K., J.Q. Zhu, D.H. Wallace, S.P. Singh & N.F. Weeden. 1998. Analysis of genes controlling photoperiod sensitivity in common bean using DNA markers. Euphytica 102: 125-132.
Guerra, M., A. Kenton & M. Bennett. 1996. rDNA sites in mitotic and polytene chromosomes of *Vigna unguiculata* (L.) Walp. and *Phaseolus coccineus* L. revealed by *in situ* hybridization. Ann. Bot. 78: 157-161.
Haley, S., L. Afanador & J.D. Kelly. 1994a. Selection for monogenic pest resistance traits with coupling- and repulsion-phase RAPD markers. Crop Sci. 34: 1061-1066.
Haley, S.D., L. Afanador & J.D. Kelly. 1994c. Identification and application of a random amplified polymorphic DNA marker for the *I* gene (Potyvirus resistance) in common bean. Phytopathology 84: 157-160.
Haley, S., L. Afanador, P.N. Miklas, J. Stavely & J.D. Kelly. 1994b. Heterogeneous inbred populations are useful as sources of near-isogenic lines for RAPD marker localization. Theor. Appl. Genet. 88: 337-342.
Haley, S.D., P.N. Miklas, L. Afanador & J.D. Kelly. 1994d. Random amplified polymorphic DNA (RAPD) marker variability between and within gene pools of common bean. J. Amer. Soc. Hort. Sci. 119: 122-125.
Haley, S.D., P.N. Miklas, J.R. Stavely, J. Byrum & J.D. Kelly. 1993. Identification of RAPD markers linked to a major rust resistance gene block in common bean. Theor. Appl. Genet. 86: 505-512.
Hammer, K. 1984. Das Domestikationssyndrom. Kulturpflanze 32: 11-34.
Hammond-Kosack, K. & J. Jones. 1997. Plant disease resistance genes. Ann. Rev. Plant Phys. Plant Molec. Biol. 48: 575-607.
He, S., Z. Yu, C. Vallejos & S. Mackenzie. 1995. Pollen fertility restoration by nuclear gene *Fr* in CMS common bean: an *Fr* linkage map and the mode of *Fr* action. Theor. Appl. Genet. 90: 1056-1062.

Hedrick, S., J. Bell, T. Boller & C. Lamb. 1988. Chitinase cDNA cloning and mRNA induction by fungal elicitor, wounding, and infection. Plant Physiol. 86: 182-186.
Hirel, B., M. Marsolier, A. Hoarau, J. Hoarau, J. Brangeon, R. Schafer & D. Verma. 1992. Forcing expression of a soybean root glutamine synthetase gene in tobacco leaves induces a native gene encoding cytosolic enzyme. PMB 20: 217-218.
Hoffman, L.M. & D.D. Donaldson. 1985. Characterization of two *Phaseolus vulgaris* phytohemagglutinin genes closely linked on the chromosome. EMBO J. 4: 883-889.
Johannsen, W. 1909. Elementen der exakten Erblichkeitslehre. Fisher, Jena, Germany.
Johnson, E., P.N. Miklas, J. Stavely & J. Martínez-Cruzado. 1995. Coupling- and repulsion-RAPDs for marker-assisted selection of PI 181996 rust resistance in common bean. Theor Appl. Genet. 90: 659-664.
Johnson, W. 1997. Improving the efficiency of common bean (*Phaseolus vulgaris* L.) breeding programs using molecular markers. Ph.D. dissertation. Univ. of California, Davis, U.S.A.
Johnson, W., P. Guzmán, D. Mandala, A. Mkandawire, S. Temple, R. Gilbertson & P. Gepts. 1997. Molecular tagging of the *bc-3* gene for introgression into Andean common bean. Crop Sci. 37: 248-254.
Johnson, W., C. Menéndez, R. Nodari, E. Koinange, S.P. Singh & P. Gepts. 1996. Association of a seed weight factor with the phaseolin seed storage protein locus across genotypes, environments, and genomes in *Phaseolus* - *Vigna* spp.: Sax (1923) revisited. J. Quant. Trait Loci 2: Article 5, http://probe.nalusda.gov/otherdocs/jqtl/jqtl1996-1905/sax1992.htm.
Jones, C., K. Edwards, S. Castaglione, M. Winfield, F. Sala, C. van de Wiel, G. Bredemeijer, B. Vosman, M. Matthes, A. Daly, R. Brettschneider, P. Bettini, M. Buiatti, E. Maestri, A. Malcevschi, N. Marmiroli, R. Aert, G. Volckaert, J. Rueda, R. Linacero, A. Vazquez & A. Karp. 1997. Reproducibility testing of RAPD, AFLP and SSR markers in plants by a network of European laboratories. Mol. Breed. 3: 381-390.
Jorgensen, R.A., R.E. Cuéllar, W.P. Thompson & T.A. Kavanagh. 1987. Structure and variation in ribosomal RNA genes of pea. Plant Mol. Biol. 8: 3-12.
Jung, G., S. Beebe, J. Nienhuis, S. Park, D.P. Coyne, J. Marita, F. Pedraza & J. Tohme. 1998. Development of SCAR markers linked to common bacterial blight resistance genes (QTL) in common bean. Annu. Rpt. Bean Improv. Coop. 41: 97-98.
Jung, G., D.P. Coyne, P. Skroch, J. Nienhuis, E. Arnaud-Santana, J. Bokosi, H. Ariyarathne, J. Steadman, J. Beaver & S. Kaeppler. 1996. Molecular markers associated with plant architecture and resistance to common blight, web blight, and rust in common beans. J. Amer. Soc. Hort. Sci. 121: 794-803.
Jung, G., P. Skroch, D.P. Coyne, J. Nienhuis, H. Ariyarathne, S. Kaeppler & M. Bassett. 1997. Molecular-marker-based genetic analysis of tepary-bean-derived common bacterial blight resistance in different developmental stages of common bean. J. Amer. Soc. Hort. Sci. 122: 329-337.
Kami, J.A. & P. Gepts. 1994. Phaseolin nucleotide sequence diversity in *Phaseolus*. I. Intraspecific diversity in *Phaseolus vulgaris*. Genome 37: 751-757.
Keller, B., N. Sauer & C. Lamb. 1988. Glycine-rich cell wall proteins in bean: gene structure and association of the protein with the vascular system. EMBO J. 7: 3625-3633.
Kelly, J.D. & P.N. Miklas. 1998. The role of RAPD markers in breeding for disease resistance in common bean. Mol. Breed. 4: 1-11.
Kelly, J.D., J. Stavely & P.N. Miklas. 1996. Proposed symbols for rust resistance genes. Annu. Rpt. Bean Improv. Coop. 39: 25-31.

Kelly, J.D. & R.A. Young. 1996. Proposed symbols for anthracnose resistance genes. Annu. Rpt. Bean Improv. Coop. 39: 20-24.
Khush, G.S., R.J. Singh, S.C. Sur & A.L. Librojo. 1984. Primary trisomics of rice: origin, morphology, cytology and use in linkage mapping. Genetics 107: 141-163.
Knight, M. & G. Jenkins. 1992. Genes encoding the small subunit of ribulose 1,5-biphosphate carboxylase/oxygenase in *Phaseolus vulgaris* L.: nucleotide sequence of cDNA clones and initial studies of expression. Plant Mol. Biol. 18: 567-579.
Koenig, R., S.P. Singh & P. Gepts. 1990. Novel phaseolin types in wild and cultivated common bean (*Phaseolus vulgaris*, Fabaceae). Econ. Bot. 44: 50-60.
Koinange, E.M.K., S.P. Singh & P. Gepts. 1996. Genetic control of the domestication syndrome in common-bean. Crop Sci. 36: 1037-1045.
Kornegay, J., C. Cardona & C.E. Posso. 1993. Inheritance of resistance to Mexican bean weevil in common bean, determined by bioassay and bichemical tests. Crop Sci. 33: 589-594.
Kyle, M.M. & M.H. Dickson. 1988. Linkage of hypersensitivity to five viruses with the *B* locus in *Phaseolus vulgaris* L. J. Hered. 79: 308-311.
Lamprecht, H. 1940. Zur Genetik von *Phaseolus vulgaris*. XVI. Weitere Beiträge zur Vererbung der Teilfarbigkeit. Hereditas 26: 277-291.
Leakey, C.L.A. 1988. Genotypic and phenotypic variation in common bean. pp. 245-327. *In* P. Gepts (ed.), Genetic Resources of *Phaseolus* Beans. Kluwer, Dordrecht, Netherlands.
Lee, J.S. & D.P.S. Verma. 1984. Structure and chromosomal arrangement of leghemoglobin genes in kidney bean suggest divergence in soybean leghemoglobin gene loci following tetraploidization. EMBO J. 3: 2745-2752.
Lee, M. 1995. DNA markers and plant breeding programs. Adv. Agron. 55: 265-344.
Llaca, V. & P. Gepts. 1996. Pulsed field gel electrophoresis analysis of the phaseolin locus region in *Phaseolus vulgaris*. Genome 39: 722-729.
Maréchal, R. 1971. Observations sur quelques hybrides dans le genre *Phaseolus*. II. Les phénomènes méiotiques. Bull. Rech. Agron. Gembloux 6: 461-489.
McClean, P.E., J.R. Myers & J.J. Hammond. 1993. Coefficient of parentage and cluster analysis of North American dry bean cultivars. Crop Sci. 33: 190-197.
McMullen, M., P. Byrne, M. Snook, B. Wiseman, E. Lee & N. Widstrom. 1998. Quantitative trait loci and metabolic pathways. Proc. Natl. Acad. Sci. U.S.A. 95: 1996-2000.
Mehdy, M.C. & C.J. Lamb. 1987. Chalcone isomerase cDNA cloning and mRNA induction by fungal elicitor, wounding and infection. EMBO J. 6: 1527-1533.
Meier, B., N. Shaw & A. Slusarenko. 1993. Spatial and temporal accumulation of defense gene transcripts in bean (*Phaseolus vulgaris*) leaves in relation to bacteria-induced hypersensitive cell death. Mol. Plant-Micr. Inter. 6: 453-466.
Melotto, M., L. Afanador & J.D. Kelly. 1996. Development of a SCAR marker linked to the *I* gene in common bean. Genome 39: 1216-1219.
Melotto, M. & J.D. Kelly. 1998. SCAR markers linked to major disease resistance genes in common bean. Annu. Rpt. Bean Improv. Coop. 41: 64-65.
Menancio-Hautea, D., C.A. Fatokun, L. Kumar, D. Danesh & N.D. Young. 1993. Comparative genome analysis of mungbean (*Vigna radiata* L. Wilczek) and cowpea (*V. unguiculata* L. Walpers) using RFLP mapping data. Theor. Appl. Genet. 86: 797-810.
Mendel, G. 1866. Experiments on plant hybrids. pp. 1-48. *In* C. Stern & E.R. Sherwood (eds.), The Origin of Genetics. W.H. Freeman, San Francisco, U.S.A.
Menéndez, C., A. Hall & P. Gepts. 1997. A genetic linkage map of cowpea (*Vigna unguiculata*) developed from a cross between two inbred, domesticated lines. Theor. Appl. Genet. 95: 1210-1217.

Mercado-Ruaro, P. & A. Delgado Salinas. 1998. Karyotypic studies on species of *Phaseolus* (Fabaceae: Phaseolinae). Amer. J. Bot. 85: 1-9.
Miao, G.H., Z.L. Hong & D.P.S. Verma. 1993. Two functional soybean genes encoding P34 (CDC2) protein kinases are regulated by different plant developmental pathways. Proc. Natl. Acad. Sci. U.S.A. 90: 943-947.
Miklas, P.N., J.R. Stavely & J.D. Kelly. 1993. Identification and potential use of a molecular marker for rust resistance in common bean. Theor. Appl. Genet. 85: 745-749.
Mirkov, T.E., J.M. Wahlstrom, K. Hagiwara, F. Finardi-Filho, S. Kjemtrup & M.J. Chrispeels. 1994. Evolutionary relationships among proteins in the phytohemagglutinin-archelin-alpha-amylase inhibitor family of the common bean and its relatives. Plant Mol. Biol. 26: 1103-1113.
Moreno, J. & M.J. Chrispeels. 1989. A lectin gene encodes the alpha-amylase inhibitor of the common bean. Proc. Natl. Acad. Sci. U.S.A. 86: 7885-7889.
Nagl, W. 1969. Banded polytene chromosomes in the legume *Phaseolus vulgaris*. Nature 221: 70-71.
Nagl, W. 1974. The *Phaseolus* suspensor and its polytene chromosomes. Z. Pflanzenphysiol. 73: 1-44.
Nagl, W. 1985. Chromosomes in differentation. pp. 135-172. *In* A.K. Sharma & A. Sharma (eds.), Advances in Chromosome and Cell Genetics. IBH, New Delhi, India.
Nagl, W., M. Jeanjour, H. Kling, S. Kühner, I. Michels, T. Müller & B. Stein. 1983. Genome and chromatin organization in higher plants. Biol. Zentr. Bl. 102: 129-148.
Nakamura, Y., M. Leppert, P. O'Connell, R. Wolff, T. Holm, M. Culver, C. Martin, E. Fujimoto, M. Hoff, E. Kumlin & R. White. 1987. Variable number of tandem repeat (VNTR) markers for human gene mapping. Science 235: 1616-1622.
Nenno, M. & W. Nagl. 1998. Identification of polytene chromosomes of *Phaseolus coccineus* on the basis of centromeric heterochromatin morphology. Annu. Rpt. Bean Improv. Coop. 41: 105-106.
Nenno, M., K. Schumann & W. Nagl. 1994. Detection of rRNA and phaseolin genes on polytene chromosomes of *Phaseolus coccineus* by fluorescence in situ hybridization after pepsin treatment. Genome 37: 1018-1021.
Nenno, M., D. Zink & W. Nagl. 1998. The *Arabidopsis* telomere sequence is highly abundant in the genome of *Phaseolus acutifolius* and preferentially located in the centromeres. Annu. Rpt. Bean Improv. Coop. 41: 103-104.
Nguyen, T., M. Zelechowska, V. Foster, H. Bergmann & D.P.S. Verma. 1985. Primary structure of the soybean nodulin-35 gene encoding uricase II localized in the peroxisomes of uninfected cells of nodules. Proc. Natl. Acad. Sci. U.S.A. 82: 5040-5044.
Nodari, R.O., E.M.K. Koinange, J.D. Kelly & P. Gepts. 1992. Towards an integrated linkage map of common bean. I. Development of genomic DNA probes and levels of restriction fragment length polymorphism. Theor. Appl. Genet. 84: 186-192.
Nodari, R.O., S.M. Tsai, R.L. Gilbertson & P. Gepts. 1993a. Towards an integrated linkage map of common bean. II. Development of an RFLP-based linkage map. Theor. Appl. Genet. 85: 513-520.
Nodari, R.O., S.M. Tsai, P. Guzmán, R.L. Gilbertson & P. Gepts. 1993b. Towards an integrated linkage map of common bean. 3. Mapping genetic factors controlling host-bacterium interactions. Genetics 134: 341-350.
Olaya, G., G. Abawi & N.F. Weeden. 1996. Inheritance of the resistance to *Macrophomina phaseolina* and identification of RAPD markers linked to the resistance genes in beans. Phytopathology 86: 674-679.

Osborn, T.C., D.C. Alexander, S.S.M. Sun, C. Cardona & F.A. Bliss. 1988. Insecticidal activity and lectin homology of arcelin seed protein. Science 240: 207-210.
Palmer, J.D. 1987. Chloroplast DNA evolution and biosystematic uses of chloroplast DNA variation. Amer. Nat. 130: S6-S29.
Paterson, A. 1995. Molecular dissection of quantitative traits: progress and prospects. Genome Res. 5: 321-333.
Pathirana, S., C. Vance, S. Miller & J. Gantt. 1992. Alfalfa root nodule phosphoenolpyruvate carboxylase: characterization of the cDNA and expression in effective and plant-controlled ineffective nodules. Plant Mol. Biol. 20: 437-450.
Prioul, J., S. Quarrie, M. Causse & D. de Vienne. 1997. Dissecting complex physiological functions through the use of molecular quantitative genetics. J. Exp. Bot. 48: 1151-1163.
Rafalski, J.A. & S.V. Tingey. 1993. Genetic diagnostics in plant breeding: RAPDs, microsatellites and machines. Trends Genet. 9: 275-280.
Rivkin, M., C.E. Vallejos & P.E. McClean. 1999. Disease-related sequences in common bean. Genome 42:41-47.
Ryder, T.B., S.A. Hedrick, J.N. Bell, X. Liang, S.D. Clouse & C.J. Lamb. 1987. Organization and differential activation of a gene family encoding the plant defense enzyme chalcone synthase in *Phaseolus vulgaris*. Mol. Gen. Genet. 210: 219-233.
Saettler, A.W. 1989. Common bacterial blight. pp. 261-283. *In* H.F. Schwartz & M.A. Pastor-Corrales (eds.), Bean Production Problems in the Tropics 2nd ed. CIAT, Cali, Colombia.
Satovic, Z., A. Torres & J. Cubero. 1997. Genetic mapping of new morphological, isozyme and RAPD markers in *Vicia faba* L. using trisomics. Theor. Appl. Genet. 93: 1130-1138.
Sax, K. 1923. The association of size differences with seed coat pattern and pigmentation in *Phaseolus vulgaris*. Genetics 8: 552-560.
Schneider, K., M. Brothers & J.D. Kelly. 1997. Marker-assisted selection to improve drought resistance in common bean. Crop Sci. 37: 51-60.
Schondelmaier, J. & C. Jung. 1997. Chromosomal assignment of the nine linkage groups of sugar beet (*Beta vulgaris* L.) using primary trisomics. Theor. Appl. Genet. 95: 590-596.
Schuster, M.L., D.P. Coyne, T. Behre & H. Leyna. 1983. Sources of *Phaseolus* species resistance and leaf and pod differential reactions to common blight. HortScience 18: 901-903.
Schwartz, H.F. & M.A. Pastor-Corrales (eds.). 1989. Bean Production Problems in the Tropics 2nd ed. CIAT, Cali, Colombia.
Shi, L., T. Zhu & P. Keim. 1996. Ribosomal RNA genes in soybean and common bean: chromosomal organization, expression & evolution. Theor. Appl. Genet. 93: 136-411.
Shippen, D. & T. McKnight. 1998. Telomeres, telomerase and plant development. Trends Plant Sci. 3: 126-130.
Singh, S.P. 1991. Bean genetics. pp. 199-286. *In* A. van Schoonhoven & O. Voysest (eds.), Common Beans: Research for Crop Improvement. C.A.B. Int., Wallingford, U.K. & CIAT, Cali, Colombia.
Singh, S.P., P. Gepts & D.G. Debouck. 1991. Races of common bean (*Phaseolus vulgaris*, Fabaceae). Econ. Bot. 45: 379-396.
Singh, S.P., A. Molina & P. Gepts. 1995. Potential of wild common bean for seed yield improvement of cultivars in the tropics. Can. J. Plant Sci. 75: 807-813.
Singh, S.P. & C.G. Muñoz. 1999. Resistance to common bacterial blight among *Phaseolus* species and common bean improvement. Crop Sci. 39:80-89.
Skroch, P., G. Jung, J. Nienhuis & D.P. Coyne. 1996. Integration of RAPD marker linkage maps and comparative mapping of QTL for disease resistance in common bean. Annu. Rpt. Bean Improv. Coop. 39: 48-49.

Slightom, J.L., S.M. Sun & T.C. Hall. 1983. Complete nuleotide sequence of a French bean storage protein gene: phaseolin. Proc. Natl. Acad. Sci. U.S.A. 80: 1897-1901.

Sonnante, G., T. Stockton, R.O. Nodari, V.L. Becerra Velásquez & P. Gepts. 1994. Evolution of genetic diversity during the domestication of common-bean (*Phaseolus vulgaris* L.). Theor. Appl. Genet. 89: 629-635.

Tanksley, S. 1993. Mapping polygenes. Annu. Rev. Genet. 27: 205-233.

Tanksley, S.D., S. Grandillo, T.M. Fulton, D. Zamir, Y. Eshed, V. Petiard, J. López & T. Beckbunn. 1996. Advanced backcross QTL analysis in a cross between an elite processing line of tomato and its wild relative *L. pimpinellifolium*. Theor. Appl. Genet. 92: 213-224.

Tautz, D. 1989. Hypervariability of simple sequences as a general source for polymorphic DNA markers. Nucl. Ac. Res. 17: 6463-6471.

Temple, S.R. & F.J. Morales. 1986. Linkage of dominant hypersensitive resistance to bean common mosaic virus to seed color in *Phaseolus vulgaris*. Euphytica 35: 331-333.

Thompson, W., M. Everett, N. Polans & R. Jorgensen. 1983. Phytochrome control of RNA levels in developing pea and mung bean leaves. Planta 158: 487-500.

Thummler, F. & D.P.S. Verma. 1987. Nodulin-100 of soybean is the subunit of sucrose synthase regulated by availability of free heme in nodules. J. Biol. Chem. 262: 14730-14736.

Toubart, P., A. Desiderio, G. Salvi, F. Cervone, L. Daroda & G. De Lorenzo. 1992. Cloning and characterization of the gene encoding the endopolygalacturonase-inhibiting protein (PGIP) of *Phaseolus vulgaris* L. Plant J. 2: 367-373.

Tucker, M.L. & S.B. Milligan. 1991. Sequence analysis and comparison of avocado fruit and bean abscission cellulases. Plant Physiol. 95: 928-933.

Uchiumi, T., R. Kuwashiro, J. Miyamoto, M. Abe & S. Higashi. 1998. Detection of the leghemoglobin gene on two chromosomes of *Phaseolus vulgaris* by *in situ* hybridization (FISH). Plant Cell Physiol. 39: 790-794.

Urrea, C., P.N. Miklas, J.S. Beaver & R. Riley. 1996. A codominant randomly amplified polymorphic DNA (RAPD) marker useful for indirect selection of bean golden mosaic virus resistance in common bean. J. Amer. Soc. Hort. Sci. 121: 1035-1039.

Valladares-Sánchez, N.E., D.P. Coyne & M.L. Schuster. 1979. Differential reactions of leaves and pods of *Phaseolus* germplasm to strains of *Xanthomonas phaseoli* and transgressive segregation for tolerance from crosses of susceptible germplasm. J. Amer. Soc. Hort. Sci. 104: 648-654.

Vallejos, E.C., N.S. Sakiyama & C.D. Chase. 1992. A molecular marker-based linkage map of *Phaseolus vulgaris* L. Genetics 131: 733-740.

Vidal, V., B. Ranty, M. Dillenschneider, M. Charpenteau & R. Ranjeva. 1993. Molecular characterization of a 70kDa heat-shock protein of bean mitochondria. Plant J. 3: 143-150.

Vos, P., R. Hogers, M. Bleeker, M. Reijans, T. Van de Lee, M. Hornes, A. Frijters, J. Pot, J. Peleman, M. Kuiper & M. Zabeau. 1995. AFLP: a new technique for DNA fingerprinting. Nucl. Ac. Res. 23: 4407-4414.

Walter, M., J. Grima-Pettenati, C. Grand, A. Boudet & C. Lamb. 1988. Cinnamyl-alcohol dehydrogenase, a molecular marker specific of lignin synthesis: cDNA cloning and mRNA induction by fungal elicitor. Proc. Natl. Acad. Sci. U.S.A. 85: 5546-5550.

Walter, M.H., J. Liu, C. Grand, C.J. Lamb & D. Hess. 1990. Bean-pathogenesis-related (PR) proteins deduced from elicitor-induced transcripts are members of a ubiquitous new class of conserved PR proteins including pollen allergens. Mol. Gen. Genet. 222: 353-360.

Walters, K.J., G.L. Hosfield, M.A. Uebersax & J.D. Kelly. 1997. Navy bean canning quality: correlations, heritability estimates, and randomly amplified polymorphic DNA markers associated with component traits. J. Amer. Soc. Hort. Sci. 122: 338-343.

Weeden, N.F., F.J. Muehlbauer & G. Ladizinsky. 1992. Extensive conservation of linkage relationships between pea and lentil genetic maps. J. Hered. 83: 123-129.

Welsh, J. & M. McClelland. 1990. Fingerprinting genomes with arbitrary primers. Nucl. Ac. Res. 18: 7213-7218.

Williams, J.G.K., A.R. Kubelik, K.J. Livak, J.A. Rafalski & S.V. Tingey. 1990. DNA polymorphisms amplified by arbitrary primers are useful as genetic markers. Nucl. Ac. Res. 18: 6531-6535.

Young, R.A. & J.D. Kelly. 1996. RAPD markers flanking the *Are* gene for anthracnose resistance in common bean. J. Amer. Soc. Hort. Sci. 121: 37-41.

Young, R.A. & J.D. Kelly. 1997. RAPD markers linked to three major anthracnose resistance genes in common bean. Crop Sci. 37: 940-946.

Young, R.A., M. Melotto, R.O. Nodari & J.D. Kelly. 1998. Marker-assisted dissection of the oligogenic anthracnose resistance in the common bean cultivar, 'G 2333'. Theor. Appl. Genet. 96: 87-94.

Yu, Z.H., R.E. Stall & C.E. Vallejos. 1998. Detection of genes for resistance to common bacterial blight of beans. Crop Sci. 38: 1290-1296.

Zietkiewicz, E., A. Rafalski & D. Labuda. 1994. Genome fingerprinting by Simple Sequence Repeat (SSR)-anchored polymerase chain reaction amplification. Genomics 20: 176-183.

Chapter 4

Marker-Assisted Selection

James D. Kelly & Phillip N. Miklas
Crop and Soil Sciences, Michigan State University, East Lansing, MI 48824, U.S.A.; USDA-ARS-IAREC, 24106 N. Bunn Rd., Prosser, WA 99350, U.S.A.

Key words: coupling-linkage, disease resistance, epistatic and hypostatic genes, indirect selection, pathogens, QTL, RAPD markers, repulsion-linkage

1. INTRODUCTION

The concept of indirect selection is attractive to plant breeders working to improve complex traits. For the successful application of indirect selection, a high level of correlation and/or linkage must exist between the trait of agronomic interest and the trait being directly selected. The selectable trait must be under simpler genetic control than the trait of interest and methodologies must be available that would favor its selection. For many years, indirect selection for yield has been practiced by breeders using selection based on yield components. With the discovery that visual polymorphisms were under simple genetic control, the concept of using these single gene traits as markers for more complex economic traits, became a reality. Basically there are three marker types: morphological (traits), biochemical (seed storage proteins and isozymes), and molecular or DNA-based polymorphisms, that can be used to establish linkages with traits of economic importance. A review of these markers will show that polymerase chain reaction (PCR) based molecular markers currently have the greatest application for marker-assisted selection (MAS) and therefore will predominate discussions within this chapter.

2. MORPHOLOGICAL MARKERS

Morphological traits represent the first markers used in linkage mapping. They include genes usually identified through mutations that affect plant morphology. Virescent foliage *vi*, unifoliolate leaves *uni*, and reclining foliage *rf-1* to *rf-3* are but a few examples of the recessive mutants characterized in common bean (*Phaseolus vulgaris* L.) (Bassett, 1996). Other traits are attributable to dominant genes such as those conferring stringless pods *St*, and curved pods *Ct*. There are also many different dominant and recessive genes that affect color of flowers, seeds, and pods. Other genes that affect responses to environment or diseases, or in fact any gene that is simply inherited are usually included in the morphological class of markers.

Generally, too few morphological markers are segregating in any given genetic population to establish tight linkages with a specific trait of economic importance. Moreover, many morphological mutations are deleterious and undesirable in breeding populations and thus have limited utility as selectable markers. When linkages between morphological markers and traits of economic importance have been found it has usually been by happenstance and the association has been undesirable. For instance, it was difficult to recover parental seed color while breeding for *I* gene resistance to bean common mosaic (BCM, caused by a potyvirus) due to linkage with the *B* locus that conditions a seed color darkening effect in certain bean types (Kyle & Dickson, 1988; Temple & Morales, 1986). Similarly, the presence of the *I* gene is often associated with a bluish background color in pinto beans thought to be due to linkage with the *B* locus (J.D. Kelly, unpublished results). The linkage between flower color (V/v^{lae}) and resistance to common bacterial blight is another example (Jung et al., 1997). As more morphological markers are placed on linkage maps, however, their relationships and utility as selectable markers for genes and traits that are difficult to breed will become more apparent. In addition, morphological markers have been useful for integrating different linkage maps so their segregation in breeding populations used primarily to map resistance traits should not be ignored (Gepts et al., 1993).

3. BIOCHEMICAL MARKERS

With the discovery of simply inherited, structural variation present in seed storage proteins (phaseolin in *P. vulgaris*) and isozymes, the potential for using this variability as a genetic marker became a reality. The relative ease of determination (starch gel electrophoresis), immediately offered the opportunity to use these genetic markers for indirect selection, provided associations and/or linkages with economic traits existed and could be established. Unlike other crops, isozymatic variability in common bean was limited and expressed primarily at the gene pool (Andean - A, Middle American - M) level and not at the genotype level (Singh et al., 1991a,b). Storage protein markers displayed a similar pattern and proved useful in the classification of both cultivated and wild germplasm into gene pools. In addition, seed storage proteins have been used to indirectly select for insect resistance and increased protein content of common bean. A unique seed storage protein, arcelin, proved to be very valuable as an insecticide effective against the storage weevil, *Zabrotes subfasciatus* Boheman. Using artificial seeds containing the purified arcelin protein, Osborn et al. (1988) demonstrated that arcelin produced valuable insecticidal properties without any negative effects on mammals. Since the arcelin protein was simply inherited (Romero-Andreas et al., 1986) and possessed a distinct electrophoretic pattern as compared to other storage proteins, direct selection based on its banding pattern became a reality and rapid progress was made introducing arcelin genes into a range of germplasm (Kornegay et al., 1993). These markers are being used to facilitate selection for resistance to specific storage weevils in multi-parent gamete selection programs (Singh et al., 1998), and have been used to alter the seed protein composition to generate higher levels of resistance to a second storage weevil (Hartweck et al., 1997).

In a similar study, Delaney & Bliss (1991) demonstrated that one cycle of selection based on specific electrophoretic bands, i.e., alleles for the seed storage protein(s), including phaseolin, was as effective as three cycles of direct selection for increasing phaseolin protein levels in the seed of common bean. In effect these markers are components of the desired traits and thus behave more as a pleiotropic marker than a linked marker having potential for recombination with the particular trait of interest.

4. DNA-BASED MARKERS

With the advent of a wide range of analytical tools for DNA analysis, two types of molecular markers became available to breeders, geneticists, and germplasm specialists. Codominant restriction fragment length polymorphisms (RFLPs), discovered first (Botstein et al., 1980), proved to be abundant and became valuable as genetic markers. Dominant, random amplified polymorphic DNA (RAPD) markers proved easier to generate (Williams et al., 1990), but were less reproducible across labs and were less informative as markers in segregating populations. Since then, newer markers, such as AFLPs (amplified fragment length polymorphisms) that combine RFLP and RAPD technologies (Vos et al., 1995), and the microsatellite simple sequence repeats (SSRs) commonly used in human genetics, have gained popularity in higher plants through comparative mapping and DNA fingerprinting applications (Mohan et al., 1997; Staub et al., 1996). Scientists immediately saw the potential for these highly abundant markers and used them extensively to expand the genetic linkage maps for most commodities. Many of these maps had previously been limited to either morphological traits or protein markers (Bassett, 1991; Gepts, 1988). Maps proved valuable in positioning traits in relation to each other but were limited to the traits segregating in the specific mapping population. For details on linkage mapping in common bean, see Chapter 3 by Gepts in this book.

Since RFLPs were more informative due to the expression of codominance in early generation populations, they became the marker of choice among geneticists interested in genetic mapping. The application to breeders, however, was restricted by the costly and sophisticated techniques required in the generation of RFLPs. Germplasm specialists likewise used RFLPs to assist in the classification of related and wild germplasm, but costs have limited their extensive deployment. In certain instances these maps are being used as a basis for the cloning of specific genes into crops for which an efficient regeneration and transformation system is available. Similarly, AFLPs amenable to high throughput marker generation (Tohme et al., 1996) and the PCR-based inter simple sequence repeats (ISSRs) (González et al., 1998) have gained popularity in mapping and germplasm comparison studies, but as of yet there are limited reports of their use for tagging genes for indirect selection. A relatively new class of markers known as resistance gene analogues (RGA), developed from sequences of cloned resistance genes may have potential for tagging resistance traits (Geffroy et al., 1998; Kanazin et al., 1996). In soybean, RGA markers clustered near previously mapped resistance genes. These markers might be less useful for MAS of specific resistance genes, however, because most of them will arise from polymorphisms generated through accumulated mutations in silenced puesdogenes (Kanazin et al., 1996). The most recent markers are the single nucleotide polymorphisms (SNPs) which appear useful for mapping, manipulation, and study of disease genes in human (Landegren et al., 1998), and will likely have comparable application in plant genomics (Cregan et al., 1998).

RAPD markers proved to be cheaper but more variable technology with regard to problems of reproducibility between labs (Jones et al., 1997; Weeden et al., 1992). Variability has been attributed to differences in PCR protocols, equipment, and the type of heat labile polymerase, all of which theoretically, can be controlled. In practice, however, variability between labs cannot be controlled due to differences in equipment, which affects portability of RAPDs. Breeders immediately saw the advantage of RAPD markers as a rapid, cost-effective tool for the indirect selection of traits once linkages between the marker and the trait have been identified. As an extension of single gene tagging, linkages between RAPD markers and quantitative trait loci (QTL) controlling complex traits such as yield, stress tolerance (Schneider et al., 1997), and canning quality (Walters et al., 1997) have also been identified.

To facilitate the identification of linked markers, mapping populations, and methods have been developed which include near isogenic lines and bulked segregant analysis (BSA) (Michelmore et al., 1991). Markers identified using these methods have been used by breeders to incorporate new resistance genes or combine genes from different sources, especially in cases where normal screening techniques are laborious and expensive. To improve the reproducibility of previously identified RAPD markers, allele specific associated primers (ASAP), and/or sequence characterized amplified region (SCAR) markers, derived from corresponding RAPD bands (Gu et al., 1995; Kesseli et al., 1992; Paran & Michelmore, 1993) have become the basis for the indirect selection of economically valuable traits in common bean (Kelly & Miklas, 1998; Melotto et al., 1996).

Although indirect selection for qualitative traits appears promising (Kelly, 1995), the comparative efficiency of direct versus indirect selection for major gene traits comes into question. Often, such traits are easier, faster, and more cost effective to select for directly rather than indirectly. In the case of agronomic traits such as plant growth habit, height, flowering, and maturity, indirect selection will likely not improve efficiency over routine field selection. However, markers for genes conditioning photoperiodism (*ppd* and *hr*; Gu et al., 1995, 1998) and growth habit (*fin*) which could be used in germplasm conversion programs (Patterson et al., 1998) represent possible exceptions. Improvement of other traits such as disease and insect resistance could be accelerated using indirect selection. Disease and insect resistance present a set of challenges different from those involved with agronomic traits. One major difference is that disease resistance is evasive and breeders must work with two biological organisms and have expertise in handling both. The limitation in disease screening methods becomes apparent when the breeder is challenged with mixed races of a pathogen or by the presence of many resistance genes in the host. When screening conditions are not ideal, escapes can occur and other disease causing organisms can and do interfere with the successful detection of resistant individuals. Indirect selection of race-specific genes offers breeders a viable alternative to ensure that favorable gene combinations are present in new breeding lines. RAPD markers tightly linked to individual race-specific resistance genes form the basis for effective indirect selection of major gene resistance. A list of available RAPD markers linked to different resistance genes in common bean is shown in Table 1. The use of linked RAPD markers and their potential role in MAS for disease resistance to facilitate the efficient pyramiding of epistatic resistance genes for different pathogens of common bean is discussed (Kelly & Miklas, 1998).

Table 1. RAPD markers linked to major disease resistance (R) genes in common bean

R genes	Germ-plasm source	Gene pool[1]	Disease[2]	RAPD marker[3]	Linkage distance (cM) and orientation[4]	References
Co-1	Michig-an DRK	A	Ant	$OF10_{530}$	1.9 repulsion-phase	Young & Kelly, 1997
Co-2	Cornell 49-242	M	Ant	$OQ4_{1440}$ $B355_{1000}$ $OH20_{450}$	2.0 - 5.5 coupling-phase[6] 5.4 - 7.7 coupling-phase[6] co-dominant SCAR	Adam-B et al., 1994; Young & Kelly, 1996a
Co-4^2	SEL 1308, G2333	M	Ant	$OAS13_{950}$ $OAL9_{740}$ $SAS13_{950}$	No recombinants 3.1 - 4.5 coupling-phase Dominant SCAR	Young et al., 1998
Co-5	TU, G2333, SEL 1360	M	Ant	$OAB3_{450}$	5.9 coupling-phase	Young & Kelly, 1997
Co-6	Catrach-ita, AB 136	M	Ant	$OAH1_{780}$ $OAK20_{890}$ $OZ4_{560}$	12.3 co-dominant, coupling[6] 7.3 repulsion-phase[6] 2.9 coupling	Alzate-Marin et al., 1997; Young & Kelly, 1997
I	Seafare-r	M	BCM	$OW13_{690}$	1.3 - 5.0 coupling-phase	Haley et al., 1994b;
	Montca-lm	A		$SW13_{690}$	Dominant SCAR	Melotto et al., 1996
bc-3	B85009	M	BCM	$OAD19_{690}$	1.9 repulsion-phase	Haley et al., 1994a;
	MCM 3031	M		$OS13_{660}$	7.1 coupling-phase	Johnson et al., 1997
	MCR 2205	A		$OC20_{460}$	No recombinants, repulsion	
	MCR 2205	A		$OC20_{460}$ $OC11_{350}$ $OC11_{420}$	Dominant SCAR No recombinants Co-dominant SCAR	

R genes	Germ-plasm source	Gene pool[1]	Disease[2]	RAPD marker[3]	Linkage distance (cM) and orientation[4]	References
bgm-1	Garrapa-to, A 429	M	BGM	$OR2_{530}$ $OR2_{570}$	4.2 repulsion-phase Coupling and co-dominant	Urrea et al., 1996
Mp-1	BAT 477	M	Mac	$B386_{900}$	Coupling-phase	Olaya et al., 1996
Mp-2	BAT 477	M	Mac	$B459_{1600}$	Repulsion-phase	
Ur-3	NEP-II	M	Rust	$OK14_{620}$	2.23 coupling-phase	Haley et al., 1994c; Nemchin & Stavely, 1998
Ur-4	Early Gallatin	A	Rust	$OA14_{1100}$	No recombinants	Miklas et al., 1993
Ur-5	Mexico 309, B-190	M	Rust	$OF10_{970}$ $OI19_{460}$	2.15 coupling-phase No recombinants, Dominant SCAR	Haley et al., 1993 Melotto & Kelly, 1998
Ur-9	PC 50	A	Rust	$OJ13_{1800}$	5.0 coupling-phase[5]	Jung et al., 1996b
Ur-11	PI 181996	M	Rust	$OAC20_{490}$ $OAE19_{890}$	No recombinants 6.2 repulsion-phase	Johnson et al., 1995

[1]A, Andean and M, Middle American (Singh et al., 1991a,b); [2] Ant, anthracnose; BCM, bean common mosaic; BGM, bean golden mosaic; Mac, Macrophomina root rot; [3] 'O' and 'B' were derived by decamer primers from Operon Technologies (Alameda, CA) and the University of British Columbia (Vancouver, Canada), respectively; [4] SCAR, sequence characterized amplified region; [5] Recombination within a RIL mapping population, all other represent linkage distances with F_2 populations; [6] Flanking markers.

4.1 RAPD Markers and Qualitative Resistance

Breeding for durable pest resistance has challenged both plant breeders and pathologists. Non-specific resistance has been reported to offer more durable resistance but its deployment creates a major challenge for the breeder since this type of resistance is often masked by epistatic major gene resistance. Compounded by complex inheritance, non-specific resistance is not easily transmitted or identified in the progeny. Improving the durability of major gene resistance would be a more attractive alternative for plant breeders since genetic manipulations are easier to accomplish and outcomes are more predictable than the inconsistent expression of minor genes.

Monogenic resistance genes are attractive to the breeder because they are easy to manipulate and can be rapidly introgressed into susceptible materials through simple backcrossing. These same race-specific genes are recognized as the least durable source of genetic resistance to highly variable plant pathogens and insects. Duvick (1996) states that even though major resistance genes have only short useful lifetimes when used one at a time, the opportunity to pyramid genes (using MAS) with some knowledge of their complementarity will make major gene resistance more useful than at present. Bean breeders have a unique opportunity to utilize resistance genes from the two major gene pools to develop complementary resistance to a range of bean pathogens. Stavely (1984) demonstrated that the Andean *Ur-4* gene conditions resistance to different races of the bean rust [caused by *Uromyces appendiculatus* (Pers.) Ung.] than those controlled by the Middle American *Ur-3* and *Ur-5* genes. Kelly & Young (1996) showed that the Andean *Co-1* gene has a different resistance spectrum from the Middle American *Co-2* gene against races of *Colletotrichum lindemuthianum* (Sacc. & Magn.) Bri. & Cav. causing bean anthracnose. When several genes are to be pyramided using MAS, Beaver & Macchiavelli (1998) suggested waiting until the F_4 generation to screen lines for linked markers. Screening in the F_4 would improve the probability of identifying lines with the desired genotype and would drastically reduce the minimum number of lines needed to be evaluated.

Pyramiding race-specific genes has been highly effective in controlling stem rust (caused by *Puccinia graminis* Pers.:Pers.) of wheat (*Triticum aestivum* L.) for 40 years in Canada and the north-central U.S.A. Epidemics of stem rust have greatly reduced the yield of wheat cultivars lacking pyramided resistance genes in this region (Schafer & Roelfs, 1985). Similar observations have been made in common bean. The landrace Colorado de Teopisca (G 2333) from Chiapas, Mexico shown to possess two independent dominant genes (*Co-4*^{2}, *Co-5*), was resistant to 380 isolates of *C. lindemuthianum* (Pastor-Corrales et al., 1994). More recently, the presence of a third independent gene, *Co-7*, in G 2333 was confirmed (Young & Kelly, 1996b; Young et al., 1998). Traditional gene pyramiding requires the incorporation of several different resistance genes into a single cultivar. Due to the race specificity of many of these genes, screening nurseries have to be systematically inoculated with different races of the pathogen in an attempt to ensure that gene combinations are being maintained. In the absence of effective selection due to a lack of informative (differentiating) races, epistatic interactions between many resistance genes prevent the identification of hypostatic genes resulting in their potential loss from breeding populations. Testcrossing to susceptible genotypes needs to be conducted each cycle to detect the presence (or absence) of the masked genes. As breeders continue to identify and incorporate more broadly-based resistance genes into their breeding populations, valuable hypostatic genes will continue to be lost. Two situations where markers are most likely to have the greatest impact in combining oligogenic disease resistance are with epistatic and hypostatic genes.

4.1.1 Epistatic Genes

In the situation where a new broadly based resistance gene is being backcrossed into susceptible germplasm, the breeder may be reluctant to work with a more virulent race of the pathogen to detect the presence of the new gene in segregating populations. Direct screening may not be desirable for other reasons such as lack of routine screening methods and informative races for discriminating new resistance genes, quarantine, host escapes, or field inoculations may not permit the precise identification of a specific gene or gene combination since mixtures (contamination) from endemic races may occur.

4.1.2 Hypostatic Genes

In the above situation the breeder may wish to maintain a defeated or hypostatic gene that may have value in gene pyramiding. Despite being defeated by a new race(s) of a pathogen, the same gene may offer resistance to other potential new races of a pathogen. Pathogen populations are never static and may change annually based on sexual or asexual recombination, mutation, migration, competition between races, fitness, and genetic composition of the host. Such factors influence the longevity of new races, so a previously defeated resistance gene may regain utility in the future. If screening for resistance is based on the most virulent race present, the hypostatic gene will not be selected and could be lost from the breeding populations. The strategies of incorporating genes that lack resistance to all races or incorporating the most broad resistance sources have resulted in the omission of defeated resistance genes in the past or the underestimation of their potential in future cultivars. Marker-assisted selection of hypostatic genes, discussed later for anthracnose, BCM, and rust, would facilitate their maintenance in breeding populations.

4.1.3 Anthracnose

RAPD markers for five independent dominant genes (*Co-1, Co-2,* $Co\text{-}4^2$, *Co-5,* and *Co-6*) conditioning resistance to anthracnose have been identified and SCAR markers for two of these genes (*Co-2* and $Co\text{-}4^2$) have been developed (Adam-Blondon et al., 1994; Alzate-Marin et al., 1997; Young & Kelly, 1996a, 1997; Young et al., 1998). Pyramiding diverse resistance genes using MAS and deploying different gene combinations in different regions is proposed as the most practical and realistic approach to provide efficient long-term control of bean anthracnose in a relatively short time (Balardin & Kelly, 1998). Highly virulent races (1473 and 1572) of *C. lindemuthianum* of Middle American origin have been observed to be avirulent to the Andean *Co-1* gene (Balardin et al., 1997; Pastor-Corrales et al., 1994, 1995; Sicard et al., 1997; Young & Kelly, 1996b). Combining the Andean *Co-1* gene with the Middle American genes (*Co-2*, *Co-4*^{2}, *Co-5*, and *Co-6*) would be the best strategy to develop complementary resistance to the predominant Middle American races of the pathogen. This combination has been suggested as a viable approach for improving the durability of anthracnose resistance genes in Central America (Young & Kelly, 1997). The *Co-1* gene has been underutilized in breeding for anthracnose resistance because it does not exhibit the broad-based resistance of certain Middle American genes and often is masked by these other genes. Using MAS, the Andean *Co-1* gene can be combined with Middle American resistance genes, which would otherwise not be possible under local field selection because of the presence of multiple races of the pathogen.

When selecting for resistance in a particular region, the breeder should carefully choose a gene pair that, if deployed, would confer resistance to all known races in that region. In North America, this would be accomplished by combining the *Co-5* and *Co-6* genes, whereas for areas of Central America the most suitable combination would be the *Co-1* and *Co-4*^{2} genes. Markers linked to these resistance genes (Young & Kelly, 1996a, 1997) allow the breeder the opportunity to effectively introgress the *Co-4*^{2}, *Co-5*, and *Co-6* genes into North American germplasm in the absence of more virulent races of the pathogen.

Markers flanking the resistance gene improve upon the use of loosely-linked markers. Young & Kelly (1996a) showed that selection efficiency increased from 94% to 99% when two loosely-linked flanking markers were used to select for the *Co-2* gene. Single markers with linkage distances in excess of 5 cM are unlikely to be useful alone as selectable markers in a breeding program.

In some instances, the presence of certain gene combinations in resistant germplasm cannot be detected, because differentiating races of anthracnose are not available to the breeder. When combined in a single genotype, those unknown resistance genes, which cannot be identified by traditional pathological techniques, can be distinguished from each other using a marker tightly linked to one of the resistance genes. RAPD and SCAR markers tightly linked to the *Co-4*^{2} gene have been used successfully to identify a third independent gene (*Co-7*) in the highly resistant G 2333 (Young et al., 1998).

4.1.4 Ashy Stem Blight

Two complementary dominant genes (*Mp-1, Mp-2*) were identified as conditioning resistance to a Colombian isolate (Mp 50) of *Macrophomina phaseolina* (Tassi) Goid., causing ashy stem blight in BAT 477 (Olaya et al., 1996). The greenhouse screening technique used was effective in categorizing progeny of the cross BAT 477/A 70 into discrete resistant and/or susceptible (9:7) categories. Using BSA, Olaya et al. (1996) identified two independent markers: $B386_{900}$ present in the resistant bulk and $B459_{1600}$ present in the susceptible bulk. The use of these markers in breeding for resistance to *Macrophomina* is speculative since their presence (or absence) was not surveyed in susceptible germplasm or in other sources of resistance to *Macrophomina*.

4.1.5 Bean Common Mosaic

In breeding for resistance to BCM, the combination of dominant and recessive genes which have distinctly different mechanisms of resistance should offer durability over a single gene for resistance (Kelly, 1997; Kelly et al., 1995). The dominant *I* gene is masked by the recessive *bc-3* gene so as efforts to incorporate the *bc-3* gene into new germplasm proceed, the risk of losing the *I* gene increases. Linked markers offer the only realistic opportunity to maintain and continue to utilize the *I* gene in future cultivars. A marker $OW13_{690}$, tightly linked to the *I* gene, has been reported (Haley et al., 1994b) and has been demonstrated to be effective across a wide range of germplasm in both gene pools.

Conversely, selection against the *I* gene may be advantageous in East Africa where necrotic strains of the potyvirus (BCMNV) exist (Silbernagel et al., 1986) or in Central America where strains of bean severe mosaic virus (BSMV) occur (Morales & Castaño, 1992). Since the *I* gene has been incorporated into a broad array of germplasm and is highly sensitive (resulting in plant death) to both BCMNV and BSMV, it poses an unacceptable risk to producers in both regions. Morales & Kornegay (1996) suggest detecting the *I* gene, when masked by the presence of *bc-3* gene, by using a strain of BSMV. However, this method has not yet become routine or available for general use by plant breeders who may be reluctant to import BSMV for testing. Markers linked to the *I* gene could be utilized to select against the *I* gene while maintaining the BCM resistance afforded by the *bc-3* gene in regions where the *I* gene is undesirable.

Markers linked to the *bc-3* gene would also be useful since the breeder may not want to import strains of BCMNV for purposes of testing. A private company is introgressing *bc-3* into snap or garden bean in the absence of the pathogen, by selecting against a coupling-phase SCAR marker ($OC20_{460}$) (Johnson et al., 1997). Markers more tightly linked to other resistance genes like *bc-u, bc-1*^{2}, and *bc-2*^{2} (Drijfhout, 1978) are needed to facilitate introgression and pyramiding of other resistance genes to combat BCM.

4.1.6 Bean Rust

The pyramiding of different resistance genes and mechanisms (specific, adult plant, slow rusting, reduced pustule size) will likely prolong the life of a cultivar by creating a more durable resistance complex against the highly variable rust pathogen. Since screening for specific rust resistance in the field or greenhouse is relatively easy, use of an informative race for detection of the hypostatic gene would be more efficient, but often discriminating races are unavailable or are too risky to use. Alternatively, RAPD markers have been useful for maintaining hypostatic rust resistance genes in the presence of epistatic resistance genes. For instance in the development of the navy bean line BelMiDak-RR-7 (Stavely et al., 1994), a RAPD marker (Miklas et al., 1993) was used to detect *Ur-4* in the presence of the *Ur-11* gene. *Ur-11* derives from PI 181996 and was briefly named *Ur-3*^{2} (Johnson et al., 1995; Kelly et al., 1996; Stavely, 1998). Due to inadequate sampling, previous progeny testing had failed to detect the *Ur- 4* gene in the presence of the *Ur-11* gene (Kelly et al., 1993).

In addition to the *Ur-4* gene, four other rust resistance genes, *Ur-3, Ur-5, Ur-9,* and *Ur-11*, have been tagged (Haley et al., 1993, 1994c; Johnson et al., 1995; Jung et al., 1996b; Kelly et al., 1996) with tightly linked RAPD markers (Table 1). The RAPD markers for the *Ur-11* gene (Johnson et al., 1995) are difficult to reproduce; therefore, will require further refinement as SCAR markers, similar to that for the *Ur-3* (Nemchinova & Stavely, 1998) and *Ur-5* genes (Melotto & Kelly, 1998), before they become useful to plant breeders in different laboratories.

Using traditional approaches, novel rust resistance from the plant introductions, namely PIs 181996 (M gene pool), 190078 (M), 151385 (A), and 151395 (A) was incorporated into some North American dry and snap bean market classes in combination with, and for protection of, other widely used sources of resistance like the *Ur-3, Ur-4, Ur-5,* and *Ur-6* genes. For more rapid development of durable rust resistance in other market classes, linked RAPD markers could be used to pyramid the *Ur-3, Ur-4, Ur-5, Ur-9,* and *Ur-11* genes with the broadly effective epistatic resistance genes found in the PI collection (Stavely et al., 1992). The most efficient use of markers linked with rust resistance will probably be realized in later generations for detection of hypostatic genes in advanced lines which already possess the epistatic resistance gene as determined by routine greenhouse screening. The importance of previously defeated resistance genes should not be underestimated. In Honduras, lines carrying the broadly effective *Ur-11* resistance gene, derived from PI 181996 (Stavely,1990), were infected by a newly identified rust pathotype (race 108) (Stavely et al., 1997), whereas lines possessing either the hypostatic *Ur-4* or *Ur-5* resistance gene in addition to *Ur-11* were not infected (Mmbaga et al., 1996).

4.1.7 Bean Golden Mosaic (BGM)

Many different types and sources of genetic resistance and/or tolerance have been identified and used, in combination, to breed BGM resistant cultivars (Beebe, 1994; Morales & Singh, 1991). Because of epistatic interactions among the different resistance sources, it becomes difficult to know exactly which sources are combined in improved lines. A partial resistance source conditioned by the recessive *bgm-1* gene derived from Garrapato is characterized by a lack of mosaic symptom development on trifoliolate leaves (Blair & Beaver, 1993), but pod deformation and reduced yield will occur under moderate to severe disease pressure. To counter the loss of pod quality and yield, the recessive *bgm-1* gene should be combined with other resistance genes conditioning non-deformed pods (Molina Castañeda & Beaver, 1998), reduced mosaic symptoms (Vélez et al., 1998), and high yield (Beebe, 1994). Dorado (DOR 364) has quantitative resistance to BGM, expressed as a reduction in yellow mosaic symptoms, which tend to be masked in the presence of *bgm-1*. Furthermore, presence of *bgm-1* appears necessary for expression of the *Bgp* gene for normal pod development (Molina Castañeda & Beaver, 1998). A codominant RAPD marker ($OR2_{570/530}$) tightly linked with *bgm-1* (Urrea et al., 1996) (Table 1), is being used by public and private breeders to hasten the development of snap bean germplasm with an initial moderate level of BGM resistance (McMillan et al., 1998; Stavely et al., 1997). The codominant RAPD marker for *bgm-1* is independent of gene pools and easy to score, but nonetheless SCARs for both codominant bands have been developed to simplify MAS for *bgm-1* (S. Beebe, personal communication, 1998).

4.2 RAPD Markers and Quantitative Resistance

Quantitative traits appear to have the most potential for MAS, because selection for them is confounded by environmental effects. Since the cost to evaluate these complex traits with low to moderate heritability is high, MAS would be more likely to show the greatest gain. However, for the very reasons given: complex inheritance, low heritability, and confounding environmental effects; markers associated with QTL have been difficult to find, and once found have exhibited limited usefulness across a range of genetic backgrounds or environments. Efforts to utilize MAS for the improvement of quantitative traits have been limited, but may improve as better quantitative data is generated and denser linkage maps become available from map co-integration across laboratories (Freyre et al., 1998; see also Chapter 3 by Gepts).

4.2.1 Ashy Stem Blight

Field resistance to *Macrophomina* was conditioned by four QTL having relatively minor effects (15% singly and 47% cumulatively) in advanced lines from a Dorado/XAN 176 population (Miklas et al., 1998b). RAPDs linked with these QTLs and the *Mp-1 and Mp-2* genes may provide a starting point from which to use markers to combine quantitative and qualitative resistance sources for control of ashy stem blight in common bean. It is unknown, however, whether these QTL will be amenable to MAS because of their relatively small effects. Independent verification in a population derived between the resistant parent XAN 176 and another susceptible cultivar is needed to assess the usefulness of these QTL for MAS.

4.2.2 Bean Golden Mosaic

Two independent QTL were identified as having major influence on the reduced mosaic resistance trait in the cultivar Dorado (Miklas et al., 1996b). Together these two QTL explained 60% of the phenotypic variation in BGM reaction for one environment, and were consistently expressed across three separate field environments where the pathogen was present. Analyses of the informative markers across BGM resistance sources revealed the progenitor Porrillo Sintetico as the likely source of the QTL. Since the reduced mosaic resistance of Dorado is conditioned by few QTL, with intermediate to major effect, a backcrossing program may be the most effective way to transfer these resistance traits via MAS. A recombinant inbred line (RI) mapping population (Dorado/G 19833) segregating for reduced mosaic resistance has been developed (Beebe et al., 1998). Comparison of markers and QTLs between the Dorado/XAN 176 and Dorado/G 19833 populations is underway to identify RAPDs with the best potential for MAS of Dorado-derived BGM resistance. To date, only the RAPD $OW12_{700}$ (Miklas et al., 1996b) has been selected for subsequent SCAR development, because it was found to explain the greatest amount of variation for BGM reaction in both populations (P.N. Miklas & S. Beebe, unpublished results).

4.2.3 Common Bacterial Blight

Genetic studies have shown resistance to common bacterial blight (CBB), caused by *Xanthomonas campestris* pv. *phaseoli* (Smith) Dye, to be quantitatively inherited, but involving a few major genes (Beebe, 1989). Phenotypic observations of complex resistance have been supported by QTL data. Using RFLPs, Nodari et al. (1993) identified four QTL which together explained 75% of the variation in greenhouse-leaf reaction (first trifoliolate leaf) to CBB in a BAT 93/Jalo EEP 558 (RI) population. One major-effect QTL explained 35% of the variation in reaction (Table 2). The CBB resistance of BAT 93 was obtained from two sources, PI 207262 and Tara. The latter was originally derived from a cross with tepary bean (Honma, 1956) and selected in bean line GN No. 1 Sel. 27 (Coyne & Schuster, 1974). Three QTL explaining 77% of the variation in greenhouse-leaf reaction were identified for the GN No. 1 Sel. 27 resistance source segregating in a different RI population (Dorado/XAN 176). One major-effect QTL, defined by the RAPD marker $OAP6_{820}$, explained 60% of the variation in resistance (Miklas et al., 1996b). A survey revealed that most lines with CBB resistance derived from GN No. 1 Sel. 27 possessed $OAP6_{820}$ indicating this marker should be useful for MAS of CBB resistance. Moreover, a SCAR recently developed for $OAP6_{820}$ (P.N. Miklas, unpublished results) was found to explain the greatest amount of variation for CBB resistance in other mapping populations segregating for the GN No. 1 Sel. 27 resistance source (G. Jung, personal communication, 1998).

Table 2. Genetic markers with potential use in MAS of major-effect QTL conditioning resistance to common bacterial blight in common bean

Resistance source/ organ[1]	Dry bean line	Marker Type	Designation/ size	R^2	Linkage group[2]	References
GN No.1 Sel 27 & PI 207262 /1st trfl	BAT 93	RFLP	D1390	0.35	D7 (B7)	Nodari et al., 1993
		Seed protein	*Phs*	0.27	D7 (B7)	
GN No.1 Sel.27 /1st trfl	XAN 176	RAPD	$OAP6_{820}$	0.60	US2(B10)	Miklas et al., 1996b, 1998a; Ariyarathne et al., 1999
		RAPD	$OG17_{540}$	0.57	US2(B10)	
	BAC 6	RAPD	$BC409_{1250}$	0.11	UM(B10)	Jung et al., 1996a; Ariyarathne et al., 1999
		RAPD	$OU16_{600}$	0.09	5(B10)	

Resistance source/ organ[1]	Dry bean line	Marker Type	Designation/ size	R^2	Linkage group[2]	References
/pods	XAN 176	RAPD	$OAJ1_{760}$	0.52	US2(B10)	Miklas et al., 1996b, 1998a; Ariyarathne et al., 1999
		RAPD	$OAS18_{500}$	0.21	US7 (B7)	
	BAC 6	RAPD	$BC409_{1250}$	0.17	UM(B10)	Jung et al., 1996a; Ariyarathne et al., 1999
		RAPD	$OK19_{450}$	0.12	5(B10)	
/field	XAN 176	RAPD	$OAA19_{600}$	0.40	US7 (B7)	Miklas et al., 1996b
/later trfl	BAC 6	RAPD	$BC409_{1250}$	0.13	UM(B10)	Jung et al., 1996a; Ariyarathne et al., 1999
PI 319433	SEL1309	SCAR	SU091	0.24	1[3] (B8)	Pedraza et al., 1997
/1st trfl	XAN 159	SCAR[4]	Y5	-	5 (B6)	Jung et al., 1997,1998
		RAPD	$C7_{900}$	0.12	1 (B8)	
/pods		Flower color	V/v^{lae}	0.30	5 (B6)	
/seeds			V/v^{lae}	0.18	5 (B6)	
/later trfl		RAPD	$BC420_{900}$	0.40	5 (B6)	
		RAPD	$C7_{900}$	0.13	1 (B8)	
PI 440795	OAC 88-1	SCAR[4]	$BC73_{700}$	0.45	H	Bai et al., 1997; Beattie et al., 1998
/field		SCAR[4]	$BC486_{1000}$	0.22	B	
Phaseolus coccineus /mid trfl	XR-235-1-1	RFLP	Bng139	0.15	F (B8)	Yu et al., 1998
		RFLP	Bng40	0.11	A (B7)	

[1] trfl = trifoliolate leaf. [2]Linkage group designations do not correspond among the different genetic maps except B7 = D7 = US7 = A because of the shared *Asp* and *Phs* loci, B8 = 1 = F because of the shared *C* locus, unassigned (BAC 6) = US2 (XAN 176) because of the shared CBB QTL (G. Jung, personal communication, 1998), and linkage group 5 (PC 50/XAN 159) is probably the same as B6 = D6 because of the shared *V* locus, and B10 = 5 = UM (unassigned marker) = US2 (Ariyaranthe et al., 1999). The corresponding linkage group in parentheses is obtained from the core map (Freyre et al., 1998). Note that linkage group 5 of PC 50/XAN 159 and linkage group 5 of BAC 6/HT 7719 do not correspond. Also, linkage groups B and H (Bai et al., 1997) do not necessarily correspond with the lettered linkage group nomenclature of Yu et al. (1998). [3]The SU091 SCAR maps to linkage group 1 = B8 in the PC 50/XAN 159 population (Jung et al., 1998). [4]The sequences for the pair of specific primers which amplify these SCARs are proprietary and have not been published.

Different QTL conditioning resistance in young and adult tissues and to different strains of the pathogen were expressed for the GN No. 1 Sel. 27 source of CBB resistance in a BAC 6/HT 7719 population (Jung et al., 1996a). For the same resistance source, Miklas et al. (1996b) observed QTL which affected expression of both leaf and pod resistance in both young (greenhouse) and mature plants (field), in addition to QTL that affected resistance in a single organ or test environment. In a PC 50/XAN 159 population, one genomic region possessed a factor(s) which influenced resistance in all three organs, seeds, leaves, and pods, while another QTL only influenced resistance within a single plant organ (Jung et al., 1997). The CBB resistance in XAN 159 was derived from *P. acutifolius* (G 40020 = PI 319443).

A SCAR marker (SU091) for the same XAN 159 source of resistance segregating in a different population (DOR 476/SEL 1309) explained 24% of the phenotypic variation in leaf reaction (Pedraza et al., 1997). In addition, the SU091 SCAR maps to a major QTL for CBB resistance in the PC 50/XAN 159 population (Jung et al., 1998). A SCAR has been developed for a second QTL in this same population. Breeding lines with both SCARs were more resistant than lines with only the SU091 SCAR marker. The marker (SU091) itself is derived from *P. acutifolius* and is likely located within the introgressed genomic fragment containing CBB resistance.

Interestingly, the SU091 and $AP6_{820}$ SCARs were observed in OAC 88-1 (P.N. Miklas, unpublished results) which has CBB resistance derived from a different tepary line, PI 440795 (Scott & Michaels, 1992). Perhaps PI 319443 and PI 440795, and the tepary in GN No. 1 Sel. 27 contributed the same CBB resistance to common bean through separate introgression events, i.e., XAN 159, OAC 88-1, and GN No. 1 Sel. 27 breeding lines. The presence of two resistance genes in OAC 88-1 is supported by the identification of two independent QTL that explained 80% of the variation for CBB resistance derived from PI 440795 (Bai et al., 1997). The potential commonality of the genes in OAC 88-1 to existing resistance sources in XAN 159 and GN No. 1 Sel. 27 is supported by the observation that one or more loci conditioning CBB resistance were in common across four different tepary lines (Urrea et al., 1999). These observations also support the presence of independent loci for CBB resistance in XAN 159 and OAC 88-1, as suggested by the 15:1 ratio for resistant to susceptible plants observed in the F_2 generation from a cross between XAN 159 and OAC 88-1 (Jung et al., 1997; Michaels, 1992). It will be interesting to see how the recently developed SCARs linked to the CBB resistance QTL from PI 440795 (Beattie et al., 1998), XAN 159, and GN No. 1 Sel. 27 associate amongst each other.

The combination of resistance QTL, as in the pyramiding of qualitative genes, will likely contribute to more effective and durable CBB resistance. The two QTL identified for CBB resistance in the BAC 6/HT 7719 and PC 50/XAN 159 populations mapped to different regions in a map integration study (Skroch et al., 1996). This finding suggests that these diverse sources of CBB resistance could be combined in an attempt to obtain superior resistant lines. Using traditional breeding, the resistance of GN No. 1 Sel. 27 and PI 207262 were combined in the highly resistant line XAN 112 (Beebe & Pastor-Corrales, 1991) and these same resistance sources were likely combined in some of the WBB and VAX lines (Zapata et al., 1991; Singh & Muñoz, 1999). We are backcrossing the major independent QTL from XAN 159 and GN No. 1 Sel. 27 into snap bean and all the major dry bean market classes in the U.S.A. by MAS for the SU091 and $OAP6_{820}$ SCARs (P.N. Miklas, unpublished results). It is unclear whether the CBB resistance QTL and associated markers derived from PI 207262 in BAT 93 (Nodari et al., 1993) and from *P. coccineus* in XR-235-1-1 (Yu et al., 1998) represent distinct genes from the tepary-derived resistance alleles in XAN 159, OAC 88-1, and GN No. 1 Sel. 27. The presence of SU091 in BAT 93 and $OAP6_{820}$ in XR-235-1-1 suggests that these lines might not contain resistance genes distinct from tepary bean (P.N. Miklas & S. Singh, unpublished results).

The identification of RAPDs and development of SCARs linked with QTL from diverse sources (GN No. 1 Sel. 27 via XAN 176, PI 319433 via XAN 159, PI 440795 via OAC 88-1) with major to moderate influence on resistance to CBB, implies that MAS will be useful in combining resistance sources to CBB in common bean. The independence of the resistance genes to be combined must be closely monitored as many different lines and cultivars have common sources of CBB resistance. During domestication of *P. acutifolius*, a pronounced bottleneck appears to have occurred, as few unique loci controlling rust (Miklas & Stavely, 1998) and CBB (Urrea et al., 1999) resistance were observed across an array of cultivated tepary bean. Similar reduction in variability in domesticated tepary bean has also been confirmed at the biochemical and molecular level (Schinkel & Gepts, 1988, 1989). A potentially interesting use for markers would be in the introgression of resistance genes across species of *Phaseolus*. Markers linked to a block of monogenically inherited CBB resistance genes segregating in a tepary bean population have been identified (Park et al., 1998) and may facilitate the recovery of CBB resistant plants from interspecific populations.

4.2.4 Web Blight

Five QTL which explained 34% of the variation in foliar reaction to web blight [caused by *Thanatephorus cucumeris* (Frank) Donk.], were observed in the field in the BAC 6/HT 7719 RI population (Jung et al., 1996a). Indirect selection of these relatively minor-effect QTL are unlikely to improve upon traditional field selection for resistance to web blight. Field resistance to web blight is confounded by expression of both physiological and morphological avoidance mechanisms. An association between QTL for web blight resistance and plant architecture (branch density) may be partially explained by disease avoidance due to the upright architecture (Jung et al., 1996a).

4.3 Linkage Between Resistance To Diseases

Since linkages between traits are not uncommon; both coupling and repulsion linkages have been reported between markers associated with disease resistance traits. A repulsion linkage between QTL affecting BGM and CBB resistance were observed on two separate linkage groups, which partially explained the low, but consistent, negative phenotypic correlation (r = -0.23) observed between those two traits in the Dorado/XAN 176 population (Miklas et al., 1996b). In the same population, the *Ur-5* marker was linked to QTL conditioning resistance to BGM and *Macrophomina*, and an unknown gene for rust resistance *Ur-X* and the *Co-2* marker $Q4_{1300}$ were linked (Miklas et al., 1998a). The gene for small pustule rust resistance mapped near a region influencing web blight resistance (Jung et al., 1996a) in the BAC 6/HT 7719 population. This gene is different from the previously reported *Ur-9*, and *Ur-10* genes (Jung et al., 1996b). Another association between web blight and CBB resistance with the RAPD marker $BC409_{1250}$ was reported by Jung et al. (1996a), but the marker was not mapped in the same population. A QTL for CBB resistance, segregating in the BAT 93/Jalo-EEP 558 RI population, has been located on linkage group 7 in a region that includes the phaseolin locus *Phs* (Nodari et al., 1993). Recently, a QTL for white mold resistance, segregating in the A 55/G 122 RI population, exposed to the straw test, has also been identified in the same general region (P.N. Miklas & P. Gepts, unpublished results). Specific markers tagging these QTL need to be developed for use in MAS. These linkages may represent initial visualization of resistance gene clusters in common bean (Miklas et al., 1998a) as has been observed in other crops (Tamulonis et al., 1997; Witsenboer et al., 1995).

From a breeder's perspective, linkages among resistance loci are beneficial when they are derived from an individual parental line (cis-configuration), because when crossed with susceptible parents these resistances will be inherited as a single unit and likely will remain intact in subsequent progeny (Geffroy et al., 1998). When linked resistance genes are donated by different parents (trans-configuration) recombination between the loci is necessary to obtain both resistance sources in a single line. For example recombination among QTL will be needed to combine parental levels of resistance to BGM, CBB, and *Macrophomina* in the Dorado/XAN 176 population (Miklas et al., 1996b, 1998a, b). Thus, breeding cultivars resistant to multiple diseases will be complicated initially by these trans-linkages, and subsequently simplified once the linked genes are recombined in cis-orientation. An appealing use for MAS would be the indirect selection of gene clusters using flanking markers, and although not yet demonstrated, the use of linked markers to select for a block of rust resistance genes (*Ur-5*) has been described (Haley et al., 1993).

4.4 Selection Efficiency Among Different Types of RAPD Markers

Indirect selection efficiency can be influenced by the number, linkage distance, and orientation of RAPD marker linkages and the breeding method used, so consideration should be given to employing one of the following strategies to improve efficiency.

- With dominant RAPD markers directly linked (on the same sister chromatid) to the susceptible allele at a locus, selection against susceptibility is possible. The increased efficiency results from the ability to select against the heterozygote and is particularly useful when selecting for recessive resistance genes. Since RAPD markers are dominant, when linked to a recessive allele - whether it conditions resistance or susceptibility, the linkage orientation is repulsion-phase; conversely, when linked to a dominant allele, regardless of its expression (resistance/susceptible), the linkage is in coupling-phase. Susceptibility-linked markers for the BCM resistance *bc-3* gene, *the Co-1* and *Co-6* anthracnose resistance genes, the *Ur-11* rust resistance gene, and the *Mp-1* ashy stem blight resistance gene have been identified. For the identification of homozygous resistant *bc-3/bc-3* individuals, Haley et al. (1994a) demonstrated an 82% selection efficiency by selecting against a susceptibility-linked marker versus 26% efficiency in selecting for the resistance-linked marker. It is noteworthy that selection against a marker linked to the susceptible allele is restricted to the F_2 and later generation populations. For rapid backcross introgression of dominant traits, resistance-linked RAPD markers are required, since they can be used to identify heterozygous BC_nF_1's for subsequent backcross cycles (Johnson et al., 1995). Resistance-linked markers may also be more useful in gamete selection programs (Singh et al., 1998) for identifying superior F_1's.

- Markers flanking the resistance gene significantly increase selection efficiency of resistant individuals when loosely linked (>5 cM) markers are utilized. Young & Kelly (1996a) showed that selection efficiency was increased from 94% to 99% when two flanking markers were used to select for the *Co-2* gene for anthracnose resistance in common bean.
- The information generated by a pair of markers, one linked in coupling and the other in repulsion to the targeted locus, is genetically equivalent to that obtained from a single codominant marker. Johnson et al. (1995) observed 19% selection efficiency of homozygous resistant *Ur-11/Ur-11* individuals using a coupling marker, 77% using a repulsion marker, and 83% using both markers as a codominant pair.
- A codominant pair of flanking markers further increases selection efficiency since they combine the advantages of flanking markers with the ability to identify heterozygotes. This marker combination allows the identification of double recombinants not possible with flanking markers alone. Young & Kelly (1997) demonstrated an increase in selection efficiency from 33% for the coupling-phase marker to 92% for the repulsion-phase marker to 95% for the codominant flanking markers in the selection of *Co-6/Co-6* homozygote individuals conferring resistance to anthracnose in common bean.
- Whereas, each dominant RAPD marker forming a codominant pair is amplified by a different primer, single primers do give rise to natural codominant RAPD markers at the relatively low frequency of about 1 out of every 20 RAPDs detected (P.N. Miklas, unpublished results). The identification of a codominant RAPD marker linked to a resistance gene is due strictly to chance. However, when using BSA to screen for linked polymorphisms, the identification of a linked codominant RAPD marker is enhanced if lines within the resistant and susceptible bulks are completely homozygous for the corresponding allele at the targeted locus. True-breeding DNA bulks enabled Urrea et al. (1996) to identify a codominant RAPD marker linked to the *bgm-1* gene conditioning resistance to BGM. In addition to using MAS to identify the homozygous recessive resistance gene, the heterozygote can be distinguished from the susceptible homozygote and be retained for additional selfing and selection if needed.
- A codominant marker (or codominant pair) is advantageous because when linked to a recessive resistance gene that exhibits epistasis it can be used to facilitate gene pyramiding (Urrea et al., 1996). For example, advanced lines are phenotypically selected for the reduced mosaic resistance to

BGM derived from Dorado (hypostatic resistance source), while simultaneously maintaining heterozygosity (*Bgm-1/bgm-1*) at the "non-mosaic" *bgm* resistance gene locus by indirect selection of both alleles of the linked codominant RAPD marker $OR2_{570/530}$. Upon selfing these lines and then selecting individual plants for the 530 bp fragment linked to the non-mosaic resistance of *bgm-1*, both resistances can be effectively pyramided.

- Recombination-facilitated RAPD markers have been identified by Miklas et al. (1996a) as a solution to the amplification of a similar sized DNA fragment in susceptible germplasm. Recombinants for the $OAD19_{690}$ and $OS13_{660}$ markers, originally linked in repulsion and coupling to the *bc-3* gene, respectively, were identified. The strategy of recombination-facilitated MAS has the potential to broaden the range of usefulness of specific markers in susceptible germplasm, improve selection efficiency by changing the linkage orientation, and indicate whether similar sized RAPD fragments assort independently or are amplified from the same genomic region. The latter information would be invaluable in deciding whether to develop a SCAR or ASAP marker from a problematic RAPD marker.
- ASAP (Gu et al., 1995) and SCAR (Paran & Michelmore, 1993) markers offer an additional refinement over RAPD markers. Generally both produce a single polymorphic band that is more reproducible across labs and is easier to score and applicable to use with low-quality DNA obtained through speedy DNA extraction procedures (Matrix Mill, N.F. Weeden, personal communication, 1998). Their development can be a laborious undertaking, however. A SCAR for the OH20 RAPD marker linked to the *Co-2* anthracnose resistance gene has been developed but it requires splicing with a restriction endonuclease to demonstrate polymorphism between resistant and susceptible genotypes (Adam-Blondon et al., 1994). The restricted fragment differed in size from the original RAPD fragment and demonstrated codominance which would be useful for the detection of heterozygotes. A SCAR marker developed for the *I* gene RAPD marker $OW13_{690}$ has proven to be less laborious and more robust. The single polymorphic band can be amplified using different heat stable polymerases and thermal cyclers in resistant lines from different gene pools (Melotto et al., 1996). However, in other crops, certain SCARs did not demonstrate the same polymorphism as did the RAPD marker (Paran & Michelmore, 1993). Similar limitations were observed by J.D. Kelly in the development of a SCAR for the RAPD marker $OAD19_{690}$ linked to the *bc-3* gene (Haley et al., 1994a) and by Johnson et al. (1997) for SCAR markers associated with the $OS13_{660}$ and $OC11_{350}$ RAPD markers. Only the SCAR for $OC11_{420}$ marker amplified a single band polymorphism

similar to the original RAPD. In those cases where SCARs are not polymorphic, the only solution is to find additional linkages to more useful RAPD markers.

Other limitations of certain SCARs is that they generate a second monomorphic band which requires the use of gel electrophoresis to separate out the polymorphic band. In order to overcome some of the limitations of SCARs, ASAPs were developed using primers of different lengths which would only amplify the polymorphic RAPD band. The use of 18mer ASAPs to selectively amplify a fragment linked to the *Ppd* (photoperiod) gene was demonstrated in common bean (Gu et al., 1995). Combining ASAPs of different lengths (18mer with a 15mer), Gu et al. (1995) successfully amplified a single polymorphic band linked to the recessive epistatic *hr* photoperiod gene (Gu et al., 1998) which interacts with the *Ppd* gene. Tailoring primers to specifically amplify a single fragment enables the use of alternative detection techniques. The amplified DNA is stained directly with ethidium bromide in microtiter plates eliminating the need for gel electrophoresis (Gu et al., 1995; Melotto et al., 1996). This modification is a considerable saving in time and money. These refinements are still dependent on tight linkages between the original RAPD marker and resistance gene and selection efficiency could still be improved by deploying flanking markers or those linked to the susceptible allele. SCARs and ASAPs are amenable to multiplex PCR reactions (Mitchell et al., 1997) where by more than one set of specific primers is included in the same reaction to expedite MAS and save costs. Multiplexed PCR will require gel electrophoresis for visualization of the multiple SCARs or ASAPs. Unless of course the multiplexed ASAPs were linked with susceptibility then, DNA negative cells within microtiter plates would be selected for without the need for gel-electrophoresis.

The use of evenly spaced markers to select against the genome of the donor parent in order to quickly recover the recurrent parent in fewer backcrosses has been used sparingly in the breeding of corn and soybean (Openshaw et al., 1994). For example, in a wide cross designed to improve yield, the highest yielding progeny with the least number of markers from the exotic donor parent would be included in the next cycle of selection. Similarly, markers linked to the trait of interest can be selected against to reduce linkage drag while selecting directly for the trait of interest.

4.5 Limitations of RAPD Markers

In some instances the application of dominant RAPD markers for indirect selection has been restricted to specific *Phaseolus* gene pools because of the presence in susceptible germplasm of DNA fragments of equal size (kb) to the markers linked to the resistance genes. This was the case with respect to the application of the dominant RAPD marker $OA14_{1100}$, tightly linked to the *Ur-4* rust resistance gene (no recombinants observed), which limited the use of the marker to the Middle American gene pool (Miklas et al., 1993). Other markers showed similar limitations. The $OF10_{970}$ marker linked to the *Ur-5* gene can only be used for indirect selection in the Andean gene pool (Haley et al., 1993), and the $OAD19_{690}$ marker linked to the *bc-3* gene can only be used in germplasm of race Mesoamerica within the Middle American gene pool (Haley et al., 1994a). Indirect selection across a broad range of germplasm is possible with gene pool non-specific RAPD markers such as $OW13_{690}$ for the *I* gene, $OK14_{420}$ for the *Ur-3* gene, $OI19_{460}$ for the *Ur-5* gene, $OR2_{570/530}$ for the *bgm-1* gene, and $OF10_{530}$ for the *Co-1* gene because in susceptible germplasm these RAPDs do not result in same-sized fragments. To overcome gene pool specificity of a RAPD marker, Miklas et al. (1996a) showed that recombinant individuals, resistant but without the linked marker, could be crossed with susceptible germplasm containing the marker. In such populations, resistant progeny are obtained by selecting against the marker. Conversely, resistant recombinants with susceptibility-linked markers can be used in crosses with susceptible lines lacking the susceptibility-linked marker to obtain resistant progeny by direct selection for the marker. This recombination-facilitated MAS strategy overcomes gene pool specificity of a dominant RAPD genetic marker by significantly broadening its range of application across gene pools. This strategy can also be used in the presence or absence of gene pool specificity to reverse the linkage orientation of a RAPD marker to fit a particular breeding program.

5. SUMMARY

With the identification of over 30 RAPD markers linked to 17 different resistance genes in *P. vulgaris*, the feasibility and potential for MAS in disease resistance breeding in common bean is clearly possible. When costs are balanced against applications, breeders may choose MAS as a practical alternative to direct selection for multiple traits. As additional markers are detected for other genes, the value of the technology will increase since MAS for more than one trait can be practiced in breeding populations. The value of MAS of quantitative traits will become more important and useful as phenotypic data and the technology for the detection of QTL improves. The direct incorporation of single gene resistance to the prevalent races of a pathogen within a particular area is currently the breeding method of choice, but this resistance is often short-lived, forcing breeders to continuously incorporate new resistance genes into their programs. Combining different genes with resistance to the prevalent races, although more difficult and expensive to achieve initially, should provide a more durable resistance for long-term protection. Marker-assisted selection provides opportunities to pyramid genes not feasible with traditional breeding methods. Genetic markers will likely be used in the development of future bean cultivars possessing durable disease resistance.

REFERENCES

Adam-Blondon, A.F., M. Sévignac, H. Bannerot & M. Dron. 1994. SCAR, RAPD, and RFLP markers linked to dominant gene (*Are*) conferring resistance to anthracnose in common bean. Theor. Appl. Genet. 88:865-870.

Alzate-Marin, A.L., G.A. de Carvalho, H. Menarim, G.S. Baia, T.J. de Paula Junior, E.G. de Barros & M.A. Moreira. 1997. Identification of RAPD markers associated with resistance to anthracnose in common bean. Annu. Rpt. Bean Improv. Coop. 40:130-131.

Ariyarathne, H.M., D.P. Coyne, G. Jung, P.W. Skroch, A.K. Vidaver, J.R. Steadman, P.N. Miklas & M.J. Bassett. 1999. Molecular mapping of disease resistance genes for halo blight, common bacterial blight, and bean common mosaic virus in a segregating population of common bean. J. Am. Soc. Hort. Sci. 124: (In press).

Bai, Y., T.E. Michaels & K.P. Pauls. 1997. Identification of RAPD markers linked to common bacterial blight resistance genes in *Phaseolus vulgaris* L. Genome 40:544-551.

Balardin, R.S., A.M. Jarosz & J.D. Kelly. 1997. Virulence and molecular diversity in *Colletotrichum lindemuthianum* from South, Central and North America. Phytopathology 87:1184-1191.

Balardin, R.S. & J.D. Kelly. 1998. Interaction between races of *Colletotrichum lindemuthianum* and gene pool diversity in *Phaseolus vulgaris*. J. Amer. Soc. Hort. Sci. 123:1038-1047.

Bassett, M.J. 1991. A revised linkage map of common bean. HortScience 26:834-836.

Bassett, M.J. 1996. List of Genes - *Phaseolus vulgaris* L. Annu. Rpt. Bean Improv. Coop. 39:1-19.

Beattie, A., T.E. Michaels & K.P. Pauls. 1998. An efficient reliable method to screen for common bacterial blight (CBB) resistance in *Phaseolus vulgaris* L. Annu. Rpt. Bean Improv. Coop. 41:53-54.

Beaver, J.S. & R. Macchiavelli. 1998. Breeding strategies for pyramiding genes for resistance. Annu. Rpt. Bean Improv. Coop. 41:141-142.

Beebe, S. 1989. Quantitative genetics in *Phaseolus vulgaris*: the example of resistance to *Xanthomonas campestris* pv. *phaseoli*. pp.213-238. *In* S. Beebe (ed.), Current Topics in Breeding of Common Bean. CIAT, Cali, Colombia.

Beebe, S. 1994. Breeding for resistance to bean golden mosaic virus: history and perspectives. pp. 148-150. *In* F. Morales (ed.), Bean Golden Mosaic: Research Advances. CIAT, Cali, Colombia.

Beebe, S. & M. Pastor-Corrales. 1991. Breeding for disease resistance. pp. 561-617. *In* A. van Schoonhoven & O. Voysest (eds.), Common Beans: Research for Crop Improvement. C.A.B. Int., Wallington, U.K. & CIAT, Cali, Colombia.

Beebe, S., F. Pedraza, M. Rojas, J. Gutiérrez & J. Tohme. 1998. A genetic map of common bean combining RFLP, RAPD, SCAR, and AFLP markers. Annu. Rpt. Bean Improv. Coop.41:95-96.

Blair, M.W. & J.S. Beaver. 1993. Inheritance of BGMV resistance from bean genotype A 429. Annu. Rpt. Bean Improv. Coop. 36:143.

Botstein, D., R.L. White, M. Skolnick & R.W. Davis. 1980. Construction of a genetic linkage map in man using restriction fragment length polymorphisms. Amer. J. Hum. Genet. 32:314-331.

Coyne, D.P. & M.L. Schuster. 1974. Breeding and genetic studies of tolerance to several bean (*Phaseolus vulgaris*) bacterial pathogens. Euphytica 231:651-656.

Cregan, P.B., T. Jarvik, A.L. Bush, R.C. Shoemaker, K.G. Lark, A.L. Kahler & J.E. Specht. 1998. The soybean simple sequence repeat map: some applications. p 80. *In* Agronomy Abstracts. Amer. Soc. Agron., Madison, U.S.A.

Delaney, D.E. & F.A. Bliss. 1991. Selection for increased percentage protein in common bean. 1. Comparison of selection for seed protein alleles and S_1 family recurrent selection. Theor. Appl. Genet. 81:301-305.

Drijfhout, E. 1978. Genetic interaction between *Phaseolus vulgaris* L. and bean common mosaic virus with implications for strain identification and breeding for resistance. Agric. Res. Rpt. 872. Centre for Agricultural Publishing and Documentation, Wageningen, Netherlands.

Duvick, D.N. 1996. Plant breeding, an evolutionary concept. Crop Sci. 36:539-548.

Freyre, R., P.W. Skroch, V. Geffroy, A.F. Adam-Blondon, A. Shirmohamadali, W.C. Johnson, V. Llaca, R.O. Nodari, P.A. Pereira, S.M. Tsai, J. Tohme, M. Dron, J. Nienhuis, C.E. Vallejos & P. Gepts. 1998. Towards an integrated linkage map of common bean. 4. Development of a core linkage map and alignment of RFLP maps. Theor. Appl. Genet. 97:847-856.

Geffroy, V., F. Creusot, J. Falquet, M. Sévignac, A.F. Adam-Blondon, H. Bannerot, P. Gepts & M. Dron. 1998. A family of LRR sequences at the *Co-2* locus for anthracnose resistance in *Phaseolus vulgaris* and its potential use in marker-assisted selection. Theor. Appl. Genet. 96:494-502.

Gepts, P. 1988. Provisional linkage map of common bean. Annu. Rpt. Bean Improv. Coop. 31:20-25.

Gepts, P., R. Nodari, S.M. Tsai, E.M.K. Koinange, V. Llaca, R. Gilbertson & P. Guzmán. 1993. Linkage mapping in common bean. Annu. Rpt. Bean Improv. Coop. 36: xxiv-xxxviii.

Gonzáles, A., J.A. Wong Leon, A. Delgado Salinas & P. Gepts. 1998. Determination of genetic diversity among wild and domesticated beans using inter simple sequence repeats (ISSRs). Annu. Rpt. Bean Improv. Coop. 41:99-100.

Gu, W.K., N.F. Weeden, J. Yu & D.H. Wallace. 1995. Large-scale, cost-effective screening of PCR products in marker-assisted selection applications. Theor. Appl. Genet. 91:465-470.

Gu, W.K., J.Q. Zhu, D.H. Wallace, S.P. Singh & N.F. Weeden. 1998. Analysis of genes controlling photoperiod sensitivity in common bean using DNA markers. Euphytica 102:125-132.

Haley, S.D., L.K. Afanador & J.D. Kelly. 1994a. Selection for monogenic resistance traits with coupling- and repulsion-phase RAPD markers. Crop Sci. 34:1061-1066.

Haley, S.D., L.K. Afanador & J.D. Kelly. 1994b. Identification and application of a Random Amplified Polymorphic DNA marker for the *I* gene (Potyvirus resistance) in common bean. Phytopathology 84:157-160.

Haley, S.D., L.K. Afanador, P.N. Miklas, J.R. Stavely & J.D. Kelly. 1994c. Heterogeneous inbred populations are useful as sources of near-isogenic lines for RAPD marker localization. Theor. Appl. Genet. 88:337-342.

Haley, S.D., P.N. Miklas, J.R. Stavely, J. Byrum & J.D. Kelly. 1993. Identification of RAPD markers linked to a major rust resistance gene block in common bean. Theor. Appl. Genet. 86:505-512.

Hartweck, L.M., C. Cardona & T.C. Osborne. 1997. Bruchid resistance of common bean lines having an altered seed protein composition. Theor. Appl. Genet. 95:1018-1023.

Honma, S. 1956. A bean interspecific hybrid. J. Hered. 47:217-220.

Johnson, E., P.N. Miklas, J.R. Stavely & J.C. Martínez-Cruzado. 1995. Coupling and repulsion RAPD markers for marker-assisted selection of a rust resistance gene in common bean. Theor. Appl. Genet. 90:659-664.

Johnson, W.C., P. Guzmán, D. Mandala, A.B.C. Mkandawire, S. Temple, R.L. Gilbertson & P. Gepts. 1997. Molecular tagging of the *bc-3* gene for introgression into Andean common bean. Crop Sci. 37:248-254.

Jones, C.J., K.J. Edwards, S. Castaglione, M.O. Winfield, F. Sala, C. Van DeWiel, G. Bredemeijer, B. Vosman, M. Matthes, A. Daly, R. Brettschneider, P. Bettini, M. Buiatti, E. Maestri, A. Malcevschi, N. Marmiroli, R. Aert, G. Volckaert, J. Rueda, R. Linacero, A. Vazquez & A. Karp. 1997. Reproducibility testing of RAPD, AFLP, and SSR markers in plants by a network of European laboratories. Mol. Breed. 3:381-390.

Jung, G., S. Beebe, J. Nienhuis, S. Park, D.P. Coyne, J. Marita, F. Pedraza & J. Tohme. 1998. Development of SCAR markers linked to common bacterial blight resistance genes (QTL) in common bean. Annu. Rpt. Bean Improv. Coop. 41:97-98.

Jung, G., D.P. Coyne, P.W. Skroch, J. Nienhuis, E. Arnaud-Santana, J. Bokosi, H.M. Ariyarathne, J.R. Steadman, J.S. Beaver & S.M. Kaeppler. 1996a. Molecular markers associated with plant architecture and resistance to common blight, web blight and rust in common bean (*Phaseolus vulgaris* L.). J. Amer. Soc. Hort. Sci. 121:794-803.

Jung, G., D.P. Coyne, P. Skroch, J. Nienhuis, J.M. Bokosi & J.R. Steadman. 1996b. RAPD marker linked to a gene for specific rust resistance in common bean. Annu. Rpt. Bean Improv. Coop. 39:59-60.

Jung, G., P.W. Skroch, D.P. Coyne, J. Nienhuis, E. Arnaud-Santana, H.M. Ariyarathne, S.M. Kaeppler & M.J. Bassett. 1997. Molecular-markers-based genetic analysis of tepary bean-derived common bacterial blight resistance in different developmental stage of common bean. J. Amer. Soc. Hort. Sci. 122:329-337.

Kanazin, V., L.F. Marek & R.C. Shoemaker. 1996. Resistance gene analogs are conserved and clustered in soybean. Proc. Natl. Acad. Sci. U.S.A. 93:11746-11750.

Kelly, J.D. 1995. Use of Random Amplified Polymorphic DNA markers in breeding for major gene resistance to plant pathogens. HortScience 30:461-465.

Kelly, J.D. 1997. A review of varietal response to bean common mosaic potyvirus in *Phaseolus vulgaris*. Plant Varieties & Seeds 10: 1-6.

Kelly, J.D., L. Afanador & S.D. Haley. 1995. Pyramiding genes for resistance to bean common mosaic virus. Euphytica 82:207-212.

Kelly, J.D. & P.N. Miklas. 1998. The role of RAPD markers in breeding for disease resistance in common bean. Mol. Breed. 4:1-11.

Kelly, J.D., J.R. Stavely & P.N. Miklas. 1996. Proposed symbols for rust resistance genes. Annu. Rpt. Bean Improv. Coop. 39:25-31.

Kelly, J.D., J.R. Stavely, P.N. Miklas, L. Afanador & S.D. Haley. 1993. Pyramiding rust resistance genes using RAPD markers. Annu. Rpt. Bean Improv. Coop. 36:166-167.

Kelly, J.D. & R.A. Young. 1996. Proposed symbols for anthracnose resistance genes. Annu. Rpt. Bean Improv. Coop. 39:20-24.

Kesseli, R.V., I. Paran & R.W. Michelmore. 1992. Efficient mapping of specifically targeted genomic regions and the tagging of these regions with reliable PCR-based genetic markers. pp. 31-36. CSSA-ASHS-AGA Joint Plant Breeding Symp. Ser., Minneapolis and Crop Sci. Soc. Amer., Madison, U.S.A.

Kornegay, J., C. Cardona & C.E. Posso. 1993. Inheritance of resistance to Mexican bean weevil in common bean, determined by bioassay and biochemical tests. Crop Sci. 33:589-594.

Kyle, M.M. & M.H. Dickson. 1988. Linkage of hypersensitivity to five potyviruses with the *B* locus for seed coat color in *Phaseolus vulgaris* L. J. Heredity 79:308-311.

Landegren, U., M. Nilsson & P.Y. Kwok. 1998. Reading bits of genetic information: methods for single-nucleotide polymorphism analysis. Genome Research 8:769-776.

McMillan Jr., R.T., M.J. Davis & H.J. McLaughlin. 1998. PCR evaluation of fourteen bean golden mosaic virus (BGMV) resistant snap bean lines for presence of the virus. Annu. Rpt. Bean Improv. Coop. 41:31-32.

Melotto, M., L. Afanador & J.D. Kelly. 1996. Development of a SCAR marker linked to the *I* gene in common bean. Genome 39:1216-1219.

Melotto, M. & J.D. Kelly. 1998. SCAR markers linked to major disease resistance genes in common bean. Annu. Rpt. Bean Improv. Coop. 41:64-65.

Michaels, T.E. 1992. Genetic control of common blight resistance in lines derived from *P. vulgaris / P. acutifolius* crosses. Annu. Rpt. Bean Improv. Coop. 35:40-41.

Michelmore, R.W., I. Paran & R.V. Kesseli. 1991. Identification of markers linked to disease resistance genes by bulked segregant analysis: A rapid method to detect markers in specific genomic regions using segregating populations. Proc. Natl. Acad. Sci. U.S.A. 9828-9832.

Miklas, P.N., L. Afanador & J.D. Kelly. 1996a. Recombination-facilitated RAPD marker-assisted selection for disease resistance in common bean. Crop Sci. 36:86-90.

Miklas, P.N., R. Delorme, V. Stone, C.A. Urrea, J.S. Beaver & J.R. Steadman. 1998a. A RAPD map of disease resistance traits in common bean. Annu. Rpt. Bean Improv. Coop. 41:93-94.

Miklas, P.N., E. Johnson, V. Stone, J.S. Beaver, C. Montoya & M. Zapata. 1996b. Selective mapping of QTL conditioning disease resistance in common bean. Crop Sci. 36:1344-1351.

Miklas, P.N. & J.R. Stavely. 1998. Incomplete dominance of rust resistance in tepary bean. HortScience 33:143-145.

Miklas, P.N., J.R. Stavely & J.D. Kelly. 1993. Identification and potential use of a molecular marker for rust resistance in common bean. Theor. Appl. Genet. 85:745-749.

Miklas, P.N., V. Stone, C.A. Urrea, E. Johnson & J.S. Beaver. 1998b. Inheritance and QTL analysis of field resistance to ashy stem blight. Crop Sci. 38: 916-921.

Mitchell, S.E., S. Kresovich, C.A. Jester, C. Javier Hernández & A.K. Szewc-McFadden. 1997. Application of multiplex PCR and fluorescence-based, semi-automated allele sizing technology for genotyping plant genetic resources. Crop Sci. 37:617-624.

Mmbaga, M, J.R. Steadman & J.R. Stavely. 1996. The use of host resistance in disease management of rust in common bean. Integrated Pest Mgt. Rev. 1:191-200.

Mohan, M., S. Nair, A. Bhaqwat, T.G. Krishna, M. Yano, C.R. Bhatia & T. Sasaki. 1997. Genome mapping, molecular markers and marker-assisted selection in crop plants. Mol. Breed. 3: 87-103.

Molina Castañeda, A. & J.S. Beaver. 1998. Inheritance of normal pod development in bean golden mosaic resistant common beans. Annu. Rpt. Bean Improv. Coop. 41:3-4.
Morales, F.J. & M. Castaño. 1992. Increased disease severity induced by some comoviruses in bean genotypes possessing monogenic dominant resistance to bean common mosaic potyvirus. Plant Dis. 76:570-573.
Morales, F.J. & J. Kornegay. 1996. The use of plant viruses as markers to detect genes for resistance to bean common mosaic and bean common mosaic necrosis viruses. Annu. Rpt. Bean Improv. Coop. 39:272-273.
Morales, F.J. & S.P. Singh. 1991. Genetics of the resistance to bean golden mosaic virus in *Phaseolus vulgaris* L. Euphytica 52:113-117.
Nemchinova, Y.P. & J.R. Stavely. 1998. Development of SCAR primers for the *Ur-3* rust resistance gene in common bean. Phytopathology 88:S67.
Nodari, R.O., S.M. Tsai, P. Guzmán, R.L. Gilbertson & P. Gepts. 1993. Towards an integrated linkage map of common bean. III. Mapping genetic factors controlling host-bacteria interactions. Genetics 134:341-350.
Olaya, G., G.S. Abawi & N.F. Weeden. 1996. Inheritance of the resistance to *Macrophomina phaseolina* and identification of RAPD markers linked to the resistance genes in beans. Phytopathology 86:674-679.
Openshaw, S.J., S.G. Jarboe & W.D. Beavis. 1994. Marker-assisted selection in backcross breeding. pp. 41-43. *In* Analysis of molecular data. CSSA-ASHS Joint Plant Breeding Symp. Ser., Corvallis and Crop Sci. Soc. Amer., Madison, U.S.A.
Osborn, T.C., D.C. Alexander, S.S.M. Sun, C. Cardona & F.A. Bliss. 1988. Insecticidal activity and lectin homology of arcelin seed protein. Science 240:207-210.
Paran, I. & R.W. Michelmore. 1993. Development of reliable PCR-based markers linked to downy mildew resistance genes in lettuce. Theor. Appl. Genet. 85:985-993.
Park, S.O., D.P. Coyne, A. Dursun & G. Jung. 1998. Identifying randomly amplified polymorphic DNA (RAPD) markers linked to major genes for common bacterial blight resistance in tepary bean. J. Amer. Soc. Hort. Sci. 123:278-282.
Pastor-Corrales, M.A., O.A. Erazo, E.I. Estrada & S.P. Singh. 1994. Inheritance of anthracnose resistance in common bean accession G 2333. Plant Dis. 78:959-962.
Pastor-Corrales, M.A., M.M. Otoya, A. Molina & S.P. Singh. 1995. Resistance to *Colletotrichum lindemuthianum* isolates from Middle America and Andean South America in different common bean races. Plant Dis. 79:63-67.
Patterson, A.H., K.F. Schertz, Y.R. Lin & Z. Li. 1998. Case history in plant domestication: Sorghum, an example of cereal evolution. pp. 187-195. *In* A.H. Patterson (ed.), Molecular Dissection of Complex Traits, CRC Press, New York, U.S.A.
Pedraza, F., G. Gallego, S. Beebe & J. Tohme. 1997. Marcadores SCAR y RAPD para la resistencia a la bacteriosis comun (CBB). pp.130-134. *In* S.P. Singh & O. Voysest (eds.), Taller de Mejoramiento de Frijol para el Siglo XXI: Bases para Una Estrategia para America Latina. CIAT, Cali, Colombia.
Romero-Andreas, J., B.S. Yandell & F.A. Bliss. 1986. Bean arcelin. 1. Inheritance of a novel seed protein of *Phaseolus vulgaris* L. and its effect on seed composition. Theor. Appl. Genet. 72:123-128.
Schafer, F.R. & A.P. Roelfs. 1985. Estimated relation between numbers of urediniospores of *Puccinia graminis* f. sp. *tritici* and rates of occurrence of virulence. Phytopathology 75:749-750.
Schinkel, C. & P. Gepts. 1988. Phaseolin diversity in the tepary bean, *Phaseolus acutifolius* A. Gray. Plant Breed. 101:292-301.

Schinkel, C. & P. Gepts. 1989. Allozyme variability in the tepary bean, *Phaseolus acutifolius* A. Gray. Plant Breed. 102:182-195.
Schneider, K.A., M.E. Brothers & J.D. Kelly. 1997. Marker assisted selection to improve drought resistance in common bean. Crop Sci. 37:51-60.
Scott, M.E. & T.E. Michaels. 1992. *Xanthomonas* resistance of *Phaseolus* interspecific cross selections confirmed by field performance. HortScience 27:348-250.
Sicard, D., Y. Michalakis, M. Dron & C. Neema. 1997. Genetic diversity and pathogenic variation of *Colletotrichum lindemuthianum* in the three centers of diversity of its host, *Phaseolus vulgaris*. Phytopathology 87:807-813.
Silbernagel, M.J., L.J. Mills & W.Y. Wang. 1986. Tanzanian strain of bean common mosaic virus. Plant Dis. 70:839-841.
Singh, S.P., C. Cardona, F.J. Morales, M.A. Pastor-Corrales & O. Voysest. 1998. Gamete selection for upright carioca bean with resistance to five diseases and a leafhopper. Crop Sci. 38:666-672.
Singh, S.P., J.A. Gutiérrez, A. Molina, C. Urrea & P. Gepts. 1991a. Genetic diversity in cultivated common bean. II. Marker-based analysis of morphological and agronomic traits. Crop Sci. 31:23-29.
Singh, S.P. & C.G. Muñoz. 1999. Resistance to common bacterial blight among *Phaseolus* species and common bean improvement. Crop Sci. 39: 80-89.
Singh, S.P., R. Nodari & P. Gepts. 1991b. Genetic diversity in cultivated common bean. I. Allozymes. Crop Sci. 31:19-23.
Skroch, P., G. Jung, J. Nienhuis & D.P. Coyne. 1996. Integration of RAPD marker linkage maps and common mapping of QTL for disease resistance in common bean. Annu. Rpt. Bean Improv. Coop. 39:48-49.
Staub, J.E., F.C. Serquen & M. Gupta. 1996. Genetic markers, map construction, and their application in plant breeding. HortScience 31:729-741.
Stavely, J.R. 1984. Genetics of resistance to *Uromyces phaseoli* in a *Phaseolus vulgaris* line resistant to most races of the pathogen. Phytopathology 74:339-344.
Stavely, J.R. 1990. Genetics of rust resistance in *Phaseolus vulgaris* plant introduction PI 181996. Phytopathology 80:1056 (abstract).
Stavely, J.R. 1998. Recombination of two major dominant rust resistance genes that are tightly linked in repulsion. Annu. Rpt. Bean Improv. Coop. 41:17-18.
Stavely, J.R., J.D. Kelly & K.F. Grafton. 1994. BelMiDak-rust-resistant navy dry bean germplasm lines. HortScience 29:709-710.
Stavely, J.R., J.D. Kelly, K.F. Grafton, C.A. Mullins, A. Straw, R.T. McMillan Jr., J.S. Beaver, P.N. Miklas, J. Steinke, J.R. Steadman, D.P. Coyne, D.T. Lindgren & M.J. Silbernagel. 1997. Rust resistant bean germplasm releases, 1994-1996. Annu. Rpt. Bean Improv. Coop. 40:120-121.
Stavely, J.R., J. Steinke, R.T. McMillan Jr., K.F. Grafton, J.R. Steadman, J.D. Kelly, D.P. Coyne, D.T. Lindgren & M.J. Silbernagel. 1992. Rust resistant bean germplasm releases. Annu. Rpt. Bean Improv. Coop. 35:228-229.
Tamulonis, J.P., B.M. Lussi, R.S. Hussey, W.A. Parrott & R. Boerma. 1997. DNA markers associated with resistance to Japanese root-knot nematode in soybean. Crop Sci. 37:783-788.
Temple, S.R. & F.J. Morales. 1986. Linkage of dominant hypersensitive resistance to bean common mosaic virus to seed color in *Phaseolus vulgaris* L. Euphytica 35:331-333.
Tohme, J., D.O. González, S. Beebe & M.C. Duque. 1996. AFLP analysis of gene pools of a wild bean core collection. Crop Sci. 36: 1375-1384.

Urrea, C.A., P.N. Miklas & J.S. Beaver. 1999. Inheritance of resistance to common bacterial blight in four tepary lines. J. Amer. Soc. Hort. Sci. 124:24-27.

Urrea, C.A., P.N. Miklas, J.S. Beaver & R.H. Riley. 1996. A codominant RAPD marker useful for indirect selection of BGMV resistance in common bean. J. Amer. Soc. Hort. Sci. 121:1035-1039.

Vélez, J.J., M.J. Bassett, J.S. Beaver & A. Molina. 1998. Inheritance of resistance to bean golden mosaic virus in common bean. J. Amer. Soc. Hort. Sci. 123: 628-631.

Vos, P., R. Hogers, M. Bleeker, M. Reijans, T. Van der Lee, M. Hornes, A. Frijters A, J. Pot, J. Peleman, M. Kuiper & M. Zabeau. 1995. AFLP: a new technique for DNA fingerprinting. Nucl. Ac. Res. 23:4407-4414.

Walters, K.J., G.L. Hosfield, M.A. Uebersax & J.D. Kelly. 1997. Navy bean canning quality: correlations, heritability estimates and randomly amplified polymorphic DNA markers associated with component traits. J. Amer. Soc. Hort. Sci. 122:338-343.

Weeden, N.F., G.M. Timmerman, M. Hemmat, B.E. Kneen & M.A. Lodhi. 1992. Inheritance and reliability of RAPD markers. pp. 12-17. *In* CSSA-ASHS-AGA Joint Plant Breeding Symp. Ser., Minneapolis and Crop Sci. Soc. Amer., Madison, U.S.A.

Williams, J.G.K., A.R. Kubelik, K.J. Livak, J.A. Rafalksi & S.V. Tingey. 1990. DNA polymorphisms amplified by arbitrary primers are useful as genetic markers. Nucl. Ac. Res. 18: 6531-6535.

Witsenboer, H., R.V. Kessili, M.G. Fortin, M. Stanghellini & R.W. Michelmore. 1995. Sources and genetic structure of a cluster of genes for resistance to three pathogens in lettuce. Theor. Appl. Genet. 91: 178-188.

Young, R.A. & J.D. Kelly. 1996a. RAPD markers flanking the *Are* gene for anthracnose resistance in common bean. J. Amer. Soc. Hort. Sci. 121:37-41.

Young, R.A. & J.D. Kelly. 1996b. Characterization of the genetic resistance to *Colletotrichum lindemuthianum* in common bean differential cultivars. Plant Dis. 80:650-654.

Young, R.A. & J.D. Kelly. 1997. RAPD markers linked to three major anthracnose resistance genes in common bean. Crop Sci. 37:940-946.

Young, R.A., M. Melotto, R.O. Nodari & J.D. Kelly. 1998. Marker assisted dissection of the oligogenic anthracnose resistance in common bean cultivar, 'G 2333'. Theor. Appl. Genet. 96:87-94.

Yu, Z.H., R.E. Stall & C.E. Vallejos. 1998. Detection of genes for resistance to common bacterial blight of beans. Crop Sci. 38:1290-1296.

Zapata, M., R. Wilkinson, G.F. Freytag, H. Vélez, F.H. Ortíz & J.H. López-Rosa. 1991. Incorporating resistance to *Xanthomonas campestris* pv. *phaseoli* into bean using the latent period as a resistance marker. J. Agric. Univ. P.R. 75:345-352.

Chapter 5

Genetic Transformation

Hans-Jörg Jacobsen
Universität Hannover, Herrenhaeuserst 2, 30419 Hannover, Germany

Key words: direct gene transfer, explant culture, plant regeneration systems, transformation, transient gene expression

1. INTRODUCTION

The large-seeded legume crops, including the common bean (*Phaseolus vulgaris* L.), are second to cereals with regard to their role in human and animal nutrition, and they are a cheap source of dietary protein, calories, fiber, vitamins, and minerals. This nutritional value is particularly important for the poorer parts of populations in developing countries in Africa, Asia, and Latin America. In the common bean, conventional breeding methods have achieved remarkable progress using sexual recombination (including successful wide crosses, Mejía-Jiménez et al., 1994). However, the lack of genetic variation for certain resistance traits and the difficulties with further interspecific hybridizations, especially with species in the quaternary gene pools and beyond (see Chapter 2 by Debouck), still make breeding a rather slow process, even when marker-assisted breeding is being used (as discussed in Chapter 4 by Kelly & Miklas). In the past, plant geneticists tried to make use of radiation- or chemically-induced mutations, but the respective low frequencies and occurrence of many recessive and loss-of-function mutations failed to convince breeders of the value of these techniques. In addition, in the era of mutation breeding, major emphasis was on yield-related parameters and not on resistance to biotic and abiotic production constraints.

The traditional goals of crop breeding therefore remain unchanged. However, the rapid advances in plant molecular genetics and plant gene transfer are potential novel tools in the genetic improvement of crop species. The number and type of available useful genes are ever-increasing, starting from model genes first (antibiotic- or herbicide-resistance genes) to genes improving the resistance against insects and bacterial, fungal, or viral diseases, or modifying the quality characteristics of crop plants (e.g., the "FlavrSavr" tomato). The DNA-transfer techniques applied were either *Agrobacterium tumefaciens*-mediated gene transfer or, in those cases where the crop plants did not belong to the host range of *Agrobacterium*, direct gene delivery methods involving protoplasts or particle bombardment techniques (for references see below). Thus, genetic transformation technology now complements conventional breeding methods. In many crops, genes for pest resistance, quality, and other traits can be added by these new methods. However, common bean and other large-seeded legumes have not been easy to transform.

In the common bean, existing tissue culture technology was applied to produce transgenic plants from transformed tissues, but the development of an efficient and reproducible protocol for stable transformation has not been achieved, although several papers have been published. In these cases, there was no follow-up. The present chapter reviews the state of the art and discusses strategies.

2. IN VITRO - BEAN REGENERATION

The major requirement for a successful transformation protocol is a fast, robust, and efficient regeneration system. Such a system has to be compatible with the protocol used for transformation. Most emphasis in the past was on crop species which were either comparably easy to regenerate and transform, or of importance for agriculture in developed countries. Compared to crops like soybean [*Glycine max* (L.) Merr.] or corn (*Zea mays* L.), the common bean, as a typical "orphan crop" was not among those species which have found great attention, particularly not from the private sector thus far. In addition, the complex biology of the large-seeded legumes, especially pulses, has hampered progress. There is, however, some success: common bean plants can be produced *in vitro* directly from excised shoot meristems (Kartha et al., 1981), shoot tips (Saam et al., 1987), leaf petioles (Malik & Saxena, 1991), seedlings (Malik & Saxena, 1992), embryonic axes (Mohamed et al., 1992a, b; Zambre et al., 1998), seed cotyledons (Mohamed et al., 1991), seedling nodes (Franklin et al., 1991; McClean & Grafton, 1989), or meristematic calli (Mejía-Jiménez et al., 1998). Dillen et al. (1996) regenerated plants from green nodular callus derived from *P. acutifolius* buds and pedicels with a frequency of 7 out of 12 genotypes responding. Regeneration via meristem culture or direct shoot regeneration from primary explant cultures thus is comparably easy to achieve and therefore may be the system of choice, since these methods are faster than protoplast- or callus-based systems and thus may have a reduced risk for undesirable variation in the resulting regenerants or transformants.

Table 1. Regeneration of *Phaseolus vulgaris* plants from various explant sources

Explant	References
Shoot meristem	Kartha et al., 1981
Shoot tip	Saam et al., 1987
Leaf petiole	Malik & Saxena, 1991
Seedlings	Malik & Saxena, 1992
Cotyledons	Mohamed et al., 1991
Embryonic axes	Mohamed et al., 1992a, b; Zambre et al., 1998
Seedling nodes	Franklin et al., 1991
Meristematic calli	Mejía-Jiménez et al., 1998

3. TRANSFORMATION METHODS

3.1 Indirect Transformation: Tumor Formation and Transient Gene Expression

As has been mentioned before, in principle two different gene delivery systems can be applied with higher plants: 1) indirect transformation using disarmed *A. tumefaciens* strains, or 2) direct transformation of protoplasts, cells or tissues. The latter technique includes forced DNA uptake into protoplast with polyethylene glycol (PEG, León et al., 1991) or electroporation, bombardment of tissues with DNA-coated gold or tungsten particles (for a review see Christou, 1995), or direct tissue electroporation (Chowrira et al., 1995; Dillen et al., 1995).

The initial phase in the development of a transformation protocol is the search for compatible *Agrobacterium* strains and the improvement of transformation efficiencies using scorable marker systems like *glucoronidase (gus* gene, Jefferson, 1987). Common beans have been found to be susceptible to *Agrobacterium*-wildtype strains and derivatives (Allavena, 1985; McClean et al., 1989, 1991). Surprisingly, several genotypes of common bean have been reported to exhibit similar levels of compatible reactions with various biotypes of *A. tumefaciens*, but younger seedlings were found to be more susceptible. In that, McClean et al. (1991) indicated that tumor formation developed 7 days after inoculating bean tissues of 1 to 3 day-old seedlings, while 4 weeks were needed before tumors began to appear in the tissues of 4 to 7 day-old seedlings. Lewis & Bliss (1994), however, reported that 3-week-old seedlings exhibited different reactions, depending on genotype, with respect to tumor formation induced by the *Agrobacterium* strain C58. After inoculating with a disarmed, *glucoronidase*-harboring C58 derivative, it was noted that strong tumor formation and *gus* expression did not correlate. In a recently published paper, Zhang et al. (1997) tested two strains, A2760 and EHA 105, with 11 common bean genotypes. They found significant *Agrobacterium* x host interactions. Other strong factors influencing the transformation rate were explant maturity, pre-culture and co-cultivation conditions. Kapila et al. (1997) reported on an infiltration system where *gus*-expressing cells covered up to 90% of the *Phaseolus* leaf area.

Agrobacterium-mediated gene transfer has further been applied to common bean by Mariotti et al. (1989), Genga & Ceriotti (1990), Franklin et al. (1991), and McClean et al. (1991), but the production of stably transformed bean plants has not been reported by these groups so far. Most bean transformation research used nodal explants in an *Agrobacterium*-mediated system. Various aspects in the process of *Agrobacterium*-mediated transformation in beans were addressed in these initial studies. McClean & Grafton (1989) demonstrated shoot organogenesis in nodal tissues of benzyladenine-preconditioned bean seedlings. McClean et al. (1991) subsequently used this regeneration protocol in *Agrobacterium*-mediated transformation, but only transformed calli and no shoots were recovered. Since callus can be induced easily from any type of bean tissues, it seems that competent cells for shoot differentiation in their regeneration system were not accessible to *Agrobacterium.* In the regeneration system from seedling nodal tissues (McClean & Grafton, 1989), the area of the existing axillary meristems was gently scraped prior to the culture on regeneration medium. *Agrobacterium* infects the wounded tissues of the disrupted meristem, while new adventitious meristems initiate around the disrupted regions.

Genga & Ceriotti (1990) used a modified technique of *Agrobacterium*-mediated transformation for bean cotyledonary node tissues in which the first set of elongated shoots was removed from 10 to 15 day-old seedlings and then the meristems around the excised shoots were inoculated. Although they regenerated shoots on selective medium, root initiation was not observed and the GUS assays were negative. This observation may be due to some problems with the construct used and it may be advisable to delay the selection against non-transformed tissues for some time. Delaying the selection phase may provide sufficient time for proliferation of transformed cells and subsequent marker gene expression. The gene expression will enable the differentiation of transformed from non-transformed tissues at a later stage. The elimination of non-regenerative tissues around the putative transformed regenerable tissues will minimize the chance for shoot escape and consequently reduce the regeneration of non-transformed shoots.

In most cases, no information was given regarding the temperature during co-cultivation. In a recent paper, Dillen et al. (1997a) report that temperature > 22°C drastically reduced the expression of the transgenes in *P. acutifolius*. It seems that the respective co-cultivation conditions require further research.

3.2 Direct Gene Transfer

First reports on the application of biolistics (acceleration of DNA-coated gold particles for delivering DNA into intact cells and tissues) are from Genga & Ceriotti (1990) and Aragão et al. (1993) reporting transient gene expression. Russell et al. (1993) reported the first stable and heritable transformation of common bean with a rate of 0.0035 germline transformations. Stable bean transformation following particle bombardment was also reported by Kim & Minamikawa (1995), but no progeny analysis has yet been published. Brasileiro et al. (1996) reported an interesting combination of *Agrobacterium*-mediated gene transfer and bombardment. They used wounded meristems obtained after bombardment for co-cultivation with disarmed *Agrobacteria* and apparently saw enhanced transformation frequencies (scorable marker gene expression). The same group found morphological factors influencing the recovery of transgenic shoots after particle bombardment (Aragão & Rech, 1997). It was found that, depending on the cultivar, the meristems are either exposed to the particles or covered by leaf primordia. Transgenic shoots were obtained only from the first morphological group, while from the other group no transgenic shoots were reported.

Other attempts to directly transform common bean include protoplast transformation using the PEG method (León et al., 1991) or tissue electroporation (Dillen et al., 1995). In both cases, transient expression of the transgenes was reported, but no fertile plants or progenies were reported.

3.3 Stable Bean Transformation

So far, only three groups have convincingly claimed to have produced stable and heritable bean transformation: Russell et al. (1993) and Aragão et al. (1998) in *P. vulgaris,* and Dillen et al. (1997b) in *P. acutifolius*. These recent papers demonstrate that bean transformation is possible, as is the case with many other crop plants. The next step is to demonstrate transgene transmission to subsequent generations. In principle, at least two different explanations for the lack of transmission have been discussed. Either the transgenes were silenced through transgene methylation or other modes of homologous inactivation, or the primary transformants were false positives, if, for instance, endophytes had been transformed.

4. CONCLUSIONS

The available literature is clearly demonstrating that common bean transformation is no longer a dream. It is a reality. However, transformation in common bean is not yet as well developed as it is with other crops, including several important large-seeded legumes. Progress will depend on the scientific expertise devoted to developing an efficient and reproducible bean transformation system. The reports available, however, also make clear that most approaches were done solely with marker or reporter genes. Further research in common bean should focus on the development of stable and efficient transformation systems, using genes with resistance to pests and abiotic stresses, among others. For the latter, however, wide crosses may offer more promise. In addition, for the sake of end user acceptance, the research should focus on: 1) the development of methods avoiding antibiotic resistance genes as selectable markers, and 2) selecting useful genes that do not interfere with seed sale and distribution systems in developing countries.

REFERENCES

Allavena, A. 1985. Infection of P. vulgaris with strains of A. tumefaciens and A. rhizogenes. Annu. Rpt. Bean Improv. Coop. 28: 90-91.

Aragão, F.J.L., M.F. Grossi de Sa, M.R. Davey, A.C.M. Brasileiro, J.C. Faria & E.L. Rech. 1993. Factors influencing transient gene expression in bean (Phaseolus vulgaris L.) using electrical particle acceleration device. Plant Cell. Rep. 12:483-490.

Aragão, F.J.L. & E.L. Rech. 1997. Morphological factors influencing the recovery of transgenic bean plants (Phaseolus vulgaris L.) of a carioca cultivar. Int. J. Plant Sci. 158: 157-163.

Aragão, F.J.L., S.G. Ribeiro, L.M.G. Barros, A.C.M. Brasileiro, D.P. Maxwell, E.L. Rech & J.C. Faria. 1998. Transgenic beans (Phaseolus vulgaris L.) engineered to express viral antisense RNAs show delayed and attenuated symptoms to bean golden mosaic virus. Molec. Breed. 4: 491-499.

Brasileiro, A.C.M., F.J.L. Aragão, S. Rossi, D.M.A. Dusi, L.M.G. Barros & E.L. Rech. 1996. Susceptibility of common and tepary beans to Agrobacterium ssp. strains and improvement of Agrobacterium-mediated transformation using microprojectile bombardement. J. Amer. Soc. Hort. Sci. 121: 810-815.

Chowrira, G.M., V. Akella & P.F. Lurquin. 1995. Electroporation-mediated gene transfer into intact nodal meristems in plants. Molec. Biotechn. 3:17-23.
Christou, P. 1995. Strategies for variety-independent genetic transformation of important cereals, legumes and woody species utilizing particle bombardement. Euphytica 85:13-27.
Dillen, W., J. de Clercq, A. Goossens, M. van Montagu & G. Angenon. 1997b. Agrobacterium-mediated transformation of Phaseolus acutifolius A. Gray. Theor. Appl. Genet. 94:151-158.
Dillen, W., J. de Clercq, J. Kapila, M. Zambre, M. van Montagu & G. Angenon. 1997a. The effect of temperature on Agrobacterium tumefaciens- mediated gene transfer to plants. The Plant J. 12:1459-1463.
Dillen, W., J. de Clercq, M. van Montagu & G. Angenon. 1996. Plant regeneration from callus in a range of Phaseolus acutifolius A.Gray genotypes.1996. Plant Sci. 118:81-88.
Dillen, W., G. Engler, M. van Montagu & G. Angenon.1995. Electroporation-mediated DNA delivery to seedling tissues of Phaseolus vulgaris L.(common bean). 1995. Plant Cell Rep. 15: 119-124.
Franklin, C.I., T.N. Trieu, R.A. Gonzáles & R.A. Dixon. 1991. Plant regeneration from seedling explants of green bean (Phaseolus vulgaris L.) via organogenesis. Plant Cell Tissue Organ Cult. 24:199-206.
Genga, A. & A. Ceriotti. 1990. Towards genetic transformation of bean by Agrobacterium tumefaciens. pp. 527-536. In Proc. lst Intl. ISHS Symp. In vitro culture and horticulture breeding, Cesena, Italy.
Jefferson, R.A. 1987. Assaying chimeric genes in plants: The GUS gene fusion system. Plant Mol. Biol. Rptr. 5:387-405.
Kapila, J., R. de Rycke, M. van Montagu & G. Angenon. 1997. An Agrobacterium-mediated transient gene expression system for intact leaves. Plant Sci. 122:101-108.
Kartha, K.K., K. Pahl, N.L. Leung & L.A. Mroginski. 1981. Plant regeneration from meristems of grain legumes: soybean, cowpea, peanut, chickpea, and bean. Can. J. Bot. 59:1671-1679.
Kim, J.W. & T. Minamikawa. 1995. Transformation and regeneration of french bean plants by the particle bombardement process. Plant Sci. 117:131-138.
León, P., F. Planckaert & V. Walbot. 1991. Transient gene expression in protoplasts of Phaseolus vulgaris isolated from a cell suspension. Plant Physiol. 95:968-972.
Lewis, M.E. & F.A. Bliss. 1994. Tumor formation and ß-glucoronidase expression in Phaseolus vulgaris inoculated with Agrobacterium tumefaciens. J. Amer. Hort. Sci. 119:361-366.
Malik, K.A. & P.K. Saxena. 1991. Regeneration in Phaseolus vulgaris L.: promotive role of N6-benzylaminopurine in cultures from juvenile leaves. Planta 184:148-150.
Malik, K.A. & P.K. Saxena. 1992. Regeneration of Phaseolus vulgaris L.: high frequency induction of direct shoot formation in intact seedlings by N6-benzylaminopurine and Thidiazuron. Planta 186:384-389.
Mariotti, D., G.S. Fontana & L. Santini. 1989. Genetic transformation of grain legumes: Phaseolus vulgaris L. and Phaseolus coccineus L. J. Genet. Breed. 43:77-82.
McClean, P. & K.F. Grafton. 1989. Regeneration of dry bean (Phaseolus vulgaris L.) via organogenesis. Plant Sci. 60: 117-122.
McClean, P., P. Chee, B. Held, J. Simental, R.F. Drong & J. Slighom. 1991. Susceptibility of dry bean (Phaseolus vulgaris L.) to Agrobacterium infection: transformation of cotyledonary and hypocotyl tissues. Plant Cell Tissue Organ Cult. 24:131-138.
Mejía-Jiménez, A., H.J. Jacobsen & W.M. Roca. 1998. Development of an in vitro system in common bean suitable for genetic transformation. pp 73-74. In A.M. de Ron (ed.), Proc. of

EUCARPIA-International Symposium on Breeding of Protein and Oil Crops. Pontevedra, Spain.

Mejía-Jiménez, A., C. Muñoz, H.J. Jacobsen, W.M. Roca & S.P. Singh. 1994. Interspecific hybridization between common and tepary beans: increased hybrid embryo growth, fertility, and efficiency of hybridization through congruity backcrossing. Theor. Appl. Genet. 88:324-331.

Mohamed, M.F., P.E. Read & D.P. Coyne. 1991. In vitro response of bean (Phaseolus vulgaris L.) cotyledonary explants to benzyladenine in the medium. Plant Growth Regulat. Soc. Amer. Quart. 19:19-26.

Mohamed, M.F., P.E. Read & D.P. Coyne. 1992a. Plant regeneration from in vitro culture of embryonic axis explants in common and tepary beans. J. Amer. Soc. Hort. Sci. 117:332-336.

Mohamed, M.F., P.E. Read & D.P. Coyne. 1992b. Dark preconditioning, CPPU, and thidiazuron promote shoot organogenesis on seedling node explants of common and faba beans. J. Amer. Soc. Hort. Sci. 117:668-672.

Russell, D.R., K.M. Wallace, J.H. Bathe, B.J. Martinell & D.E. McCabe. 1993. Stable transformation of Phaseolus vulgaris via electric-dicharge mediated particle acceleration. Plant Cell Rep. 12: 165-169.

Saam, M.M., G.L. Hosfield & J.W. Saunders. 1987. In vitro propagation of dry bean seedling shoot tips. J. Amer. Soc. Hort. Sci. 112:852-855.

Zambre, M.A., J. de Clerq, E. Vranová, M. van Montagu, G. Angenon & W. Dillen. 1998. Plant regeneration from embryo-derived callus in Phaseolus vulgaris (common bean) and P. acutifolius A.Gray (tepary bean). Plant Cell Rep. 17: 626-630.

Zhang, Z., D. P. Coyne & A. Mitra. 1997. Factors affecting Agrobacterium- mediated transformation of common bean. J. Amer. Soc. Hort. Sci. 122:300-305.

Chapter 6

Integrated Genetic Improvement[1]

Shree P. Singh
Plant, Soil and Entomological Sciences, University of Idaho, 3793 North 3600 East, Kimberly, Idaho 83341-5076, U.S.A.

Key words: cultivar development, gamete selection, gene introgression, gene pools, gene pyramiding, multiple-parent crosses, parental development, selection in F_1 and F_1-derived families, selection methods

1. INTRODUCTION

Since the 1950's, common bean (*Phaseolus vulgaris* L.) researchers have developed an improved understanding of the origin (Berglund-Brücher & Brücher, 1976; Brücher, 1988; Delgado Salinas et al., 1988; Gentry, 1969; Kami et al., 1995; Kaplan, 1981; Khairallah et al., 1992; Miranda Colín, 1967; Weiseth, 1954), domestication (Gepts et al., 1986; Kaplan, 1965; Koenig et al., 1990), evolution under domestication (Evans, 1980; Gepts & Debouck, 1991; Koenig & Gepts, 1989a; Koinange et al., 1996; Smartt, 1969, 1988), and the patterns of variation and organization of genetic diversity in the crop (Becerra Velásquez & Gepts, 1994; Evans, 1973; Gepts & Bliss, 1985; Khairallah et al., 1990; Koenig & Gepts, 1989b; Singh, 1989; Singh & Gutiérrez, 1984; Singh et al., 1991a, b, c) and related species (Debouck & Smartt, 1995; see also Chapter 2 by Debouck). Significant advances have also been made in germplasm collection, characterization, and screening methods, including identification of useful genes and different types of markers for their direct and indirect selection. Alternative recombination and selection methods have also been developed. On the other hand, there has been a growing concern for conservation and efficient use of

[1] Published as Idaho Agricultural Experiment Station Journal Article no. 99704. University of Idaho, College of Agriculture, Moscow, ID.

natural resources including agrobiodiversity, sustainable agriculture, and production and consumption of healthy and so-called organic food. Moreover, in order to sustain families on the farm and keep pace with ever-growing demographic development, yield per unit of cropped land must be maximized, production costs reduced, and farming made attractive to increasingly demanding younger generations.

Thus, it is essential to develop broadly adapted high yielding, high-quality common bean cultivars that are less dependent on water, fertilizers, pesticides, and labor in the shortest time possible. This will be possible only by doing the following: 1) maximizing genetic gains from selection for specific traits, and 2)simultaneously improving the maximum number of agronomic traits by accumulating all the favorable alleles from the crop's cultivated races, gene pools, and wild populations forming its primary, secondary, and tertiary gene pools. Moreover, a thorough understanding of the patterns of variation in and the organization of genetic diversity and evolution under domestication within the species in particular and other species of interest in the genus at large is essential.

The organization of diversity among the different *Phaseolus* species (i.e., taxonomy and phylogenetic relationship or crossability) in relation to the common bean was discussed by Debouck in Chapter 2. I shall briefly examine the traits and genes deficient or lacking in common bean in relation to its major production problems (Graham, 1978; Schwartz & Pastor-Corrales, 1989; Thung & Oliveira, 1998; Wortmann et al., 1998; Zaumeyer & Thomas, 1957; see also Chapter 1 by Singh). This chapter will also cover useful genes or traits available in different alien germplasms (Table 1) to improve common bean. Finally, for systematic accumulation of favorable alleles for simultaneous improvement of the maximum number of agronomic traits from across the crop's cultivated races, gene pools, and wild populations forming its primary, secondary, and tertiary gene pools, I shall discuss the integrated genetic improvement of common bean cultivars. For this purpose, a three-tiered breeding approach will be described (Kelly et al., 1997, 1998; see also Chapter 8). The three-tiered approach involves: 1) broadening the genetic base by introgression of useful genes from alien cultivated and wild relatives, 2) parental development for specific traits by pyramiding useful genes, and 3) cultivar development by simultaneous selection for multiple desirable traits. In the past, each of these major activities has been carried out, often by different groups of researchers and without all the necessary attention and care that is required for the development of truly superior cultivars for growers, processors, traders, agricultural industries, and consumers. Integrated genetic improvement can maximize the use of genetic diversity and gains from selection, and reduce the bean crop's dependence on water, fertilizers, pesticides, and labor for the benefit of humankind.

2. TRAITS DEFICIENT IN COMMON BEAN CULTIGENS AND THEIR SOURCES IN RELATED SPECIES

Cultivated and wild common bean germplasm accessions have been evaluated at the International Center for Tropical Agriculture (CIAT, Cali, Colombia) and elsewhere for the last several decades for their response to biotic and abiotic stresses. These stresses include major diseases, insects, water stress, and low soil fertility (Bannerot, 1965; Beebe et al., 1981; Menezes & Dianese, 1988; Morales & Niessen, 1988; Pastor-Corrales et al., 1995, 1998; Schoonhoven & Cardona, 1982; Schoonhoven et al., 1983; Schuster et al., 1983; Schwartz et al., 1982; Singh & Muñoz, 1999; Stavely, 1984). Plant architectural and phenological traits have also been evaluated (Nienhuis & Singh, 1985; Singh & Gutiérrez, 1982; White & Laing, 1989). Considering common bean production problems, it may be safe to conclude that while adequate levels of resistance to some factors, such as rust [caused by *Uromyces appendiculatus* (Pers.) Ung.], bean common mosaic (BCM, caused by a potyvirus), and anthracnose [caused by *Colletotrichum lindemuthianum* (Sacc. & Magn.) Bri. & Cav.] are found in cultigens, the levels of resistance to other factors are not adequate. The diseases and pests for which present resistance is not adequate include the following: common bacterial blight [caused by *Xanthomonas campestris* pv. *phaseoli* (Smith) Dye], halo blight [caused by *Pseudomonas syringae* var. *phaseolicola* (Burkh.) Young et al.], bacterial brown spot [(caused by *Pseudomonas syringae* pv. *syringae* (van Hall)], ascochyta blight [(caused by *Phoma exigua* var. *diversispora* (Bub.) Boerma], web blight [caused by *Thanatephorus cucumeris* (Frank) Donk.], white mold [caused by *Sclerotinia sclerotiorum* (Lib.) de Bary], bean fly (*Ophiomyia phaseoli* Tryon), leafhoppers (*Empoasca kraemeri* Ross & Moore), and bruchids [*Zabrotes subfasciatus* (Boheman) and *Acanthoscelides obtectus* (Say)].

Systematic evaluation of wild common bean and wild and cultivated germplasm of alien species for resistance to pests and diseases and for other useful traits has so far been limited. The main reasons are probably: 1) lack of available germplasm, 2) special growing requirements and evaluation protocols for wild common bean and alien species, and 3) reluctance by bean researchers to look beyond the commercial classes.

Despite the fact that limited evaluations have been done for resistance to pests and diseases and abiotic stresses, alien germplasm seems promising for common bean improvement (Table 1). For example, resistance to bruchids was found in wild *P. vulgaris* (Acosta Gallegos et al., 1998; Cardona & Kornegay, 1989; Schoonhoven et al., 1983). Variation was also reported for a few photosynthesis traits in wild common bean (González et al., 1995; Lynch et al., 1992). *P. polyanthus* is particularly known for its resistance to ascochyta blight (Schmit & Baudoin, 1992) as well as to white mold (Hunter et al., 1982). *P. coccineus* has long been known as source of resistance to anthracnose (Hubbeling, 1957), root rots (Wilkinson, 1983), white mold (Abawi et al., 1978), bean yellow mosaic (Baggett, 1956), and bean golden mosaic (BGM, caused by a geminivirus) (F. Morales & S. Singh, unpublished). *P. costaricensis* might be interesting for its resistance to BGM (Singh et al., 1997b).

Table 1. Some useful traits deficient in common bean and their sources in closely related Phaseolus species

Traits	Source species	References
Cold	*P. coccineus*	Bannerot, 1979
Drought	*P. acutifolius*	Markhart, 1985; Parsons & Howe, 1984
Heat	*P. acutifolius*	Garver, 1934; Petersen & Davis, 1982
Leafhopper	*P. acutifolius*	C. Cardona (unpublished)
Bruchid	Wild *P. vulgaris*	Schoonhoven et al., 1983
	P. acutifolius	C. Cardona (unpublished)
Ascochyta blight	*P. polyanthus*	Schmit & Baudoin, 1992
Common bacterial blight	*P. acutifolius*	Coyne et al., 1963; Singh & Muñoz, 1999
	P. coccineus	Mohan, 1982
White mold	*P. coccineus*	Abawi et al., 1978; Hunter et al., 1982
Bean golden mosaic	*P. coccineus*	F. Morales & S. Singh (unpublished)
	P. polyanthus	F. Morales & S. Singh (unpublished)
	P. costaricencis	Singh et al., 1997a
Root rots	*P. coccineus*	Wilkinson, 1983
Anthracnose	*P. coccineus*	Hubbeling, 1957
Bean yellow mosaic	*P. coccineus*	Baggett, 1956

Good sources of tolerance to leafhoppers exist in *P. acutifolius* (C. Cardona, unpublished). Also, high levels of resistance to common bacterial blight (Coyne et al., 1963; Schuster et al., 1983; Singh & Muñoz, 1999), and bruchids (C. Cardona, unpublished) are found in some accessions of *P. acutifolius*.

3. INTEGRATED GENETIC IMPROVEMENT OF COMMON BEAN CULTIVARS

3.1 Introduction

The pedigree of common bean cultivars released around the world, and the list of germplasm accessions available for its cultivated and wild members of the primary, secondary, and tertiary gene pools (see Debouck in Chapter 2) are available. Analysis of this information shows that the genetic basis of cultivars within each market class (Voysest & Dessert, 1991; see also Chapter 1 by Singh) and race of common bean (Singh et al., 1991a) is extremely narrow (Adams, 1977; McClean et al., 1993; Voysest et al., 1994). Only a small fragment (<5%) of the available genetic diversity has been used globally (Singh, 1992) despite nearly a century of organized common bean improvement (Robertson & Frazier, 1978).

Useful genes are scattered across cultivated and wild populations in the primary, secondary, and tertiary gene pools of the common bean (Table 1; Singh, 1989; Singh et al., 1991a). Extremely large differences are found in seed characteristics (e.g., size, shape, and color), growth habit, and adaptation traits. There is also a deficiency of other useful genes including resistance to major biotic and abiotic production constraints in common bean cultivars belonging to different market classes within and among different races and gene pools (Singh, 1989; Singh et al., 1991a). Moreover, there are some problems of recombination and gene exchange (Koinange & Gepts, 1992; Kornegay et al., 1992; Singh & Gutiérrez, 1984; Singh & Molina, 1996) between races and gene pools. It would be extremely difficult, if not impossible, to accumulate at once all the necessary favorable alleles from across races and gene pools into one high-yielding high-quality commercial cultivar resistant to major biotic and abiotic stresses. Thus, a three-tiered breeding strategy (or its modification) as suggested by Kelly et al. (1997, 1998, see also Chapter 8) and summarized in Figure 1 is likely to have more success for integrated genetic improvement of common bean. For clarity and convenience I shall discuss: 1) useful gene introgression from alien germplasm, 2) pyramiding genes for specific traits and parental development, and 3) cultivar development. Before I do so, different selection methods used in common bean improvement will be briefly reviewed.

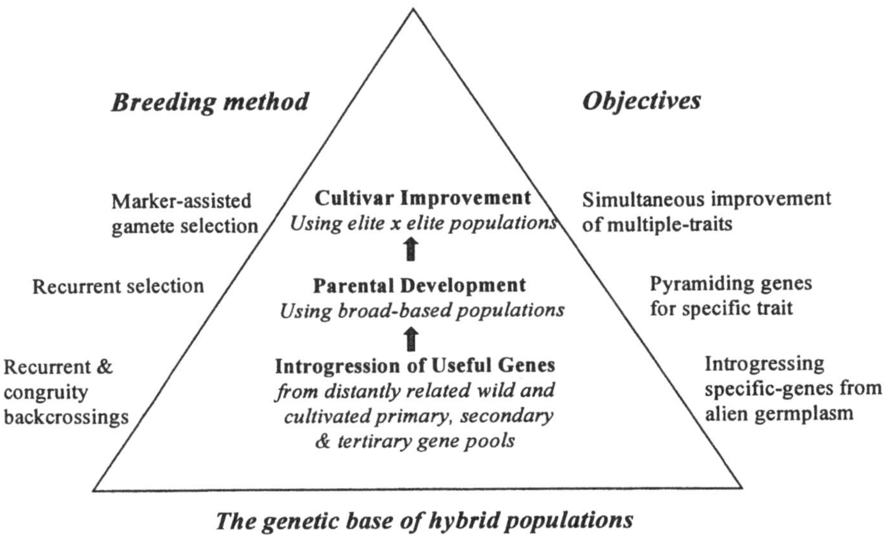

Figure 1. Summary of integrated genetic improvement of common bean. The breeding pyramid is adopted from Kelly et al. (1998).

3.2 Selection Methods Used

Until the first quarter of this century in Canada and the U.S.A., and the fifties and sixties in Latin America and Africa, selection between and within locally available and introduced landraces of common bean was probably the method of choice. This procedure resulted in the release of many important cultivars. These cultivars had a significant impact on bean production for many years. For example, the cultivar Robust was released in Michigan (U.S.A.) in 1915 (Robertson & Frazier, 1978); Actopan, Antigua, and Jamapa in Mexico; Porrillos in El Salvador; and Rio Tibagi and Carioca in Brazil (Voysest, 1983). Later, the recurrent backcross (Miranda et al., 1979; Pompeu, 1980, 1982), pedigree (Kelly et al., 1994a, b), and mass-pedigree (Beebe et al., 1993; Grafton et al., 1993; Singh et al., 1989, 1993) methods and their modifications were used. These procedures resulted in the development and release of hundreds of successful cultivars in the Americas, Europe, Africa, and Asia. More recently, congruity backcrossing (Haghighi & Ascher, 1988; Mejía-Jiménez et al., 1994; Urrea & Singh, 1995), single seed descent (Kelly et al., 1989; Urrea & Singh, 1994), recurrent (Beaver & Kelly, 1994; Duarte, 1966; Kelly & Adams, 1987; Singh et al., 1999; Sullivan & Bliss, 1983), and gamete (Singh, 1994, 1998; Singh et al., 1998a) selection methods have been used for common bean improvement. Thus, most if not all commonly used crop breeding methods have been employed in common bean. In spite of this work, objective data comparing the efficiency of different selection methods, with some exceptions (Beaver & Kelly, 1994; Gutiérrez & Singh, 1992; Singh & Terán, 1998; Urrea & Singh, 1994, 1995), are scarce. Given the immense diversity in bean types (market classes), growing environments, production problems, breeding objectives, and resources available to common bean researchers, no breeding method will probably be equally suitable for all circumstances. Instead, breeders must use a separate selection method or combine two or more methods to suit their needs.

No attempts will be made to describe or review various breeding methods currently in use in common bean because their details can be found in standard plant breeding books and research articles (e.g., Fouilloux & Bannerot, 1988). Based on my experience in common bean improvement, the recurrent and congruity inbred-backcrossing, recurrent, and gamete selection methods or their modifications could be suggested for the introgression of useful genes from alien germplasm, for parental development and gene pyramiding, and for cultivar development, respectively (Figure 1). Because gamete selection (Singh, 1994, 1998) facilitates the integration of marker-assisted and conventional selection methods for simultaneous improvement of multiple qualitative and quantitative agronomic traits, and it is comparatively less known, a brief account of this technique will be given later in the chapter.

3.3 Gene Introgression from Alien Germplasm

Because of the differences in genetic distance between different *Phaseolus* species and *P. vulgaris* (Debouck & Smartt, 1995; see also Chapter 2 by Debouck), between gene pools and races within the common bean cultigens (Gepts & Bliss, 1985; Singh, 1989; Singh et al., 1991a), and the breeding methods and strategies that are required, introgression of useful genes from each major distantly related cultivated race, gene pool, wild populations, and alien species from the secondary and tertiary gene pools must be accomplished separately. Thus, each of these types of introgression will be covered separately here, keeping in mind the organization of genetic diversity in *P. vulgaris* and its related species.

3.3.1 Introgression Between Cultivated Races Within the Primary Gene Pool

In addition to differences in response to biotic and abiotic stresses, the knowledge of parental performance *per se* for seed yield and other quantitative characters, combining ability, occurrence of incompatibility genes, and undesirable linkages is essential while introgressing desirable alleles from alien germplasm. The problem of gene recombination often magnifies with increase in genetic distance between races and gene pools. Nonetheless, in general, there is good complementation and positive combining ability occurs between members of different cultivated races of common bean within the Middle American gene pool. These occur with traits such as seed yield (Nienhuis & Singh, 1988a; Singh et al., 1992, 1993) and tolerance of water stress (Singh, 1995b; White et al., 1994b). Thus, despite considerable differences in seed, plant, and adaptation characteristics, it is not too difficult to introgress useful genes among and across the three races, namely, Durango, Jalisco, and Mesoamerica. Introgression is made easier when the differences in photoperiod response and phenological traits are not large, especially when the crop is grown during the long summer days at higher latitudes. Even these difficulties can be overcome by growing parents under shorter daylengths (approximately 12 hr photoperiod) for hybridization and assuring >70% genetic contribution of the adapted cultivars and elite lines under improvement. An elaborate backcrossing program is not required until and unless one parent is early maturing determinate and another is highly photoperiod sensitive or a late maturing extreme climber (i.e., growth habit type IV, Singh, 1982b) belonging to another race. Often a three-way or modified-double cross (Singh, 1982a) ensuring >70% genetic contribution of the parents of the same race and market class under improvement suffices.

Examples of gene introgression achieved between races within a gene pool include: 1) upright plant architectural traits introgressed from race Mesoamerica to race Durango, 2) transfer of *A. godmani* resistance from race Jalisco to race Mesoamerica, and 3) introgression of recessive *bgm-1* gene resistance to BGM from race Durango to race Mesoamerica.

3.3.1.1 Upright plant type

Most landraces and improved cultivars belonging to race Durango possess the prostrate semiclimbing growth habit type III (Singh, 1989; Singh et al., 1991a). In semiarid production regions of the central highlands of Mexico and the northwestern U.S.A. this prostrate type III growth habit often may not impose serious problems. But in comparatively wet or humid growing environments (e.g., Michigan, Minnesota, Nebraska, and North

Dakota, U.S.A.) such cultivars are prone to white mold, anthracnose, and seed discoloration because of their closed canopy and pods touching the ground. Thus, upright erect cultivars that permit better air movement through the canopy and keep pods away from the soil are preferred. These traits were introgressed from upright type II cultivars belonging to race Mesoamerica using phenotypic recurrent selection (Kelly & Adams, 1987). Similarly, within race Mesoamerica, carioca beans with upright growth habit type II and resistance to leafhopper and five diseases were recently developed using gamete selection (Singh et al., 1998a). These types are also required for large scale mechanized commercial farming to reduce dependence on hand labor and production costs. For further details on breeding for plant type, refer to Chapter 7 by Vandenberg & Nleya and Chapter 8 by Kelly et al.

3.3.1.2 Resistance to bean pod weevil

Extremely high levels of resistance to Mexican pod weevil (*Apion godman* Boheman) are found in medium-seeded race Jalisco climbing bean germplasm from the highlands of Mexico (Beebe et al., 1993; Garza et al., 1996). The insect is also a severe problem in the warmer regions of Guatemala, El Salvador, Honduras, and northern Nicaragua where small-seeded cultivars belonging to race Mesoamerica are grown (Singh, 1989, 1992; Singh et al., 1991a). Most of this latter group of cultivars are highly susceptible to this insect. Over the past several years high levels of the*Apion* resistance from the race Jalisco has been transferred into small red- and black-seeded race Mesoamerica cultivars of bush growth habits II and III, using the mass-pedigree method (Beebe et al., 1993).

3.3.1.3 Resistance to bean golden mosaic

Resistance to BGM can be expressed as percent infection, plant dwarfing, leaf chlorosis, and pod deformation (Morales & Nissen, 1988). Resistance to each of these component traits is controlled by different genes (Molina Castañeda & Beaver, 1998; Morales & Singh, 1991; Urrea et al., 1996; Vélez et al., 1998). Moreover, at least two different sources of resistance to leaf chlorosis are known to occur. One of the sources found in race Durango (e.g., landrace cultivar Garrapato or G 2402) is controlled by the recessive gene *bgm-1* (Urrea et al., 1996). This gene was inadvertently transferred into common bean line A 429 (Morales & Singh, 1993). Since then, the line A 429 or Garrapato has been used to develop highly resistant small red (DOR 482, MD 30-75), black (Turbo III, DOR 500), and carioca (Singh et al., 1998a) seeded cultivars for Central America and Brazil, using pedigree, mass-pedigree or gamete selection methods. Moreover, availability of a RAPD and SCAR marker (Urrea et al., 1996) should minimize the need for and complement direct screening in presence of the virus. Introgression of the *bgm-1* gene from line A 429 into snap bean cultivars for the state of Florida (U.S.A.) was accomplished using the marker (McMillan et al., 1998).

3.3.2 Gene Introgression Between Two Cultivated Primary Gene Pools

Because of increasing genetic distance (Singh et al., 1991a, b), differences in yielding ability (Singh, 1989; White & González, 1990; White et al., 1992), poor combining ability (Kornegay et al., 1992; Singh & Urrea, 1995; Welsh et al., 1995), and occasional incompatibility problems (Gepts & Bliss, 1985; Singh & Gutiérrez, 1984; Singh & Molina, 1996), it is often difficult to recover essential agronomic characteristics from inter-gene pool crosses. Some of these characteristics include adaptation, seed traits, and yielding ability. These desirable characteristics may be difficult to maintain while introgressing favorable alleles between Andean and Middle American gene pools, using biparental crosses followed by pedigree (Kornegay et al., 1992), single seed descent (Welsh et al., 1995), or mass selection methods (Singh et al., 1989). More elaborate programs of recurrent or congruity inbred-backcrossings (Urrea & Singh, 1995) and recurrent selection (Beaver & Kelly, 1994; Singh et al., 1999) are required. Moreover, there may be an occasional need to use bridging-parents (Singh & Gutiérrez, 1984), if *Dl-1* and *Dl-2* incompatibility alleles are found.

Useful genes have only rarely been transferred between the Andean and Middle American gene pools. Nonetheless, there are a few recent examples of introgression of useful genes between these two gene pools, including resistance to BCM, BGM, rust, anthracnose, and leafhoppers. BCM, an aphid and seed-transmitted viral disease of common bean (Morales & Castaño, 1987), causes severe yield losses in susceptible cultivars in most bean growing areas of the world. There are several strains of the virus, some causing necrosis of foliar or aerial plant parts, especially at higher temperatures (Burke & Silbernagel, 1974; Drijfhout et al., 1978). Both strain-specific and non-specific resistance to BCM have been reported (Drijfhout, 1978; Drijfhout et al., 1978). Also, molecular markers for both the dominant *I* as well as recessive *bc-3* genes have been reported (Haley et al., 1994; Johnson et al., 1997; Melotto et al., 1996). Breeders have or are pyramiding these dominant and recessive resistant genes with the aid of molecular markers.

Both dominant *I* and recessive *bc-3* gene resistances for BCM have been transferred from small-seeded race Mesoamerica cultivars to large-seeded Nueva Granada types (F. Morales, personal communication, 1996). A clearer understanding of the evolutionary origin (Andean versus Middle American) of the *I* gene and its undesirable linkages with certain seedcoat color genes (Cáceres O. & Morales, 1985; Kyle & Dickson, 1988; Temple & Morales, 1986) should assist with breeding. Knowledge of the necrosis-inducing strains of the virus (Drijfhout, 1978), and the specificity of resistance covered by other recessive genes should facilitate pyramiding of resistant genes for this important disease. To date, no one has searched for new sources of resistance to BCM in the wild bean populations and related species belonging to the primary, secondary, and tertiary gene pools of *P. vulgaris*.

Resistance to leafhoppers has been introgressed from race Mesoamerica to Nueva Granada (Kornegay & Cardona, 1990). High levels of resistance to leaf chlorosis caused by BGM and controlled by the *bgm-1* gene were initially found in race Durango. This resistance has been transferred into large-seeded red and red mottled Andean beans of race Nueva Granada for the Caribbean countries (J. Beaver, personal communication, 1998) and into snap bean (McMillan et al., 1998).

3.3.3 Gene Introgression from Wild Populations of the Primary Gene Pool

Except for the occurrence of *Dl-1* and *Dl-2* genes in inter-gene-pool crosses of wild populations (Koinange & Gepts, 1992), there are no known barriers for transferring genes from wild populations of either gene pool into cultigens. The F_1 hybrids between cultivated x wild, as well as their progenies in subsequent generations, are fully fertile. Consequently, such crosses have been used to study inheritance of seed size (Motto et al., 1978), determine yield potential (Singh et al., 1995), and to study genetic differences between the wild and cultivated phenotypes, including mapping of major genes involved in domestication (Koinange et al., 1996). In addition, extremely high levels of resistance to bruchids (*Z. subfasciatus*) have been successfully transferred from the Mexican wild bean populations to a range of cultivars (Cardona & Kornegay, 1989; Cardona et al., 1989, 1990).

The resistance to *Z. subfasciatus* was not found in thousands of accessions of cultivated common bean that have been screened (Schoonhoven & Cardona, 1982). Only a few wild bean populations from the highlands of Mexico were found to be resistant (Schoonhoven et al., 1983). By extracting purified arcelin seed protein and using a range of concentrations in artificial seeds, Osborn et al. (1988) demonstrated that the bruchid resistance was due to the presence of this protein in the seed cotyledons of wild bean. The presence of arcelin is controlled by a single dominant gene (Osborn et al., 1986) and there are over half a dozen alleles with varying effects (Acosta Gallegos, 1998; Cardona et al., 1990). The presence of arcelin can easily be detected by SDS-PAGE electrophoresis. Now, since antiserum to the protein is available, it is easier, faster, and cheaper to use the ELISA (enzyme linked immunosorbent assay) technique for large scale screening of segregating populations, families, and derived advance-generation lines. This development has facilitated and expedited the breeding for bruchid resistance and minimized dependence on screening using insect infestation that is time-consuming and affected by the environment. A recurrent backcross program was used for this purpose. Arcelin, like lectin and other related proteins, is heat labile and hence its antinutritional and harmful effects are lost during prolonged (>30 minutes) cooking.

It is not known, however, what mating or recombination and selection scheme would be the most appropriate for transferring quantitatively inherited traits (such as yield potential) from wild populations to cultigens should there be such a need. Knowing that only a small number of genes with large phenotypic effects, clustered in a few regions of the genome, control the inheritance of traits involved in domestication (Koinange et al., 1996), it should not be difficult to recover the cultivated phenotype from wild x cultivated crosses. On the contrary, a more directed germplasm conversion approach could be followed by which the major genes involved in domestication could easily be introgressed into desired wild populations by backcrossing to domesticate them and realize their full agronomic potential in comparatively short time. Dependable molecular markers tightly linked with major genes involved in domestication and other useful traits should aid in this task, especially in temperate North American and European environments.

3.3.4 Introgression from the Secondary Gene Pool

As noted earlier, extremely high levels of resistance to ascochyta blight (Schmit & Baudoin, 1992) and BGM (F.J. Morales & S.P. Singh, unpublished results; Singh et al., 1997b) are among the useful traits found in some accessions of *P. costaricensis, P. polyanthus,* and *P. coccineus* comprising the secondary gene pool of common bean. Moderate levels of resistance to common bacterial blight (Mohan, 1982; Schuster et al., 1983; Singh & Muñoz, 1999), anthracnose (Hubbeling, 1957), halo blight, white mold (Abawi et al., 1978; Hunter et al., 1982), root rots (Wilkinson, 1983), and/or cold (Bannerot, 1979) would also be expected in members of this group because of their natural habitat (Table 1).

Crosses among the three species of the secondary gene pool as well as their crosses with common bean are easily effected without any need of embryo rescue (Baggett, 1956; Camarena & Baudoin, 1987; Cheng et al., 1981; Park & Dhanvantari, 1987; Singh et al., 1997a), especially when *P. vulgaris* is used as the female parent. However, hybrid progenies between crosses of common bean and any of the three species forming the secondary gene pool may be partially sterile and it may be difficult to recover the desired stable common bean phenotypes (Wall, 1970). Thus, difficulties may be encountered in developing true breeding common bean lines with introgressed desirable genes from members of the secondary gene pool. Researchers in the past have successfully used backcrossing to the recurrent common bean parent for introgression of resistance to common bacterial blight (Freytag et al., 1982; Miklas et al., 1994; Park & Dhanvantary, 1987). In my limited experience thus far I observed a tendency to revert to the female-parent genotype, and that recombinants are unstable. Consequently, the congruity backcrossing (Haghighi & Ascher, 1988; Mejía-Jiménez et al., 1994; Urrea & Singh, 1995) was also used at CIAT to maximize recombination between species, increase the fertility of progenies, and recover more useful recombinants (S.P. Singh, unpublished results).

3.3.5 Introgression from the Tertiary Gene Pool

Since the initial *P. vulgaris* x *P. acutifolius* interspecific cross (Honma, 1956) other researchers have repeatedly made this interspecies hybridization (Haghighi & Ascher, 1988; Mejía-Jiménez et al., 1994; Scott & Michaels, 1992; Thomas & Waines, 1984). Due to the great genetic distance, the hybrid embryos must be rescued between two to three weeks after pollination. Moreover, since the hybrids are male sterile, one or more backcrosses to the recurrent common bean parent are often required. Using *P. acutifolius* as the female parent of the initial F_1 cross and/or the first backcrossing of the *P. vulgaris* x *P. acutifolius* hybrid on to *P. acutifolius* is often more difficult than using *P. vulgaris* as the female parent of the initial and first backcrossing the interspecies hybrid on to it (Mejía-Jiménez et al., 1994). Haghighi & Ascher (1988) and Mejía-Jiménez et al. (1994) demonstrated the usefulness of the congruity over recurrent backcrossing for recovery of fertility and high number of hybrid progenies between interspecific crosses of common and tepary beans.

High level of resistance to common bacterial blight, a seed-transmitted disease (Aggour et al., 1989) from the initial common x tepary bean cross (Coyne et al., 1963), as well as from the subsequent interspecies crosses (McElroy, 1985; Scott & Michaels, 1992; Singh & Muñoz, 1999) has been transferred to common bean. However, introgression of resistance to leafhoppers, bruchids, heat, and water stress still needs to be achieved. It is likely that lack of thorough and specific parental germplasm screening for each of these traits and use of the most resistant *P. acutifolius* accession in interspecies hybridization has slowed progress on these projects. Use of a comparatively small population size of interspecific hybrid progenies and/or inappropriate screening techniques have contributed to the problem. As a result, these supposedly quantitatively inherited traits have not been introgressed. To overcome some of difficulties associated with introgression of quantitatively inherited traits from *P. acutifolius* to *P. vulgaris*, the following may be helpful: 1) use of congruity backcrossing, 2) production of large interspecies hybrid progenies from plant-to-plant paired pollinations at each step of crossing, and 3) development of a large number of inbred lines before subjecting them to appropriate screenings.

P. parvifolius, a member of the tertiary gene pool, crosses easily with *P. acutifolius*, and has been crossed with *P. vulgaris*, using embryo rescue (Singh et al., 1998b). In spite of this work, its potential usefulness for common bean improvement has yet to be determined.

Useful genes, when and if found in *P. angustissimus*, *P. filiformis*, *P. lunatus*, and 30 other species of the section *Phaseolus* and all members of sections *Chiapasana*, *Minkelersia*, and *Xanthotricha* (see Chapter 2 by Debouck) cannot be accessed for common bean improvement at the moment. Identification and cloning of useful alien genes and successful regeneration and transformation of common bean may facilitate these interspecies gene introgressions. Similarly, for the immediate needs, molecular marker-assisted recurrent and congruity inbred-backcrossings would expedite and permit more focussed introgression of useful genes from alien cultivated and wild germplasm forming the common bean's primary, secondary, and tertiary gene pools.

3.4 Pyramiding Genes and Parental Development for Specific Traits

3.4.1 Introduction

Traits for which there is convincing and unequivocal evidence for the existence of two or more complementary useful genes, pyramiding those genes into a common genotype would help maximize the character expression or gains from selection. Pyramiding such genes would also broaden the genetic base of cultivars, and increase the durability of resistance, especially when it is caused by variable pathogens (e.g., rust, angular leaf spot, and anthracnose diseases; see Chapter 4 by Kelly & Miklas). In such circumstances, it is essential to pyramid useful genes from within and across cultivated races and gene pools and wild populations of common bean, and from its secondary (*P. coccineus*, *P. costaricensis*, and *P.*

polyanthus) and tertiary (*P. acutifolius* and *P. parvifolius*) gene pools. While under certain circumstances it may be feasible to achieve both the simultaneous introgression and pyramiding of useful genes (Singh & Muñoz, 1999), often gene pyramiding into a common bean genotype follows after successful gene introgression from an alien germplasm has been accomplished.

The task of gene pyramiding for specific traits independently or simultaneously for two or more traits is carried out depending upon the availability of resources and compatibility of screening and selection methods. While pyramiding genes for specific traits it is essential to keep in mind the adaptive features, growth habit, maturity, and seed characteristics of the principal market class or bean type that needs to be improved. For example, if small-seeded black, red or white beans from the race Mesoamerica predominate in a region then the pyramiding of useful genes should be accomplished in those types or something very similar to them. On the other hand, in some regions large-seeded red mottled Andean common bean from race Nueva Granada or medium-seeded pinto from Middle American race Durango are cultivated. For these regions, it is advisable to develop elite parents with pyramided genes for specific traits that are similar or close to their respective seed types, growth habit, and adaptation.

Pyramiding multiple genes for specific traits has been used in these situations: 1) from within a given cultivated race of common bean (e.g., leafhopper resistance), 2) from among races within a primary gene pool (e.g., seed yield and water stress tolerance), 3) from across the two cultivated primary gene pools within *P. vulgaris* (e.g., rust resistance), and 4) from across *Phaseolus* species (e.g., common bacterial blight). We shall discuss each of these in the following paragraphs.

3.4.2 Leafhopper Resistance

Heritable differences in cultivated common bean for ovipositional non-preference and tolerance to feeding by leafhoppers (*E. kraemeri*) have been found only in small-seeded common bean cultivars belonging to race Mesoamerica (Kornegay et al., 1986). Both traits are inherited quantitatively (Galwey & Evans, 1982; Kornegay & Temple, 1986). Through recurring cycles of breeding, using the bulk-pedigree method of selection, significant levels of tolerance to leafhopper have been accumulated in common bean lines and cultivars (Kornegay et al., 1989). But at best this is still an intermediate level of resistance which may not be adequate under high insect population pressure (Singh et al., 1998a). Thus, leafhopper resistance in alien species needs to be investigated. For example, much higher levels of resistance to leafhoppers have been found in tepary bean (e.g., accession G

40036). This resistance seems to be controlled by two recessive genes (C. Cardona, unpublished). These and other sources of resistance need to be introgressed and combined with the leafhopper tolerance available in common bean.

3.4.3 Seed Yield

In Chapter 8, Kelly et al. provide a thorough and detailed account of breeding for high seed yield in common bean. For further details also see Adams (1982), Kelly et al. (1997, 1998), Singh (1991, 1992), and Wallace et al. (1993). Differences in yield potential are much larger between than within races and gene pools of cultivated common bean (Singh, 1989; Singh et al., 1991a). Heritability of seed yield varies from low to moderately high (Nienhuis & Singh, 1988b; Singh et al., 1991d; Welsh et al., 1995). Moreover, the knowledge of parental yield *per se*, their combining ability, and evolutionary origins (Middle American versus Andean) are essential for breeding for high yielding common bean cultivars (Singh, 1991, 1992). For example, in general, good combining ability occurs among the three races of Middle American common bean (Nienhuis & Singh, 1986, 1988a). High yielding genotypes have been developed, using mass-pedigree (Singh, 1995a; Singh et al., 1989, 1993) and recurrent (Singh et al., 1999) selection methods from interracial populations within the Middle American gene pool. While Kornegay et al. (1992) and Singh et al. (1989) were unable to develop high yielding lines, Beaver and Kelly (1994) and Singh et al. (1999) were also able to select for high yield in Andean x Middle American populations, using recurrent selection.

3.4.4 Water Stress Tolerance

Water stress or drought is among the most widely distributed and endemic abiotic problems affecting bean production in many regions of the world, especially northeastern Brazil and the central and northern highlands of Mexico (Acosta G. et al., 1989; White & Singh, 1991). Among physiological and agronomic traits, seed yield still appears to be the most useful and reliable selection criterion for breeding for water stress tolerance in common bean (White et al., 1994a). Heritability for seed yield under water stress may range from low to moderately high (Singh, 1995b). Also, large differences in water stress tolerance occur in races of common bean. Cultivars from race Durango seem to possess the highest levels of water stress tolerance (H. Terán & S.P. Singh, unpublished) as would be expected because of their origin and domestication in the semiarid harsh environments of the Mexican highlands (Singh, 1989; Singh et al., 1991a). White et al.

(1994b) observed large differences in combining ability for water stress tolerance in common bean cultivars of different races of Middle American origin. Singh (1995b), using this information, developed lines possessing higher levels of water stress tolerance from a Durango x Mesoamerica interracial population. More recently, other newer sources possessing higher levels of water stress tolerance have been identified in races Durango, Jalisco, and Mesoamerica (H. Terán & S. P. Singh, unpublished) which need to be combined. However, the highest levels of water stress tolerance reported in *P. acutifolius* (Markhart, 1985; Thomas et al., 1983) has yet to be introgressed into and combined with the highest levels of tolerance available in *P. vulgaris*. Search for and use of reliable molecular markers for traits associated with drought tolerance should complement and facilitate breeding for this important trait (Schneider et al., 1997).

3.4.5 Bean Rust

The most variable pathogen causing disease in common bean may be the bean rust pathogen, *U. appendiculatus* (CIAT, 1985; Stavely, 1984). Being wind-borne, it is highly mobile and thus possesses great potential for changing its population composition within and between growing seasons and production regions. Fortunately, from judicious field screening over the years and across countries through international nurseries (CIAT, 1985) and screening in greenhouses with known races (Stavely, 1984), high levels of resistance have been identified in a range of cultivated bean types of both Andean and Middle American origins. While the genetics of resistance found in some important sources of germplasm are still not fully understood, nearly a dozen different genes carrying race-specific resistance have been identified (Carvalho et al., 1978; Christ & Groth, 1982; Grafton et al., 1985; Kolmer & Groth, 1984; Stavely, 1990). RAPD markers (Haley et al., 1993; Jung et al., 1996) are being used to incorporate rust resistance into common bean cultivars and to combine rust resistance with resistance to other diseases such as BCM, BGM, anthracnose, and/or common bacterial blight. However, the soundest strategy for breeding for rust resistance would likely include a concerted effort using gene pyramiding (see Chapter 4 by Kelly & Miklas). Complementary genes from both Andean and Middle American gene pools conferring resistance to the broadest spectrum of pathogenic races would be systematically combined with such traits as presence of trichomes on abaxial leaf surfaces, imparting race nonspecific resistance (Shaik, 1984).

3.4.6 Common Bacterial Blight (CBB)

The CBB (a seed-transmitted disease), the most important bacterial disease of common bean, affecting all aerial plant parts, may cause yield losses of more than 40% in susceptible cultivars (Serracin et al., 1991; Yoshii et al., 1976). Only low to intermediate levels of resistance are found in common and scarlet runner (*P. coccineus*) beans, but the tepary bean possesses the highest levels of resistance, imparting near immunity (Coyne et al., 1963; Schuster et al., 1983; Singh & Muñoz, 1999; Zapata et al., 1985). Inheritance of race-specific resistance to CBB in tepary bean is controlled by several dominant genes (Drijfhout & Blok, 1987; Dursun et al., 1995; Freytag, 1989). However, both the resistance found in common bean and that introgressed from the scarlet runner and tepary beans are known to be quantitatively inherited. These resistances involve one or more genes with major effects and five to eight genes or QTLs with small effects (Jung et al., 1996; McElroy, 1985; Nodari et al., 1993).

Over the years the resistances found in landraces of common (Beebe & Pastor-Corrales, 1991), scarlet runner (Freytag et al., 1982; Miklas et al., 1994; Park & Dhanvantari, 1987), and tepary beans (Coyne et al., 1963; McElroy, 1985; Scott & Michaels, 1992) have been introgressed independently. Attempts have also recently been made to both introgress and pyramid CBB resistance (Singh & Muñoz, 1999). This is a singular example where useful complementary genes have been pyramided from across different *Phaseolus* species belonging to the primary, secondary, and tertiary gene pools of common bean. Consequently, this work has resulted into the highest levels of CBB resistance ever developed (Singh & Muñoz, 1999).

Because different levels of CBB resistance are found in common, scarlet runner, and tepary beans (Singh & Muñoz, 1999) different CBB resistance genes within each species may need to be combined first. Next, CBB resistance should be pyramided between common and scarlet runner beans because they possess comparatively lower levels of CBB resistance than the tepary beans. Eventually, the pyramided CBB resistance thus obtained must be combined with that of the tepary bean. Use of tightly linked molecular markers for each of the CBB resistance genes found across these three and other *Phaseolus* species in the secondary and tertiary gene pools, access to the widest range of pathogenic populations of the bacterium, and molecular marker-assisted recurrent selection would expedite this breeding process.

3.5 Cultivar Development

3.5.1 Introduction

As noted earlier, large differences in seed characteristics and consumer preferences exist for each of the major market classes of beans (e.g., black, navy, pinto, red kidney, and cranberry) (Voysest, 1983; Voysest & Dessert, 1991; see also Chapter 1 by Singh). Different plant types (growth habit type I, II, III, or IV; Singh, 1982b), adaptive traits (sensitivity or insensitivity to photoperiod, White & Laing, 1989), and combinations of resistance to biotic and abiotic stresses are also often required for each market class (Singh, 1992; see also Chapter 1). For these reasons, a separate integrated genetic improvement program for cultivars of each major market class and production region is justified. In order to bring maximum health, social, economic, and environmental benefits to the growers, processors, traders, and consumers, high yielding cultivars with the maximum expression of each trait and a combination of the maximum number of desirable traits must be sought in each successive breeding cycle.

Breeding for cultivars of major market classes of the three most widely grown races, namely, small-seeded Mesoamerica, medium-seeded Durango, and large-seeded Nueva Granada covering over 80% of the production of dry bean worldwide is dealt with in separate chapters elsewhere in this book. However, the salient features common to the integrated genetic improvement of cultivars of all market classes, using the gamete selection method (Singh, 1994, 1998; Singh et al., 1998a) will be briefly discussed here.

3.5.2 Parental Selection and Programming Crosses

All commercial cultivars, elite lines, and reliable donor parents of necessary genes (including those obtained from introgression from alien germplasm and pyramiding useful complementary genes) are selected. These selections are based on their evaluation information for specific traits, their general agronomic worth, adaptation or overall performance in the bean growing areas of concern, and their combining ability. It is not often customary to test parental performance per se and combining ability in common bean. However, these tests are needed to assure that only the most popular and leading landraces, cultivars, and elite lines of proven performance of the market class under improvement are used as the recipient parents. Parental evaluation across sites and through complementary nurseries is a prerequisite for choosing parents, testing their combining ability, and their eventual use in hybridization. The tests are also needed to assure that the necessary genes are present in the donor parents, and the virtues and defects of all the parents to be used in crosses are known. The necessary facilities and resources must exist to simultaneously improve or select for resistance to specific biotic and abiotic stresses, seed yield, and other agronomic characters of interest in bean types of the preferred market classes.

While selecting parents for specific cross combinations, due consideration is given for seed, plant, and adaptation characteristics of the parents, in addition to the resistance to biotic and abiotic stresses and other useful traits that they might possess. For each market class, all parents or the maximum number of possible parents must be similar to the specific cultivar groups under improvement in growth habit, seed color and size, and all must be well adapted. Thus, each cross is made among only high yielding well-adapted elite recipient and donor parents. Distantly related, poorly adapted, or noncommercial parents, especially belonging to another gene pool, for the most part, are not used for immediate development of superior cultivars of specific market classes.

The other factor that must be considered is that the simultaneous improvement of the maximum number of traits is sought in each selection cycle. Thus, when the necessary genes for each of the major traits of interest are found in separate parents, a few multiple-parent crosses are preferred over a large number of single crosses. Although comparatively more time is spent during hybridization to generate multiple-parent crosses, the process allows production of recombinants with favorable alleles for multiple traits, something that is not possible through single crosses without repeated cycles of selection for specific traits, one at a time. For example, if simultaneous selection for resistance to anthracnose, BCM, bean rust, and bruchids is sought, a four-way cross involving all four donor parents is made first. The double-cross F_1 hybrid thus developed then serves as the pollinator parent for the cultivar or elite lines to be improved. Often it is advisable to assure between 10 to 25% genetic contribution of each donor-parent of useful genes in a final multiple-parent cross. Moreover, a large number of plant-to-plant pollinations are made at each step of multiple-parent cross development to assure adequate sampling of gametes and genetic contribution of each parent involved in the final crosses. If this procedure is followed, there will be enough (>30 plants) selected F_1 plants for subsequent evaluations and line development.

3.5.3 Gamete Selection Using Dominant and Codominant Markers During Hybridization and Development of Final Multiple-Parent F_1 Crosses

When multiple-parent crosses are made, there is an opportunity to evaluate and select for dominant and codominant traits or markers in heterogametic and heterogeneous F_1 hybrids. For example, the F_1 of a four-way cross will be segregating into 1 resistant: 1 susceptible for each of anthracnose, BCM, bruchids, and rust if each is controlled by a separate dominant gene or linked to such a marker, provided only one of the four parents contributed such a gene. Thus, only 1/16 of F_1 plants would contribute gametes possessing all four desirable alleles, which should then be used for pollinating the cultivar or elite line under improvement. The process is repeated as long as necessary and includes evaluation of the final F_1 multiple-parent crosses.

Gamete selection as proposed by Singh (1994, 1998), using dominant and codominant morphological, biochemical, and DNA-based markers in heterogametic and heterogeneous crosses during hybridization and development of multiple-parent F_1 crosses helps assure accumulation of all necessary alleles early on. This process reduces the population size in the subsequent segregating generations (i.e., F_2 onwards). Also, considerable time and resources are saved for improving multiple traits. The selected F_1 plants from the final multiple-parent crosses are each harvested separately to develop F_1-derived F_2 families.

3.5.4 Selection Within and Between F_1-Derived Families in Early Generations and Identification of Promising Families

Gamete selection during hybridization and development of the multiple-parent crosses and in the final F_1, when combined with early generation (F_2 to F_4) evaluation and selection helps identify promising populations and families within populations. These are then used to develop superior lines for subsequent evaluations and new cultivar identification. Both qualitative and quantitative traits between and within F_1-derived families are selected, using complementary nurseries for each major biotic and abiotic stress. Seed yield is determined in contrasting environments representative of the bean production regions. The effectiveness of this selection strategy among F_1-derived families for resistance to angular leaf spot, anthracnose, BGM, CBB, and leafhoppers, in F_2 to F_4 of multiple-parent populations was demonstrated by Singh et al. (1998a). All 127 F_1-derived families from one population (BZ 9780) were discarded by the F_4. Out of 460 F_1-derived families from another population (GX 9792) only 17 families that segregated for all five resistances were left by the F_4. Although evaluation and selection for seed yield was not practiced in that study, Singh & Terán (1998) demonstrated the feasibility and effectiveness of yield testing in early generations of the F_1-derived families between and within populations of common bean. Thus, by combining gamete selection in multiple-parent F_1 crosses and judicious evaluation and selection in F_1-derived families in early generations, the probability of developing new, high-yielding superior cultivars possessing the maximum number of desirable traits is maximized.

3.5.5 Development and Evaluation of Lines from Promising Families

To develop advanced breeding lines possessing multiple desirable traits and conform seed, plant, and adaptation characteristics of common bean for growers, processors, traders, and consumers only selected promising families are used. The F_5 seed harvested from the F_4 plants of promising families, after discarding all non-commercial seeds, is space planted under appropriate disease pressure (e.g., anthracnose and/or rust) for evaluation and selection of the maximum number of resistant plants with commercially acceptable seeds of good culinary quality.

All selected plants in the F_5 are harvested individually for progeny tests in the F_6 and seed increase in F_7. These experimental lines are evaluated through adaptation and complementary nurseries for each major biotic and abiotic stress. Evaluation also includes seed quality characteristics and yield trials across contrasting sites, including in the growers' fields, to identify new cultivars (Singh, 1992). This process generally takes five years.

4. CONCLUSIONS AND FUTURE PROSPECTS

There are over 29,000 cultivated and more than 1,300 wild accessions of *P. vulgaris* housed in the germplasm bank at CIAT, Cali, Colombia, and elsewhere. At CIAT the number of accessions belonging to its secondary and tertiary gene pools, respectively, are 1,049 and 335 (see Chapter 2 by Debouck). In spite of this diversity, the genetic base of commercial cultivars of specific market classes is narrow (Adams, 1977; McClean et al., 1993; Voysest et al., 1994). Moreover, the average global yield of common bean remains low (<900 kg/ha) and its production suffers from a wide range of both biotic and abiotic constraints, some causing up to 100% yield losses (see Chapter 1 by Singh).

In the Latin American tropics and sub-tropics, more than 22,000 accessions of cultivated common bean have been evaluated systematically for their reaction to important diseases, insects, and response to long photoperiod, drought, and low soil fertility. Similar evaluations of wild accessions of common bean and all other related species has lagged behind. The situation may be more alarming in the temperate environments of North America, Europe, Africa, and Asia. Nonetheless, it is widely known that common bean cultigens are deficient in many agronomic traits including resistance to storage insects, leafhoppers, ascochyta blight, common bacterial blight, and white mold. Useful genes imparting resistance to these factors have been identified in wild beans (primary gene pool) and related species forming the secondary (*P. coccineus, P. costaricensis*, and *P. polyanthus*) and tertiary (*P. acutifolius* and *P. parvifolius*) gene pools (Table 1).

Until recently, common bean breeders and geneticists have tended to confine hybridization and selection within a specific market class. Moreover, despite nearly a century of breeding history (Robertson & Frazier, 1978), less than 5% of the available germplasm has been used in hybridization programs (Singh, 1992). In order to maximize and sustain bean production it is essential to develop high yielding, high quality cultivars that are less dependent on water, fertilizer, pesticides, and manual labor. This need warrants a comprehensive integrated genetic improvement program in which all favorable alleles from cultivated and wild populations of common bean's primary, secondary, and tertiary gene pools are accumulated in superior cultivars of each major market class. To facilitate this task it will be essential to carry out three major breeding activities: 1) introgression of useful genes from distantly related races and gene pools of cultivated common bean and from its wild populations and related species, 2) accumulation or pyramiding of favorable alleles for specific traits from across different species (i.e., parental development), and 3) cultivar development for specific market classes of beans. Germplasm recombination and selection methods will vary depending upon the genetic distance between parents, breeding objectives, and available resources. Nonetheless, recurrent and congruity inbred-backcrossing, recurrent, and gamete selection methods are suggested, respectively, for each of these breeding activities.

Availability of an efficient and repeatable transformation system for *P. vulgaris* and integrated linkage maps, use of the knowledge of genetics of domestication and evolution, and development and use of marker-assisted selection should expedite and facilitate need-based integrated genetic improvement of common bean.

REFERENCES

Abawi, G.S., R. Provvidenti, D.C. Crosier & J.E. Hunter. 1978. Inheritance of resistance to white mold disease in *Phaseolus coccineus*. J. Hered. 69:200-202.

Acosta G., J.A., F.J. Ibarra-Pérez, A. Pajarito R., R. Ochoa M., R. Lépiz I. & M.W. Adams. 1989. Breeding for tolerance to drought in bean. pp. 416-417. *In* S. Beebe (ed.), Current Topics in Breeding of Common Bean. CIAT, Cali, Colombia.

Acosta Gallegos, J., C. Quintero, J. Vargas, O. Toro, J. Tohme & C. Cardona. 1998. A new variant of arcelin in wild common bean, *Phaseolus vulgaris* L., from southern Mexico. Genet. Resources & Crop Evol. 45:235-242.

Adams, M.W. 1977. An estimation of homogeneity in crop plants with specific reference to genetic vulnerability in the dry bean *Phaseolus vulgaris* L. Euphytica 26:665-679.

Adams, M.W. 1982. Plant architecture and yield breeding in *Phaseolus vulgaris* L. Iowa State J. Res. 56:225-254.

Aggour, A.R., D.P. Coyne, A.K. Vidaver & K.M. Eskridge. 1989. Transmission of the common blight pathogen in bean seed. J. Amer. Soc. Hort. Sci. 114:1002-1008.

Baggett, J.R. 1956. The inheritance of resistance to strains of bean yellow mosaic virus in the interspecific cross *Phaseolus vulgaris* x *P. coccineus*. Plant Dis. Rptr. 40:702-707.

Bannerot, H. 1965. Résultats de l'infection d'une collection de haricots par six races physiologiques d'anthracnose. Ann. Amélior. Plant. 15:201-222.

Bannerot, H. 1979. Cold tolerance in beans. Annu. Rpt. Bean Improv. Coop. 22:81-84.

Beaver, J.S. & J.D. Kelly. 1994. Comparison of two selection methods for the improvement of dry bean populations derived from crosses between gene pools. Crop Sci. 34:34-37.

Becerra Velásquez, V.L. & P. Gepts. 1994. RFLP diversity of common bean (*Phaseolus vulgaris*) in its centres of origin. Genome 37:256-263.

Beebe, S.E., F.A. Bliss & H.F. Schwartz. 1981. Root rot resistance in common bean germplasm of Latin American origin. Plant Dis. 65:485-489.

Beebe, S., C. Cardona, O. Díaz, F. Rodríguez, E. Mancía & S. Ajquejay. 1993. Development of common bean (*Phaseolus vulgaris* L.) lines resistant to the pod weevil, *Apion godmani* Wagner, in Central America. Euphytica 69:83-88.

Beebe, S.E. & M.A. Pastor-Corrales. 1991. Breeding for disease resistance. pp. 561-617. *In* A. van Schoonhoven & O. Voysest (eds.), Common Beans: Research for Crop Improvement. C.A.B. Int., Wallingford, U.K. & CIAT, Cali, Colombia.

Berglund-Brücher, O. & H. Brücher. 1976. The South American wild bean (*Phaseolus aborigineus* Burk) as ancestor of the common bean. Econ. Bot. 30:257-272.

Brücher, H. 1988. The wild ancestor of *Phaseolus vulgaris* in South America. pp. 185-214. *In* P. Gepts (ed.), Genetic Resources of *Phaseolus* Beans. Kluwer, Dordrecht, Netherlands.

Burke, D.W. & M.J. Silbernagel. 1974. Hypersensitive resistance to bean common mosaic virus in dry field beans. Annu. Rpt. Bean Improv. Coop. 17:22.

Cáceres O., S.R. & F.J. Morales. 1985. Incorporación de genes de resistencia dominante y recesiva a cepas del virus del mosaico común en dos variedades comerciales de frijol (*Phaseolus vulgaris* L). Acta Agr. 35:7-20.

Camarena, F. & J.P. Baudoin. 1987. Obtention des premiers hybrides interspécifiques entre *Phaseolus vulgaris* et *Phaseolus polyanthus* avec le cytoplasme de cette dernière forme. Bull. Rech. Agron. Gembloux 22:43-55.

Cardona, C. & J. Kornegay. 1989. Use of wild *Phaseolus vulgaris* to improve beans for resistance to bruchids. pp. 90-98. *In* S. Beebe (ed.), Current Topics in Breeding of Common Beans. CIAT, Cali, Colombia.

Cardona, C., J. Kornegay, C.E. Posso, F. Morales & H. Ramírez. 1990. Comparative value of four arcelin variants in the development of dry bean lines resistant to the Mexican bean weevil. Entomol. Expt. Appl. 56:197-206.

Cardona, C., C.E. Posso, J. Kornegay, J. Valor & M. Serrano. 1989. Antibiosis effects of wild dry bean accessions on the Mexican bean weevil and the bean weevil (Coleoptera: Bruchidae). J. Econ. Entomol. 82:310-315.

Carvalho, L.P. de, C. Vieira & G.M. Chaves. 1978. Hereditariedade da resistència a cinco raças fisiológicas de *Uromyces phaseoli* var. typica Arth. em *Phaseolus vulgaris* L. Fitopatol. Bras. 3:181-185.

Cheng, S.S., M.J. Bassett & K.H. Quesenberry. 1981. Cytogenetic analysis of interspecific hybrids between common bean and scarlet runner bean. Crop Sci. 21:75-79.

Christ, B.J. & J.V. Groth. 1982. Inheritance of resistance in three cultivars of beans to the bean rust pathogen and the interaction of virulence and resistance genes. Phytopathology 72:771-773.

CIAT (Centro Internacional de Agricultura Tropical). 1985. International bean rust nursery: results 1983-1984. CIAT, Cali, Colombia.

Coyne, D.P., M.L. Schuster & S. Al-Yasiri. 1963. Reaction studies of bean species and varieties to common blight and bacterial wilt. Plant. Dis. Rptr. 47:534-537.

Debouck, D.G. & J. Smartt. 1995. Beans, *Phaseolus* spp. (Leguminosae-Papilionoideae). pp. 287-294. *In* J. Smartt & N. W. Simmonds (eds.), Evolution of Crop Plants. 2nd ed. Longman, London, U.K.

Delgado Salinas, A., A. Bonet & P. Gepts. 1988. The wild relative of *Phaseolus vulgaris* in Middle America. pp. 163-184. *In* P. Gepts (ed.), Genetic Resources of *Phaseolus* Beans. Kluwer, Dordrecht, Netherlands.

Drijfhout, E. 1978. Genetic interaction between *Phaseolus vulgaris* and bean common mosaic virus with implications for strain identification and breeding for resistance. Agric. Res. Rpt. 872. Centre for Agricultural Publishing and Documentation, Wageningen, Netherlands.

Drijfhout, E. & W.J. Blok. 1987. Inheritance of resistance to *Xanthomonas campestris* pv. *phaseoli* in tepary bean (*Phaseolus acutifolius*). Euphytica 36:803-808.

Drijfhout, E., M.J. Silbernagel & D.W. Burke. 1978. Differentiation of strains of bean common mosaic virus. Neth. J. Plant Path. 84:13-26.

Duarte, R.A. 1966. Responses in yield and yield components from recurrent selection practiced in a bean hybrid population in three locations in North and South America. Diss. Abrst. 27:1339B-1340B.

Dursun, A., D.P. Coyne, M.F. Mohamed & G. Jung. 1995. Inheritance of resistance to common bacterial blight in tepary beans. Annu. Rpt. Bean Improv. Coop. 38:162-163.

Evans, A.M. 1973. Plant architecture and physiological efficiency in the field bean. pp. 279-284. *In* D. Wall (ed.), Potentials of Field Beans and Other Food Legumes in Latin America. CIAT, Cali, Colombia.

Evans, A.M. 1980. Structure, variation, evolution, and classification in *Phaseolus*. pp. 337-347. *In* R.J. Summerfield & A.H. Bunting (eds.), Advances in Legume Science. Royal Botanic Gardens, Kew, U.K.

Fouilloux, G. & H. Bannerot. 1988. Selection methods in the common bean (*Phaseolus vulgaris*). pp. 503-542. *In* P. Gepts (ed.), Genetic Resources of *Phaseolus* Beans. Kluwer, Dordrecht, Netherlands.

Freytag, G.F. 1989. Inheritance of resistance to three strains of common bacterial blight (*Xanthomonas campestris*) in the cultivated tepary bean (*Phaseolus acutifolius* var. *latifolius*). Annu. Rpt. Bean Improv. Coop. 32:101-102.

Freytag, G.F., M.J. Bassett & M. Zapata. 1982. Registration of XR-235-1-1 bean germplasm. Crop Sci. 22:1268-1269.
Galwey, N.W. & A.M. Evans. 1982. The inheritance of resistance to *Empoasca kraemeri* Ross & Moore in the common bean, *Phaseolus vulgaris* L. Euphytica 31:933-952.
Garver, S. 1934. The Redfield tepary bean, an early maturing variety. J. Amer. Soc. Agron. 3:397-403.
Garza, R., C. Cardona & S.P. Singh. 1996. Inheritance of resistance to the bean-pod weevil (*Apion godmani* Wagner) in common beans from Mexico. Theor. Appl. Genet. 92:357-362.
Gentry, H.S. 1969. Origin of the common bean, *Phaseolus vulgaris*. Econ. Bot. 23:55-69.
Gepts, P. & F.A. Bliss. 1985. F_1 hybrid weakness in the common bean: differential geographic origin suggests two gene pools in cultivated bean germplasm. J. Hered. 76:447-450.
Gepts, P. & D.G. Debouck. 1991. Origin, domestication, and evolution of the common bean (*Phaseolus vulgaris* L.). pp. 7-53. *In* A. van Schoonhoven & O. Voysest (eds.), Common Beans: Research for Crop Improvement. C.A.B. Int., Wallingford, U.K. & CIAT, Cali, Colombia.
Gepts, P., T.C. Osborn, K. Raskha & F.A. Bliss. 1986. Phaseolin-protein variability in wild forms and landraces of the common bean (*Phaseolus vulgaris*): evidence for multiple centers of domestication. Econ. Bot. 40:451-468.
González, A., J. Lynch, J.M. Tohme, S.E. Beebe & R.E. Macchiaveli. 1995. Characters related to leaf photosynthesis in wild populations and landraces of common bean. Crop Sci. 35:1468-1476.
Grafton, K.F., K.C. Chang, J.R. Venette & A.J. Vander Wal. 1993. Registration of 'Norstar' navy bean. Crop Sci. 33:1405-1406.
Grafton, K.F., G.C. Weiser, L.J. Littlefield & J.R. Stavely. 1985. Inheritance of resistance to two races of leaf rust in dry edible bean. Crop Sci. 25:537-539.
Graham, P.H. 1978. Problems and potentials of field beans (*Phaseolus vulgaris* L.) in Latin America. Field Crops Res. 1:259-317.
Gutiérrez, J.A. & S.P. Singh. 1992. Effect of alternative methods of advancing early generation populations of common bean in the tropics. Turrialba 42:482-486.
Haghighi, K.R. & P.D. Ascher. 1988. Fertile intermediate hybrids between *Phaseolus vulgaris* and *P. acutifolius* from conguity backcrossing. Sex. Plant Reprod. 1:51-58.
Haley, S.D., L. Afanador & J.D. Kelly. 1994. Identification and application of a random amplified polymorphic DNA marker for the *I* gene (potyvirus resistance) in common bean. Phytopathology 84:157-160.
Haley, S.D., P.N. Miklas, J.R. Stavely, J. Byrum & J.D. Kelly. 1993. Identification of RAPD markers linked to a major rust resistance gene block in common bean. Theor. Appl. Genet. 86:505-512.
Honma, S. 1956. A bean interspecific hybrid. J. Hered. 47:217-220.
Hubbeling, N. 1957. New aspects of breeding for disease resistance in beans (*Phaseolus vulgaris* L.). Euphytica 6:111-141.
Hunter, J.E., M.H. Dickson, M.A. Boettger & J.A. Cigna. 1982. Evaluation of plant introductions of *Phaseolus* spp. for resistance to white mold. Plant Dis. 66:320-322.
Johnson, W.C., P. Guzmán, D. Mandala, A.B.C. Mkandawire, S. Temple, R.L. Gilbertson & P. Gepts. 1997. Molecular tagging of the *bc-3* gene for introgression into Andean common bean. Crop Sci. 37: 248-254.
Jung, G., D.P. Coyne, P.W. Skroch, J. Nienhuis, E. Arnaud-Santana, J. Bokosi, H.M. Ariyarathne, S.R. Steadman, J.S. Beaver & S.M. Kaeppler. 1996. Molecular markers

associated with plant architecture and resistance to common blight, web blight, and rust in common beans. J. Amer. Soc. Hort. Sci. 121:794-803.

Kami, J., V. Becerra Velásquez, D.G. Debouck & P. Gepts. 1995. Identification of presumed ancestral DNA sequences of phaseolin in *Phaseolus vulgaris*. Proc. Natl. Acad. Sci. U.S.A. 92:1101-1104.

Kaplan, L. 1965. Archeology and domestication in American *Phaseolus* (beans). Econ. Bot. 19:358-368.

Kaplan, L. 1981. What is the origin of the common bean? Econ. Bot. 35:240-254.

Kelly, J.D. & M.W. Adams. 1987. Phenotypic recurrent selection in ideotype breeding of pinto beans. Euphytica 36:69-80.

Kelly, J.D., M.W. Adams, A.W. Saettler, G.S.Hosfield, G.V. Varner, J.S. Beaver, M.A. Uebersax & J. Taylor. 1989. Registration of 'Mayflower' navy bean. Crop Sci. 29:1571-1572.

Kelly, J.D., G.L. Hosfield, G.V. Varner, M.A. Uebersax, M.E. Brothers & J. Taylor. 1994a. Registration of 'Huron' navy bean. Crop Sci. 34:1408.

Kelly, J.D., G.L. Hosfield, G.V. Varner, M.A. Uebersax, S.D. Haley & J. Taylor. 1994b. Registration of 'Raven' black bean. Crop Sci. 34:1406-1407.

Kelly, J.D., J.M. Kolkman & K.A. Schneider. 1997. Breeding for high yield in common bean. pp. 187-204. *In* S.P. Singh & O. Voysest (eds.), Taller de Mejoramiento de Frijol para el Siglo XXI: Bases para Una Estrategia para América Latina. CIAT, Cali, Colombia.

Kelly, J.D., J.M. Kolkman & K. Schneider. 1998. Breeding for yield in dry bean (*Phaseolus vulgaris* L.). Euphytica 102:343-356.

Khairallah, M.M., M.W. Adams & B.B. Sears. 1990. Mitochondrial DNA polymorphisms of Malawian bean lines: further evidence for two major gene pools. Theor. Appl. Genet. 80:753-761.

Khairallah, M.M., B.B. Sears & M.W. Adams. 1992. Mitochondrial restriction fragment length polymorphisms in wild *Phaseolus vulgaris* L.: insights on the domestication of the common bean. Theor. Appl. Genet. 84:915-922.

Koenig, R. & P. Gepts. 1989a. Segregation and linkage of genes for seed proteins, isozymes, and morphological traits in common bean (*Phaseolus vulgaris*). J. Hered. 80:455-459.

Koenig, R. & P. Gepts. 1989b. Allozyme diversity in wild *Phaseolus vulgaris*: further evidence for two major centers of genetic diversity. Theor. Appl. Genet. 78:809-817.

Koenig, R.L., S.P. Singh & P. Gepts. 1990. Novel phaseolin types in wild and cultivated common bean (*Phaseolus vulgaris*, Fabaceae). Econ. Bot. 44:50-60.

Koinange, E.M.K. & P. Gepts. 1992. Hybrid weakness in wild *Phaseolus vulgaris* L. J. Hered. 83:135-139.

Koinange, E.M.K., S.P. Singh & P. Gepts. 1996. Genetic control of the domestication syndrome in common bean. Crop Sci. 36:1037-1045.

Kolmer, J.A. & J.V. Groth. 1984. Inheritance of a minute uredinium infection type of bean rust in bean breeding line 914. Phytopathology 74:205-207.

Kornegay, J.L. & C. Cardona. 1990. Development of an appropriate breeding scheme for tolerance to *Empoasca kraemeri* in common bean. Euphytica 47:223-231.

Kornegay, J.L., C. Cardona, J.V. Esch & M. Alvarado. 1989. Identification of common bean lines with ovipositional resistance to *Empoasca kraemeri* (Homoptera: Cicadellidae). J. Econ. Entomol. 82:649-654.

Kornegay, J.L., C. Cardona & A. van Schoonhoven. 1986. The mechanisms of resistance in common beans to the leafhopper *Empoasca kraemeri*. Entomol. Exp. Appl. 40:273-279.

Kornegay, J.L. & S.R. Temple. 1986. Inheritance and combining ability of leafhopper defense mechanisms in common bean. Crop Sci. 26:1153-1158.

Kornegay, J.L., J.W. White & O. Ortíz de la Cruz. 1992. Growth habit and gene pool effects on inheritance of yield in common bean. Euphytica 62:171-180.
Kyle, M.M. & M.H. Dickson. 1988. Linkage of hypersensitivity to five viruses with the *B* locus in *Phaseolus vulgaris* L. J. Hered. 79:308-311.
Lynch, J., A. González, J.M. Tohme & J. García. 1992. Variation in characters related to leaf photosynthesis in wild bean populations. Crop Sci. 32:633-640.
Markhart III, A.H. 1985. Comparative water relations of *Phaseous vulgaris* L. and *Phaseolus acutifolius* A. Gray. Plant Physiol. 77:113-117.
McClean, P.E., J.R. Myers & J.J. Hammond. 1993. Coefficient of parentage and cluster analysis of North American dry bean cultivars. Crop Sci. 33:190-197.
McElroy, J.B. 1985. Breeding for dry beans, *P. vulgaris* L., for common bacterial blight resistance derived from *Phaseolus acutifolius* A. Gray. Ph.D. dissertation. Cornell Univ., Ithaca, U.S.A.
McMillan Jr., R.T., M.J. Davis, H.J. McLaughlin & J.R. Stavely. 1998. PCR evaluation of fourteen bean golden mosaic virus (BGMV) resistant snap bean germplasm lines for the presence of the virus. Annu. Rpt. Bean Improv. Coop. 41: 31-32.
Mejía-Jiménez, A., C. Muñoz, H.J. Jacobsen, W.M. Roca & S.P. Singh. 1994. Interspecific hybridization between common and tepary beans: increased hybrid embryo growth, fertility, and efficiency of hybridization through recurrent and congruity backcrossing. Theor. Appl. Genet. 88:324-331.
Melotto, M., L. Afanador & J.D. Kelly. 1996. Development of a SCAR marker linked to the *I* gene in common bean. Genome 39: 1216-1219.
Menezes, J.R. & J.C. Dianese. 1988. Race characterization of Brazilian isolates of *Colletotrichum lindemuthianum* and detection of resistance to anthracnose of *Phaseolus vulgaris*. Phytopathology 78:650-655.
Miklas, P.N., J.S. Beaver, K.F. Grafton & G.F. Freytag. 1994. Registration of TARS VCI-4B multilple disease resistant dry bean germplasm. Crop Sci. 34:1415.
Miranda Colín, S. 1967. Origén de Phaseolus *vulgaris* L. (frijol común). Agrociencia 1:99-109.
Miranda, P., R.C. Mafra, E.B. Correia & M.A. De-Queiroz. 1979. "IPA-79-19" uma nova variedade de feijão "mulatinho" (*Phaseolus vulgaris* L.) para Pernambuco. Pes. Agrop. Pernamb. 3:105-111.
Mohan, S.T. 1982. Evaluation of *Phaseolus coccineus* Lam. germplasm for resistance to common bacterial blight of bean. Turrialba 32:489-490.
Molina Castañeda, A. & J.S. Beaver. 1998. Inheritance of normal pod development in bean golden mosaic resistant common beans. Annu. Rpt. Bean Improv. Coop. 41:3-4.
Morales, F.J. & M. Castaño. 1987. Seed transmission characteristics of selected bean common mosaic virus strains in differential bean cultivars. Plant Dis. 71:51-53.
Morales, F.J. & A.I. Niessen. 1988. Comparative responses of selected *Phaseolus vulgaris* germplasm inoculated artificially and naturally with bean golden mosaic virus. Plant Dis. 72:1020-1023.
Morales, F.J. & S.P. Singh. 1991. Genetics of resistance to bean golden mosaic virus in *Phaseolus vulgaris* L. Euphytica 52:113-117.
Morales, F.J. & S.P. Singh. 1993. Breeding for resistance to bean golden mosaic virus in an interracial population of *Phaseolus vulgaris* L. Euphytica 67:59-63.
Motto, M., G.P. Sorresi & F. Salamini. 1978. Seed size inheritance in a cross between wild and cultivated common beans (*Phaseolus vulgaris* L.). Genetica 49:31-36.
Nienhuis, J. & S.P. Singh. 1985. Effects of location and plant density on yield and architectural traits in dry beans. Crop Sci. 25:579-584.

Nienhuis, J. & S.P. Singh. 1986. Combining ability analyses and relationships among yield, yield components, and architectural traits in dry bean. Crop Sci. 26:21-27.

Nienhuis, J. & S.P. Singh. 1988a. Genetics of seed yield and its components in common bean (*Phaseolus vulgaris* L.) of Middle American origin. I. General combining ability. Plant Breed. 101:143-154.

Nienhuis, J. & S.P. Singh. 1988b. Genetics of seed yield and its components in common bean (*Phaseolus vulgaris* L.) of Middle American origin. II. Genetic variance, heritability and expected response from selection. Plant Breed. 101:155-163.

Nodari, R.O., S.M. Tsai, R.L. Gilbertson & P. Gepts. 1993. Towards an integrated linkage map of common bean. 2. Development of an RFLP-based linkage map. Theor. Appl. Genet. 85:513-520.

Osborn, T.C., D.C. Alexander, S.S.M. Sun, C. Cardona & F.A. Bliss. 1988. Insecticidal activity and lectin homology of arcelin seed protein. Science (Washington DC) 240:207-210.

Osborn, T.C., T. Blake, P. Gepts & F.A. Bliss. 1986. Bean arcelin. 2. Genetic variation, inheritance and linkage relationships of a novel seed protein of *Phaseolus vulgaris* L. Theor. Appl. Genet. 71:847-855.

Park, S.J. & B.N. Dhanvantari. 1987. Transfer of common blight (*Xanthomonas campestris* pv. *phaseoli*) resistance from *Phaseolus coccineus* Lam. to *P. vulgaris* L. through interspecific hybridization. Can. J. Plant Sci. 67:685-695.

Parsons, L.R. & T.K. Howe. 1984. Effects of water stress on the water relations of *Phaseolus vulgaris* and the drought resistant *Phaseolus acutifolius*. Physiol. Plant. 60:197-202.

Pastor-Corrales, M.A., C. Jara & S.P. Singh. 1998. Pathogenic variation in, sources of, and breeding for resistance to *Phaeoisariopsis griseola* causing angular leaf spot in common bean. Euphytica 103:161-171.

Pastor-Corrales, M.A., M.M. Otoya, A. Molina & S.P. Singh. 1995. Resistance to *Colletotrichum lindemuthianum* isolates from Middle America and Andean South America in different common bean races. Plant Dis. 79:63-67.

Petersen, A.C. & D.D. Davis. 1982. Yield response of *Phaseolus vulgaris* L. and *Phaseolus acutifolius* subjected to water stress. Annu. Rpt. Bean Improv. Coop. 25:53-54.

Pompeu, A.S. 1980. Yields of French bean lines (*Phaseolus vulgaris* L.) of the rosinha and roxinho groups resistant to *Colletotrichum lindemuthianum*. Bragantia 39:89-97.

Pompeu, A.S. 1982. Catu, Aete-3, Aroana-80, Moruna-80, Carioca-80 e Ayso: novos cultivares de feijoeiro. Bragantia 41:213-218.

Robertson, L.S. & R.D. Frazier (eds.). 1978. Dry Bean Production-Principles & Practices. Michigan State Univ. Ext. Bull. E-1251. East Lansing, U.S.A.

Schmit, V. & J.P. Baudoin. 1992. Screening for resistance to *Ascochyta* blight in populations of *Phaseolus coccineus* L. and *P. polyanthus* Greenman. Field Crops Res. 30:155-165.

Schneider, K., M.E. Brothers & J.D. Kelly. 1997. Marker-assisted selection to improve drought resistance in common bean. Crop Sci. 37:51-60.

Schoonhoven, A. van & C. Cardona. 1982. Low levels of resistance to the Mexican bean weevil in dry beans. J. Econ. Entomol. 75:567-569.

Schoonhoven, A. van, C. Cardona & J. Valor. 1983. Resistance to the bean weevil and the Mexican bean weevil (Coleopter: Bruchidae) in noncultivated common bean accessions. J. Econ. Entomol. 76:1255-1259.

Schuster, M.L., D.P. Coyne, T. Behre & H. Leyna. 1983. Sources of *Phaseolus* species resistance and leaf and pod differential reactions to common blight. HortScience 18:901-903.

Schwartz, H.F. & M.A.Pastor-Corrales (eds.). 1989. Bean Production Problems in the Tropics. 2nd ed. CIAT, Cali, Colombia.
Schwartz, H.F., M.A. Pastor-Corrales & S.P. Singh. 1982. New sources of resistance to anthracnose and angular leaf spot of beans (*Phaseolus vulgaris* L.). Euphytica 31:741-754.
Scott, M.E. & T.E. Michaels. 1992. *Xanthomonas* resistance of *Phaseolus* interspecific cross selections confirmed by field performance. HortScience 27:348-350.
Serracin, J., R.A. Young, J.C. Rosas & J. Cáceres. 1991. Daños causados por *Xanthomonas campestris* pv. *phaseoli* y su efecto en el rendimiento del frijol común (habichuela; *Phaseolus vulgaris*). J. Agric. Univ. PR 75:353-361.
Shaik, M. 1984. Race-nonspecific resistance in bean cultivars to races of *Uromyces appendiculatus* pv. *appendiculatus* and its correlation with leaf epidermal characteristics. Phytopathology 75:478-481.
Singh, S.P. 1982a. Alternative methods to backcross breeding. Annu. Rpt. Bean Improv. Coop. 25:11-12.
Singh, S.P. 1982b. A key for identification of different growth habits of *Phaseolus vulgaris* L. Annu. Rpt. Bean Improv. Coop. 25:92-95.
Singh, S.P. 1989. Patterns of variation in cultivated common bean (*Phaseolus vulgaris*, Fabaceae). Econ. Bot. 43:39-57.
Singh, S.P. 1991. Breeding for seed yield. pp. 383-443. *In* A. van Schoonhoven & O. Voysest (eds.), Common Beans: Research for Crop Improvement. C.A.B. Int., Wallingford, U.K. & CIAT, Cali, Colombia.
Singh, S.P. 1992. Common bean improvement in the tropics. Plant. Breed. Rev. 10:199-269.
Singh, S.P. 1994. Gamete selection for simultaneous improvement of multiple traits in common bean. Crop Sci. 34:352-355.
Singh, S.P. 1995a. Selection for seed yield in Middle American versus Andean x Middle American interracial common-bean populations. Plant Breed. 114:269-271.
Singh, S.P. 1995b. Selection for water-stress tolerance in interracial populations of common bean. Crop Sci. 35:118-124.
Singh, S.P. 1998. Uso de marcadores y selección de gametos para el mejoramiento simultáneo de caracteres múltiples de frijol (*Phaseolus vulgaris* L.) para Mesoamérica y el Caribe. Agron. Mesoam. 9: 1-9.
Singh, S.P., C. Cajiao, J.A. Gutiérrez, J. García, M.A. Pastor-Corrales & F.J. Morales. 1989. Selection for seed yield in inter-gene pool crosses of common bean. Crop Sci. 29:1126-1131.
Singh, S.P., C. Cardona, F.J. Morales, M.A. Pastor-Corrales & O. Voysest. 1998a. Gamete selection for upright carioca bean with resistance to five diseases and a leafhopper. Crop Sci. 38: 666-672.
Singh, S.P., D.G. Debouck & W.M. Roca. 1997a. Successful interspecific hybridization between *Phaseolus vulgaris* L. and *P. costaricensis* Freytag & Debouck. Annu. Rpt. Bean Improv. Coop. 40:40-41.
Singh, S.P., D.G. Debouck & W.M. Roca. 1998b. Interspecific hybridization between *Phaseolus vulgaris* L. and *P. parvifolius* Freytag. Annu. Rpt. Bean Improv. Coop. 41:7-8.
Singh, S.P., P. Gepts & D.G. Debouck. 1991a. Races of common bean (*Phaseolus vulgaris*, Fabaceae). Econ. Bot. 45:379-396.
Singh, S.P. & J.A. Gutiérrez. 1982. Sources of some architectural traits in dry bush beans *Phaseolus vulgaris* L. Annu. Rpt. Bean Improv. Coop. 25:6-8.
Singh, S.P. & J.A. Gutiérrez. 1984. Geographical distribution of DL_1 and DL_2 genes causing hybrid dwarfism in *Phaseolus vulgaris* L., their association with seed size, and their significance to breeding. Euphytica 33:337-345.

Singh, S.P., J.A. Gutiérrez, A. Molina, C. Urrea & P. Gepts. 1991b. Genetic diversity in cultivated common bean. II. Marker-based analyses of morphological and agronomic traits. Crop Sci. 31:23-29.

Singh, S.P. & A. Molina. 1996. Inheritance of crippled trifoliolate leaves occurring in interracial crosses of common bean and its relationship with hybrid dwarfism. J. Hered. 87:464-469.

Singh, S.P., A. Molina & P. Gepts. 1995. Potential of wild common bean for seed yield improvement of cultivars in the tropics. Can. J. Plant Sci. 75:807-813.

Singh, S.P., A. Molina, C.A. Urrea & J.A. Gutiérrez. 1993. Use of interracial hybridization in breeding the race Durango common bean. Can. J. Plant Sci. 73:785-793.

Singh, S.P. & C.G. Muñoz. 1999. Resistance to common bacterial blight among *Phaseolus* species and common bean improvement. Crop Sci. 39:80-89.

Singh, S.P., R. Nodari & P. Gepts. 1991c. Genetic diversity in cultivated common bean. I. Allozymes. Crop Sci. 31:19-23.

Singh, S.P., W.M. Roca & D.G. Debouck. 1997b. Ampliación de la base genética de los cultivares de frijol: hibridación interespecífica en especies de *Phaseolus*. pp. 9-19. *In* S.P. Singh & O. Voysest (eds.), Taller de Mejoramiento de Frijol para el Siglo XXI: Bases para Una Estrategia para América Latina. CIAT, Cali, Colombia.

Singh, S.P. & H. Terán. 1998. Population bulk versus F_1-derived family methods of yield testing in early generations of multiple-parent interracial and inter-gene pool crosses of common bean. Can J. Plant Sci. 78:417-421.

Singh, S.P., H. Terán, A. Molina & J.A. Gutiérrez. 1991d. Genetics of seed yield and its components in common beans (*Phaseolus vulgaris* L.) of Andean origin. Plant Breed. 107:254-257.

Singh, S.P., H. Terán, A. Molina & J.A. Gutiérrez. 1992. Combining ability for seed yield and its components in comon bean of Andean origin. Crop Sci. 32:81-84.

Singh, S.P., H. Terán, C.G. Muñoz & J.C. Takegami. 1999. Two cycles of recurrent selection for seed yield in common bean. Crop Sci. 39:391-397.

Singh, S.P. & C. Urrea. 1995. Inter-and intraracial hybridization and selection for seed yield in early generations of common bean, *Phaseolus vulgaris* L. Euphytica 81:131-137.

Smartt, J. 1969. Evolution of American *Phaseolus* beans under domestication. pp. 451-461. *In* P.J. Ucko & G.W. Dimbleby (eds.), The Domestication and Exploitation of Plants and Animals. Duckworth, London, U.K.

Smartt, J. 1988. Morphological, physiological and biochemical changes in *Phaseolus* beans under domestication. pp. 143-161. *In* P. Gepts (ed.), Genetic Resources of *Phaseolus* Beans. Kluwer, Dordrecht, Netherlands.

Stavely, J.R. 1984. Pathogenic specialization in *Uromyces phaseoli* in the United States and rust resistance in beans. Plant Dis. 68:95-99.

Stavely, J.R. 1990. Genetics of rust resistance in *Phaseolus vulgaris* plant introduction PI 181996. Phytopathology 80: 1056.

Sullivan, J.G. & F.A. Bliss. 1983. Recurrent mass selection for increased seed yield and seed protein percentage in the common bean (*Phaseolus vulgaris* L.) using a selection index. J. Amer. Soc. Hort. Sci. 108:42-46.

Temple, S.R. & F.J. Morales. 1986. Linkage of dominant hypersensitive resistance to bean common mosaic virus to seed color in *Phaseolus vulgaris* L. Euphytica 35:331-333.

Thomas, C.V., R.M. Manshardt & J.G. Waines. 1983. Teparies as a source of useful traits for improving common beans (*Phaseolus acutifolius* hybridization with *Phaseolus vulgaris*). Desert Plants 5:43.48.

Thomas, C.V. & J.G. Waines. 1984. Fertile backcross and allotetraploid plants from crosses between tepary beans and common beans. J. Hered. 75:93-98.
Thung, M.D.T. & I.P. Oliveira. 1998. Problemas Abioticos que Afectam a Producão do Feijoeiro e seus Metodós de Controle. EMBRAPA-CNPAF, Santo Antonio de Goias, Brazil.
Urrea, C.A., P.N. Miklas, J.S. Beaver & R.H. Riley. 1996. A codominant randomly amplified polymorphic DNA (RAPD) marker useful for indirect selection of bean golden mosaic virus resistance in common bean. J. Amer. Soc. Hort. Sci. 121:1035-1039.
Urrea, C.A. & S.P. Singh. 1994. Comparison of mass, F_2-derived family, and single-seed-descent selection methods in an interracial population of common bean. Can. J. Plant Sci. 74:461-464.
Urrea, C.A. & S.P. Singh. 1995. Comparison of recurrent and congruity backcrossing for interracial hybridization in common bean. Euphytica 81:21-26.
Vélez, J.J., M.J. Bassett, J.S. Beaver & A. Molina. 1998. Inheritance of resistance to bean golden mosaic virus in common bean. J. Amer. Soc. Hort. Sci. 123:628-631.
Voysest, O. 1983. Variedades de frijol en América Latina y su origén. CIAT, Cali, Colombia.
Voysest, O. & M. Dessert. 1991. Bean cultivars: classes and commercial seed types. pp. 119-162. *In* A. van Schoonhoven & O. Voysest (eds.), Common Beans: Research for Crop Improvement. C.A.B. Int., Wallingford, U.K. & CIAT, Cali, Colombia.
Voysest, O., M.C. Valencia & M.C. Amézquita. 1994. Genetic diversity among Latin American Andean and Mesoamerican common bean cultivars. Crop Sci. 34:1100-1110.
Wall, J.R. 1970. Experimental introgression in the genus *Phaseolus*. 1. Effect of mating systems on interspecific gene flow. Evolution 24:356-366.
Wallace, D.H., J.P. Baudoin, J.S. Beaver, D.P. Coyne, D.E. Halseth, P.N. Masaya, H.M. Munger, J.R. Myers, M. Silbernagel, K.S. Yourstone & R.W. Zobel. 1993. Improving efficiency of breeding for higher crop yield. Theor. Appl. Genet. 86:27-40.
Weiseth, G. 1954. Una variedad silvestre del poroto común (*Phaseolus vulgaris*), autóctona del Noroeste Argentino y su relación genética con variedades cultivadas. Rev. Agron. Noroeste Argentino 1:71-81.
Welsh, W., W. Bushuk, W. Roca & S.P. Singh. 1995. Characterization of agronomic traits and markers of recombinant inbred lines from intra- and interracial populations of *Phaseolus vulgaris* L. Theor. Appl. Genet. 91:169-177.
White, J.W., J.A. Castillo, J.R. Ehleringer, J.A. García-C. & S.P. Singh. 1994a. Relations of carbon isotope discrimination and other physiological traits to yield in common bean (*Phaseolus vulgaris*) under rainfed conditions. J. Agric. Sci. (Cambridge) 122:275-284.
White, J.W. & A. González. 1990. Characterization of the negative association between seed yield and seed size among genotypes of common bean. Field Crops Res. 23:159-175.
White, J.W. & D.R. Laing. 1989. Photoperiod response of flowering in diverse genotypes of common bean (*Phaseolus vulgaris*). Field Crops Res. 22:113-128.
White, J.W., R. Ochoa M., F.J. Ibarra-Pérez & S.P. Singh. 1994b. Inheritance of seed yield, maturity and seed weight of common bean (*Phaseolus vulgaris*) under semi-arid rainfed conditions. J. Agric. Sci. (Cambridge) 122:265-273.
White, J.W. & S.P. Singh. 1991. Breeding for adaptation to drought. pp. 501-560. *In* A. van Schoonhoven & O. Voysest (eds.), Common Beans: Research for Crop Improvement. C.A.B. Int., Wallingford, U.K. & CIAT, Cali, Colombia.
White, J.W., S.P. Singh, C. Pino, M.J. Ríos & I. Buddenhagen. 1992. Effect of seed size and photoperiod response on crop growth and yield of common bean. Field Crops Res. 28:295-307.

Wilkinson, R.E. 1983. Incorporation of *Phaseolus coccineus* germplasm may facilitate production of high yielding *P. vulgaris* lines. Annu. Rpt. Bean Improv. Coop. 26:28-29.

Wortmann, C.S., R.A. Kirkby, C.A. Eledu & D.J. Allen. 1998. Atlas of Common Bean (*Phaseolus vulgaris* L.) Production in Africa. CIAT, Cali, Colombia.

Yoshii, K., G.E. Gálvez & G. Alvarez. 1976. Estimation of yield losses in beans caused by common blight. Proc. Amer. Phytopathol. Soc. 3:298-299 (abstract).

Zapata, M., G.F. Freytag & R.E. Wilkinson. 1985. Evaluation for bacterial blight resistance in beans. Phytopathology 75:1032-1039.

Zaumeyer, W.J. & H.R. Thomas. 1957. A monographic study of bean diseases and methods of their control. USDA Tech. Bull. 868. US Govrnt. Print. Off., Washington DC, U.S.A.

Chapter 7

Breeding to Improve Plant Type

Albert Vandenberg & Thandiwe Nleya
Crop Development Centre, University Of Saskatchewan, 51 Campus Drive, Saskatoon, Saskatchewan, Canada, S7N 5A8

Key words: breeding pyramid, common bean, ideotype breeding, lentil, mechanized harvesting, pea, plant type modification, pulse crops, soybean

1. INTRODUCTION

Common bean (*Phaseolus vulgaris* L.) is the most widely cultivated species in a group of more than 60 legume species referred to as pulse crops. All pulse crops have gone through a process of agricultural domestication and adaptation to farming systems. Early crop improvements were made by selecting desirable plant and seed types. Over time, a rich diversity of landraces evolved for most pulses. In the early part of this century, private and public plant breeding programs were established for most of the economically important pulse crops.

Many pulses are now grown in regions far from the geographical origins of the crop. Common bean originated in Andean South and Middle America, but is now cultivated worldwide in diverse environments. Pulse crop producers often have sought new plant types to suit local needs. In the case of common bean, a rich diversity of growth habit was available to suit various farming systems (Evans, 1973a; Gepts & Debouck, 1991). Climbing types were selected for co-cultivation with maize (*Zea mays* L.), prostrate indeterminate types were selected for adaptation to semiarid environments with variable moisture, and upright determinate and indeterminate types were selected for cultivation in humid tropical and subtropical environments.

Common bean has been introduced into diverse production regions at much higher latitudes than the centers of origin. The process of matching growth habit to changing environment, economy, and technology is a challenge for all scientists involved in common bean improvement. Plant breeders must anticipate and adapt to changes that occur in specific production areas based on economic and technological trends.

The introduction of large-scale mechanized farming systems in the past several decades accelerated the development of new plant types in all major field crops. The world's major cereal crops, such as rice (*Oryza sativa* L.), wheat (*Triticum aestivum* L.), maize, and sorghum [*Sorghum bicolor* (L.) Moench], have all undergone modification of plant type through plant breeding to make them more suitable for highly mechanized farming systems. For example, long-stemmed wheat cultivars, originally grown for a combination of uses as grain, forage, and straw, are now largely replaced by short and stiff-stemmed types designed to suit intensive production and direct combining systems. Similarly, dwarf sorghum cultivars were developed for mechanized farming systems. Agricultural economies are never static, and the application of scientific plant breeding methods to effect changes in plant type to accommodate changes in farming systems is an accelerated version of the historical domestication process for all crop plants.

Donald (1968) formally proposed the concept of ideotype plant breeding to describe the efforts that were underway in cereal breeding at that time. Ideotype breeding differs from traditional breeding in that breeders set out to modify individual plant traits to increase yield and/or adapt plants to specific agronomic production systems. An attempt is made to provide a physiological rationale for breeding objectives. The choice of specific plant traits for modification depends on environmental conditions and the production system targeted for the crop. This approach has been used successfully to improve many crop plants, including common bean (Acquaah et al., 1991; Adams, 1973, 1982; Brothers & Kelly, 1993; Kelly & Adams, 1987; Nienhuis & Singh, 1985; Singh & Gutiérrez, 1982). The demand for new plant types increases with the need to adapt plants to new production environments with different climates and different management technologies. In some cases, economic or technological changes in farming systems may occur at a faster pace than changes in breeding methodology or in the biological capacity to introduce changes through plant breeding.

Historical efforts to modify plant type and canopy structure in other pulse crops and in common bean are reviewed in this chapter. Following that, consideration is given to how plant type modification could be approached for various common bean production systems. A fundamental concept is that efforts to improve plant type must involve clear objectives based on clear understanding of the agronomic, technologic, climatic, and economic environment of the production region.

2. BREEDING FOR IMPROVED PLANT TYPES IN GRAIN LEGUMES

2.1 Soybean

Breeders have altered plant growth habit in many crops. The two outstanding examples among the grain legumes are the soybean [*Glycine max* (L.) Merr.] and the field pea (*Pisum sativum* L.). The soybean was originally introduced to North America from Asia as a forage crop. As its potential as a grain for oil and protein was exploited, the production area expanded to large portions of North and South America. Modification of the soybean plant and canopy characteristics occurred at the same time that other technological innovations in production were developed. For instance, the availability of herbicides made it possible to replace row cropping with solid-seeded production systems. This introduced more flexibility into the production system, allowing the crop to expand into new production regions.

Cooper (1975) outlined approaches by which breeders could modify morphological and physiological characteristics of soybean to maximize yield. Breeding efforts focused on developing short-statured early maturing soybeans to suit specific midwestern U.S.A. cropping systems (Cooper, 1981; Walker & Cooper, 1982). Developing and introducing new plant types for specific production systems may create constraints in plant breeding systems during the phase-in period of a new set of breeding objectives. For soybean, breeding methods involving early generation testing were developed to increase efficiency so that more crosses could be screened annually (Cooper, 1990).

Soybean is now a valuable crop for the high latitude cooler regions of northern Europe, the northern U.S.A., and Canada. In the 1960s and 1970s, major efforts in research went into adapting soybean to high latitude areas. Initially the major limiting factor for soybean production in these areas was considered to be photoperiod sensitivity. The short-day requirement was eliminated by identifying daylength-insensitive genotypes.

Low temperature remains a major problem for soybean production in cool regions of Europe and Canada. Research conducted in Switzerland (Soldati & Keller, 1984) revealed a strong influence of climatic conditions on yield and yield components of soybean. Cool temperatures lengthen the vegetative period and increase the number of nodes and branches developed per plant. Flower abortion rate is much higher in soybean grown at cool temperatures, reducing the yield relative to soybean growing in warm conditions. Poor partitioning of assimilates to the grain seems to be a major problem in cool areas and is apparently related to interaction between temperature and photoperiod (Soldati & Keller, 1984). Good germination and growth under cold temperature conditions, tolerance to cold stress before and during flowering, and high photosynthetic activity under moderate temperatures for maximum grain filling were identified as important characteristics of soybean cultivars adapted to cool regions.

Development of extra-early cultivars has allowed production of the soybean crop to move to northern latitudes. Jingling & Fanglan (1984) developed super-early soybean cultivars from crosses between two early maturing cultivars. They noted that using extremely early parents from different geographical and ecological environments increased the chance of obtaining offspring earlier than their parents and that transgressive segregation is a common phenomenon in such soybean hybrid populations.

Breeders have developed soybean cultivars with semi-determinate and determinate stem termination. The determinate growth habit is important in reducing lodging (Cober & Tanner, 1995) and is suited to the narrow- row direct harvest production systems (Ablett et al., 1991). The problem of excessive dwarfing in early maturing determinate cultivars was overcome by breeding for high yielding, tall determinate lines with high pod clearance to suit mechanized harvesting.

In addition to breeding for more upright plant types, considerable research effort was devoted to designing harvest equipment specific to soybean. Bichel & Hengen (1978) reported that more than 90 patents on devices for reducing soybean-gathering losses were granted, though most of them proved ineffective. Some successful innovations were the row-crop head (Bichel & Hengen, 1978), the integral flexible cutterbar cutting platform (Bichel et al., 1976), and the narrow-pitch combine cutterbar (Quick & Mills, 1978). Bichel & Hengen (1978) concluded that reducing harvest losses in soybean to within a 2-4% range required specifically designed equipment for the soybean crop in combination with improved cultivars more suitable for direct harvesting.

2.2 Pea

In dry pea, plant breeders have attempted to increase yield and improve yield stability by developing new plant types (Heath et al., 1994). The major change in plant architecture was the development of the semi-leafless pea, with leaflets replaced by tendrils. This morphological change is due to expression of the single recessive *af* (afila) gene. While the crop canopy of semi-leafless pea is as efficient photosynthetically as conventional pea (Heath et al., 1994), semi-leafless types have some agronomic advantages. Successful direct harvesting of pea requires that the cultivars have good standing ability; otherwise yield losses during combining are high. The major advantage of the semi-leafless growth habit is improved standing ability. The tendrils bind stems together, reducing lodging. The more open plant canopy with reduced foliage alters the canopy microclimate of the crop, leading to reduced humidity and reduced disease incidence. The yield advantage of the new plant types is due to better standing ability and easier harvesting. The more open canopy of semi-leafless types is a disadvantage from the standpoint of weed control. The yield potential of these new plant types was achieved through development of better herbicides.

Other modifications to plant type in field pea include development of determinate growth habit and improved stem strength. In production systems for which herbicides were not available, the traditional sprawling indeterminate growth habit was helpful for weed control but impeded disease control. The determinate growth habit, combined with herbicide use, slows the development of foliar diseases in the lower canopy during wet periods. Lodging tolerance in field pea is determined by a combination of genetic and environmental factors, including stem strength, internode length, and lower canopy disease potential. Steady improvement in lodging tolerance through breeding has occurred over the past twenty years, although the genetic basis for the improvement is not clearly understood.

Improvements in canopy structure and the associated benefits in agronomic systems were fundamental to the rapid expansion of field pea production in Europe and western Canada over the past 20 years. Improvements in lodging tolerance reduced the need to purchase specialized harvest equipment. This change paralleled the expansion of soybean production in the 1960s and 1970s, though on a smaller scale. In western Canada, the area devoted to field pea production expanded from 50,000 ha to more than 1,000,000 ha over the past 10 years. As yield increased and production became easier, the volume of field pea production increased while the cost of production declined. This change resulted in the expansion of field pea use in compound animal feeds and in the expansion of markets for human consumption. Then further expansion and the realization of rotational benefits of pulse crop production occurred.

The successful plant breeding efforts for canopy modification for soybean and field pea are in part due to availability of a wide range of morphology, adaptation, and germplasm. The genetic basis of many morphological and physiological traits is well understood for these crops. In this situation, breeders can target specific changes, develop clear objectives, and actively pursue those objectives based on numerous breeding strategies that include specific morphological traits and single plant selections.

2.3 Lentil

In the case of lentil (*Lens culinaris* L.), potential improvements in canopy structure are less obvious and single plant observations and selections are difficult and probably ineffective. The lentil plant is short, highly indeterminate, and branches profusely. In this case, a passive improvement strategy for canopy modification is used. The basic concept is to ensure that the agronomic techniques used in field screening closely resemble the techniques used in commercial production. At the Crop Development Centre at the University of Saskatchewan, lentil plots are desiccated at physiological maturity and then mechanically harvested starting at the F_3 generation. In this way, indirect selection occurs for reduced shattering, increased canopy height, improved stem strength, and erect branching. High yield is the major selection criterion, but selection occurs within specific commercial seed classes that are based on seed shape, cotyledon color, and seedcoat color and pattern. More than 10,000 F_3 plots are planted each year. Visual selection of F_2-derived families is used to eliminate poor crosses on the basis of canopy architecture and disease resistance. This relatively inexpensive and effective method for developing improvements in canopy structure in lentil has been used since 1991.

3. PREVIOUS EXPERIENCES WITH NEW PLANT TYPES IN COMMON BEAN

The domestication process in common bean led to new plant types (Evans, 1973a, b; Gepts & Debouck, 1991). The wild species were climbers that, over time, were developed into determinate bush, upright indeterminate, and prostrate vine types. One of the earliest examples of a technological attempt to improve plant type in common bean was the development of the determinate navy bean Sanilac from an X-ray mutation breeding program in Michigan. The determinate trait was subsequently used in other bean breeding programs in North America. After reviewing literature on architectural and physiological features important to yield in grain legumes, Adams (1973, 1982) proposed breeding for an ideotype in common bean. The main features of the ideotype were: 1) a narrow plant profile, 2) thick main stem, 3) indeterminate growth habit, 4) tall (50-55 cm) with a large number of main stem nodes (12 to 15), 5) leaf area index of four at flowering, and 6) high rate of seed-filling. The concept was used to develop a bean ideotype using traditional breeding methods and mutation techniques. The success of the ideotype breeding concept in common bean was first realized in the navy commercial class and later in the black, cream, cream stripe, pinto, purple mottled, and red market classes (Adams, 1982; Kelly & Adams, 1987; Nienhuis & Singh, 1985; Singh & Gutiérrez, 1982; Singh et al., 1998), both from Andean and Middle American gene pools. Besides high yield, the other main advantage of the new plant type was the erect growth habit suited for mechanized harvesting. The success of the ideotype concept in the small-seeded beans prompted the adoption of this breeding approach in other market classes of bean (Coyne et al., 1991; Grafton et al., 1997; Kelly & Adams, 1987; Kelly et al., 1990, 1992; Singh & Gutiérrez, 1982).

4. THE NEED TO DEVELOP NEW PLANT TYPES

4.1 Introduction

If the common bean is to remain an economically viable crop, particularly in regions with high land prices, the cost of production must be decreased or local production will decline. This trend has occurred in pulse production regions everywhere in the world. When land and labor values rise, production shifts to easily mechanized cereal crops, to higher value horticultural crops, or to crops that can be processed industrially. In the case of common bean, the production of climbing types is declining in the Andean and Middle American highlands. An economic shift is taking place, and large scale mechanized production is increasing in Argentina, Bolivia, and Brazil.

Based on the historical experiences with most grain legumes, the underlying reason for modifying plant type is to facilitate mechanization of harvest. Breeding efforts for all crops must take account of the production system. In traditional production systems, which use a low level of mechanization and harvest by hand, plant type is not a problem except in cases where the morphology of the plant may have a negative association with productivity (Nienhuis & Singh, 1985). Culinary traits, absolute yield, and disease resistance may often be the main concern in traditional production systems. In regions where the price of land used for common bean production has become very expensive, for example in eastern North America or the Central Valley of Chile, the bean crop is in decline. The main reason for this is that the price of land has risen faster than the potential return from the bean crop in comparison to alternatives. Lower cost mechanized production systems that replace traditional row cropping systems may become predominant in these areas, but this might be only a temporary adjustment.

Canopy architecture is a major consideration in breeding common bean for direct harvesting systems. In the northern temperate zone, environmental conditions may influence growth habit to the extent that indeterminate types (type II and type III) develop short lower internodes, causing pods to hang close to the ground. The type III growth habit in particular is characterized by a canopy in which pods are produced on the lower half of the stem (Kelly & Adams, 1987). Losses during direct harvest can be unacceptably high if type III cultivars are grown (Vandenberg et al., 1994).

4.2 The Humid Midwest U.S.A. and Southeastern Canada

The pinto and great northern bean classes are favored for production in the Midwest U.S.A. However, the prostrate indeterminate growth habit (type III) increases the risk of losing yield to foliar diseases and increased harvest losses. The challenge was combining the architectural characteristics of the small-seeded ideotype with the seed size, shape, and color of the pinto bean (Brothers & Kelly, 1993). The pinto class belongs to race Durango (medium-seeded) while the small-seeded types are race Mesoamerica (Singh et al., 1991). The difficulty of transferring upright characteristics from small to medium seeded pinto germplasm lead Kelly & Adams (1987) to use a system of phenotypic recurrent selection to recombine desirable traits from the two diverse germplasm pools. Due to yield component compensation, the pinto bean ideotype that was developed had fewer seeds per pod and fewer pods per plant than the navy bean ideotype. Cultivars with improved plant type are now available in pinto (Grafton et al., 1997; Kelly et al., 1990), great northern (Coyne et al., 1991; Kelly et al., 1992), and other medium-seeded commercial classes.

The research described in the previous section resulted in the development of specific plant types to suit bean production systems in the humid Midwest U.S.A. These new plant types are not universally adaptable. Different plant types may be required for different agronomic and climatic conditions. For example, the emphasis on taller, more erect, narrower plant profile, and fewer basal branches is largely based on the requirement for bean production under humid conditions and direct harvest systems. In the non-mechanized bean production systems in semiarid tropical and sub-tropical areas highly indeterminate sprawling plant types may be required. In such environments, indeterminate type III growth habit typical of the Durango race may be beneficial to yield stability since the risk of disease is much lower under arid conditions. The indeterminate growth habit has been shown to be more productive under unpredictable environmental conditions (Kelly et al., 1987; Nleya et al., 1999) typical of semiarid areas. Race

Durango is considered an important source of germplasm for breeders intending to develop improved high yielding indeterminate cultivars for semiarid tropics. This gene pool is a good source of early maturity, resistance to drought, high harvest index, and seed yield (Singh, 1989; Singh et al., 1991).

4.3 Semiarid North America

As bean production moves into the semiarid regions typical of the prairies of North America west of the Red River Valley, new plant types may be required. In this production environment, emphasis is placed on developing early maturing, high yielding, erect plant types suitable for direct harvesting systems. The frost-free growing season is short (typically 100 days) and is characterized by cool nights during the beginning and end of the season. It is important to develop bean cultivars that maximize yield and maturity within this growing season. Source limitation can be critical in early maturing materials and therefore rapid establishment of leaf area should be sought. Since duration of pod filling is positively associated with yield (White & Izqueirdo, 1991), cultivars maximizing duration of pod filling within the growth cycle of early cultivars should be developed. Common bean cultivars for northern temperate zones should have the following characteristics: 1) early maturity, 2) rapid germination and good growth in cool temperatures, 3) rapid establishment of leaf area, 4) high photosynthetic activity under moderate temperatures, 5) long period of pod filling within the growth cycle of early cultivars, 6) erect plant type with indeterminate growth habit, and 7) consistent development of a canopy in which the pods are held high. A major abiotic constraint in the northern prairie region is low temperature during sowing and harvesting periods. Quick emergence in cool spring soils is a desirable trait. The ideal soil temperature for germination of common bean is 18°C. Seeding when soil temperature is too far below this level results in slow emergence and weak plants. Even more desirable would be cold soil tolerance combined with some degree of frost tolerance.

4.4 Irrigated Pacific Northwest U.S.A.

The irrigated production regions in the U.S.A. Pacific Northwest present an interesting case. Commercial bean production is in decline because the crop is less profitable than processing vegetables or potatoes. Bean cultivars with later maturity may be associated with high yield potential in some environments (Welsh et al., 1995; White, 1981; White & Singh, 1991). Research also shows that at higher latitudes, late maturity reduces yield because of poor adaptation (Wallace, 1985). Earlier maturing cultivars may in fact be more profitable in the Pacific Northwest region if the bean crop can fit into double cropping systems where they can be seeded late following a crop of processing vegetables. Some bean producers in this region now grow two crops per year and are interested in low cost bean production to compensate for the higher risk associated with late seeding. One of the ways to cut costs is to eliminate row cropping.

4.5 The Latin American Tropics and Subtropics

Some regions in southern Brazil are now growing irrigated common bean in winter using direct seeding with mechanized direct harvest systems (Thung, 1997). Northwestern Argentina also has developed direct harvest systems for common bean (Medina & de Calva, 1997). In these long season areas where frost risk is lower, it may be possible to achieve high yield potential in determinate plants by increasing the number of nodes per plant. Combined with improvements in stem strength, it may be possible to extend the pod filling period in comparison to the traditional type I growth habit. Evans (1973a, b) reported that some tropical determinate types from Central America may produce 7-15 nodes. Another source of higher node number may be the climbing determinate types (Debouck et al., 1988). In humid regions and under irrigation, root rots and lower canopy foliar diseases will eventually limit production, so improvements in plant type in these regions could extend to improvement in root systems and canopy structures. Stem strength may play a role, and the case of field pea suggests that major improvements in stem morphology and strength are possible with recurrent selection.

5. STRATEGIES FOR DEVELOPMENT OF NEW PLANT TYPES

Bean breeding programs strive to develop improved cultivars through the most efficient use of available resources. The specific crossing and selection programs are influenced by resource availability and also by local experience and previous training. In an ideal situation, most bean breeding programs would benefit from implementing the breeding pyramid approach suggested by Kelly et al. (1997, 1998; see also Chapter 8) to increase yield. A germplasm development hierarchy with three levels is proposed. In this scheme, the top of the pyramid is the genetic pool from which cultivars are developed using elite x elite crosses within market classes. The intermediate level is focussed on germplasm development using adapted genotypes within gene pools but with no emphasis on market class or growth habit. The objective is to develop diversity and select the top lines for use as parents in the elite x elite crosses for cultivar development. The base of the pyramid represents the germplasm that involves interspecific and intraspecific broad-based crosses from which lines will eventually be selected for use in the intermediate level.

This conceptual model for breeding can be adapted more specifically for development of new plant types in common bean. In this case the overall objective is to improve the germplasm base for traits that could lead to improvements in canopy structure for direct harvesting systems while maximizing yield. Breeding programs could focus on development of germplasm that consistently develops a canopy in which all the pods are held more than 40 mm above the ground surface. This is an acceptable level at which the cutterbar floats above the soil surface for a floating flexible combine header (Zyla, 1993). Germplasm suitable for direct harvest systems could be developed by introducing parents that are capable of contributing genetic enhancement of pod distribution in the overall plant canopy. Harvestable economic yield would eventually become the selection criterion for development of new cultivars.

Traits that might optimize canopy structure at harvest could be: 1) long internodes in the lower stem, 2) consistent internode elongation under a wide range of environmental conditions, 3) reduction of stem stunting during early season growth, 4) increased stem length, 5) increased stem strength, particularly in the more basal internodes, 6) reduction of pod length without decreasing seed size, 7) increase in pod curvature so that pod tips do not extend below the cutterbar, 8) long upright peduncles, 9) commencement of flowering at the upper nodes, 10) high fertility at the upper nodes, and 11) sufficient number of main stem nodes to maximize productivity in the available growing season.

Genetic studies have been conducted on many of these individual traits. Singh (1982) classified growth habit on the basis of terminal bud growth, length of guide, stem strength, and fruiting pattern. A review of genetics of stem characters by Singh (1991) showed that various single, double or polygenic systems are reported to control terminal growth, climbing ability, plant height (stem length), height of pod formation, main stem internode length, hypocotyl and basal internode diameter, number of branches, and length of inflorescence. The same review revealed that numerous studies have reported on genetic systems that control pod number, pod length, pod curvature, flowering, and maturity. Expression of many of these developmental characteristics may also be influenced by complex daylength and temperature interactions (Coyne, 1966; Masaya & Wallace, 1984;

Masaya & White, 1986). Developmental responses to environment may be influenced by plant hormones. For example, gibberellic acid mutants are documented in many crop plants. Mislevy et al. (1989) used gibberellic acid application to attempt to control stem and pod position traits in soybean. It may be possible to select for specific genetic combinations that affect gibberellic acid physiology by selecting for consistency in stem and pod canopy character expression over a wide range of environments.

This list of traits useful for development of stable new plant types can be amended to suit the needs of local breeding programs. These needs include specific tolerances to soil conditions or other environmental influences that affect canopy structure. At this base level, genotypes could be introduced regardless of gene pool. The main interest for including genotypes would be specific traits that might contribute to long term improvement. Challenges at this level would include genetic barriers caused by wide hybridization and poor adaptation. By combining diverse sources of these traits, it might be possible to exploit diverse germplasm to develop unique combinations for use at the intermediate level, and eventually, the apex level. Selection of improved stable genotypes would form the basis for further recombination or potential advancement to the intermediate level.

At the intermediate level, the recombination and selection procedure would focus on development of stable improved canopy structure within an adapted background, and if possible, separation into market classes. Very large and diverse F_2 nurseries may be required to reveal new plant types. At this level, it is feasible to introduce mechanical harvesting in narrow rows as a selection tool at the F_3 level. Spaced F_2 plants produce sufficient seed for an F_3 microplot of three rows up to 2 meters in length. If sufficient land is available, it is possible to seed 10,000 microplots in two days. Pre-harvest visual selection can be used to eliminate families with poor plant type. A small plot combine can be used to direct harvest up to 1000 microplots per day. The microplots with the highest harvestable yield will produce sufficient seed for replicated preliminary yield trials or re-introduction into spaced plant nurseries. Further selection of lines can be based on high harvestable yield. This procedure will automatically integrate selection for consistent adaptation and consistent expression of improved canopy structure. Some selections may be useful as parents in the apex of the pyramid where the objectives would become more specific for seed type and disease resistance. Development of new plant types can be relatively easily focussed on introduction of desirable architectural types. The greatest challenge occurs in production regions where the crop is poorly adapted.

The concept of harvestable yield is critical for the development of new plant types because harvestability is linked to both agronomy and harvest technology. Bean breeders may have to introduce new harvest technologies as soon as possible into breeding programs in order to make selection progress using the new technology. Agronomy experiments investigating row spacing x genotype interaction often show that some genotypes have a positive yield response to narrow row spacing (Grafton et al., 1988). The same type of interactions should exist for genotype by harvest system.

When upright navy bean cultivars became available in Michigan, research into harvesting with six-bat reel with pickup finger-type guards was initiated (McColly, 1958). Losses were greater than 10%. Gunkel & Anstee (1962) developed a row crop header consisting of rubber fingers attached to the v-belt for stripping pods onto an elevating canvas. The technology reduced losses for the upright plant types but was unsuitable for type III cultivars. High shatter losses were recorded. Gunkel & Anstee (1962) tested two other harvesting systems. A spiral cone-shaped brush system for pulling pods and plants and a flat belt puller system that clamped onto ripe plants in the row were tested. These were not successful commercially and virtually no research in engineering new harvest systems for common bean took place for almost 30 years following this attempt. In the past ten years, harvest system research has increased and some systems have been adapted from soybean and lentil production, particularly in the navy class where much of the germplasm is now erect.

The current challenge in harvest technology is to develop harvest systems better suited to indeterminate high yielding germplasm of race Durango. New lifter and reel systems are under development in western Canada. The objective of the brush lifter system is to lift the basal pod clusters out of the way of the knife of the combine or swather (B. Reed, R. L. Kushwaha & L. Zyla, unpublished). This type of harvest technology may be best suited to type I and II cultivars for which pod clearance above the cutterbar is most likely to be consistent. Another harvesting procedure developed to the commercial trial level uses curved reel fingers and an eccentric reel orbit to lift the canopy just prior to cutting the stems. This harvest technology shows particular promise for cultivars with indeterminate type II and III growth habit. If these emerging harvest technologies are successfully introduced at the commercial level, bean breeders should consider incorporating the most promising harvest technologies into screening and selection programs in order to maximize economic yield potential for common bean. By keeping pace with harvest technology, breeders will be able to select the most appropriate plant types for common bean production.

6. CHALLENGES AND FUTURE PROSPECTS

Improvement of plant type may be highly site specific. Adaptation to daylength and temperature may set the gross limits to potential productivity of common bean in a given agronomic and economic environment. In short season areas, it may not be possible to increase the number of nodes on the main stem without adding to risk of late season frost injury. From a physiological perspective, many soil and plant factors may influence final productivity. From an economic perspective, net profit may be higher if lower cost harvest systems can be used successfully for cultivars with improved plant type. The frequency of successful profitable production ultimately determines whether or not common bean remains a viable crop in a given geographic region. The successful production of common bean depends as much on available technology, local market structure, competition from other crops, and competition from other economic sectors as it does on genetic characteristics. Breeding programs must be designed to fit the current and future production system of the region in order for the crop to remain economically viable in the long term. The specific plant characteristics will vary greatly depending on how the crop fits into local cropping patterns.

One of the most difficult challenges may be developing improved high yielding plant types for large-seeded beans of Andean origin with type I growth habit. Most of the useful germplasm for introducing higher productivity while maintaining harvestability (type II growth habit) is derived from small-seeded beans of Middle America. This represents a large genetic barrier.

Using genes from other species to influence plant type in common bean is a possible future strategy. Use of transgenic technology, however, will be possible only if repeatable transformation protocols became accessible at reasonable costs to breeders, and if their use does not interfere with marketing of beans from the standpoint of end user perception of genetic modification or transgenics.

Common bean will remain an important crop worldwide for the foreseeable future. Indeterminate types will continue to predominate in subsistence farming systems, and in semiarid tropical and temperate environments. In these environments the challenge will be to develop high yielding, well-adapted cultivars with seed types suited to local demands. Drought tolerance and plasticity are important traits under semiarid environments.

As was the case with soybean and pea, the greatest changes in plant types in the bean crop will occur in temperate regions. In these regions major efforts are already underway to develop more upright bean cultivars suited for mechanized harvesting systems. In the more northern latitudes of western Canada where the growing season is extremely short, the upright growth habit will need to be combined with early maturity.

REFERENCES

Ablett, G.R., W.D. Beversdorf & V.A. Dirks. 1991. Row width and seeding rate performance of indeterminate, semideterminate and determinate soybean. J. Prod. Agric. 4:391-395.
Adams, M.W. 1973. Plant architecture and physiological efficiency. pp. 226-278. *In* D. Wall (ed.), Potential of Field Beans and Other Food Legumes in Latin America. CIAT, Cali, Colombia.
Adams, M.W. 1982. Plant architecture and yield breeding. Iowa State J. Res. 52: 225-254.
Acquaah, G., M.W. Adams & J.D. Kelly. 1991. Identification of effective indicators of erect plant architecture in dry beans. Crop Sci. 36:261-264.
Bichel, D.C. & E.J. Hengen. 1978. Development of soybean harvesting equipment in the U.S.A. pp. 200-208. *In* Grain and Forage Harvesting: Proc. Amer. Soc. Agric. Engineers, St. Joseph, U.S.A.
Bichel, D.C., E.J. Hengen & R.E. Mott. 1976. Designing the new concept header. Agricultural Engineering 57: 21-23.
Brothers, M.E. & J.D. Kelly. 1993. Interrelationship of plant architecture and yield components in the pinto bean ideotype. Crop Sci. 33: 1234-1238.
Cober, E.R. & J.W. Tanner. 1995. Performance of related indeterminate and tall determinate soybean lines in short-season areas. Crop Sci. 35: 361-364.
Cooper, R.L. 1975. Modifying morphological and physiological characteristics of soybeans to maximize yields. pp. 230-236. *In* D. Hill (ed.), Proceedings of the World Soybean Research Conference. Interstate Printers & Publishers, Danville, U.S.A.
Cooper, R.L. 1981. Development of short-statured soybean cultivars. Crop Sci. 21:127-131.
Cooper, R.L. 1990. Modified early generation testing procedure for yield selection in soybean. Crop Sci. 30: 417-419.
Coyne, D.P. 1966. The genetics of photoperiodism and the effect of temperature on the photoperiodic response for time to flowering in *Phaseolus vulgaris* L. varieties. Proc. Amer. Soc. Hort. Sci. 89:350-360.
Coyne, D.P., J.R. Steadman, D.T. Lindgren & D.S. Nuland. 1991. 'Starlight'-Great northern dry bean. HortScience 26:441-442.
Debouck, D.G., H. Rubiano, S. Menéndez & M. del Carmen. 1988. Determinate climbers among Argentinian materials of *Phaseolus vulgaris* L. Annu. Rpt. Bean Improv. Coop. 31: 189-190.
Donald, C.M. 1968. The breeding of crop ideotypes. Euphytica 17: 385-403.
Evans, A.M. 1973a. Commentary upon plant architecture and physiological efficiency in the field bean. pp.279-286. *In* D. Wall (ed.), Potential of Field Beans and other Food legumes in Latin America. CIAT, Cali, Colombia.
Evans, A.M. 1973b. Genetic improvement of *Phaseolus vulgaris*. pp. 107-115. *In* M. Milner (ed.), Nutritional Improvement of Food Legumes by Breeding: Proceedings of a symposium sponsored by Protein Advisory Group (PAG) held at the FAO, Rome, Italy.
Gepts, P. & D. Debouck. 1991. Origin, domestication and evolution of the common bean. pp.7-54. *In* A. van Schoonhoven & O. Voysest (eds.), Common Beans: Research for Crop Improvement. C.A.B. Int., Wallingford, U.K. & CIAT, Cali, Colombia.

Grafton, K.F., A.A. Schneiter & B.J. Nagle. 1988. Row spacing, plant population and genotype x row spacing interaction on yield and yield components of dry bean. Agron. J. 80: 631-634.

Grafton, K.F., J.R. Venette & K.C. Chang. 1997. Registration of 'Maverick' pinto bean. Crop Sci. 37:1672.

Gunkel, W.W. & L.L. Anstee. 1962. Direct harvesting of dry beans. Agricultural Engineering 43: 694-697.

Heath, M.C., C.J. Pilbeam, B.A. McKenzie & P.D. Hebblethwaite. 1994. Plant architecture, competitive ability and crop productivity in food legumes with particular emphasis on pea (*Pisum sativum* L.) and faba bean (*Vicia faba* L.). pp. 771-790. *In* F.J. Muehlbauer & W.J. Kaiser (ed), Expanding the Production and Use of Cool Season Food Legumes. Proc 2nd Int. Food Legume Res. Conf. Cairo, Egypt. Kluwer, Dordrecht, Netherlands.

Jingling, W. & G. Fanglan. 1984. Breeding super-early soybean cultivars. pp. 439-445. *In* R. Shibles (ed.), World Soybean Research Conference III: Proceedings. Westview Press, Boulder, U.S.A.

Kelly, J.D. & M.W. Adams. 1987. Phenotypic recurrent selection in ideotype breeding of pinto beans. Euphytica 36:69-80.

Kelly, J.D., M.W. Adams, A.W. Saettler, G.L. Hosfield, G.V. Varner, M.A. Uebersax & J. Taylor. 1990. Registration of 'Sierra' pinto bean. Crop Sci. 30:745-746.

Kelly, J.D., M.W. Adams & G.V.Varner. 1987. Yield stability of determinate and indeterminate dry bean cultivars. Theor. Appl. Genet. 74:516-521.

Kelly, J.D., G.S. Hosfield, G.V. Varner, M.A. Uebersax, P.N. Miklas & J. Taylor. 1992. Registration of 'Alpine' great northern bean. Crop Sci. 32:1509-1510.

Kelly, J.D., J.M. Kolkman & K.A. Schneider. 1997. Breeding for high yield in common bean. pp.187-204. *In* S.P. Singh & O. Voysest (eds.), Taller de Mejoramiento de Frijol para el Siglo XX1: Bases para Una Estrategia para America Latina. CIAT, Cali, Colombia.

Kelly, J.D., J.M. Kolkman & K. Schneider. 1998. Breeding for yield in dry bean (*Phaseolus vulgaris* L.). Euphytica 102:343-356.

Masaya, P. & D.H. Wallace. 1984. The effect of elevation (temperature) on number of days to flower and node of flowering in beans. Annu. Rpt. Bean Improv. Coop. 27:199-202.

Masaya, P. & J. White. 1986. Effects of short days on stem elongation in some indeterminate dry bean cultivars adapted to the tropics. Annu. Rpt. Bean Improv. Coop. 29:1-3.

McColly, H.F. 1958. Harvesting edible beans in Michigan. Transactions Amer. Soc. Agric. Engineers 1: 68-71.

Medina, S.G. & V.F. de Calva. 1997. Logros del programa nacional de investigacion en mejoramiento y tecnologia del cultivo de poroto de Argentina. pp.235-238. *In* S.P. Singh & O. Voysest (eds.), Taller de Mejoramiento de Frijol para el Siglo XX1: Bases para Una Estrategia para America Latina. CIAT, Cali, Colombia.

Mislevy, P., K.J. Boote & F.C. Margin. 1989. Soybean response to gibberellic acid treaments. J. Plant Growth Regul. 8:11-18.

Nienhuis, J. & S.P. Singh. 1985. Effects of location and plant density on yield and architectural traits in dry beans. Crop Sci. 25: 579-584.

Nleya, T.M., A.E. Slinkard & A. Vandenberg. 1999. Evaluation of determinate and indeterminate pinto bean cultivars under an available soil moisture gradient. Can. J. Plant Sci. 79:27-34.

Quick, G.R. & W.M. Mills. 1978. Narrow-pitch combine cutterbar design and appraisal. pp. 96-100. *In* Grain and Forage Harvesting: Proc. Amer. Soc. Agric. Engineers, St Joseph, U.S.A.

Singh, S.P. 1982. A key for identification of different growth habits of *Phaseolus vulgaris* L. Annu. Rpt. Bean Improv. Coop. 25:92-95.
Singh, S.P. 1989. Patterns of variation in cultivated common bean (*Phaseolus vulgaris, Fabaceae*). Econ. Bot. 43: 39-57.
Singh, S.P. 1991. Breeding for seed yield. pp. 383-443. *In* A. van Schoonhoven & O. Voysest (eds.), Common Beans: Research for Crop Improvement. C.A.B. Int., Wallingford, U.K. & CIAT, Cali, Colombia.
Singh, S.P., C. Cardona, F.J. Morales, M.A. Pastor-Corrales & O. Voysest. 1998. Gamete selection for upright carioca bean with resistance to five diseases and a leafhopper. Crop Sci. 38:666-672.
Singh, S.P., P. Gepts & D.G. Debouck. 1991. Races of common bean (*Phaseolus vulgaris*, Fabaceae). Econ. Bot. 45: 379-396.
Singh, S.P. & J.A. Gutiérrez. 1982. Sources of some architectural traits in dry bush beans (*Phaseolus vulgaris* L.). Annu.Rpt. Bean Improv. Coop. 25:6-8.
Soldati, A. & E. Keller. 1984. Soybeans adapted to cooler regions. pp. 460-467. *In* R. Shibles (ed.), World Soybean Research Conference III: Proceedings. Westview Press, Boulder, U.S.A.
Thung, M. 1997. Soil fertility conservation and enhancement in bean production systems. pp. 470-486. *In* S.P. Singh & O. Voysest (eds.), Taller de Mejoramiento de Frijol para el Siglo XX1: Bases para Una Estrategia para America Latina. CIAT, Cali, Colombia.
Walker, A.K. & R.L. Cooper. 1982. Adaptation of soybean cultivars to low yield environments. Crop Sci. 22:678-680.
Wallace, D.H. 1985. Physiological genetics of plant maturity, adaptation and yield. Plant Breed Rev. 3:21-167.
Welsh, W., W. Bushuk, W. Roca & S.P. Singh. 1995. Characterization of agronomic traits and markers of recombinant inbred lines from intra- and interracial populations of *Phaseolus vulgaris* L. Theor. Appl. Genet. 91:169-177.
White, J.W. 1981. A quantitative analysis of the growth and development of bean plants (*Phaseolus vulgaris* L.). Ph.D. dissertation. Univ. of California, Berkeley, U.S.A.
White, J.W. & J. Izquierdo. 1991. Physiology of yield potential and stress tolerance. pp. 287-382. *In* A. van Schoonhoven & O. Voysest (eds), Common Beans: Research for Crop Improvement. C.A.B. Int., Wallingford, U.K. & CIAT, Cali, Colombia.
White, J.W. & S.P. Singh. 1991. Sources and inheritance of earliness in tropically adapted indeterminate common bean. Euphytica 55:15-19.
Vandenberg, A., L. Zyla & P.J. Hucl. 1994. Progress in reduction of dry bean harvesting losses. pp.107-113. *In* Soils and Crops '94 Proceedings. Extension Division, Univ. of Saskatchewan, Saskatoon, Canada.
Zyla, L. 1993. Development of direct-cut harvesting attachments for dry bean. MSc. Thesis, Univ. of Saskatchewan, Saskatoon, Canada.

Chapter 8

Breeding to Improve Yield

James D. Kelly, Kristin A. Schneider & Judith M. Kolkman
Crop and Soil Sciences, Michigan State University, East Lansing, MI 48824, U.S.A.

Key words: breeding pyramid, combining ability, gene pools, heterosis and hybrid bean, ideotype, partitioning, wild species, yield constraints

1. INTRODUCTION

World pulse production peaked at 58 million MT in 1990 and has since shown a decline in more than 100 countries. Highly fertile land in many countries is being devoted to more profitable cereals, oilseeds, and horticultural crops, thus displacing pulses to less favorable agricultural areas. In less developed countries, the lack of mechanized production, particularly for harvest, has favored a shift to other crops, whereas in developed nations, dry bean (*Phaseolus vulgaris* L.) production has shifted toward regions of lower-priced land, such as the Canadian prairie (Parker, 1997). For example, Manitoba replaced Ontario as the leading dry bean producer in Canada in 1998. Over the last 25 years, the area planted to dry bean in the U.S.A. has expanded by 30% and while yields have increased by a similar margin, major production shifts have occurred. For example, the acreage planted to dry bean in North Dakota has increased from 13,000 ha in 1971 to 300,000 ha in 1998, whereas the area in Michigan has decreased from 220,000 ha in 1976 to 120,000 ha in 1998. Bean must be regarded by the agricultural sector as a consistently profitable commodity if production is to be maintained in these more traditional areas. Stable prices, consistent consumer demand, and high yield are all essential goals to curtail downward production trends. Dramatic annual shifts in demand among commercial classes in both national and international markets contribute to the instability associated with bean production and increase economic risks to the producer. In North

America, bean production has moved to more northern climes where a shorter growing season, compounded with unpredictable fluctuations in precipitation and temperature, result in increased production risks. In addition, less adapted cultivars, grower inexperience, and crop losses all contribute to lower yields.

In the first half of the century, dry bean production in Brazil expanded in associated plantings with maize (*Zea mays* L.) and coffee (*Coffea arabica* L.). The arrival of soybean [*Glycine max* (L.) Merr.] in the 1960's and the economic loss of coffee to frost in the 1970's forced the displacement of bean production from fertile soils to regions of sloping, infertile, acidic soils where mechanization was not possible and the adoption of new technologies was resisted. Although annual bean production remained constant (2.5 MT), average yield during the same period (1970-1990) fell 30%. In other Latin American countries, bean production has shifted to less desirable, more marginal land as land resources in the fertile highland areas are limited. As a consequence of growing beans in these sub-optimum areas, the challenge of increasing yield is substantially more difficult. For example, in the semiarid state of Zacatecas, Mexico, where only 7% of 700,000 ha planted annually to dry bean (1990-1995) is irrigated, yield under irrigation averaged 3.6 times higher than non-irrigated. In other states, where rainfall is more abundant or irrigation is available, productivity is double or triple that of Zacatecas. In many countries, maintaining current levels of productivity under harsher environmental conditions could be viewed as progress. Displacement of bean to marginal lands with shallow, infertile, low pH, low phosphorus soils, or to areas where short seasons, early frosts, inadequate precipitation, and/or high temperature prevail, contributes to low bean yields worldwide. The discussion in this chapter attempts to address the improvement of dry bean yield under both stress and non-stress environments.

2. CONSTRAINTS TO IMPROVING YIELD

Common bean is a short-season crop, usually maturing from 65-105 days after planting. It is often grown as a secondary crop in areas with more than one growing cycle per year, or on marginal land in higher altitudes or latitudes where productivity is restricted by early frosts. These sub-optimal soil conditions and less than ideal climates predispose the bean crop to abiotic stresses such as drought, high temperatures, and low soil fertility. Fortunately, genetic variation exists for tolerance to abiotic stresses (Acosta Gallegos & Kohashi-Shibata, 1989; Graham & Ranalli, 1997; Schneider et al., 1997; Shonnard & Gepts, 1994; Singh, 1995a). From a breeding and production perspective, tolerance to stress is measured by improved yield performance under sub-optimum environmental conditions. Thus, many of the considerations for improving yield addressed in this chapter apply to the improvement of stress tolerance. Yield under stress is a complex trait controlled by many genes and strongly influenced by the environment. Heritability estimates for yield under stress are low (Schneider et al., 1997; Singh, 1995a), and testing requires that genotypes be compared in both non-stressed and stressed environments to measure yield potential under different degrees of stress.

2.1 Water Stress

Wallace & Yan (1998) emphasized the need for defining the environments for which the improvement was desired. Drought environments can be classified into two groups: terminal and intermittent (Ludlow & Muchow, 1990). Terminal drought stress is typical of the lowland tropics where the bean crop is planted at the end of the rainy season, when the available moisture level is at field capacity. As the season progresses, moisture reserves are depleted resulting in drought stress during the critical reproductive period. High temperatures also confound drought stress in lowland areas, since water demand by the plant is higher. The semiarid highlands (1,800-2,500 masl) and semiarid tropics are characterized by intermittent drought with annual rainfall of only 250-500 mm. Drought or precipitation can occur at any time during the growing season. The unpredictability of this type of environment demands indeterminate genotypes that can withstand stress and recover quickly following precipitation.

Specific adaptation of bean has made certain genotypes more suitable for one type of drought environment than the other. Early maturing, determinate genotypes can avoid terminal drought stress better than indeterminate genotypes. The precise delineation between vegetative and reproductive stages in determinate genotypes, however, does not allow sustainable recovery of reproductive growth following precipitation. Determinate types do not yield well in an intermittent drought environment. Indeterminate type II and type III genotypes perform better under intermittent drought stress since the vegetative and reproductive stages overlap, permitting plants to resume growth and recommence flowering after short periods of stress. Maturity is an issue in both types of environments since the growing season for bean in many tropical lowland growing areas is cut short by the necessity to plant primary crops, whereas in highland areas, late maturing cultivars are adversely affected by early frosts. In the semiarid highlands of Mexico, phenological plasticity, present in photoperiod-sensitive pinto beans, has been a valuable trait in yield improvement (Acosta Gallegos & White, 1995).

This trait is characterized by an alteration in the length of the growing cycle in response to environmental cues, such as daylength, temperature, and moisture. Plasticity is a mechanism for avoiding mid-season drought conditions or low temperatures late in the season. Normally full-season genotypes mature in fewer days in response to later planting dates. This characteristic is not widely utilized by bean breeders since selection for daylength neutrality is more common in tropical or high latitude regions. Other traits, such as root morphology, can also be contrasted between differing drought environments (Rojas et al., 1998). The upright type II growth habit is characterized by a prominent tap root which can penetrate greater soil depths and exploit stored water reserves, whereas shallow-rooted, early-maturing type I genotypes evade moisture stress better in terminal drought conditions (Kelly, 1998). The type III habit has a sprawling, fibrous root system and is more productive than other growth habits in intermittent drought situations, since more surface roots are available to absorb water from infrequent rainfall events. Knowledge of the environmental conditions is an important aspect of breeding for stress-prone environments.

Easily identifiable traits associated with drought tolerance mechanisms have eluded both breeders and physiologists. Consequently, costly replicated field trials, conducted in both stress and non-stress environments, are still required to understand performance under drought. Although management practices can promote higher yield, genetic improvement has greater potential to increase long-term yield (Singh, 1995a; White et al., 1994). Traits contributing to drought tolerance in bean include morphological, physiological, and phenological characteristics. Heliotropism (the movement of leaves perpendicular to incident sunlight) and loss of leaf area jointly influence yield under drought stress, but taken alone neither is a good indicator of performance (Acosta Gallegos & Adams, 1991; Kao et al., 1994; Subbarao et al., 1995). Likewise, drought-affected physiological traits such as osmotic adjustment, stomatal closure, water use efficiency, and transpiration rates cannot be used alone to predict higher yield in bean (Ramírez-Vallejo & Kelly, 1998). Measurements of yield and biomass under stress and non-stress and various indexes derived from these data, including harvest index (HI), percentage reduction of yield, and the drought susceptibility index (*S*), may be good indicators of drought tolerance (Fischer & Maurer, 1978; Ramírez-Vallejo & Kelly, 1998).

The most practical measure of drought tolerance is yield under drought. In certain seasons, it may be beneficial to have an equally high-yielding cultivar if the stress is minimal. The ideal genotype for drought-prone environments, therefore, is one that is high-yielding in both stress and non-stress environments. A reasonable alternative is to identify genotypes that do not vary greatly in yield performance over environments using the *S* index (percentage reduction in yield adjusted for drought intensity). These calculations can be misleading since genotypes intrinsically low yielding in both environments will be selected as exhibiting low *S* values. Furthermore, genotypes exhibiting large reductions in yield will be discarded despite possessing superior yielding capacity under non-stress environments (Clarke et al., 1992). A breeding strategy was postulated (Ramírez-Vallejo & Kelly, 1998) where genotypes were first selected based on the geometric mean yield under stress and non-stress environments. The group exhibiting high geometric mean yield would then be selected for low *S* values. Schneider et al. (1997), however, demonstrated that the top 10% genotypes, ranked for highest geometric mean yield, were not identified using the *S* index. Misclassifications among genotypes have been reported when rankings were based on *S* alone (White & Singh, 1991b). Since yield under both stress and non-stress environments must be measured to calculate these indexes, it would be practical and less time consuming to base selection first on geometric mean yield and then select only those genotypes with high yield under stress environments. Breeding lines derived from the best drought tolerant genotypes identified by Schneider et al. (1997) exhibited superior yield performance under drought stress in Michigan in 1998 (Kolkman & Kelly, 1999). These breeding lines significantly out-yielded the adapted cultivars from which they were derived, suggesting that selection for improved drought tolerance can be achieved in bean (Singh, 1995a).

2.2 High Temperatures

Many bean production areas are limited by more than just insufficient moisture. In tropical areas, drought stress is confounded with high temperature stress to which bean is susceptible (Shonnard & Gepts, 1994). Yield losses under high temperature result from damage to reproductive organs, accentuated by bud loss and seed abortion. Reduced photosynthetic rate, caused by damage to enzymes and membranes, also contributes to yield reduction (Chen et al., 1982). Genetic variation for heat stress tolerance exists in bean, but is sometimes confused with drought tolerance. Shonnard & Gepts (1994) studied bud abortion rates and percentage pod fill as an indicator of heat tolerance. Bud abortion tolerance was correlated with indeterminateness, suggesting that a component of this trait was either tightly linked to the *fin* gene for determinacy or affected by the pleiotropic action of this gene (Wallace & Yan, 1998). Other factors must also contribute to heat tolerance, since differences in tolerance to bud abortion were also observed in crosses between two determinate parents. Diallel analysis demonstrated the presence of significant additive genetic variance as well as maternal or cytoplasmic contributions. Dominance effects were only significant in populations from crosses involving an indeterminate parent. Percentage pod fill was controlled by additive x additive epistatic interactions, suggesting that selection should be delayed until later generations. Heat tolerance in bean is a complex, quantitative trait to which more breeding efforts should be targeted (Shonnard & Gepts, 1994). Progress in selection for heat tolerance in determinate kidney bean has been reported only in crosses with indeterminate genotypes such as Indeterminate Jamaica Red adapted to the lowland tropics (P.N. Miklas, personal communication, 1998). Contrary to drought experiments where yield under both stress and non-stress environments can be measured at one site, non-stress experiments for heat must be conducted at separate locations. Heat-tolerant genotypes should be identified in irrigated trials where moisture stress cannot be confounded with heat stress. Commonly, drought stress and heat stress are inseparable environmental conditions, and combining tolerance to both of these pressures in a single, high-yielding genotype will be a formidable task. Although the tepary bean (*P. acutifolius*) is recognized as possessing tolerance for both heat and moisture stress, success in transferring these traits to common bean has been limited (Thomas et al., 1983).

2.3 Soil Fertility

Low soil fertility is a major constraint to bean production, particularly in subsistence agricultural areas where fertilizer is unavailable or unaffordable. These regions possess a variety of soil fertility problems including low P, N, and Fe availability, sodic and/or saline conditions, among others (Aggarwal et al., 1997; Escalante Estrada et al., 1998; Gonzáles et al., 1998). It is estimated that 23 % of all cultivated soil is classified as saline and 37% as sodic (Escalante Estrada et al., 1998). In Latin America, 89% of the soils are deficient in N and 82% are deficient in P (Graham & Ranalli, 1997). Phosphorus deficiency in Africa is reported to decrease bean yield by 355,000 MT year^{-1} (Rachier et al., 1998). Improved soil management combined with genetic tolerance is proposed as a solution to enhance bean production in problematic soils. Genetic variation exists in bean for tolerance to saline and sodic soils, and low Fe and low P (Aggarwal et al., 1997; Escalante Estrada et al., 1998; Gonzáles et al., 1998). In Mexico, the black bean genotype, Negro Precoz, was the top yielding genotype under sodic/saline conditions, whereas the Ojo de Cabra and Flor de Mayo seed types yielded 60 and 70 % less, respectively (Escalante Estrada et al., 1998). The yield component most sensitive to sodic/saline conditions in the study was pod number. Under low P availability, large-seeded Andean genotypes demonstrated superior tolerance compared with small-seeded Middle American genotypes (Yan et al., 1995). Middle American genotypes, however, were shown to be more responsive to added P. Low P tolerance in bean may be associated with root architecture. Bean genotypes possessing increased root weight and root length performed better under limited P (Rachier et al., 1998).

Enhancing the biological N_2 fixation capacity of bean has been suggested as an alternative means to improve yield under low N fertility (Bliss, 1993a, b; Buttery et al., 1992). Common bean is widely regarded as weak in nodulation and N_2 fixation with fixation rates ranging from 3 to 125 kg N ha^{-1} (Graham & Ranalli, 1997). Modest fixation levels are important, however, since only 30-50% of applied N is recovered by the plant (Bliss, 1993b). An application of 100 kg N ha^{-1} equates to 50 kg N ha^{-1} fixed by the plant, assuming that 50% of the N is recovered by the plant. Under optimum environmental conditions, bean plants nodulated with efficient *Rhizobia* have yielded up to 2 MT ha^{-1} (Bliss, 1993b). However, bean is a relatively inefficient nitrogen fixer in comparison to cowpea [*Vigna unguiculata* L. (Walp.)] and soybean, especially when produced under environmental stresses such as low P availability (Pereira & Bliss, 1987; Piha & Munns, 1987; Whiteaker et al., 1976).

Variability in N_2 fixation exists in bean, particularly in plants with a larger photosynthetic area such as type IV genotypes (Bliss, 1993a; Graham, 1981; Graham & Rosas, 1977). Enhanced N_2 fixation in bean is inherited quantitatively and exhibits low heritability, which explains the limited progress toward improving this trait (Bliss, 1993a). Nodule dry weight and mean nodule weight are both controlled by additive gene effects, suggesting that early generation selection for N_2 fixation would be ineffective (Franco et al., 1998). Chaverra & Graham (1992) identified bean genotypes with contrasting differences in N_2 fixation. The cultivar Puebla 152 possessed excellent early nodule development up to 25 days after planting, whereas the genotype RIZ 21 accumulated more N and performed better during the period from 25 to 51 days after planting. Combining the positive attributes from these genotypes was suggested as a strategy to improve N_2 fixation (Chaverra & Graham, 1992). Due to screening difficulties, studies to improve N_2 fixation in bean have been problematic (Bliss, 1993a; Buttery et al., 1992; Rennie, 1984). Soil N levels can influence N_2 fixation ability in bean and confound the selection of improved genotypes. When N was not limiting, nodulating and non-nodulating mutant genotypes accumulated the same levels of N whereas, under N-limiting conditions, the nodulating genotypes performed better (Shirtliffe et al., 1996). Yield testing in low N soils should favor selection of breeding lines with higher N_2 fixation. Selection in low N environments, however, is not common since yield trials are conducted under optimal growing conditions in order to control environmental variation (Bliss, 1993a). For further details on integrated management of abiotic stresses, see Chapter 13 by Thung & Rao in this book.

2.4 Growth Habit

Given the obvious limitations to productivity from abiotic stresses, breeding for high yield in bean must be conducted within the major constraints of growth habit, maturity (adaptation), disease resistance, and seed size-shape-color (quality) preferences. Breeding for yield outside these major constraints will result in germplasm with limited commercial value. If high yield is used as the only basis for selection in bean and growth habit is ignored, only genotypes with type IV growth habit would be selected. Determinate genotypes are lower yielding than indeterminate genotypes (Nienhuis & Singh, 1985; White et al., 1992). But a mere change in stem type from determinate to indeterminate did not increase yield of large-seeded near-isogenic lines differing in growth habit (White et al., 1992). Other traits, such as node and branch number, must be increased to change yield. In a cross between genotypes from the Nueva Granada and Durango races, lines possessing indeterminate growth habit had significantly higher seed yield than lines with determinate growth habit (Welsh et al., 1995).

In bean, like other crops, major qualitative genes controlling plant height and maturity exert a large control over expression of quantitative traits. The dwarf genes of cereals reduce internode length and the unused metabolites are partitioned to the seed. The *fin* gene in bean reduces plant height and increases HI. The recessive allele of the *ppd* gene autonomously partitions available photosynthate toward growth of reproductive organs, resulting in faster growth, higher HI, and yield per day, whereas the dominant allele inhibits partitioning to reproductive organs, increases biomass, and lengthens days to harvest (Wallace & Yan, 1998). In reality, the pleiotropic effects of the *fin* gene, expressed in genotypes with larger seed size and early maturity, may enhance the inherent inability of the determinate plant to recover from stress, and contribute to the lower yield in type I cultivars. Determinate beans were inherently less stable and lower yielding in trials conducted in the American Midwest over six years (Kelly et al., 1987). Yield and stability were higher among 15 indeterminate lines evaluated across nine locations in Puerto Rico and the Dominican Republic, when compared to 15 determinate lines (Beaver et al., 1985, 1996). In an interracial cross (Mesoamerica/ Jalisco), Kornegay et al. (1992) suggested that a growth habit modification toward semi-climbing was the major reason for increased yield in this type of cross. The ideotype for large-seeded indeterminate beans has not yet been elucidated, according to White et al. (1992). The data suggest that indeterminate growth habit permits a greater expression of yield potential than the determinate growth habit. Major modifications of growth habit, if undertaken to increase yield, may not be readily accepted by growers.

2.5 Maturity

Maturity is another major constraint to yield since it restricts yield potential of germplasm selected for climatic extremes. In temperate regions and Mexican highlands (above 2,000 masl), the growing season is shortened by early killing frosts, whereas cropping systems and rainfall patterns similarly restrict the growing season in tropical zones. Selection for early maturity has resulted in lower yield (White & Singh, 1991a). Yield potential was decreased by 72 kg ha^{-1} for each day of reduction in maturity. In a cross of Redkloud/MAM 4, an increase in days to maturity resulted in higher yield (Welsh et al., 1995). A negative association between yield and seed size, however, suggested that large-seeded Andean beans achieve maximum yields through different physiological phenomena than small- or medium-sized beans. This implies that bean breeders will not be successful in exploiting yield potential from wide crosses differing in genetic background and maturity, since yield is realized differently between gene pools. Maturity must always be factored into a breeding program for yield in bean (Wallace et al., 1993, 1995).

2.6 Seed Traits

Quality in beans is a critical consumer requirement, given the wide range of pod and seed types differing in size, shape, seedcoat color, and cooking quality. Quality traits restrict the amount of germplasm available to the breeder by limiting the use of wide crosses and dramatically reducing the number of progeny suitable for advance to yield trials. Certain color combinations are inherently difficult to breed, limiting genetic progress within these seed types. In addition, White & Gonzáles (1990) demonstrated a 280 kg ha^{-1} yield reduction for every 100 mg increase in seed weight. In fact, parents from the Middle American gene pool exhibit negative combining ability for seed size, which restricts their value as parents for improving the yield of large-seeded Andean beans (Singh et al., 1992a). Similar reports confirmed an inverse relationship between seed size and yield in crosses between gene pools (Welsh et al., 1995). Breeders working with the large Nueva Granada type bean must find alternative genetic sources for increasing yield rather than through inter-gene pool crosses.

High-yielding, dark-colored red beans had limited acceptance by growers and consumers in Central America. The impact of high-yielding germplasm has been negated in those areas where light-colored red beans are preferred. Similar experiences have been observed in North America, where high-yielding kidney bean cultivars (Isabella, Charlevoix) were not accepted by the processing industry due to poor canning quality. Also, a new kidney bean genotype, CN-4109, with a 20% yield increase, was abandoned since it produced an unacceptable canned product (Wallace & Yan, 1998; Wallace et al., 1993). Farmers continue to grow contemporary cultivars with lower yield potential in order to meet the desired quality traits demanded by the consumer. Recently, with the release of the cultivar Redkanner, growers have access to a high-yielding full-season kidney bean with improved canning quality (Wallace & Yan, 1998). The expenditure of resources by North American breeders in selecting for processing quality is large relative to the expenditure for other traits because of the expense of canning all potential new breeding lines. This need limits progress in breeding for improved yield. A recent consumer survey conducted in Mexico (Castellanos et al., 1997), showed that the consumer considered cooking time when selecting bean types for consumption. This information has forced bean breeders in Mexico to evaluate all breeding material entering yield trials for water uptake and cooking time (Castellanos et al., 1995). High yielding cultivars, such as Pinto Villa (Acosta Gallegos et al., 1995b) and Bayo Victoria, have dramatically longer cooking times compared to the higher priced Flor de Mayo Bajio cultivar. Additional selection for shorter cooking time will certainly have a negative effect on breeding for high yield since a portion of the selected progeny derived from crosses with the high-yielding Pinto Villa parent must be discarded on the basis of cooking time.

2.7 Diseases and Insects

Disease and insect resistance is a critically important objective in many bean breeding programs. The lack of resistance can dramatically lower the productivity of new cultivars and limit grower acceptance. Yield variation from 0.95 to 2.26 MT ha^{-1} in dry bean produced in Ontario can be attributed in part to pests such as the potato leafhopper [*Empoasca fabae* (Harris)] (Schaafsma et al., 1998). Elsewhere in the humid American Midwest, vigorous, high-yielding type III Durango race pinto beans are very susceptible to white mold, which can greatly reduce yield potential. High-yielding genotypes may help promote specific diseases, such as white mold, because of their prolific foliage (biomass). Morphological avoidance mechanisms must be incorporated which may prevent the optimization of yield potential. Disease resistance does not directly increase yield, but in the absence of adequate levels of resistance to diseases, such as bean common mosaic, common bacterial blight, bean golden mosaic, and anthracnose, yield will be reduced. Adequate resistance to a number of pathogens is needed to help stabilize yield of common bean. Despite evidence to the contrary, some breeders believe that certain resistance genes carry a physiological load that results in lower yield. Near-isogenic lines carrying the different alleles of the *Are/are* anthracnose resistance gene showed no difference in yield and/or maturity (Park & Tu, 1987). Breeders must recognize the disease constraints within their production zones and restrict resistance breeding to these specific pathogens. In the semiarid western production states of the U.S.A., bean common mosaic and root rots are major diseases. Incorporating resistance to a wide array of additional diseases that are not primary constraints will delay progress toward improving yield and offer little advantage to growers in those regions. Similarly, in the humid midwestern U.S.A., where a broader range of pathogens limits productivity, certain seed-borne bacterial pathogens can be adequately controlled through preventive measures such as sanitation and seed certification programs. An over-emphasis on resistance breeding may hinder breeding for yield (Park et al., 1998) by restricting variability and limiting resources, but it may be essential in certain areas where disease management programs are inadequate.

2.8 Other Limitations

Bean breeders are further restricted by size and resource limitations since many work with a number of different commercial types. Even with current levels of mechanization, the number of yield trials that can be conducted annually is substantially less than the number normally conducted in oilseed crops such as soybean, where direct harvest is possible. In North America, where soybean, the major grain legume, is bred almost exclusively for yield by both the private and public sectors, it is estimated that some commercial companies have the capability to evaluate up to 250,000 lines annually in preliminary yield trials. A comparison of resource allocation for the bean and soybean breeding programs at Michigan State University is presented in Table 1. Both programs operate during a similar time frame for planting and harvest with equivalent personnel. Soybean breeders, however, have greater flexibility for harvesting later in the season since soybean does not suffer as great a reduction in quality due to delayed harvest as does bean. A single individual can direct-harvest soybean, whereas bean harvest requires a team approach. Perhaps, with an increase in grower interest in direct harvest of upright bean cultivars in certain regions of North America (T. Michaels, personal communication, 1998), this gap will gradually shrink. In addition to the physical limitations between the two programs, the number of market classes targeted in each program is quite different. Soybean breeders typically focus on one market class, whereas bean breeders must improve a number of contrasting market classes, each with its own specific quality issues such as seed size and seedcoat color. On a per market class basis, therefore, soybean breeders can exert more selection pressure to identify elite lines for release, whereas the selection pressure exercised by bean breeders is much less in any one commercial class. If one assumes similar levels of heritability and genetic variance for yield in both commodities, the selection pressure in soybean would be nine times that in dry bean (Table 1). The ability of the soybean breeder to make larger gains in quantitatively-inherited traits, such as yield, is therefore enhanced primarily by the greater number of lines available for selection. As a result of the ability to manage large numbers of lines in yield trials in soybean, breeding has resulted in a steady gain in yield of 1.5 g m^{-2} $year^{-1}$ over 75 years. This gain can be attributed solely to genetic improvement (Fehr, 1987). Annual yield gain in other grain legumes, where harvest is difficult to mechanize, initially appears to be equivalent. Over the last half century, peanut (*Arachis hypogea* L.) yield has improved by 1.8 g m^{-2} $year^{-1}$ (Wells et al., 1991), whereas commercial bean yield in the U.S.A. has increased by 1.6 g m^{-2} $year^{-1}$ over the last quarter century. Estimates for yield in peanut and bean are confounded with genetic and management components. If management contributes half to the total yield improvement, progress in breeding for yield in soybean is approximately twice that of peanut and bean.

Table 1. Comparison of soybean and dry bean breeding programs at Michigan State University

Annual number of	Soybean	Dry bean
Replicated yield plots	5,000	1,000
Commercial seed types being tested	1	5
Yield plots per seed type	5,000	200
New lines entered in yield trials	700	400
Potential cultivar releases	1	5
Selection pressure (%)[1]	0.14	1.25
Comparative selection pressure	9	1

[1]Ratio of released cultivars to lines entered in yield trial.

3. APPROACHES TO IMPROVING YIELD

The narrow genetic base available within cultivated germplasm is a major limitation to improved yield. This limitation is commonly recognized by breeders of autogamous species. Examples appear in bean (Adams, 1977; McClean et al., 1993; Voysest et al., 1994), peanut (Holley & Wynne, 1986), and soybean (Fehr, 1987; Kisha et al., 1997). Soybean has the most sophisticated of the grain legume breeding programs, and breeders often choose parents that are genetically diverse, based on measures such as the coefficient of parentage (CP). Limitations to CP are recognized since the effect of selection is not included in this measurement. Genetic distance, based on molecular markers, is becoming increasingly important as an independent method to aid in the selection of genetically diverse parents. Studies that compared genetic distance, based on CP or molecular markers, with genetic variance for yield have shown the comparative advantage of the information generated by molecular markers (Kisha et al., 1997). The major strategy employed by soybean breeders in North America is to cross elite, genetically-distinct, high-yielding lines that exhibit agronomic similarity for seed size, growth habit, and maturity. Expected gain from a cross can be estimated by the mean parent value (MPV) for yield or by using the best linear unbiased predictions (BLUP), which provides higher rank correlations and lower standard errors than MPV. Using the BLUP method, breeders have identified higher percentages of the superior crosses than with MPV (Panter & Allen, 1995).

The use of plant introductions (PI) as novel sources of variability to increase genetic gain for yield in soybean was not effective, based on three cycles of recurrent selection (Ininda et al., 1996). The increased variability anticipated in the PIs was not demonstrated in marker studies (B. Diers, personal communication, 1998). Many of the PIs showed no greater variability than that observed among cultivated materials and their use as parents has not resulted in major gains in yield (Ininda et al., 1996). If variability in this germplasm is limited, as suggested by the marker studies, then its potential value in yield improvement will be similarly restricted. Potential progress may be achieved by using the wild species, *G. soja*, in crosses with the cultivated species. Since *G. soja* has a weedy, low-yielding phenotype, single crosses to *G. max* exhibit unacceptable variability. Three-way crosses and backcrosses offer greater opportunity to exploit useful genetic variability from *G. soja* in the agronomic background of *G. max* (B. Diers, personal communication, 1998). Similar approaches using molecular markers, linkage maps, and wild species, must be evaluated in *P. vulgaris* as an alternative to conventional breeding strategies.

3.1 Early Generation Yield Testing

Breeding strategies to improve bean yield have included early generation testing (EGT) and selection based on yield components. Adoption of either approach has had limited success because the distinct market classes in bean differ greatly in seed size, which restricts improvement due to yield component compensation. In addition, the need to select for disease resistance and quality traits competes in efficiency with EGT for yield, particularly when wide crosses are made. Selection for yield, based on individual components, has not been successful (Nienhuis & Singh, 1985). It is not unexpected that selection based on individual components failed to result in overall yield improvement, given that the theory of yield component compensation in crop plants was demonstrated first in bean (Adams, 1967). The use of EGT to identify F_2-derived lines with the highest yield potential was successful in developing semidwarf soybean cultivars suitable for high-input production systems in the American midwest (Cooper, 1988). The greatest impact of semidwarf soybean cultivars is on more productive soils, where lodging constitutes a yield barrier to taller, indeterminate cultivars (Cooper, 1985). F_2-derived EGT schemes, however, have not been widely accepted by soybean breeders except for specific management situations (semidwarf soybean) or for the introgression of determinate growth habit into indeterminate cultivars grown in the northern U.S.A. (Cooper, 1985). EGT was successful in improving yield for chickpea (*Cicer arietinum* L.) and cowpea, although qualitative agronomic traits were not simultaneously examined (Dahiya et al., 1984; Ntare et al., 1984b). Stratified mass selection in the F_2 was successful in identifying high-yielding genotypes when compared to a selected group of low-yielding genotypes in peanut (Holley & Wynne, 1986). The results could be misleading, however, since a few undesirable, simply inherited traits could be responsible for the expression of low yield in the latter group.

In bean and cowpea, visual selection for yield in the F_2 and F_3 was ineffective (Dahiya et al., 1984; Patiño & Singh, 1989), whereas among the individual yield components tested, the best selection criterion in F_3 to F_7 materials was direct selection for yield (Singh & Gutitérrez, 1990). EGT was effective for yield across locations, but no comparison of efficiency with other breeding systems was demonstrated (Singh et al., 1990). Selection for other key traits in this study could be conducted in later generations, but the value of EGT is doubtful in breeding programs where resources for yield evaluation are limiting. Yield testing is the most costly and labor intensive operation breeders perform. Financial limitations delay yield testing until later generations when numbers have been reduced, based on other criteria. EGT, however, could be combined with other breeding systems [pedigree, single seed descent (SSD), gamete selection] to evaluate F_2 and F_3 populations across contrasting locations (Singh & Terán, 1998; Singh & Urrea, 1995). In cowpea, SSD was recommended over EGT, since it was equally effective and required fewer resources (Ntare et al., 1984a; Obisesan, 1992). An early generation selection method based on the performance of F_2 plants, equidistantly spaced in a honeycomb pattern, was the most effective for improved yield of mungbean [*Vigna radiata* (L.) Wilczek] (Gill et al., 1995). Gamete selection for the simultaneous improvement of multiple traits in common bean has been proposed (Singh, 1994). Results suggest that gamete selection permits the rapid identification of lines with a combination of disease and plant architecture traits. A relatively low proportion of the resistant lines, however, significantly out-yielded the highest yielding commercial cultivar Carioca (Singh et al., 1998).

3.2 Ideotype Approach

In addition to traditional empirical approaches to selecting for yield in common bean, ideotype breeding methods have been proposed and implemented (Adams, 1973). The ideotype concept was first proposed in cereals (Donald, 1968), and was later applied to bean (Adams, 1973, 1982) and other grain legumes such as pigeonpea (*Cajanus cajan* L. Millsp.), pea (*Pisum sativum* L.), and mungbean (Hovinen, 1988; Reddy & Rao, 1980; Williams et al., 1988). Breeding for a small-seeded navy bean ideotype resulted in the development of high-yielding genotypes with later maturity and a growth habit different from previous commercial cultivars. Type II cultivars exhibited the highest yield and yield stability across both rainfed and irrigated environments in Michigan (Kelly et al., 1987). Type III indeterminate cultivars exhibited high yield, but were less stable across locations because of disease problems with white mold, and harvest risks associated with prostrate growth habit. In general, cultivars with determinate growth habit were lower yielding and less stable across a broad range of environments than indeterminate cultivars. Lower-yielding, small-seeded, determinate cultivars, however, were usually more stable. It was not until small-seeded cultivars with higher yield potential were developed that this instability was observed. Data from the Cooperative Dry Bean Nursery, however, indicated the overall superior performance of type III genotypes across North America, despite potential production problems with this growth habit in the humid Midwest (Stewart-Williams & Myers, 1995). Selection for specific morphological traits, such as plant height, node and leaf number, and branching characteristics, has not been a successful strategy to increase yield of bean in tropical regions (Nienhuis & Singh, 1985). Such traits, however, are positively correlated with yield in growth types I, II, and III. Recognizing the important role that growth habit plays in yield, adaptation, maturity, and harvest mechanization, most bean breeders in North America continue to work within the constraints of the growth habit of current commercial seed types.

Based on the success of ideotype breeding in small-seeded race Mesoamerica bean, an interracial recurrent selection system was used to improve plant type of Durango race cultivars (Kelly & Adams, 1987). Currently, the type II growth habit from tropical black bean has been successfully introduced into commercial pinto and great northern beans. Pinto bean cultivars such as Sierra (Kelly et al., 1990) and Maverick (Grafton et al., 1997b), and great northern cultivars like Alpine (Kelly et al., 1992a) and Starlight (Coyne et al., 1991) were among the first Durango race cultivars to possess a type II growth habit originating from race Mesoamerica. Since these improved cultivars were adopted by growers, current commercial bean yields in Michigan for the 1990-1995 period are 25% higher than any previous five-year period. Similar cyclic breeding programs have been used to develop high-yielding Andean beans for the tropics. Recurrent selection, based on S_4 testing, was used to develop indeterminate, large-seeded Caribbean red beans that yielded 30% more than the determinate checks, but these genotypes were 10 days later maturing (Beaver & Kelly, 1994). The delay in maturity is unacceptable to growers in the Dominican Republic who prefer early type I cultivars (D.P. Coyne, personal communication, 1999). Inter-gene-pool crosses between races Nueva Granada and Mesoamerica have been used to develop large-seeded, upright type II beans with high yield but longer maturity (Wallace & Yan, 1998).

The use of ideotype breeding in other crops has met with only limited success. In barley (*Hordeum vulgare* L.), for example, the requirement for symmetry among plant parts, yield component compensation, pleiotropy, and differences in genetic backgrounds are major factors which limit progress in ideotype breeding (Rasmusson, 1987). He recommended that ideotype breeding be used to augment, but not replace, traditional breeding. Similar limitations can occur with bean, given major differences in growth habit and seed size, and may contribute to the lack of acceptance of the ideotype concept by breeders working with different seed types. Different ideotypes must be developed for market classes that differ widely in seed size (Brothers & Kelly, 1993). The pinto bean ideotype differed from the small-seeded navy bean ideotype in having fewer pods per plant and fewer seeds per pod. In the semiarid regions of the U.S.A., where a more rigid upright stem offers no agronomic advantage, J. Myers (personal communication, 1997) believes that, in comparison with the type III plants, the type II growth habit may actually result in lower-yielding plants. The type II plant may partition more energy into building a rigid, lodging-resistant, thick stem that competes with seed for the available photosynthate. Breeders must develop specific ideotypes for each major market class and production zone.

3.3 Physiological Efficiency

Breeding for physiological efficiency has been proposed as an approach to improve yield in bean (Wallace et al., 1993). Adaptation is the major constraint to breeding for yield, since a cultivar must fit the environment in which it will be grown. Days to maturity is the most important physiological trait affecting that outcome. All bean plants have a specific growth rate. Through genetic manipulation, this growth rate can be increased. Increasing growth rate does not necessarily translate into higher seed yield, however, unless greater amounts of the photosynthate from biomass are partitioned to the seed. Indirect selection for the three major physiological traits affecting yield, namely biomass, HI, and days to maturity, should result in improved yield (Wallace et al., 1993). Simultaneous selection is required because genetically established interrelationships occur among these three traits. An increase in maturity results in an increase in biomass, whereas an increase in HI results in a decrease in maturity and a decrease in biomass. All three physiological traits and the correlations among them can be quantified by a yield system analysis (YSA) of yield trial data (Wallace & Yan, 1998; Wallace et al., 1993). Long-term direct selection for yield will gradually exploit all useful genetic variability for HI and maturity. Ultimately, selection will lead to an excessively high HI because a smaller vegetative structure must support larger yield without lodging. This will either reduce canopy photosynthesis or result in an increase in disease. Lower planting densities of type III medium-seed size cultivars in the western U.S.A., where disease pressure is minimal, can offset this disadvantage. Indirect selection for higher biomass can occur only if genes for superior biomass are introduced. A YSA approach will insure that biomass is measured, since long term benefits will not be accrued if this trait is ignored (Wallace et al., 1993). Selection for biomass within the constraints of maturity is equivalent to actual selection for a higher rate of biomass accumulation. Selection for yield per day should help compensate for the tendency for higher biomass to result in longer duration of growth. Selection solely for greater biomass will give later maturity, lower HI, and higher yield, whereas selection solely for HI results in early maturity with an accompanying reduction in yield. In an analysis of yield breeding in peanut, newer, high-yielding cultivars exhibited improved allocation of photosynthate (Wells et al., 1991). From 1943-1977, photosynthate allocation in peanut has increased 57%. This improved partitioning also reflects improved HI, earlier maturity, and more upright architecture. Breeders, however, recognize that HI will plateau at some cultivar-specific value, and yield will be improved only by increasing biomass. Selection for early maturity will reduce both yield and biomass. Superior performance can be achieved, based on selection for yield and the three physiological traits, since these negative indirect effects result from exclusive selection of individual traits. Many breeders do not collect data on biomass. Without these data or an indirect measure of biomass, yield will stagnate as breeders eliminate larger, less efficient genotypes from their nurseries. Yield in Sierra pinto is a direct result of an increase in biomass with an extension in the seed-filling period, which lowered HI. Further progress in breeding for yield in North America can be achieved by combining traits from the high HI, early season type III cultivars from the West with the high biomass and full season type II cultivars from the Midwest U.S.A. Pinto bean cultivars Aztec (Kelly et al., 1992b), Maverick, and Hatton (Grafton et al., 1997a, b) have resulted from this strategy. Similar success has been achieved in the Mexican highlands by crossing tropically adapted race Mesoamerica genotypes with photoperiod-sensitive germplasm of Durango (type III) and Jalisco (IV) races adapted to highland conditions (Acosta Gallegos et al., 1995a; Singh et al., 1989a, 1993).

Indeterminate genotypes have different degrees of partitioning. Navy (race Mesoamerica) and pinto (race Durango) beans have similar growth rates (Stewart-Williams & Myers, 1995), and exhibit similar biomass accumulation. Navy beans are less complete in the transition from vegetative to reproductive growth than are pinto beans. As a result, navy beans continue to add vegetative growth (biomass) and the rate of seed fill is relatively slow, resulting in delayed maturity (incomplete partitioning types). Since pinto beans make a rapid and complete transition to reproductive growth, seed fill is rapid, resulting in faster progression to maturity (complete partitioning types). Because the transition is more complete in pinto beans, a higher HI is achieved. The rapid and complete transition from vegetative to reproductive growth is essential to achieve high yields in duration-limiting environments. Without an efficient method to collect biomass data, vital information critical for the YSA is lost. Breeders must be convinced that the gain in information is worth the effort to collect the biomass data. Finally, breeders should select genotypes that exhibit an efficient partitioning switch from vegetative to reproductive growth for specific local environments.

3.4 Parental Performance and Combining Ability

Breeding for yield, based on combining ability and genotypic performance, was suggested by Singh (1991, 1992). The narrow genetic base of all bean market classes was revealed in a survey of the parentage of North American and Latin American dry bean cultivars (Adams, 1977; McClean et al., 1993; Voysest et al., 1994). The lack of variability may be one explanation for the limited yield increases in bean. Progeny generated from crosses within the small-seeded Middle American gene pool and between Andean and Middle American gene pools did not out-yield the best check cultivars. Interracial crosses between small- and medium-seeded genotypes within the Middle American gene pool, however, produced lines that out-yielded the best parent and check cultivars (Singh, 1995b; Singh et al., 1989a). In general, limited gains in yield were achieved from crosses within races and gene pools, due to insufficient levels of genetic variation (Singh et al., 1989b). Progeny from Middle American interracial crosses outyielded the best parent, whereas none of the progeny of the intraracial cross outyielded the best parent, regardless of planting density (Singh & Gutiérrez, 1990). Substantial yield gains can be expected from interracial crosses within the Middle American gene pool, whereas selection for higher yield in Andean x Middle American crosses seem problematic (Singh, 1995b). Large differences in general combining ability (GCA) were found among bean cultivars from different races and gene pools (Nienhuis & Singh, 1986, 1988; Singh et al. 1989a, 1992a). The mean of interracial F_1 hybrids was higher than that of intraracial hybrids (Singh & Urrea, 1995). Knowledge of parental performance and GCA among and within races and selection for yield *per se* are essential in breeding for higher yield (Singh, 1991, 1992). Using this information, high-yielding lines from interracial bean populations have been developed within gene pools (Singh, 1995a; Singh et al., 1989a, 1992b, 1993) and from interracial populations between gene pools (Beaver & Kelly, 1994; Singh et al., 1992b, 1999). Despite the information available on the yield potential of specific races and the favorable combining ability for yield present in specific interracial crosses, bean breeders in North America have not utilized other gene pools or races to effect change in yield potential of bean, with few exceptions such as pinto Chase (Coyne et al., 1994).

3.5 Other Considerations

Yield relative to a production system is critical. Breeding bean for yield can be accomplished only within the framework of specific constraints. Work conducted at CIAT suggests that breeding for increased yield in bean must consider the interrelated effects of growth habit, seed size, maturity, and gene pool. Improved yield in segregating populations was associated with increased expression of climbing bean growth habit, which involves traits such as vine length, climbing ability, node number on the main stem, and plant height (Kornegay et al., 1992). Genetic progress within one growth habit type does not ensure progress in another type, nor does it apply universally across seed types. High yields in the small-seeded, long-season, highland, type IV climbing beans have limited application to the improvement of determinate, short-season beans for lowland production in the Caribbean. Support for individual physiological processes that combine differentially to generate yield in genotypes from the different gene pools is based on data from interracial crosses between gene pools (Welsh et al., 1995). Pathways to yield accumulation in cultivated bean follow those observed in landrace cultivars from the two domestication centers. For example, yield in early maturing genotypes with few pods and larger seeds is best combined in determinate, bush Andean beans in the tropics, whereas yield in late-maturing genotypes is best expressed in the indeterminate Middle American small- and medium-seeded genotypes that have more pods and greater biomass. Many of the physiological processes that control yield differ among diverse genotypes, so genetic gains in yield are not directly transferable. Rather than apply yield improvement to specific genotypes or seed types, the strategy for yield improvement in bean must be better integrated.

In any analysis as complex as breeding for yield, the common approach is to dissect the problem into its major constituents and analyze each individually. Strengths can be separated from weaknesses and new approaches can be devised to exploit those strengths while minimizing the weaknesses. The synergistic interaction of components that contribute exponentially when combined may be lost by this approach. In our analysis of yield in bean, breeding on the basis of physiological traits, combining ability, and ideotype were discussed independently as major approaches to improve yield. In bean, however, all three approaches are highly interrelated. Ideotypes infer a desirable combination of morphological traits, but these must also have a useful physiological basis, whereas combining ability is key in all aspects of breeding. Parents with low combining ability for any trait are quickly eliminated from breeding programs, as their progeny fail to meet the desired expectations.

4. BREEDING PYRAMID: THREE-TIERED APPROACH TO YIELD BREEDING

A more generalized approach to incorporating diverse germplasm into elite cultivars of bean, while maintaining superior cultivar development, was recently proposed to address both short- and long-term breeding objectives (Kelly et al., 1998). The breeding pyramid, consisting of three levels, the base, intermediate, and apex, is illustrated in Figure 1. Each level of the pyramid represents three distinct levels of germplasm development and improvement similar to a hierarchical open-ended population enrichment (HOPE) breeding system (Cramer & Kannenberg, 1992). HOPE was proposed and used in maize breeding at CIMMYT by E. W. Sprague since the early seventies (S. Singh, personal communication, 1999). Specific strategies deployed at each level, however, involve the utilization of different breeding methodologies. The approach to breeding for yield is different at each level of the breeding pyramid. Improved materials developed at the lower levels would be designed to move upward sequentially, recognizing that a strong continuity among all three levels is necessary. The genetic diversity and the development time would be greatest at the lowest level. The exploitation and maintenance of this diversity would be optimized at the intermediate level while uniformity and yield performance would be maximized at the top of the pyramid. Suggestions for appropriate recombination strategies and breeding methodologies are discussed to exploit the special resources at each level.

4.1 The Apex

The apex of the breeding pyramid primarily involves elite by elite crosses, and short-term breeding strategies that target high yield. Although an attempt should be made to cross parents with maximum diversity, the constraints faced in bean breeding must limit breeders to work within the limits of market class. Crosses are restricted to within gene pools, within races, and within market classes. Genotypes at the apex level should be relatively uniform for a majority of traits. Parents should be selected, however, for maximum diversity, based either on CP or molecular marker information. Diversity can also be increased by using three-way crosses among genetically diverse parents. For instance, crossing P_1/ P_2// P_3 (where P_1 = an adapted line that needs improvement, P_2 = an elite donor, that perhaps lacks specific adaptation, and P_3 = an adapted high-performing parent with superior combining ability) could be an approach. The focus on breeding for yield should be most evident at this level. Single seed descent would be an optimal method to use at this level of the program. In a comparison of SSD, pedigree, and bulk breeding methods, SSD was the most efficient when selecting for traits of low (10%) heritability (Casali & Tigchelaar, 1975). The main objectives of SSD are to maintain maximum genetic variability while increasing homozygosity. Once an adequate level of homozygosity is achieved, seed can be bulked and lines tested for improved yield. Since many of the constraints (traits) listed above are fixed for germplasm at the apex, pedigree selection for highly heritable agronomic

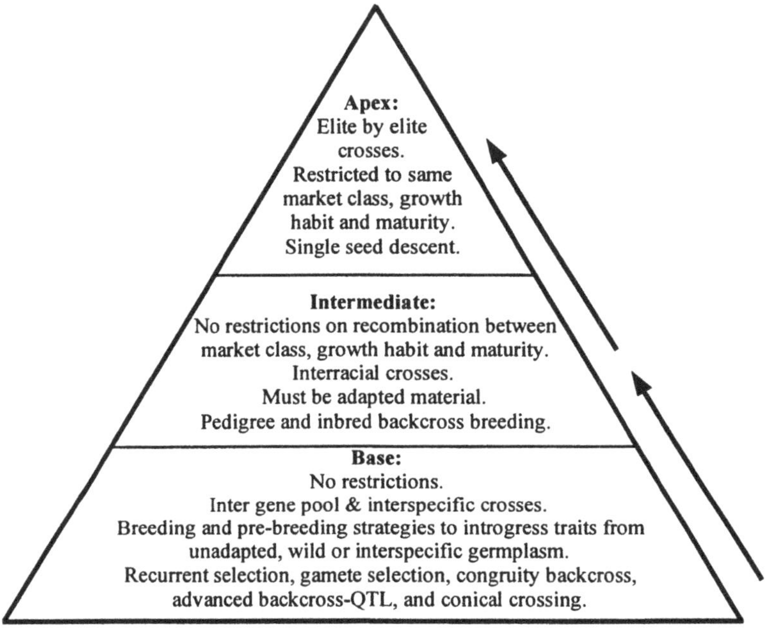

Figure 1. Breeding pyramid. A three tiered approach to breeding for yield (Kelly et al., 1998).

traits is not required. Genetic variation will not be evident for many visually scored traits in early generations, so selection should be delayed until the advanced generations where additive genetic variance predominates and quantitative traits can be measured effectively. Selection, based on YSA, could also be used at this level since seed size and maturity differences between parents is limited and emphasis is placed on yield data as the basis for selection.

Soybean breeders have made annual progress in yield through early generation selection (F_4) of large numbers of lines derived from elite by elite crosses. Yield improvement in soybean is achieved almost exclusively by the use of SSD, which is facilitated by multi-location testing, mechanized direct harvest, and electronic field measurements. Similar successes have been observed in bean where private breeders, out of necessity, concentrate their efforts on the apex of the breeding pyramid and some neglect other constraints, such as disease resistance. These private programs utilize germplasm from other programs as a source of variability without which progress would be limited. Similar models could be envisioned for national programs, where a limited number of market classes are important. If breeders agree that newly released cultivars represent the highest available yield potential at the time of release, a review of the pedigree information of contemporary cultivars demonstrates that they are the product of elite by elite crosses (McClean et al., 1993). Finally, the limited success of increasing yield through the use of PIs in soybean (Ininda et al., 1996), strongly suggests the need to continue elite by elite crosses in bean within the apex of the breeding pyramid to ensure the short-term improvement of yield in the different commercial classes and gene pools.

4.2 Intermediate Level

The intermediate level of the breeding pyramid addresses the introgression of greater levels of diversity into breeding materials. The top performing intermediate lines may be yield tested and crossed with elite lines when appropriate. Crosses in this level should be restricted within gene pools and races, whereas crosses between market classes of similar seed size, growth habits, architectural types, and maturity are encouraged. Larger number of crosses are required, since no prior estimate of combining ability among genotypes is available. Single plant or pedigree selection in early generations (F_2) is rigorously practiced for highly heritable agronomic and seed traits which can be easily scored and fixed. Selected individuals are then advanced rapidly to the F_5 by SSD. A combination of pedigree and SSD may be useful, taking advantage of the ability to select effectively for traits with high and low heritability. Inbred-backcross breeding (Bliss, 1993a) including 3- and 4-way crosses, would also be suitable methods to incorporate diversity from less adapted materials into locally adapted lines. At least two backcrosses may be necessary; the choice depends upon the variability for agronomic and quality traits that was introduced in the initial cross. Consideration should also be given to the use of the conical cross (Fouilloux & Bannerot, 1988), provided the number and choice of parents is restricted to the limits defined for this level. Agronomically acceptable high-yielding lines would enter the elite group at the apex to be tested and utilized accordingly. More flexibility in the choice of parents is possible at the intermediate level. Breeders should consider developing an ideotype for the different seed classes in their respective production zones. Flexibility in breeding for an ideotype or plant architecture is possible at this level. In addition, based on data collected from wider crosses or information reported in the literature, breeding should emphasize combining ability. Breeders must know the combining ability of elite parental material, since favorable combining ability can be more readily exploited at the intermediate level where a broader array of crosses is possible.

4.3 The Base

The base of the breeding pyramid facilitates the development of long-term, complex objectives, through the incorporation and recombination of diverse germplasm. Hybridization between races, gene pools, and species would invoke opportunities for increased diversity. Opportunities at this level attempt to capitalize on yield potential, such as that demonstrated by the Durango and Jalisco races of bean (Kornegay et al., 1992; Singh et al., 1989a, 1993). Introgression of these races into other gene pools or races is critical. Desirable yield traits found in other *Phaseolus* species could be introgressed into a base *P. vulgaris* type. Once incorporated into *P. vulgaris*, efforts should be focused on bringing these traits into the intermediate level and finally to the elite level. The success of the pyramid depends on the utilization of these new diverse populations in the upper levels of the breeding pyramid. In order to meet this challenge, the selection and advance of progeny from these wide crosses must be rigidly structured. Breeding methods for maximizing beneficial recombination between adapted and donor germplasm would best suit the long-term objectives of breeding for yield. Systems that involve cyclic intermating such as recurrent selection (Beaver & Kelly, 1994; Kelly & Adams, 1987; Singh et al., 1999; Wallace & Yan, 1998), inbred-backcross (Bliss, 1993a), congruity backcross (Urrea & Singh, 1995), or conical crossing (Bett & Michaels, 1994; Fouilloux & Bannerot, 1988) should be the most effective. In each of these crossing systems, success will be dependent upon the level of initial diversity going into the cross, the combining ability of favorable allele combinations, minimal incorporation of undesirable traits or linkage drag, and maintenance of favorable linkage groups. Depending upon these factors, material may be incorporated back into a higher level of the breeding pyramid.

Recurrent selection should create new combinations of desirable traits from the different gene pools because the different genetic and developmental associations between germplasm from different gene pools have evolved over millennia (Welsh et al., 1995). One example of such a program was a recurrent selection program designed to improve the yield of Caribbean red beans (Beaver & Kelly, 1994). Yield of determinate Andean red beans was improved through a system of S_4 recurrent selection designed to identify high-yielding progeny that were recombined in additional cycles of recurrent selection. The system demands patience and resources, but it offers a realistic avenue to effectively combine diverse germplasm sources without targeting single genes. Additional support for the effectiveness of recurrent selection in improving yield was demonstrated in both inter-gene pool and interracial broad based populations, using S_1 testing (Singh et al., 1999). Singh (1996) cautioned that the use of random intermating to simultaneously improve yield and seed size in interracial populations will result in mostly small-seeded progeny. The need to maintain some large-seeded recombinants for additional intermatings was demonstrated in interracial crosses used to improve Durango race pinto bean (Kelly & Adams, 1987). A similar recurrent selection strategy to improve yield was successful in peanut (Monteverde-Penso & Wynne, 1988; Monteverde-Penso et al., 1987). Recurrent selection is difficult to establish in a self-pollinated crop, since hybridization is relatively demanding compared to cross-pollinated species. Inbreeding after each recombination cycle, however, makes recurrent selection applicable to self-pollinated crops, especially for complex traits such as yield, where replicated trials must be conducted (Singh et al., 1999).

Cyclic breeding systems, other than a structured recurrent selection system can be used effectively to achieve similar results. The inbred-backcross line method is well suited for the transfer of genes controlling a quantitative trait from an unadapted or exotic source into a commercially acceptable type (Bliss, 1993a). Congruity backcrossing was preferred over recurrent backcrossing to maximize recombination between and retain desirable traits from distantly related parents used in interracial crosses (Urrea & Singh, 1995). The conical crossing system has the potential to combine genes for quantitatively inherited traits from several different sources (Fouilloux & Bannerot, 1988). In this system, breeders may choose the degree of diversity by influencing the number and choice of elite or non-elite parents in the design. A lack of combining ability among diverse parents entering a conical cross could negatively impact the outcome. Combining ability can be lower in the wide crosses conducted at the base of the breeding pyramid. Breeders need patience to give these programs time to achieve some level of recombination and assortment before identifying lines for advancement to the intermediate level of the breeding pyramid. The breeding strategies at the base of the pyramid typically involve more time and effort with regard to hybridization and evaluation. The amount of diversity incorporated at the base level should reflect the amount of resources available for long-term yield objectives. Ideally, all breeders should be working to some extent at each level of the breeding pyramid.

Bean breeders must recognize the level of the pyramid at which they are currently working and to which level they must look for further improvement. These decisions can vary dramatically among breeding programs, but the model should be constant. If a breeder works exclusively with elite lines, the lack of diversity in one's germplasm base may limit further progress. The development of an intermediate level program to find more diversity or collaboration with other breeders may become critical. If the breeder is working on germplasm enhancement through wide crossing, the breeding program must be adequately structured to identify high-yielding lines that may be useful at a higher level in the pyramid or be of value to other breeders. Many breeders may recognize the level of the breeding pyramid at which they expend the most effort with the realization that they are not achieving their desired objective. Our challenge to those breeders is to become involved in all three levels of the pyramid, since many programs may be restricted at one or two levels. Collaboration and germplasm exchange among programs would be one way in which breeders, focusing on the apex, could continue to maximize the diversity entering their programs. Similar collaboration would be encouraged for more basic programs.

5. FUTURE OPPORTUNITIES FOR IMPROVING YIELD

Current trends in genetic improvement emphasize the need to identify, transfer, and maintain novel sources of genetic variation, particularly the variability found in the wild relatives of crop species. Recognizing that improvement of a quantitative trait, such as yield, will not be realized by incorporating single genes found in alternative organisms such as bacteria or viruses, species that readily hybridize with *P. vulgaris* must be examined. Wild relatives are available in *P. vulgaris*, but natural populations are threatened by encroachment, and concern is widespread that much of this untapped variation may be lost. Meanwhile, the practical limitation faced by breeders, when using this germplasm source, is identifying desirable alleles within genotypes exhibiting largely undesirable agronomic characters. The genetic control of the "domestication syndrome" (marked phenotypic differences between the cultivated and wild bean) in cultivated bean involves few genes with large effects (25-30%) which account for a substantial (40-50%) part of the phenotypic variation (Koinange et al., 1996). In addition, these same genes are localized in three regions of the genome, which suggests that variability was greatly reduced by domestication and additional useful (novel) variability should exist within the wild species. The introgression of diversity from the wild to the cultivated *P. vulgaris* should present no special difficulties since these phenotypic differences are under simple genetic control (Koinange et al., 1996). The favorable alleles, however, are masked by unfavorable genes and linkages in the unadapted wild species, restricting their use for the genetic improvement of quantitative traits. Alternative mating schemes involving wild relatives should be investigated, while increasing the recovery of progenies with a cultivated phenotype (Singh et al., 1995).

Wild relatives, however, must first be collected, maintained, and characterized. Meaningful characterization of all exotic germplasm is difficult since many collections contain redundant variation and require special evaluation environments or protocols. As a solution, the core concept was proposed (Brown, 1989). A core collection or representative sample of genotypes is identified based on simple objective criteria, such as different geographical origin, seed color, and size. The sample population is tested for the trait of interest, and further sampling of accessions from the original collection is based on passport criteria similar to the selected genotypes. For example, the core concept has been used to identify nine white mold-tolerant bean genotypes from a core collection of 202 accessions (Miklas et al., 1998). Screening for improved yield traits in unadapted genotypes is illogical, since any desirable alleles will be masked by the inherently unadapted background of most exotic germplasm. To use germplasm better, sorghum [*Sorghum bicolor* (L.) Moench] breeders in North America proposed a conversion program (Stephens et al., 1967) with the Tropical Agriculture Research Station (TARS) in Mayaguez, Puerto Rico. A similar collaborative conversion program in common bean focuses on the tropical Andean germplasm which, because of photoperiod sensitivity and type IV growth habit, makes both crossing and evaluation difficult in temperate regions. The simple introgression of determinacy (*fin* gene) and photoperiod insensitivity (*ppd, hr* genes) will allow for the testing and evaluation of introgressed tropical germplasm in high latitudes. The TARS program is responsible for crossing and selfing F_1 progeny, while selection in F_2 populations for growth habit and photoperiod insensitivity is conducted in temperate regions. Photoperiod insensitive F_3 progeny are returned to the tropics for additional crossing with tropical materials. The cyclic nature of this breeding program is essential to enhance useful genetic variation.

Another breeding scheme to simultaneously identify and transfer positive alleles for yield was developed in tomato (*Lycopersicon esculentum* Mill., Tanksley & McCouch, 1997;Tanksley et al., 1996). The potential of exploiting a wild relative (*L. pimpinellifolium*) of tomato was demonstrated, using advanced-backcross QTL analysis to enhance elite cultivars for a number of quantitative traits (Tanksley et al., 1996). Superior lines were selected in the BC_2, at which stage adequate adaptation is available but some wild background remains. Molecular markers associated with QTL for yield introgressed from the wild can be identified and further used to transfer alleles to other cultivated germplasm. Using a high-density molecular map in rice (*Oryza sativa* L.), Xiao et al. (1997) showed that 15% of the BC_2 interspecific (*O. rufipogon*) test cross populations out-yielded the high-yielding commercial hybrid. Despite the low yield potential of the wild species (*O. rufipogon*), two QTL, each associated with a 17% increase in yield, have been introduced into the cultivated hybrid. This method is a long-term strategy especially for self-pollinated crops, where four generations of inbreeding must follow the three generations of hybridization. It may, however, succeed in exploiting wild germplasm where other methods have failed. Similar approaches must be considered in *Phaseolus* breeding. Use of the wild species is essential since the genetic variability from one cultivated gene pool has limited use for improving quantitative traits in the other gene pool. A point of consideration for bean breeders attempting the advanced-backcross QTL method concerns the choice of wild relatives to use in the program. The selection of suitable wild parents for crossing is complex, since valuable genes or QTL for yield are not expressed. Selection can be facilitated using molecular marker methodologies to identify only unique parental germplasm within the wild gene pools (Tohme et al., 1996).

Marker assisted selection (MAS), originally postulated to facilitate identification of genotypes possessing desired single gene combinations, was eventually applied to facilitate selection of QTL. In this way, genotypes that are superior in performance for complex traits can be selected in earlier generations (Eathington et al., 1997). Although many reports have identified molecular markers associated with QTL for yield, very few studies have addressed the practical effectiveness of MAS for quantitative traits. In bean, Schneider et al. (1997) identified markers associated with improved performance under drought stress. Although these markers proved effective in selecting superior genotypes within the population from which they were identified, selection using these same markers, conducted in a different population, was not successful. Apparently, markers associated with QTL may be population-specific, demonstrating different effects in different populations. Furthermore, significant marker-QTL associations for yield are often location-specific (Eathington et al., 1997). Despite this restrictive application, many of the same parents are continuously used for cultivar development, so markers associated with QTL specific to a valuable parent may still prove useful. Likewise, markers associated with alleles desirable in a particular environment may also be beneficial.

6. HETEROSIS AND FUTURE POTENTIAL OF HYBRID BEANS

Breeding for yield in many crops involves the exploitation of hybrid vigor. While this concept has been an integral part of breeding programs for cross-pollinated crop species such as maize and sunflower (*Helianthus annuus* L.), self-pollinating crops have relied primarily on breeding programs designed to develop inbred cultivars. Hybrid development in bean may be regarded as a potential future opportunity to increase yield (Bannerot, 1989). Heterosis in bean has been described for various traits such as plant height (Coyne, 1965), seed yield, and yield components (Chung & Stevenson, 1973; Gutiérrez & Singh, 1985; Singh & Urrea, 1995; Wassimi et al., 1986). For example, the level of heterosis exhibited in a cross of UI 114 pinto by California Light Red Kidney cultivars reached 105% relative to the high-parent value for a single yield component, whereas other yield components within the same cross had either varying levels of heterosis or absence of heterosis (Foolad & Bassiri, 1983). Twenty interracial F_1crosses yielded significantly higher (22 to 53%) than the high parent (Singh & Urrea, 1995). F_1 heterosis for yield, measured as gain above the mid-parent value, is dependent upon the level of genetic diversity of the parents. Parents of divergent growth habit have various levels of F_1 heterosis for yield (Ghaderi et al., 1984). For example, F_1 heterosis for yield was 174% above the mid-parent yield value and 146% above the high parent for a cross between parents with growth habit type I and growth habit type III (Nienhuis & Singh, 1986). In another study, parents that were divergent for growth habit, seed size, and geographical origin displayed heterosis between 28% to 47% above the mid-parent value (Gutiérrez & Singh, 1985). No progeny, however, yielded significantly higher than the high-parent value. Genetic distance was correlated to heterotic effect for seed yield, and two yield components (number of pods plant^{-1} and number of seeds pod^{-1}), but not all yield components displayed heterosis (Ghaderi et al., 1984).

In order to produce hybrids that exploit heterosis, a system must be developed to facilitate hybrid production. Cytoplasmic male sterility (CMS) has been identified in bean (Singh et al., 1980). The male sterility found in G08063 (Singh et al., 1980) is maternally inherited, resulting in high pollen abortion rates and the subsequent failure to set seeds or pods (Bassett & Shuh, 1982). A single gene (*Fr*) exhibits incomplete dominance, with partially restored fertility in the F_1, and fully restored fertility in the F_2 (Mackenzie & Bassett, 1987). A second, non-allelic restorer gene, *Fr2*, suppressed the expression of CMS (Mackenzie, 1991). Problems exist in restoring fertility to CMS genotypes, since dominant restoring ability of the known restorer genes is incomplete (Bannerot, 1989; Mackenzie & Bassett, 1987). Restorer genes are crucial to hybrid development, as they are necessary to restore self-fertility in the commercial hybrid seed.

Common bean is typically highly self-pollinating (>95%). Outcrossing rates are low and vary from 0 to 0.007% (Tucker & Harding, 1975), 0 to 0.09% (Park et al., 1996), and 0 to 78.0% (Ibarra-Pérez et al., 1997). Bumblebees, honeybees, and wild bees have been suggested as the possible vehicles of pollen transfer (Ibarra-Pérez et al., 1997). A dwarf outcrossing mutant, *do*, had outcrossing rates of 5 to 47% (Nagata & Bassett, 1985) and up to 39% at another location (Brunner & Beaver, 1989). The dwarf mutant had no abnormal pollen development. A delay in the dehiscence of the anthers facilitated the high outcrossing rate (Nagata & Bassett, 1985). In order to ensure effective cross-pollination, higher outcrossing rates must be developed in a number of different bean seed types. Many obstacles must be overcome before a hybrid development program in bean is feasible. The diversity that exists within bean cannot be fully utilized for heterotic potential, since breeders must work within a single seed type for commercial hybrid development. Unfortunately, the amount of diversity within a market class of dry bean is limited (Adams, 1977; McClean et al., 1993). For CMS to be economically feasible, restorer genes that restore fertility to 100% in the F_1 plants will be necessary. In addition, the outcrossing potential in bean is extremely limited. Finally, the ability to produce hybrids through manual manipulation would be logistically unreasonable due to the high labor intensity for hybridization and low fecundity rates in common bean.

7. CONCLUSIONS

Most plant breeders will agree that breeding for yield can be conducted effectively only at the apex of the pyramid. Why discuss a three-tiered pyramid to meet equivalent objectives? Yield is the ultimate culmination of a multitude of physiological and morphological processes that must be effectively accumulated in an additive and developmental sequence to optimize performance and maximize yield. Performance genes must be built into a new cultivar and the structure of the pyramid provides a mechanism to achieve that objective. Managing the complex interaction of the 30,000 genes estimated to collectively contribute to biomass and yield in bean is a formidable task (Wallace & Yan, 1998). The constant flow of new germplasm from the lower levels offsets the inevitable loss of genetic variability caused by apex-level (elite x elite) breeding which culminates in a narrow genetic base. Breeders recognize that the pyramid can be used for other breeding purposes, but generally the most successful cultivars are yield-based and require a structured system to effectively accumulate minor genes. Defensive traits such as pest resistance or quality traits can be extracted from more exotic germplasm at the lower levels of the pyramid, but these progeny ultimately must be advanced and selected on the basis of yield. In the absence of a structured system, breeding efforts for yield will be less organized and likely less productive.

Yield is the coordinate expression of many genes, so elite parents should possess the greatest accumulation of favorable alleles. Certain parents are recognized for their breeding value, the ability to pass on the greatest proportion of those accumulated genes to their progeny. Other elite lines lack this capacity, despite exhibiting high yield potential. Identifying and utilizing the former as parents is critical in breeding for yield. Bean breeders must facilitate communication of information on combining ability for yield. Information concerning pest resistance characteristics of germplasm is regularly disseminated without any mention of combining ability for yield. Cultivars like Jamapa and Porrillo black beans, Sierra pinto, and C-20 navy bean dominate the pedigrees of new cultivars in their respective seed types, not as sources of disease resistance traits, but for exceptional breeding value for yield. Early identification and sharing of information on combining ability, rather than a preoccupation with pest resistance, is vital to cultivar improvement.

Since the final step in breeding for yield must be focused at the apex of the pyramid, the reader may query the significance of the contributions funneled from the lower levels of the pyramid. Breeders must rely on the identification of new genes and gene combinations that will contribute to future yield improvement. Lower levels of the pyramid allow for a recombinational process, utilizing cyclic breeding systems to create unique gene combinations, combined with an organized method to cross among diverse seed types, gene pools, growth habits, and highly adapted genotypes. As breeders, we also need to look further than the cultivated species of bean for future improvement. Colleagues working in other crops have demonstrated the potential for finding unique genes or QTL masked within the wild species. Identifying and incorporating such QTL into a cultivated bean background will require time and effort. Ignoring this potential would be shortsighted. The opportunity to use the wild relatives from diverse gene pools of bean is unique. The pyramid is the structure that provides breeders with an opportunity to move improved genotypes upward, for ultimate utilization in cultivars. Since vision is important to achieve this end result, the pyramid offers a valuable visual representation of the breeding process that is sometimes lost in detail. Recognizing the many constraints that limit progress in bean breeding programs, breeders must still accomplish the objective of improving yield without neglecting the other non-yield-related traits vital for the success of future bean cultivars.

REFERENCES

Acosta Gallegos, J.A. & M.W. Adams. 1991. Plant traits and yield stability of dry bean (*Phaseolus vulgaris*) cultivars under drought stress. J. Agric. Sci. (Cambridge) 117:213-219.

Acosta Gallegos, J.A., J.Z. Castellanos, S. Nuñez-Gonzáles, R. Ochoa-Márquez, R. Rosales-Serna & S.P Singh. 1995a. Registration of 'Flor de Mayo M38' common bean. Crop Sci. 35:941-942.

Acosta Gallegos, J.A. & J. Kohashi-Shibata. 1989. Effect of water stress on growth and yield of indeterminate dry-bean (*Phaseolus vulgaris*) cultivars. Field Crops Res. 20:81-93.

Acosta Gallegos, J.A., R. Ochoa-Márquez, F.J. Ibarra-Pérez , A. Pajarito-Ravelero & I. Sánchez-Valdez. 1995b. Registration of 'Pinto Villa' common bean. Crop Sci. 35:1211.

Acosta Gallegos, J.A. & J.W. White. 1995. Phenological plasticity as an adaptation by common bean to rainfed environments. Crop Sci. 35:199-204.

Adams, M.W. 1967. Basis of yield component compensation in crop plants with special reference for field beans, *Phaseolus vulgaris*. Crop Sci. 7:505-510.

Adams, M.W. 1973. Plant architecture and physiological efficiency. pp. 226-278. *In* D. Wall (ed.), Potential of Field Beans and Other Food Legumes in Latin America. CIAT, Cali, Colombia.

Adams, M.W. 1977. An estimation of homogeneity in crop plants, with special reference to genetic vulnerability in dry bean, *Phaseolus vulgaris* L. Euphytica 26:665-679.

Adams, M.W. 1982. Plant architecture and yield breeding in *Phaseolus vulgaris* L. Iowa State J. Res. 56:225-254.

Aggarwal, V.D., S.K. Mughogho, R.M. Chirwa & S. Snapp. 1997. Field-based screening methodology to improve tolerance of common bean to low-P soils. Commun. Soil. Sci. Plant Anal. 28:1623-1632.

Bannerot, H. 1989. The potential of hybrid beans. pp. 111-134. *In* S. Beebe (ed.), Current Topics in Breeding of Common Bean. CIAT, Cali, Colombia.

Bassett, M.J. & D.M. Shuh. 1982. Cytoplasmic male sterility in common bean. J. Amer. Soc. Hort. Sci. 107:791-793.

Beaver, J.S., E. Arnaud-Santana & D.P. Coyne. 1996. Yield stability of determinate and indeterminate red mottled beans. J. Agric. Univ. Puerto Rico 80:187-189.

Beaver, J.S. & J.D. Kelly. 1994. Comparison of two selection methods for the improvement of dry bean populations derived from crosses between gene pools. Crop Sci. 34:34-37.

Beaver, J.S., C. Paniagua, D.P. Coyne & G. Freytag. 1985. Yield stability of dry bean genotypes in the Dominican Republic. Crop Sci. 25:923-926.

Bett, K.E. & T.E. Michaels. 1994. Accumulation of blight resistance using a conical cross. Annu. Rpt. Bean Improv. Coop. 37:75-76.

Bliss, F.A. 1993a. Breeding common bean for improved nitrogen fixation. Plant & Soil 152:71-79.

Bliss, F.A. 1993b. Utilizing the potential for increased nitrogen fixation in common bean. Plant & Soil 152:157-160.

Brothers, M.E. & J.D. Kelly. 1993. Interrelationship of plant architecture and yield components in the pinto bean ideotype. Crop Sci. 33:1234-1238.

Brown, A.H.D. 1989. Core collections: A practical approach to genetic resource management. Genome 31:818-824.

Brunner, B.R. & J.S. Beaver. 1989. Estimation of outcrossing of the common bean in Puerto Rico. HortScience 24:669-671.

Buttery, B.R., S.J. Park & D.J. Hume. 1992. Potential for increasing nitrogen fixation in grain legumes. Can. J. Plant Sci. 72:323-349.
Casali, V.W.D. & E.C. Tigchelaar. 1975 Computer simulation studies comparing pedigree, bulk and single seed descent selection in self pollinated populations. J. Amer. Soc. Hort. Sci. 100:364-367.
Castellanos, J.Z., H. Guzmán-Maldonado, J.A. Acosta Gallegos & J.D. Kelly. 1995. Effects of hardshell character on cooking time of common bean grown in the semiarid highlands of Mexico. J. Sci. Food Agric. 69:437:443.
Castellanos, J.Z., S.H. Guzmán-Maldonado, A. Jiménez, C. Mejía, J. Muñoz-Ramos, J.A. Acosta Gallegos, G. Hoyos, E. López-Salinas, D. Gonzáles-Eguiarte, R. Salinas-Pérez, J. Gonzáles-Acuña, J. A. Muñoz-Villalobos, P. Fernández-Hernández & B. Cáceres. 1997. Preferential habits of consumers of common bean (*Phaseolus vulgaris* L.) in Mexico. Arch. LatinoAmericanos de Nutricion 47:163-167.
Chaverra, M.H. & P.H. Graham. 1992. Cultivar variation in traits affecting early nodulation of common bean. Crop Sci. 32:1432-1436.
Chen, H., Z. Shen & P.H. Li. 1982. Adaptability of crop plants to high temperature stress. Crop Sci. 22:719-725.
Chung, J.H. & E. Stevenson. 1973. Diallel analysis of the genetic variation in some quantitative traits in dry beans. N.Z. J. Agric. Res. 16:223-231.
Clarke, J.M., R.M. Depauw & T.F. Townley-Smith. 1992. Evaluation of methods for quantification of drought tolerance in wheat. Crop Sci. 32:723-728.
Cooper, R.L. 1985. Breeding semidwarf soybeans. Plant Breed. Rev. 3:289-311.
Cooper, R.L. 1988. Breeding for yield in soybeans: Early generation testing and specific adaptation to high versus low yield environments. pp. 287-306. *In* S. Beebe (ed.), Current Topics in Breeding of Common Bean. CIAT, Cali, Colombia.
Coyne, D.P. 1965. Component interaction in relation to heterosis for plant height in *Phaseolus vulgaris* L. variety crosses. Crop Sci. 5:17-18.
Coyne, D.P., D.S. Nuland, D.T. Lindgren & J.R. Steadman. 1994. 'Chase' pinto dry bean. HortScience 29:44-45.
Coyne, D.P., J.R. Steadman, D.T. Lindgren & D.S. Nuland. 1991. 'Starlight' - Great northern dry bean. HortScience 26:441-442.
Cramer, M.M. & L.W. Kannenberg. 1992. Five years of HOPE: the hierarchical open-ended corn breeding system. Crop Sci. 32:1163-1171.
Dahiya, B.S., R.S. Waldia, L.S. Kaushik & I.S. Solanki. 1984. Early generation yield testing versus visual selection in chickpea (*Cicer arietinum* L.). Theor. Appl. Genet. 68:525-529.
Donald, C.M. 1968. The breeding of crop ideotypes. Euphytica 17:385-403.
Eathington, S.R., J.W. Dudley & G.K. Rufener II. 1997. Usefulness of marker-QTL associations in early generation selection. Crop Sci. 37:1686-1693.
Escalente Estrada., J.A., L. Aguilar García & M.T. Rodríguez-G. 1998. Seed yield and its components of common beans in sodic saline soils. Annu. Rpt. Bean Improv. Coop. 41:202-203.
Fehr, W. 1987. Breeding methods for cultivar development. pp.250-293. *In* J.R. Wilcox (ed.), Soybeans: Improvement, production and uses. Amer. Soc. Agron. Madison, U.S.A.
Fischer, R.A. & R. Maurer. 1978. Drought resistance in spring wheat cultivars. I. Grain yield responses. Aust. J. Agric. Res. 29:897-912.
Foolad, M.R. & A. Bassiri. 1983. Estimates of combining ability, reciprocal effects and heterosis for yield and yield components in a common bean diallel cross. J. Agric. Sci. 100:103-108.

Fouilloux, G. & H. Bannerot. 1988. Selection methods in the common bean (*Phaseolus vulgaris*). pp. 503-542. *In* P. Gepts (ed.), Genetic Resources of *Phaseolus* Beans. Kluwer, Dordrecht, Netherlands.
Franco, M.C., S.T. A. Cassini, V.R. Oliveira, J.M. Diniz, S.M. Tsai & C. Vieira. 1998. Nodulation capacity and combinatory ability between cultivars from Mesoamerican and Andean genic pools of common bean (*Phaseolus vulgaris* L.). Annu. Rpt. Bean Improv. Coop. 41:133-134.
Ghaderi, A., M.W. Adams & A.M.Nassib. 1984. Relationship between genetic distance and heterosis for yield and morphological traits in dry edible bean and faba bean. Crop Sci. 24:37-42.
Gill, J.S., M.M. Verma, R.K. Gumber & J.S. Brar. 1995. Comparative efficiency of four selection methods for deriving high-yielding lines in mungbean (*Vigna radiata* (L.) Wilczek). Theor. Appl. Genet. 90:554-560.
Gonzáles, J.W., D.P. Coyne, W.W. Stroup & C. Carlson. 1998. Variation for iron induced deficiency chlorosis and response to foliar sprays in dry beans grown on high pH calcareous soils. Annu. Rpt. Bean Improv. Coop. 41:200-201.
Grafton, K.F., J.R. Venette & K.C. Chang. 1997a. Registration of 'Hatton' pinto bean. Crop Sci.37:288.
Grafton, K.F., J.R. Venette & K.C. Chang. 1997b. Registration of 'Maverick' pinto bean. Crop Sci.37:1672.
Graham, P.H. 1981. Some problems of nodulation and symbiotic N fixation in *Phaseolus vulgaris* L.: A review. Field Crops Res. 4:93-112.
Graham, P.H. & P. Ranalli. 1997. Common bean (*Phaseolus vulgaris* L.). Field Crops Res. 53:131-146.
Graham, P.H. & J.H. Rosas. 1977. Growth and development of indeterminate bush and climbing cultivars of *Phaseolus vulgaris* L. inoculated with *Rhizobium*. J. Agric Sci. (Cambridge) 88:503-508.
Gutiérrez, J.A. & S.P. Singh. 1985. Heterosis and inbreeding depression in dry bush beans, *Phaseolus vulgaris* L. Can. J. Plant Sci. 65:243-249.
Holley, R.N. & J.C. Wynne. 1986. Effectiveness of stratified mass selection for yield in intrasubspecific and intersubspecific crosses of peanut. Peanut Sci. 13:33-35.
Hovinen, S. 1988. Breeding of a protein pea ideotype for Finnish conditions. J. Agric. Sci. Finl. Maataloustieteellinen Aikak. 60:1-72.
Ibarra-Pérez, F.J., B. Ehdaie & J.G. Waines. 1997. Estimation of outcrossing rate in common bean. Crop Sci. 37:60-65.
Ininda, J., W.R. Fehr, S.R. Cianzio & S.R. Schnebly. 1996. Genetic gain in soybean populations with different percentages of plant introduction parentage. Crop Sci. 36:1470-1472.
Kao, W.Y., M.P. Comstock & J.R. Ehleringer. 1994. Variation in leaf movements among common bean cultivars. Crop Sci. 34:1273-1278.
Kelly, J.D. 1998. Bean roots -- A plant breeders perspective. Annu. Rpt. Bean Improv. Coop. 41:214-215.
Kelly, J.D. & M.W. Adams. 1987. Phenotypic recurrent selection in ideotype breeding of pinto beans. Euphytica 36:69-80.
Kelly, J.D., M.W. Adams, A.W. Saettler, G.L. Hosfield, G.V. Varner, M.A. Uebersax & J. Taylor. 1990. Registration of 'Sierra' pinto bean. Crop Sci. 30:745-746.
Kelly, J.D., M.W. Adams & G.V. Varner. 1987. Yield stability of determinate and indeterminate dry bean cultivars. Theor. Appl. Genet. 74:516-21.

Kelly, J.D., G.L. Hosfield, G.V. Varner, M.A. Uebersax, P.N. Miklas & J. Taylor. 1992a. Registration of 'Alpine' great northern bean. Crop Sci. 32:1509-1510.

Kelly, J.D., G.L. Hosfield, G.V. Varner, M.A. Uebersax, N. Wassimi & J. Taylor. 1992b. Registration of 'Aztec' pinto bean. Crop Sci. 32:1509.

Kelly, J.D., J.M. Kolkman & K. Schneider. 1998. Breeding for yield in dry bean (*Phaseolus vulgaris* L.). Euphytica 102:343-356.

Kisha, T.J., C.H. Sneller & B.W. Diers. 1997. Relationship between genetic distance among parents and genetic variances in populations of soybean. Crop Sci. 37:1317-1325.

Koinange, E.M.K., S.P. Singh & P. Gepts. 1996. Genetic control of the domestication syndrome in common bean. Crop Sci. 36:1037-1045.

Kolkman, J.M. & J.D. Kelly. 1999. Drought stress and bean productivity during the 1998 growing season. Michigan Dry Bean Digest 23(2):2-3.

Kornegay, J., J.W. White & O. Ortíz de la Cruz. 1992. Growth habit and gene pool effects on inheritance of yield in common bean. Euphytica 62:171-180.

Ludlow, M.M. & R.C. Muchow. 1990. A critical evaluation of traits for improving crop yields in water-limited environments. Adv. Agron. 43:107-153.

Mackenzie, S.A. 1991. Identification of a sterility-inducing cytoplasm in a fertile accession line of *Phaseolus vulgaris* L. Genetics 127:411-416.

Mackenzie, S.A. & M.J. Bassett. 1987. Genetics of fertility restoration in cytoplasmic male sterile *Phaseolus vulgaris* L. I. Cytoplasmic alteration by a nuclear restorer gene. Theor. Appl. Genet. 74:642-645.

McClean, P.E., J.R. Myers & J.J. Hammond. 1993. Coefficient of parentage and cluster analysis of North American dry bean cultivars. Crop Sci. 33:190-197.

Miklas, P., R. Delorme, R. Hannan & M. Dickson. 1998. Using the core concept to identify new white mold resistance sources in common bean. Annu. Rpt. Bean Improv. Coop. 41:82-84.

Monteverde-Penso, E.J. & J.C. Wynne. 1988. Evaluation of three cycles of recurrent selection for fruit yield within a population of Virginia-type peanut. Crop. Sci. 28:75-78.

Monteverde-Penso, E.J., J.C. Wynne, T.G. Isleib & R.W. Mozingo. 1987. A comprehensive breeding procedure utilizing recurrent selection for peanuts. Peanut Sci. 14:1-3.

Nagata, R.T. & M.J. Bassett. 1985. A dwarf outcrossing mutant in common bean. Crop Sci. 25:949-954.

Nienhuis, J. & S.P. Singh. 1985. Effects of location and plant density on yield and architectural traits of common bean. Crop Sci. 25:579-584.

Nienhuis, J. & S.P. Singh. 1986. Combining ability analysis and relationships among yield, yield components and architectural traits in dry bean. Crop Sci. 26:21-27.

Nienhuis, J. & S.P. Singh. 1988. Genetics of seed yield and its components in common bean (*Phaseolus vulgaris* L.) of Middle American origin. I. General combing ability. Plant Breed. 101:141-154.

Ntare, B.R., M.E. Akenova, R.J. Redden & B.B. Singh. 1984a. The effectiveness of early generation (F_3) yield testing and the single seed descent procedures in two cowpea (*Vigna unguiculata* (L.) Walp.) crosses. Euphytica 33:539-547.

Ntare, B.R., R.J. Redden, B.B. Singh & M.E. Akenova. 1984b. Evaluation of early generation selection procedures for yield in cowpea (*Vigna unguiculata* (L.) Walp.). Field Crops Res. 9:91-100.

Obisesan, I.O. 1992. Evaluation of pedigree and single seed descent selection methods for cultivar development in cowpea (*Vigna unguiculata* L. Walp). Plant Breed. 108:162-168.

Panter, D.M. & F.L. Allen. 1995. Using best linear unbiased predictions to enhance breeding for yield in soybean: I. Choosing parents. Crop Sci. 35:397-405.

Park, S.J., T.E. Michaels & B.N. Dhanvantari. 1998. Breeding for resistance to common bacterial blight and its effects on agronomic performance and processing quality in dry bean. Annu. Rpt. Bean Improv. Coop. 39:90-91.

Park, S.J., T.E. Michaels, J.R. Myers, D.W.A. Hunt & K. Stewart-Williams. 1996. Outcrossing rates of common beans grown in Ontario and Idaho. Annu. Rpt. Bean Improv. Coop. 39:90-91.

Park, S.J. & J.C. Tu. 1987. Comparison of near-isogenic lines for anthracnose *Are/are* genes in common bean, *Phaseolus vulgaris* L. Euphytica 36:251-256.

Parker, J.B. 1997. World production of pulses declining. Michigan Dry Bean Digest 21(2):10-16.

Patiño, H. & S.P. Singh. 1989. Visual selection for seed yield in the F_2 and F_3 generations of nine common bean crosses. Annu. Rpt. Bean Improv. Coop. 32:79-80.

Pereira, P.A.A. & F.A. Bliss. 1987. Nitrogen fixation and plant growth of common bean (*Phaseolus vulgaris* L.) at different levels of phosphorus availability. Plant & Soil 104:79-84.

Piha, M.I. & D.N. Munns. 1987. Nitrogen fixation capacity of field-grown bean compared to other grain legumes. Agron. J. 79:690-696.

Rachier, G.O., C.S. Wortmann & J.S. Tenywa. 1998. Low soil phosphorus tolerance in common bean as affected by root architecture. Annu. Rpt. Bean Improv. Coop. 41:206-207.

Ramírez-Vallejo, P. & J.D. Kelly. 1998. Traits related to drought resistance in common bean. Euphytica 99:127-136.

Rasmusson, D.C. 1987. An evaluation of ideotype breeding. Crop Sci. 27:1140-1146.

Reddy, R.K. & S.K. Rao. 1980. Component analysis of yield factors in segregating population of pigeonpea (*Cajanus cajan* L. Millsp*)* ideotype breeding. Legume Res. 3:33-37.

Rennie, R.J. 1984. Comparison of N balance and ^{15}N isotope dilution to quantify N_2 fixation in field-grown legumes. Agron. J. 76:785-790.

Rojas, M., S. Beebe, F. Pedraza, F. Muñoz, D. Beck, J. Lynch, S. Snapp & J. Tohme. 1998. Mapping genes for root characters in common bean. Annu. Rpt. Bean Improv. Coop. 41:1-2.

Schaafsma, A.W., C. Cardona, J.L. Kornegay, A.M. Wylde & T.E. Michaels. 1998. Resistance of common bean lines to potato leafhopper (Homoptera: Cicadellidae). J. Econ. Entomol. 91:981-986.

Schneider, K.A., R. Rosales-Serna, F.J. Ibarra-Pérez , B. Cazares-Enríquez, J.A. Acosta Gallegos, P. Ramírez-Vallejo, N. Wassimi & J.D. Kelly. 1997. Improving common bean performance under drought stress. Crop Sci. 37:43-50.

Shirtliffe, S.J., J.K. Vessey, B.R. Buttery & S.J. Park. 1996. Comparison of growth and N accumulation of common bean (*Phaseolus vulgaris* L.) cv. OAC Rico and its two nodulation mutants, R69 and R99. Can. J. Plant Sci. 76:73-83.

Shonnard, G. & P. Gepts. 1994. Genetics of heat tolerance during reproductive development in common bean. Crop Sci. 34:1168.

Singh, S.P. 1991. Breeding for seed yield. pp. 383-443. *In* A. Van Schoonhoven & O. Voysest (eds.), Common Beans. Research for Crop Improvement. C.A.B. Int., Wallingford, U.K. & CIAT, Cali, Colombia.

Singh, S.P. 1992. Common bean improvement in the tropics. Plant Breed. Rev. 10:199-269.

Singh, S.P. 1994. Gamete selection for simultaneous improvement of multiple traits in common bean. Crop Sci. 34:352-355.

Singh, S.P. 1995a. Selection for water-stress tolerance in interracial populations of common bean. Crop Sci. 35:118-124.

Singh, S.P. 1995b. Selection for seed yield in Middle American versus Andean x Middle American interracial common-bean populations. Plant Breed. 114:269-271.
Singh, S.P. 1996. Effect of random intermating in early generations on seed yield in interracial populations of common bean. Can. J. Plant Sci. 76:683-688.
Singh, S.P., C. Cajiao, J.A. Gutiérrez, J. García, M.A. Pastor-Corrales & F.J. Morales. 1989a. Selection for seed yield in inter-gene pool crosses of common bean. Crop Sci. 29:1126-1131.
Singh, S.P., C. Cardona, F.J. Morales, M.A. Pastor-Corrales & O. Voysest. 1998. Gamete selection for upright carioca bean with resistance to five diseases and a leafhopper. Crop Sci. 38:666-672.
Singh, S.P. & J.A. Gutiérrez. 1990. Effect of plant density on selection for seed yield in two population types of *Phaseolus vulgaris* L. Euphytica 51:173-178.
Singh, S.P., R. Lépiz, J.A. Gutiérrez, C. Urrea, A. Molina & H. Terán. 1990. Yield testing of early generation populations of common bean. Crop Sci. 30:874-878.
Singh, S.P., A. Molina & P. Gepts. 1995. Potential of wild common bean for seed yield improvement of cultivars in the tropics. Can. J. Plant Sci. 75:807-813.
Singh, S.P., A. Molina, C.A. Urrea & J.A. Gutiérrez. 1993. Use of interracial hybridization in breeding the race Durango common bean. Can. J. Plant Sci. 73:785-793.
Singh, S.P. & H. Terán. 1998. Population bulk versus F_1-derived family methods of yield testing in early generations of multiple-parent interracial and inter-gene pool crosses of common bean. Can. J. Plant Sci. 78:417-421.
Singh, S.P., H. Terán, A. Molina & J.A. Gutiérrez. 1992a. Combining ability for seed yield and its components in common bean of Andean origin. Crop Sci. 32:81-84.
Singh, S.P., H. Terán, C.G. Muñoz & J.C. Takegami. 1999. Two cycles of recurrent selection for seed yield in common bean. Crop Sci. 39:391-397.
Singh, S.P. & C.A. Urrea. 1995. Inter- and intraracial hybridization and selection for seed yield in early generations of common bean. Euphytica 81:131-137.
Singh, S.P., C.A. Urrea, J.A. Gutiérrez & J. García. 1989b. Selection for yield at two fertilizer levels in small-seeded common bean. Can. J. Plant Sci. 69:1011-1017.
Singh, S.P., C.A. Urrea, A. Molina & J.A. Gutiérrez. 1992b. Performance of small-seeded common bean from the second selection cycle and multiple-cross intra- and interracial populations. Can. J. Plant Sci. 72:735-741.
Singh, S.P., J.W. White & J.A. Gutiérrez. 1980. Male sterility in dry beans. Annu. Rpt. Bean Improv. Coop. 23:55-57.
Stephens, J.C., F.R. Miller & D.T. Rosenow. 1967. Conversion of alien sorghum to early combine genotypes. Crop Sci. 7:396.
Stewart-Williams, K. & J. Myers. 1995. 46th Annual report of the National Cooperative Dry Bean Nurseries. Univ. of Idaho, Kimberly, U.S.A.
Subbarao, G.V., C. Johansen, A.E. Slinkard, R.C. Nageswara Rao, N. P. Saxena & Y.S. Chauhan. 1995. Strategies for improving drought tolerance in grain legumes. Critical Rev. in Plant Sci. 14:469-523.
Tanksley, S.D., S. Grandillo, T.M. Fulton, D. Zamir, Y. Eshed, V. Petiard, J. López & T. Beck-Bunn. 1996. Advanced backcross QTL analysis in a cross between an elite processing line of tomato and its wild relative *L. pimpinellifolium*. Theor. Appl. Genet. 92:213-224.
Tanksley, S.D. & S.R. McCouch. 1997. Seed banks and molecular maps: unlocking genetic potential from the wild. Science 277:1063-1066.
Thomas, C.V., R.M. Manshardt & J.G. Waines. 1983. Teparies as a source of useful traits for improving common bean. Desert Plants 5:43-48.

Tohme, J., D. Orlando-Gonzáles, S. Beebe & M.C. Duque. 1996. AFLP analysis of gene pools of a wild bean core collection. Crop Sci. 36:1375-1384.

Tucker, C.L. & J. Harding. 1975. Outcrossing in common bean *Phaseolus vulgaris* L. J. Amer. Soc. Hort. Sci. 100:283-285.

Urrea, C.A. & S.P. Singh. 1995. Comparison of recurrent and congruity backcrossing for interracial hybridization in common bean. Euphytica 81:21-26.

Voysest, O., M.C. Valencia & M.C. Amézquita. 1994. Genetic diversity among Latin American, Andean and Mesoamerican common bean cultivars. Crop Sci. 34:1100-1110.

Wallace, D.H., J.P. Baudoin, J.S. Beaver, D.P. Coyne, D.E. Halseth, P.N. Masaya, H.M. Munger, J.R. Myers, M. Silbernagel, K.S. Yourstone & R.W. Zobel. 1993. Improving efficiency of breeding for higher crop yield. Theor. Appl. Genet. 86:27-40.

Wallace, D.H., P.N. Masaya, R. Rodríquez & R.W. Zobel. 1995. Genotype, temperature and genotype x temperature interaction effects on yield of bean (*Phaseolus vulgaris* L.). pp. 893-915. *In* M. Pessarakli (ed.), Handbook of Plant and Crop Physiology. Marcel Dekker, New York, U.S.A.

Wallace, D.H. & W. Yan. 1998. Plant breeding and whole system crop physiology, improving adaptation, maturity and yield. C.A.B. Int., Wallingford, U.K.

Wassimi, N.N., T.C. Isleib & G.L. Hosfield. 1986. Fixed effect genetic analysis of a diallel cross in dry beans (*Phaseolus vulgaris* L.). Theor. Appl. Genet. 72:449-454.

Wells, R., T. Bi, W.F. Anderson & J.C. Wynne. 1991. Peanut yield as a result of fifty years of breeding. Agron. J. 83:957-961.

Welsh, W., W. Bushuk, W. Roca & S.P. Singh. 1995. Characterization of agronomic traits and markers of recombinant inbred lines from intra- and interracial populations of *Phaseolus vulgaris* L. Theor. Appl. Genet. 91:169-177.

White, J. & A. Gonzáles. 1990. Characterization of the negative association between seed yield and seed size among genotypes of common bean. Field Crops Res. 23:159-175.

White, J.W., J.L. Kornegay, J. Castillo, C.H. Cajiao & G. Tejada. 1992. Effect of growth habit on yield of large-seeded bush cultivars of common bean. Field Crops Res. 29:151-161.

White, J.W., R. Ochoa M., F.J. Ibarra-Pérez & S.P. Singh. 1994. Inheritance of seed yield, maturity and seed weight of common bean (*Phaseolus vulgaris)* under semi-arid rainfed conditions. J. Agric. Sci. (Cambridge) 122:265-273.

White, J.W. & S.P. Singh. 1991a. Sources and inheritance of earliness in tropically adapted indeterminate common bean. Euphytica 55:15-19.

White, J.W. & S.P. Singh. 1991b. Breeding for adaptation to drought. pp.501-551. *In* A. van Schoonhoven & O. Voysest (eds.), Common Beans: Research for Crop Improvement. C.A.B. Int., Wallingford, U.K. & CIAT, Cali, Colombia.

Whiteaker, G., G.C. Gerloff, W.B. Gabelman & D. Lindgren. 1976. Intraspecific differences in growth of beans at stress levels of phosphorus. J. Amer. Soc. Hort. Sci. 10:1471-1475.

Williams, R.W., R.J. Lawn, B.C. Imrie & D.E. Bythe. 1988. Development of a weather resistant ideotype for mungbean (*Vigna radiata* (L.) Wilzcek). pp. 223-224. *In* K.S. McWhirter, R.W. Downes & B.J. Read (eds.), Proceedings of the Ninth Australian Plant Breeding Conference. Agr. Res. Institute, Wagga Wagga, Australia.

Xiao, J., J. Li, S. Grandillo, S.N. Ahn, L. Yuan, S.D. Tanksley & S.R. McCouch. 1997. Identification of heterosis-enhancing QTL alleles from a wild rice relative, *Oryza rufipogon*. Plant Genome V. p147 (abstract).

Yan, X., J.P. Lynch & S.E. Beebe. 1995. Genetic variation for phosphorus efficiency of common bean in contrasting soil types. I. Vegetative response. Crop Sci. 35:1086-1093.

Chapter 9

Improvement of Medium-Seeded Race Durango Cultivars

Mark A. Brick & Kenneth F. Grafton
Soil and Crop Science, Colorado State University, Fort Collins, CO 80523-1170, U.S.A.; Plant Sciences, North Dakota State University, Fargo, ND 58105, U.S.A.

Key words: breeding history, breeding for specific constraints, diversity among cultivars, improvement programs, production and consumption

1. INTRODUCTION

This chapter will focus on improvement strategies for dry beans (*Phaseolus vulgaris* L.) that represent race Durango from the Middle American (M) gene pool. Landraces and cultivars from this race have medium seed size (25 to 40 g/100 seeds), predominantly indeterminate growth habit III, small to medium ovate or cordate leaflets, thin stems and branches, and short internodes, with fruiting commencing from and concentrated in basal nodes (Singh et al., 1991a). Four to five flattened rhombohedric seeds are produced in flattened pods. The Durango race originated in the semiarid highlands of Mexico.

Seed color varies among accessions of race Durango and includes beige (Bayo), yellow, cream, gray, black, white, red, and pink. Some accessions have irregular, spotted, speckled or striped patterns, such as Anasazi (white with red blotch), Ojo de Cabra (cream striped), and pinto (cream spotted). Commercial market classes in the U.S.A. and Canada include Anasazi, pinto, great northern, pink, and small red (a.k.a. red Mexican), and in Mexico include pinto, pink, Bayo, Ojo de Cabra, black, and others. Of these, pinto, great northern, pink, and small red types also are grown in Chile, Europe,

Africa, and western Asia (Singh, 1992). Small cream striped (Carioca), pink, red, black, beige, and white are also found in race Mesoamerica, but these often possess about one third size smaller seeds and are distinct from race Durango types. Pinto is the largest race Durango class produced in the U.S.A., Bayo is the largest in Mexico, and great northern is the largest in Europe and West Asia.

2. HISTORY OF BEAN CULTIVATION IN NORTH AMERICA

Race Durango beans were cultivated in Mexico and the U.S.A. in pre-Columbian times. Kaplan (1965) reported that beans found at an archaeological site in the southwestern U.S.A. were cultivated 2,300 years ago and that they likely originated from Mexico and Central America. New World settlers cultivated dry beans in the western U.S.A. from selections and landraces of small red, pink, pinto, and great northern beans that were cultivated by Native Americans or introduced from the semiarid highlands of Mexico. The early introduced landraces were grown on small acreage in the U.S.A. until state and federal governments initiated improvement programs during the late 1800's and early 1900's. Among the early bean researchers, R. A. Emerson worked at the University of Nebraska from 1898 until 1912. During the early 20th century, research programs introduced new germplasm and determined that seed stocks produced in the semiarid western U.S.A. had lower levels of seed-borne pathogens, notably bean common mosaic (BCM, caused by a potyvirus), bacterial blights, and anthracnose [caused by *Colletotrichum lindemuthianum* (Sacc. & Magn.) Bri. & Cav.] (Adams, 1996). By 1917, seed production shifted from the central and eastern to the semiarid western U.S.A., where most commercial bean seed is produced today (Dean, 1994). In 1906, the Michigan Agricultural College (Michigan State University) was the first institution in the U.S.A. to employ a full-time bean breeder (Adams, 1996). Subsequently, other universities and the USDA initiated breeding programs and efforts concentrated on improving disease resistance in introduced germplasm and selection of resistant plants in heterogeneous stocks.

3. GENETIC DIVERSITY AMONG RACE DURANGO CULTIVARS

The categorization of cultivated common bean germplasm into six races within two primary centers of origin has contributed immensely to our knowledge about the level of genetic diversity in the primary gene pool (Singh et al., 1991a). For example, Singh et al. (1991b, c) were able to distinguish medium-seeded landraces and cultivars of race Durango from other races of Middle American and Andean gene pools, based on variation among seed proteins, allozymes, and morphological and agronomic traits. This categorization also provided us with a better understanding of combining ability among races and genetic incompatibility factors among germplasm from different centers of origin that can result in lethality of F_1 hybrids and/or give rise to crippling of trifoliolate leaves (Singh & Gutiérrez, 1984; Singh & Molina, 1996).

The level of genetic diversity in dry bean cultivars from race Durango has been evaluated extensively using coefficient of parentage, allozymes, morphological characteristics, and cluster analyses (McClean et al., 1993; Singh et al., 1991a, b; Voysest et al., 1994). McClean et al. (1993) used coefficient of parentage and reported three major cluster groups among 143 cultivars/lines grown in the U.S.A. corresponding to small-, medium-, and large-seeded market classes. The medium-seeded cluster was composed of cultivars or lines from race Durango, including, pinto, great northern, small red, and pink types. They further reported that among the three major cluster groups, the medium-sized seed group had the broadest genetic diversity both within and among clusters. The introgression of genes from the Andean gene pool for resistance to rust [caused by *Uromyces appendiculatus* (Pers.) Ung.] and BCM accounts for much of the diversity seen in the medium-seeded group. Introgression of genes for resistance to pathogens and upright plant architecture from Mesoamerica race has also contributed to diversity in these bean cultivars. Current breeding programs in North America are using broad-based parental germplasm to improve disease resistance and agronomic performance.

In Latin America, approximately 600 dry bean cultivars have been released from 1930 to 1964 (Voysest et al., 1994). Of these, Mexico released 13 cultivars of Bayo, pinto, and Ojo de Cabra types, while Chile released four pinto cultivars. Germplasm from race Durango contributed 43% to the development of these cultivars, with races Nueva Granada and Mesoamerica contributing 18 and 13%, respectively. Apparently, the race Nueva Granada, from a more distant center of domestication, contributed higher portion of genetic diversity in the race Durango cultivars than its more closely related races Jalisco or Mesoamerica.

4. CURRENT PRODUCTION AND CONSUMPTION

In 1996, dry beans were planted on approximately 85, 700, and 1,100 thousand ha in Canada, the U.S.A., and the central highlands of Mexico, respectively. Total production is usually 1.1 to 1.3 million MT in the U.S.A. and Mexico, and 200,000 to 250,000 MT in Canada. Production of pinto, great northern, pink, and small red beans constituted 25 and 50% of the total production in Canada and the U.S.A., respectively. In 1998, the USDA/AMS National Statistics Service reported that the leading U.S.A. states in order of production were North Dakota (437,410 MT), Michigan (197,544 MT), Nebraska (163,660 MT), Colorado (135,714 MT), Minnesota (104,876 MT), Idaho (94,285 MT), California (90,000 MT), Washington (36,776 MT), Wyoming (39,285 MT), and New York (19,017 MT). Average yields range from 1,100 to 1,300 kg/ha under rainfed conditions in the Midwest to about 2,000 to 2,200 kg/ha under irrigated production in California and the Pacific Northwest (Anonymous, 1998). The U.S.A. production of race Durango beans is concentrated in North Dakota, Colorado, Nebraska, and Idaho; however, states such as California, Michigan, Kansas, New Mexico, Texas, Washington, and Wyoming also contribute to the total. Approximately 316,000 ha of race Durango beans were planted in North Dakota and Colorado in 1998. In Canada, the primary market class of race Durango bean is pinto in Manitoba, while Alberta also produces pinks, great northern, and small red beans.

In Mexico, most of the dry beans from race Durango are grown in the semiarid highlands with over 1.1 million ha, and 95% of this is produced under rainfed conditions (Kelly & Acosta Gallegos, 1997). The main season for bean production in the semiarid highlands is spring to autumn and the crop comprises 65% of total national area planted. The state of Chihuahua produces almost exclusively pinto and Ojo de Cabra, while Aguas Calientes, Durango, and Zacatecas produce pinto, Bayo, black, and other classes of bean. The major producers are Zacatecas (600-700,000 ha), Durango (280-300,000 ha), Chihuahua (220-250,000 ha), and Aguas Calientes (120-150,000). Annual fluctuations in yield occur because of environmental and disease problems. The average yield in 1995 varied from 250 kg/ha in Aguas Calientes to 490 kg/ha in Durango and averaged 460 kg/ha with total national production at 0.7 million MT. During 1996, the weather was more favorable and the average spring-summer yield was 600 kg/ha, with total national production at 1.1 million MT.

In the U.S.A., more than 80% of the pinto and great northern bean production is sold in dry packages, with the remainder sold as canned or processed products. Most pink and small red beans are sold as canned products. Consumption of dry beans has increased in the U.S.A. and Canada since 1980, partly due to the recognition that dry beans in the diet lower blood cholesterol and partly due to increased consumption of ethnic food from Latin America. Today, annual consumption of dry bean in the U.S.A. is 3.5 kg/person compared to approximately 15 kg/person in Mexico (Kelly & Acosta Gallegos, 1997). Common bean is second only to corn (*Zea mays* L.) as a source of protein, supplying 20 to 40% of the protein and carbohydrates in the diet of the Mexican population (Avila-Rodríguez et al., 1996). Mexican consumers show a preference for certain beans such as Flor de Mayo, Peruano or Azufrado, and Bayo, and an aversion for some seed types such as Ojo de Cabra, despite a substantial difference in market prices (Castellanos et al., 1996).

5. PRODUCTION CONSTRAINTS

Seasonally, the U.S.A. and Canadian bean production is limited by many environmental conditions, including heat and drought stress, early and late season frost, soil compaction, hail, and others. In addition, pathogens that cause anthracnose, rust, common bacterial blight [*Xanthomonas campestris* pv. *phaseoli* (Smith) Dye], bacterial brown spot [*Pseudomonas syringae* pv. *syringae* (van Hall)], halo blight [*Pseudomonas syringae* pv. *phaseolicola* (Burkh.)], white mold [*Sclerotinia sclerotiorum* (Lib.) de Bary], and soil borne fungi that cause root rots further reduce yield. In Mexico, additional pests such as angular leaf spot [caused by *Phaeoisariopsis griseola* (Sacc.) Ferr.], Mexican bean beetle (*Epilachna varivestis* Mulsant), and bean pod weevil (*Apion godmani* Wagner), and constraints such as inadequate use of fertilizer and pesticides, and use of poor quality seed exacerbate problems.

6. BREEDING FOR SPECIFIC CONSTRAINTS

6.1 Bean Common Mosaic

Bean common mosaic (BCM) virus is a nonpersistent, stylet-borne, aphid-transmitted potyvirus that is found in nearly every country where beans are grown. Primary transmission of this viral pathogen occurs through infected seed that ensures the wide dispersal of the disease and forces the implementation of rigid constraints on seed production programs in many countries. BCM remains a problem where both aphid vectors are present and locally grown seed of susceptible cultivars is used (Gálvez & Morales,

1989). Natural infection of susceptible cultivars can range from 0 to 100% (Kaiser et al., 1968; Zaumeyer & Thomas, 1957).

According to Drijfhout (1978), BCM was first reported in 1899 by Iwanoski in Russia. BCM was first reported as a disease of bean in the U.S.A. in 1916 (Stewart & Reddick, 1917). The first virus (BCMV) strain found was labeled "virus 1", now termed the "Type strain" assigned to pathogroup I by Drijfhout (1978). Reddick & Stewart (1919) were the first to report that BCMV was transmitted through the seed and Nelson & Down (1933) reported pollen transmission. Seed transmission is dependent on the age of a plant at time of infection. Infection of young plants increases the chance of seed transmission, while seed transmission is less probable when older plants are infected. Seven pathogenicity groups of BCMV have been identified (Drijfhout, 1978) and classified broadly into two serogroups. Serogroup A is classified as bean common mosaic necrosis virus (BCMNV) and interacts with the major dominant resistant *I* gene, and serogroup B is classified as BCMV (Mink et al., 1994).

A series of recessive strain specific resistance (*bc*) genes exist; all require a nonspecific *bc-u* gene to be functional. The *bc-u* allele is independent of the three other loci (*bc-1, bc-2, and bc-3*). The *bc-1* and *bc-2* have multiple alleles, two of which are functional (Drijfhout, 1978). These recessive resistance alleles are present in some genotypes of race Durango and provide resistance to a wide range of BCMV and BCMNV strains (Drijfhout, 1978; Forster et al., 1994; Richards & Burkholder, 1943; Silbernagel, 1969). The notable exception is the *bc-3* allele found in IVT 7214, which conditions resistance to all known strains of BCMV and BCMNV (Drijfhout, 1978). This and other resistance genes are being introgressed into race Durango cultivars (Kelly, 1997). The *I* gene is not commonly found in race Durango. But this dominant gene is present in some genotypes of races Mesoamerica and Nueva Granada and has been used by breeders worldwide to provide resistance to a wide range of BCMV strains. Kelly (1997) described the typical hypersensitive resistant reaction, referred to as systemic top necrosis, when resistant plants were challenged with strains of BCMNV. The use of this dominant, hypersensitive necrosis-inducing gene has become increasingly important since 1980, when strain NL 8 in pathogroup 3 and strain US 10 in pathogroup 7 were reported in Idaho (Forster et al., 1994). The appearance of strains in pathogroup 3 caused concern because of their ability to elicit the black root reaction at normal growing temperatures in cultivars possessing the unprotected *I* gene. Identification of strains in pathogroup 7 also caused concern because these were not known to occur in the U.S.A. (Forster et al., 1994).

The occurrence of strain differences in BCMV and BCMNV has prompted breeders to look for new sources of resistance. The dominant *I* gene confers resistance to all strains but causes black root or systemic necrosis due to a hypersensitive reaction with BCMNV, depending on the environment and host-strain combination. Many breeding programs are combining the recessive *bc* alleles with *I* gene, termed "protected *I* gene", to provide resistance to all known strains of BCMV or BCMNV. Cultivars with gene combinations such as *Ibc-1*^{2}, *Ibc-2*^{2}, and *Ibc-3* are resistant to a wide range of BCM pathogroups and prevent expression of the hypersensitive necrosis. Many pinto breeding programs are presently incorporating the *bc-3* allele alone or in combination with the *I* gene for cultivar development. Because the *bc-3* allele is epistatic over all other resistance alleles, molecular markers are used to combine the recessive and dominant resistance alleles (Kelly et al., 1995). Markers that are linked to the *I* and *bc-3* alleles are available (Haley et al., 1994; Johnson et al., 1997; Miklas et al., 1996). Breeders should be cautious about using *bc-3* in the absence of the *I* gene, because recently it has been observed that some lines with putative *bc-3* gave apparent resistance to BCMNV but susceptibility to BCMV (Miklas et al., 1998b). It is now recommended that immune resistance to a set of representative strains, one from each pathogroup, be used to select for and confirm presence of the *bc-3* allele (P.N. Miklas, personal communication, 1998). Immune resistance to an individual strain, NL-3 of pathogroup VI, has been the traditional way to select for *bc-3* (Kelly, 1997).

Only recently have the *I* and the *bc-3* genes been introgressed into Durango race (Kelly et al., 1998a, b; Stavely et al., 1997). Genotypes possessing the *I* gene alone, or in combination with a recessive resistance gene, include pintos Kodiak (Kelly et al., 1998b), BelDakMi-RMR-10 and -14 (Stavely et al., 1997), 92-US-1006 (Silbernagel, 1994), and USWA-19 and -20 (Silbernagel & Hang, 1997b); great northerns Matterhorn (Kelly et al., 1998a), BelMiNeb-RMR-3 (Stavely et al., 1997), and USWA-12 and -13 (Silbernagel & Hang, 1997c); small reds USWA-6, USWA-8, and USWA-9 (Silbernagel & Hang, 1997a) and Anasazi type (Silbernagel et al., 1998a). Forster et al. (1994) reported BCMV reaction of a wide range of cultivars and breeding lines.

6.2 Bean Rust

In the U.S.A., rust damage occurs east of the continental divide and can reduce bean yield from 40 to 100% (Stavely & Pastor-Corrales, 1989) depending on time of infection, cultivars grown, environmental conditions, pathogen race, and severity of the disease. More than 89 races of the pathogen occur in the U.S.A. (Stavely, 1984b, 1998; Stavely et al., 1989; Steadman et al., 1995). Most older cultivars of race Durango in the U.S.A. are highly susceptible to rust.

Zaumeyer & Harter (1941) were among the first to study and report sources of resistance to bean rust. Since 1979, thousands of bean lines, mainly plant introductions, have been evaluated for reactions to different races of rust, and sources of resistance are available in both the Andean and Middle American gene pools (Gonzáles-García & Grafton, 1996; Stavely, 1988, 1989). Resistance is controlled by several dominant genes (Grafton et al., 1985; Stavely, 1984a), and resistance genes may condition reactions to one or more strains (Kelly, 1995). The gene symbol, adapted from Ballantyne (1978), uses *Ur-* followed by a number to identify each gene (Kelly et al., 1996).

Recent work in the U.S.A. has focused on using the *Ur-3* gene as a source of resistance (Coyne et al., 1994; Stavely & Grafton, 1989; Stavely et al., 1992). In 1979, the pinto cultivar Olathe (Wood & Keenan, 1982) was released with resistance to the prevalent rust races in the Midwest U.S.A. Resistance genes in Olathe were obtained from a dark red kidney cultivar of race Nueva Granada. Other cultivars such as Topaz, Sierra, Frontier, Chase, Maverick, Alpine, Matterhorn, Kodiak, UI 320, and UI 465 provide resistance to rust races occurring in different production areas. Lately, breeders are also using other genes from both the Andean and Middle American gene pools (Kelly et al., 1994, 1996). Consequently, several germplasm lines that possess two or more rust resistance genes have been released (Stavely & Grafton, 1989; Stavely et al., 1992, 1994, 1997). These lines, identified by prefixes that indicate the collaborating location or state, are pintos, BelDak-RR-1 and -2 and BelDakMi -3 through BelDakMi-RMR-14 and great northern, BelNeb-RR-1 and -2 and BelMiNeb-RMR-3.

To facilitate selection and pyramiding genes for resistance to rust, molecular markers such as random amplified polymorphic DNA (RAPD) have been used (Haley et al., 1993; Kelly et al., 1994; Miklas et al., 1993; see also Chapter 4 by Kelly & Miklas). Molecular markers enable breeders to combine desired resistance genes without the need to use multiple inoculations with different rust races or test-crossing to susceptible stocks to identify genotypes that carry all desired resistant alleles.

6.3 Common Bacterial Blight

Common bacterial blight (CBB) is an important disease of dry beans worldwide (Saettler, 1989; Singh, 1992; Zaumeyer & Thomas, 1957). The disease is favored by warm temperatures and high relative humidity, with greater damage to plants at 28°C than at lower temperatures (Saettler, 1989). Initial symptoms of the disease are small irregular lesions along the edge of the leaflet. The leaflet may have water-soaked margins, and eventually dark brown lesions may form. A scalded brownish-yellow appearance develops on the canopy as the disease progresses (Schwartz, 1994). The pathogen can overwinter in plant debris, resulting in infected volunteer beans early in the next growing season (Saettler, 1989; Schwartz, 1994).

Many sources of resistance to CBB have been identified (Arnaud-Santana et al., 1993, 1994; Coyne & Schuster, 1969a, 1983; Zapata et al., 1998). Recently, new sources of resistance to CBB were identified in tepary and runner bean (*P. coccineus* L.) (Miklas et al., 1994a, b; Park & Dhanvantari, 1987; Singh & Muñoz, 1999) as well as in wild *P. vulgaris* (Navarrete-Maya & Acosta Gallegos, 1997).

Resistance to CBB can be quantitatively and qualitatively controlled depending on source germplasm, with pod and leaf resistance controlled by different genes (Aggour & Coyne, 1989; Arnaud-Santana et al., 1994; Coyne & Shuster, 1974c, 1983; Coyne et al., 1965; Jung et al., 1996, 1997; Park et al., 1998b; Silva et al., 1989; Valladares-Sánchez et al., 1979, 1983). Welsh & Grafton (1998) reported that resistance to CBB from a *P. coccineus* source introgressed into *P. vulgaris* was controlled by a single recessive gene and minor genes that modified the resistant reaction. One to four QTL have been found to account for a significant portion of the variation in leaf and pod resistance derived from great northern Nebraska 1 Selection 27 (Dursun et al., 1995; Jung et al., 1996, 1997; Miklas et al., 1996; Park et al., 1996, 1998a). Differential reaction in plant organs and low heritabilities for some sources interfere with the transfer of resistance genes into improved cultivars.

Resistance from tepary bean has been introgressed into common bean using interspecific hybridization (Honma, 1956; McElroy, 1985; Scott & Michaels, 1988; Singh & Muñoz, 1999). The original interspecific cross was made by Honma between Tepary 4 and the susceptible great northern line, Montana 5. GN Nebraska #1 Sel. 27 derived from this cross (Coyne & Schuster, 1974d) has been extensively used in the parentage of other cultivars, including Valley, Starlight, Jules, Tara, and Chase (Coyne & Schuster, 1969a, 1970, 1974a, b; Coyne et al., 1994). Tolerance to CBB is also found in cultivars Harris, Ivory, Beryl, and Marquis. Reaction to CBB among other pinto, great northern, pink, and small reds has been reported (Lindgren et al., 1993, 1994, 1995, 1996,1997). Recently, three new lines with resistance to CBB were released (Miklas, 1998b).

Molecular markers linked to resistance genes derived from tepary bean now assist in selection and offer unique opportunities to incorporate both quantitative and qualitative resistance (Park et al., 1996, 1998a; see also Chapters 3 by Gepts and 4 by Kelly & Miklas). Other novel molecular techniques such as the use of P-32 labeled DNA can quantify bacterial populations on infected plants and assist in the selection of resistant germplasm and may also improve selection response to both qualitative and quantitative resistance (Michaels et al., 1996)

6.4 Bacterial Brown Spot

Bacterial brown spot (BBS) is a sporadic problem in dry bean production areas of the High Plains, central and eastern U.S.A., and Canada. The bacterium has a very wide host range, including lima bean (*P. lunatus* L.), soybean [*Glycine max* L. (Merr.)], broad bean (*Vicia faba* L.), cowpea (*Vigna unguiculata* L. Walp.), and other species (Zaumeyer & Thomas, 1957). However, only isolates recovered from common bean are virulent to common bean (Saad & Hagedorn, 1971). Burkholder (1930) was the first to report BBS on common bean. Symptoms of the disease appear as small water-soaked spots most visible on the underside of the young foliage (Schwartz, 1994). A narrow, greenish-yellow border may surround some of the lesions; however, when the lesion matures, it typically develops a brown spot appearance. Necrotic tissue in the center of the lesion may fall out, producing a shot-hole appearance. Pods also can be infected and may show numerous, small water-soaked lesions and pod distortion.

Outbreaks of BBS are most serious when temperatures are moderately cool (below 24°C) and relative humidity is greater than 95% for 24 h or longer (Schwartz, 1994). The bacteria can overwinter on weeds, in the soil, and on volunteer beans and bean residue following an infection (Legard & Hunter, 1990; Legard & Schwartz, 1987). The bacteria can survive on the surface of bean leaves as epiphytes without causing disease, or may incite lesions under favorable environmental conditions. The pathogen typically infects young foliage and symptoms are rarely seen on older leaves. Rain or hailstorms with high winds that wound plants favor infection and spread of the pathogen if a source of inoculum is present.

Antonius (1982) reported that correlations between reactions on pods and leaves in the F_2 plants and their progenies suggested that a common genetic system controlled both reactions. Screening techniques have been shown to identify useful germplasm (Antonius & Hagedorn, 1983; Silbernagel, 1980). Bean germplasm that possess resistance include Wisconsin BBSR 17, BBSR 28, and BBSR 130 (Hagedorn & Rand, 1977, 1980); WBR 133 (Daub & Hagedorn, 1979); Earliwax, PI 326353, PI 326419, and PI 339377 (Hagedorn et al., 1972); PI 313390, PI 313416, PI 313297, and PI 313404 (Antonius & Hagedorn, 1979). BBS resistance in race Durango germplasm is not common. Coyne & Schuster (1969b) reported that GN 1140 showed moderately high tolerance to BBS in both field and greenhouse tests. Other cultivars that possess BBS tolerance include Harris, Tara, and Chase (Coyne et al., 1994; Lindgren et al., 1993).

6.5 Halo Blight

Halo bacterial blight is distributed throughout the world and is a problem in regions that have moderate temperatures during the growing season (Schwartz, 1989). Like BBS, the disease is seed transmitted and is favored by moderately cool temperatures and high humidity (Schwartz, 1994). Initial symptoms of the disease are small, greasy, water-soaked spots most visible on the undersurface of the young leaflets. The water-soaked spots later become surrounded by a broad greenish-yellow halo caused by the bacterial toxin. Size and development of the halo is variable and it may not be produced under high temperatures. During severe infection, the disease can become systemic and cause yellowing and death of new foliage if the temperature is favorable. Halo blight symptoms also appear on pods as small, circular, water-soaked spots or streaks. Cream or silver-colored bacterial ooze often is associated with these spots. Pod infection often causes discoloration, shriveling, and contamination of seeds.

Pathogenic variation in the bacteria occurs and is categorized into two major race groups (Race 1 and 2) in the Americas and Europe and up to nine races in Africa (Schwartz, 1989; Taylor et al., 1996). Because different genes control pod, leaf, and systemic reactions (Coyne & Schuster, 1974d; Coyne et al., 1967, 1971; Hill et al., 1972), it is important to incorporate genes for each of these sources of resistance in a breeding program. Resistance to race 1 is found in the small red cultivars UI 3 and UI 35, which provide a hypersensitive reaction (Taylor et al., 1978; Walker & Patel, 1964). The small red PI 150414 is also resistant to races 1 and 2 (Patel & Walker, 1965). Other sources of resistance have been identified (Baggett & Frazier, 1967; Coyne & Cambell, 1984; Keenan, 1988; Lindgren et al., 1993; Schwartz, 1989). The pinto cultivar Chase (Coyne et al., 1994) and great northern cultivar Weihing, released by the University of Nebraska in 1998, have high resistance to halo blight.

6.6 White Mold

White mold is a major concern to dry bean growers in most production areas in the U.S.A. (Steadman, 1983). A survey of dry bean growers in North Dakota and Minnesota found that 76% of the respondents ranked white mold as the most serious problem, with fungicide use on approximately 30% of the acreage in an attempt to control the disease (Lamey et al., 1998). Schwartz & Steadman (1989) cited reports of crop losses averaging 30% in Nebraska with individual field losses as high as 92%. In North Dakota, yield loss and cost of fungicide applications can exceed $20 million annually (J. R. Venette, personal communication, 1998).

The fungus produces irregularly shaped sclerotia that act as overwintering structures and can survive in the soil for at least three years (Cook et al., 1975). Schwartz & Steadman (1978) found that one sclerotium in five kg of soil could result in 46% disease severity in Nebraska. The primary infection sites for bean are decaying blossoms, and a wet period of 16-24 hours for successful infection is required (Abawi & Grogan, 1975). In a bean field, sclerotia germinate to form apothecia, giving rise to ascospores, when the plant canopy has formed and when soil moisture is adequate. Also, sclerotia in cereal grain fields the year following bean will germinate and form apothecia; these apothecia produce ascospores that infect adjacent bean fields. The importance of these spore showers on dry bean infection is unknown, but they probably allow infection of bean flowers in the upper surface of the canopy.

Reducing or eliminating late-season irrigation of fields infested with white mold may reduce disease damage and incidence (Steadman et al., 1976). Canopy development and other architectural traits, including upright plant growth, may allow for less disease development (Fuller et al., 1984b, c; Schwartz et al., 1987). However, erect architecture alone does not necessarily result in avoiding white mold damage (K. Grafton, unpublished data). For example, the relatively erect, indeterminate pinto cultivars Aztec and Ouray exhibit considerable damage to white mold in spite of plant type, while the more prostrate cultivar Chase, with a more open canopy, tends to avoid the disease (Coyne et al., 1994).

Greenhouse and laboratory evaluation techniques for white mold have met with limited success (Dickson et al., 1982; Hunter et al., 1982; Miklas et al., 1992a, b). In general, these tests are labor intensive and time consuming; they require several replications to achieve accurate results, limiting their usefulness for early generation selection. Petzoldt & Dickson (1996) developed an inoculation test called the "straw test", which is quick, easy, and highly repeatable. This test promises to be of major importance in evaluating dry bean germplasm for physiological resistance in the greenhouse and field (Hall & Phillips, 1997, 1998). Miklas et al. (1999) used the straw test to evaluate a subset of the *P. vulgaris* core collection. They evaluated 89 of 201 entries from the core collection and identified several new sources of resistance to white mold.

Fuller et al. (1984a) reported that genetic variation accounted for 74 and 76% of the variation in reaction to white mold in F_2 populations in the greenhouse and field, respectively. They concluded that resistance was quantitatively inherited and due primarily to additive gene action. Lyons et al. (1987) and Miklas & Grafton (1992) reported that physiological resistance to white mold in *P. vulgaris* is low. The latter researchers found that mean lesion length and a disease incidence index were normally distributed, and transgressive segregation occurred for both traits. Heritability estimates for lesion length (0.27, 0.38, and 0.66) were lower than the disease incidence index (0.77, 0.58, and 0.70). Low genetic correlations between lesion length and disease incidence index suggested that selection for both traits was warranted.

Breeders have attempted to improve architecture to avoid or reduce white mold disease development (Coyne et al., 1994; Grafton et el., 1997; Kelly et al., 1990, 1992a, b). Physiological resistance has been identified in navy bean cultivars Laser, Ex Rico 23 (Bunsi), and ND88-106-04 (Steadman et al., 1998). The line G 122, a cream mottled type from Nueva Granada race, also was reported to have physiological resistance (Kmiecik & Nienhuis, 1998). New sources of resistance such as I9365-3, I9365-5, I-9365-31, and 92BG-7 released by Miklas et al. (1998a) have shown superior field performance for white mold (Steadman et al., 1998). Resistance in these lines was derived from interspecific hybridization with *P. coccineus*.

6.7 Root Rots

Root diseases occur throughout the bean growing regions of North America (Abawi, 1989; Silbernagel, 1990; Zaumeyer & Thomas, 1957). Continuous bean cropping, environmental conditions, and soil compaction contribute to problems associated with root diseases. The most important root pathogens are species of *Aphanomyces*, *Fusarium*, *Macrophomina Phymatrotricyhum*, *Pythium*, *Rhizoctonia*, *Sclerotinia*, and *Thielaviopsis* (Abawi, 1989). However, pathogens that predominate and limit production differ from one bean-growing region to another.

Root rot [caused by *F. solani* (Mart.) Sacc. f. sp. *phaseoli* (Burkh.) W.C. Snyder & H.M. Hans] is one of the most prevalent diseases in the world (Abawi & Pastor-Corrales, 1990). The first symptoms of *Fusarium* root rot are longitudinal red to brown streaks on the hypocotyl and taproot on 7 to 10 day old seedlings (Burke & Hall, 1991). The cortex of the hypocotyl becomes progressively streaked and generally necrosis is largely confined to the cortex cells. The upper root system often regenerates adventitious roots on the hypocotyl near the soil surface if conditions are appropriate (Burke & Barker, 1966). Yield reductions depend on disease severity and the ability of the plant to regenerate new healthy roots near the soil surface (Burke & Barker, 1966). Keenan et al. (1974) found 72% of the fields heavily infested with *Fusarium* root rot in the San Juan Basin region of Southwest Colorado, causing yield losses of up to 84%.

Fusarium wilt or *Fusarium* yellows [caused by *F. oxysporum* Schlechtend:Fr. f. sp. *phaseoli* J.B. Kendrick & W.C. Snyder (*Fop*)] can cause rapid yellowing and foliage defoliation, and ultimately result in severe yield loss. Infected plants have reddish purple to reddish orange discoloration in the root and stem vascular tissues. The discoloration can proceed up to the third or fourth node above ground (Schwartz et al., 1989). Destruction is most severe in regions where high soil temperatures, poor water drainage, soil compaction, and drought prevail during part of the growing season. The disease occurs in Mexico (Abawi & Pastor-Corrales, 1990) and the U.S.A. (Armstrong & Armstrong, 1963; de Jensen et al., 1998; Kendrick & Snyder, 1942; Silbernagel & Schwartz, 1988).

Differential reactions among bean cultivars and lines to the pathogen have been reported (Ribeiro & Hagedorn, 1979a; Salgado & Schwartz, 1993). Five pathogenic races of *Fop* have been characterized based upon molecular markers (RAPD, RFLP) and vegetative compatibility groups (Woo et al., 1996). The races were 1 and 4 from the U.S.A., race 2 from Brazil, race 3 from Colombia, and race 5 from Greece. Races 1 and 4 from the U.S.A. correspond to strain ATCC 18131, isolated by Armstrong & Armstrong (1963) in South Carolina, and the Colorado race recovered by Schwartz in 1990 (Salgado & Schwartz, 1993). Buruchara & Bua (1997) reported that among 55 sources from the CIAT core collection, that had resistant or intermediate reactions, two-thirds originated from Mexico.

Early studies indicated that resistance to root rots occurred in *P. coccineus* (Yerkes & Freytag, 1956) and in a black seeded *P. vulgaris* accession, N-203, introduced by Oliver Norvell and later assigned PI 203958 (Azzam, 1957; Wallace & Wilkinson, 1965). Resistance in both of these sources was inherited quantitatively (Boomstra & Bliss, 1977; Hassan et al., 1971). Resistance genes from PI 203958 have been incorporated into pink, pinto, and small red cultivars. Pink cultivars that have utilized this source of resistance include Rosa, Viva, and Gloria, as well as the small red cultivars Rufus (Burke, 1975; Burke & Miller, 1983) and NW 59 and NW 63 (Burke, 1982). Pinto cultivars NW 410 and NW 590 (Burke, 1982), Pindak (Schneiter et al., 1982), and Holberg (Schneiter et al., 1983) also have resistance derived from PI 203958. The pinto germplasm USWA 19 (Silbernagel & Hang, 1997b) combines resistance to root rot, rust, and BCM. The dryland pinto Cahone (Wood et al., 1983) is also known to have resistance to *Fusarium* root rot but is not derived from PI 203958.

Salgado & Schwartz (1993) developed a greenhouse procedure to screen dry bean cultivars and germplasm for resistance to *Fop*. Several pinto lines and tropical accessions, as well as tepary beans were found to have resistance to *Fop* (Salgado et al., 1990, 1994). Ribeiro & Hagedorn (1979b) reported that resistance to the Brazilian race (race 2) of *Fop* was controlled by a single dominant gene and resistance to race 1 from the U.S.A. was controlled by an incompletely dominant gene. Salgado et al. (1995) concluded that the resistant lines from race Durango possessed a single dominant gene, while resistant lines of race Mesoamerica possessed multiple recessive-gene resistance to the Colorado race of *Fop*. Recently, Velásquez-Valle et al. (1997) and Velásquez-Valle & Schwartz (1997) published the reactions to a wide range of dry bean germplasm, including, black, navy, cranberry, small red, pinto, Flor de Mayo, Bayo, and others adapted to Mexico, the U.S.A., and Canada. Among the most resistant lines were pinto Sierra (Kelly et al., 1990), CO 59196, released as pinto Fisher (Fisher et al., 1995), and an Ojo de Cabra line, LEF-2RB, developed by personnel at the National Research Institute for Forestry and Agriculture (INIFAP) in Mexico. Pinto germplasm USWA-20 possesses high resistance to *Fop* in Colorado and Nebraska (Silbernagel et al., 1998b).

6.8 Plant Architecture

Upright type II (Singh, 1982) architecture has been an important component of breeding great northern, pinto, and small red beans in the U.S.A. Upright growth habit provides some disease avoidance for white mold by creating an open canopy that slows the development of the pathogen (Park, 1993) and enables the crop to be direct harvested, thus reducing production costs. Upright growth habit can also improve efficiency of pesticide application by reducing the amount of pesticide that is required to control disease outbreaks.

Adams (1982) proposed the development of the ideal bean plant architecture. Acquaah et al. (1991, 1992) studied plant characteristics that were associated with upright architecture to serve as indicators of desirable upright habit in both small-seeded navy and medium-seeded pinto beans. They used a stepwise regression model to select plant traits that had strong relationships with upright architecture. The traits hypocotyl diameter, plant height, branch angle, and pod number on the main stem were found to be important to identify upright lines. Broad sense heritability estimates for these characteristics were 31 to 38% for hypocotyl diameter, 10 to 80% for height, 51 to 64% for branch angle, and 10 to 80% for pod number on the main stem. Brothers & Kelly (1993) reported that narrow-sense heritability for upright growth habit ranged from 0.42 to 0.62 and indicated that genes controlling plant architecture and seed weight were not linked.

Singh & Gutiérrez (1982) also reported the importance of yield components, erect branching, and internode length to improve architecture of dry bean. Coyne (1982) proposed using crosses between parents with superior morphological traits with high yielding lines in a conventional breeding program to improve architecture.

A negative relationship between erect plant architecture and bean yield was reported by Singh (1992). Nienhuis & Singh (1985) reported that type

III cultivars had higher yield than type I or II, regardless of plant density or environment. They compared lines bred for architecture with commercial checks and found that none outyielded the checks. Factors that cause the negative relationship between upright architecture and yield in medium-seeded beans have not been well documented. One theory suggests that the energy required by the plant to produce the large stem diameter and stiff stems necessary for upright growth habit reduces the amount of energy available for pod and seed production.

Among the first medium-seeded cultivars in North America to combine upright architecture and early maturity were Ouray (Wood, 1982) and Topaz. Kelly & Adams (1987) used recurrent selection to develop pinto bean cultivar Sierra with type II architecture (Kelly et al., 1990). Other cultivars or lines that possess type II architecture include the great northern Alpine (Kelly et al., 1992a), Weihing (D.P. Coyne, unpublished), and Matterhorn (Kelly et al., 1998a); the pinto Aztec (Kelly et al., 1992b), Arapaho (Brick et al., 1995), Maverick (Grafton et al., 1997), Vision (Novartis Seed Co.), Kodiak (Kelly et al., 1998b), CO-81-12001 (Brick et al., 1991), BelDakMi-1, 2, 3, 4, 5, 6, 7, 8, and 9 (Stavely et al., 1992, 1994), and BelDakMi-RMR 10, 11, and 13 (Stavely et al., 1997); small red ICB-6 (Miklas, 1998b); and pink USWA 61 (Silbernagel & Hang, 1997d). Most of these combine type II growth habit and resistance to rust, BCM, and other foliar pathogens. For additional details on breeding for plant type see Chapter 7 by Vandenberg & Nleya and Chapter 8 by Kelly et al.

6.9 Breeding for Yield

From 1940 to 1994, mean seed yield of dry beans increased an average of 10.4 kg/ha per year due to the introduction of new cultivars and technological changes in commercial bean production (Nuland & Taylor, 1996). The introduction of new cultivars during this period accounted for 70% of the annual yield increase. In Colorado, a yield increase of almost 300 kg/ha since 1980 can be attributed to the release of cultivars Olathe and Bill Z.

Bean breeders have not emphasized breeding for yield potential *per se* because they have stressed improved disease resistance and environmental stress tolerance. In general, small and medium-seeded germplasm from the M gene pool has higher yield potential than those from the Andean region (Singh, 1992; Singh & Urrea, 1994). Today, exotic germplasm has been integrated into many programs to broaden the genetic base of the crop and for yield-improvement. Most of the success in broadening the genetic base of pinto beans has come as a result of combining genes from races within the M gene pool. However, Andean germplasm has made important contributions (Voysest et al., 1994). This is despite the fact that population bulks and lines resulting from crosses between different races within the M gene pool have higher mean seed yield than those resulting from crosses between the A and M gene pools (Singh & Urrea, 1994; Singh et al., 1989, 1993; Urrea & Singh, 1995). Furthermore, mean seed yield of lines from crosses between races within the M gene pool were higher than those that resulted from crosses within the race. These results suggest that breeders should continue to broaden the genetic base of race Durango beans by crossing to races Mesoamerica and Jalisco, but must use caution when crossing to Andean races for yield gains.

7. CURRENT CULTIVAR IMPROVEMENT PROGRAMS

7.1 Canada and the U.S.A.

Private and public state and federal research programs are actively involved in dry bean improvement throughout North America. In the U.S.A., the public sector has breeding programs to develop cultivars of race Durango at Colorado State University, Michigan State University, North Dakota State University, University of Idaho, University of Nebraska, and the USDA at Prosser, WA. In Canada, programs working on race Durango are located at the University of Saskatchewan and Agriculture Canada Research Stations at Lethbridge, Morden, and Ontario. Each breeding group has a primary focus to improve production and reduce yield constraints in their regions. Involvement in the national Cooperative Dry Bean Nursery, a yield trial coordinated by the University of Idaho and grown at more than 20 locations, allows breeders to evaluate improved germplasm across a wide geographical area. The Midwest Regional Performance Nursery, coordinated by North Dakota State University, allows breeders to evaluate advanced to mid-generation material before large-scale seed increase.

Development of cultivars and lines with high yield and good agronomic traits is a uniform objective of all breeding programs. In contrast, breeding efforts to minimize pest and environmental constraints in the U.S.A. and Canada are unique to each region. While high levels of resistance to certain diseases, including rust, white mold, bacterial blights, BCM, and root rots is a frequent objective, breeding for anthracnose resistance is a major component only in the Michigan and Ontario programs. Similarly, Michigan, North Dakota, and Manitoba have a major interest in performance under rainfed conditions, while the remainder of the programs evaluate performance under irrigated production. These differences in production regions often dictate the types of disease resistance required and differences in ideotype of the cultivars that are developed, whether erect type II or prostrate type III, early maturing or full season, and adaptation to direct combining. For additional information, see Chapter 7 by Vandenberg & Nleya.

7.2 Semiarid Highlands of Mexico

Public dry bean breeding programs throughout Mexico are organized and conducted by INIFAP. Improvement programs that focus on race Durango beans are located in and around the semiarid central highlands. Intermittent drought stress is the single most important environmental factor limiting bean yield (Acosta-Díaz, 1998; Acosta-Díaz et al., 1997; Acosta Gallegos et al., 1998). In addition, yields are severely reduced by anthracnose, common blight, rust, and root rots (Acosta Gallegos & Navarrete-Maya, 1996a, b; Araya et al., 1996).

Cultivars and lines from race Durango have been shown to possess high levels of resistance to drought stress (Acosta-Díaz et al., 1994; Acosta Gallegos et al., 1997; Singh, 1995). Michigan State University and INIFAP collaborate in an ongoing Bean/Cowpea CSRP project (Kelly & Acosta Gallegos, 1997). Breeding for resistance to drought is a major focus of the project. A study was conducted to determine an effective selection strategy for drought resistance using lines from two recombinant inbred (RI) populations derived from drought-resistant Mexican breeding lines and pinto Sierra (Schneider et al., 1997b). The study examined yield under drought, yield potential, drought susceptibility index, harvest index, and geometric mean as possible selection criteria to identify drought resistant genotypes. Results of the study suggested that a breeding strategy should involve selection based first on the geometric mean yield across stress and non-stress environments, followed by selection of lines based on yield under stress environments. Abebe et al. (1998) also compared selection criteria among dry bean lines and concluded that the geometric mean was the most useful selection criterion to identify high yielding lines under stress and non-stress environments.

Schneider et al. (1997a) utilized marker-assisted selection (MAS) in two RI populations to determine if MAS could be used to identify high yielding lines under drought stress in early generations. Experiments were conducted at seven locations from 1990 to 1994 in Michigan and Mexico. Heritability estimates for yield in one population ranged from 0.55 to 0.59 for stress and non-stress environments, respectively, and from 0.20 to 0.19 for stress and non-stress, respectively, in another population. They found MAS to be effective in Michigan under severe stress and ineffective in Mexico under moderate stress. Five RAPD markers used for MAS in the population with low heritability improved yield by 11% under stress and by 8% under non-stress, whereas conventional selection based on yield was ineffective. But response to conventional selection was three times greater in the population with high heritability, supporting the conclusion that the effectiveness of MAS was inversely proportional to the heritability of the trait under examination. Goytia-Jiménez et al. (1998) also studied the inheritance of yield under stress and non-stress conditions at three locations in central Mexico. They reported that heritability estimates ranged from 0.06 to 0.54 among environments. They concluded that breeding programs aimed at drought resistance should delay selection until advanced generations, and that drought resistant lines selected at one location would not likely be adapted at other locations due to strong genotype by environment interactions.

Phenological plasticity of bean lines has also been shown to be an important adaptive response related to productivity in the semiarid highlands. Plasticity allows cultivars to mature prior to cold temperatures in the fall even when planting occurs later in the season. A range in plasticity was found among breeding lines and wild accessions from the region (Acosta Gallegos & White, 1995). They showed that plasticity was associated with photoperiod sensitivity, but sensitivity alone does not confer plasticity. Retention of this trait appears vital in all new breeding lines under development for drought stress areas, given that the parentage of new lines includes tropically-adapted, photoperiod insensitive germplasm. The high-yielding cultivar Pinto Villa has shown broad adaptation and yield stability in the semiarid highlands, partially due to phenological plasticity and tolerance to low night temperatures during seed filling (Acosta Gallegos et al., 1995). Moreover, Acosta Gallegos & Navarrete-Maya (1996a) suggested pyramiding genes resistant to a broad array of pathogens into locally adapted cultivars.

8. CONCLUSIONS AND FURURE PROSPECTS

Production of dry beans of race Durango is limited by many environmental conditions, including heat and drought stress, early and late season frost, soil compaction, hail, and others. In addition, pathogens that cause anthracnose, rust, common bacterial blight, bacterial brown spot, halo blight, white mold, and root rots further reduce yield potential in Canada and the U.S.A. In Mexico, additional pests such as angular leaf spot, Mexican bean beetle, and bean pod weevil are important.

The level of genetic diversity in dry bean cultivars from race Durango is somewhat narrow in Canada and the U.S.A. Much of the diversity in this group results from introgression of genes from the races Mesoamerica and Nueva Granada for resistance to rust and BCM, and for upright plant architecture. Breeding programs continue to broaden the genetic base of cultivars and breeders in North America have developed cultivars with high yield, disease resistance, good agronomic traits, and excellent nutritional and seed quality. Bean yields have increased linearly an average of 10.4 kg/ha per year during the past 50 years due to the introduction of new cultivars and technological changes in commercial bean production.

Breeding for physiological resistance to white mold has been more problematic because race Durango germplasm contains neither good sources of physiological resistance nor avoidance mechanisms such as upright architecture. Resistant genes from other races and gene pools of common bean and from *P. coccineus* and other species of the secondary and tertiary gene pools should be introgressed into race Durango cultivars. Avoidance mechanisms, including upright and open plant structure, less dense canopies and branching patterns, elevated pod set, and reduced lodging, will continue to be used to reduce white mold damage. Upright architecture also has agronomically desirable features for application of furrow irrigation and direct harvest.

Breeding for yield potential *per se* has not been emphasized thus far because breeders have stressed disease resistance. Breeding programs in the future will expand the use of molecular-marker-assisted selection, diversify the genetic base of cultivars, pyramid resistance genes for both single and multiple pests, and improve cooking and canning qualities. The ideal cultivar will allow producers to harvest early, have a type II growth habit to facilitate direct mechanical harvest, and have disease resistance to prevalent pathogens to reduce the reliance on pesticides. It will also have high yield potential and possess excellent seed quality.

REFERENCES

Abawi, G.S. 1989. Root rot. pp.105-157. *In* H.F. Schwartz & M.A. Pastor-Corrales (eds.), Bean Production Problems in the Tropics. 2nd ed. CIAT, Cali, Colombia.

Abawi, G.S. & R.G. Grogan. 1975. Source of primary inoculum and effects of temperature and moisture on infection of beans by *Sclerotinia sclerotiorum*. Phytopathology 65:300-309.

Abawi, G.S. & M.A. Pastor-Corrales. 1990. Root rots of beans in Latin America and Africa: Diagnosis, research methodologies, and management strategies. CIAT, Cali, Colombia.

Abebe, A., M.A. Brick & R. Kirkby. 1998. Comparison of selection indices to identify productive dry bean lines under diverse environmental conditions. Field Crops Res. 58:15-23.

Acosta-Díaz, E. 1998. Early vigor as a selection criteria for adaptation to drought stress in common bean. Annu. Rpt. Bean Improv. Coop. 41:153-154.

Acosta-Díaz, E., J.A. Acosta Gallegos & R. Rosales-Serna. 1997. Abscission of reproductive structures in common bean under drought stress. Annu. Rpt. Bean Improv. Coop. 40:65-66.

Acosta-Díaz, E., J. Kohashi-Shibata, A. Excalante-Estradá, J.A. Acosta Gallegos & T. Nava-Sánchez. 1994. Yield response to drought in common bean (*Phaseolus vulgaris* L.) varieties. Annu. Rpt. Bean Improv. Coop. 37:165-166.

Acosta Gallegos, J.A., E. Acosta-Díaz, S. Padilla-Ramírez, E. López-Salinas, R.A. Salinas-Pérez, N. Mayek-Pérez & J.D. Kelly. 1998. Seed yield of dry bean cultivars under drought stress. Annu. Rpt. Bean Improv. Coop. 41:151-152.

Acosta Gallegos, J.A., E. Acosta-Díaz, R. Rosales-Serna, S. Padilla-Ramírez, E. López-Salinas, R.A. Salinas-Pérez & J.D. Kelly. 1997. Yield response of dry bean cultivars form different races under drought stress. Annu. Rpt. Bean Improv. Coop. 40:75-76.

Acosta Gallegos, J.A. & R. Navarrete-Maya. 1996a. Search for multiple disease resistance in common bean in the highlands of Mexico. Annu. Rpt. Bean Improv. Coop. 39:180-181.

Acosta Gallegos, J.A. &. R. Navarrete-Maya. 1996b. Identificacion de genotipos de frijol con resistencia multiple a enfermedades en la Mesa Central de Mexico. Agric. Tec. Mex. 22 :187-197.

Acosta Gallegos, J.A., R. Ochoa-Márquez, M.P. Arrieta-Montiel, F.J. Ibarra-Pérez , A. Pajarito Ravelero & I. Sánchez Valdez. 1995. Registration of "Pinto Villa" common bean. Crop Sci. 35:1211.

Acosta Gallegos, J.A. & J.W. White. 1995. Phenological plasticity as and adaptation by common bean to rainfed environments. Crop Sci. 35:199-204.

Acquaah, G., M.W. Adams & J.D. Kelly. 1991. Identification of effective indicators of erect plant architecture in dry beans. Crop Sci. 31:261-264.

Acquaah, G., M.W. Adams & J.D. Kelly. 1992. A factor analysis of plant variables associated with architecture and seed size in dry bean. Euphytica 60:171-177.

Adams, M.W. 1982. Plant architecture and yield breeding. Iowa State J. Res. 56:225-54.

Adams, M.W. 1996. An historical perspective on significant accomplishments in dry bean research. The first 100 years. Annu. Rpt. Bean Improv. Coop. 39:32-45.

Aggour, A.R. & D.P. Coyne. 1989. Heritability, phenotypic correlations, and associations of the common blight disease reactions in beans. J. Amer. Soc. Hort. Sci. 114:828-833.

Anonymous. 1998. USDA/AMS National Agricultural Statistics Service dry bean reports. http://usda.nass.gov.

Antonius, S.H. 1982. Identification and inheritance of resistance to bacterial brown spot (*Pseudomonas syringae*) in *Phaseolus vulgaris*. Ph.D. dissertation. Univ. of Wisconsin, Madison, U.S.A.

Antonius, S.H. & D.J. Hagedorn. 1979. New sources of resistance to *Pseudomonas syringae* in *Phaseolus vulgaris*. Annu. Rpt. Bean Improv. Coop. 22:89-90.

Antonius, S.H. & D.J. Hagedorn. 1983. Genotype assay: number of effective factors (genes) controlling reaction to *Pseudomonas syringae* on snap beans. Phytopathology 73:824 (abstract).

Araya, C.M., J.R. Steadman & J. Acosta Gallegos. 1996. Pathogenic variability of *Uromyces appendiculatus* on dry edible beans in Mexico. Annu. Rpt. Bean Improv. Coop. 39:150-151.

Armstrong, G.M. & J.K. Armstrong. 1963. Fusarium wilt of bean in South Carolina and some host relations of the bean *Fusarium*. Plant Dis. Rptr. 47:1088-1091.

Arnaud-Santana, E., D.P. Coyne, K.M. Eskridge & A.K. Vidaver. 1994. Inheritance, low correlations of leaf, pod, and seed reactions to common blight disease in common beans, and implications for selection. J. Amer. Soc. Hort. Sci. 119:116-121.

Arnaud-Santana, E.T., T. Mmbaga, D.P. Coyne & J.R. Steadman. 1993. Sources of resistance to common bacterial blight and rust in elite *Phaseolus vulgaris* L. germplasm. HortScience 28:644-46.

Avila-Rodríguez, E.H., F.J. Ibarra-Pérez, M.P. Intraiago-Ortega & M.L. Valderrama-Chairez. 1996. Genetic variation of grain quality and nutritional characteristics in common bean (*Phaseolus vulgaris* L.) cultivars of Mesoamerican and Andean origin. Annu. Rpt. Bean Improv. Coop. 39:258-259.

Azzam, H.A. 1957. Inheritance of resistance to *Fusarium* root rot in *Phaseolus vulgaris* L. and *Phaseolus coccineus* L. Diss. Abstr. 18:32-33.

Baggett, J.R. & W.A. Frazier. 1967. Sources of resistance to halo blight in *Phaseolus vulgaris*. Plant Dis. Rptr. 51:661-665.

Ballantyne, B. 1978. The genetic basis of resistance to rust caused by *Uromyces appendiculatus* in bean (*Phaseolus vulgaris*). Ph.D. dissertation. Univ. of Sidney, Sidney, Australia.

Boomstra, A.G. & F.A. Bliss. 1977. Inheritance of resistance to *Fusarium solani* f. sp. *phaseoli* in beans (*Phaseolus vulgaris* L.) and breeding strategy to transfer resistance. J. Amer. Soc. Hort. Sci. 102:186-188.

Brick, M.A., D.R. Wood, M. Ballarin, H.F. Schwartz, C.H. Pearson, J.B. Ogg & J.W. Echols. 1991. Registration of three germplasm lines of pinto bean. Crop Sci. 31:1100.

Brick, M.A., D.R. Wood, H.F. Schwartz, C.H. Pearson, J.B. Ogg, M. Ballarin & J.F. Shanahan. 1995. Registration of Arapaho pinto bean. Crop Sci. 35:1511-1512.

Brothers, M.E. & J.D. Kelly. 1993. Interrelationship of plant architecture and yield components in the pinto bean ideotype. Crop Sci. 33:1234-38.

Burke, D.W. 1975. New red and pink beans resistant to *Fusarium* root rot. Annu. Rpt. Bean Improv. Coop. 18:19-20.

Burke, D.W. 1982. Registration of red Mexican beans Rufus, NW-59 and NW-63. Crop Sci. 22:685-686.

Burke, D.W. & A.W. Barker. 1966. Importance of lateral roots in *Fusarium* root rot of beans. Phytopathology 56:292-294.

Burke, D.W. & R. Hall. 1991. *Fusarium* root rot. pp 9-10. *In* Compendium of Bean Diseases. Amer. Phytopath. Soc., St. Paul, U.S.A.

Burke, D.W. & D.E. Miller. 1983. Control of *Fusarium* root rot with resistant beans and cultural management. Plant Dis. 67:1312-1317.

Burkholder, W.H. 1930. The bacterial diseases of the bean: a comparative study. Cornell Univ. Agr. Exp. Sta. Mem. 127. Ithaca, U.S.A.

Buruchara, R.A. & B. Bua. 1997. Characterizing the core collection for resistance to *Pythium* and *Rhizoctonia* root rots. Annu. Rpt. Bean Improv. Coop. 40:89-90.

Castellanos, F.Z., S.H. Gusmán-Maldonado, A. Jinenix & J.A. Acosta Gallegos. 1996. Preferential habits of common bean (*Phaseolus vulgaris* L.) consumers in Mexico. Annu. Rpt. Bean Improv. Coop. 39:182-183.

Cook, G.E., J.R. Steadman & M.G. Boosalis. 1975. Survival of *Whetzelinia sclerotiorum* and initial infection of dry edible beans in western Nebraska. Phytopathology 65:250-255.

Coyne, D.P. 1982. Modification of plant architecture and crop yield by breeding. HortScience 15:244-247.

Coyne, D.P. & C. Cambell. 1984. Reaction of *Phaseolus vulgaris* germplasm to western Nebraska isolates of halo blight bacterium. Annu. Rpt. Bean Improv. Coop. 27:161.

Coyne, D.P., D.S. Nuland, D.T. Lindgren & J.R. Steadman. 1994. 'Chase' pinto dry bean. HortScience 29:44-45.

Coyne, D.P. & M.L. Schuster. 1969a. A new great northern dry bean variety tolerant to common blight-"Tara". Annu. Rpt. Bean Improv. Coop. 12:12.

Coyne, D.P. & M.L. Schuster. 1969b. Moderate varietal tolerance of beans to brown spot (*Pseudomonas syringae*). Annu. Rpt. Bean Improv. Coop. 12:13.

Coyne, D.P. & M.L. Schuster. 1970. 'Jules' a great northern dry bean variety tolerant to common blight bacterial disease. SB 506. Agric. Exp. Sta., College of Agric. and Home Econ., Univ. of Nebraska, Lincoln, U.S.A.

Coyne, D.P. & M.L. Schuster. 1974a. 'Great Northern Star' dry bean tolerant to bacterial diseases. HortScience 11:621.

Coyne, D.P. & M.L. Schuster. 1974b. 'Great Northern Valley' dry bean. HortScience 9:482.

Coyne, D.P. & M.L. Schuster. 1974c. Differential reaction of pods and foliage of beans (*Phaseolus vulgaris*) to *Xanthomonas phaseoli*. Plant Dis. Rptr. 58:270-81.

Coyne, D.P. & M.L. Schuster. 1974d. Breeding and genetic studies of tolerance to several bean (*Phaseolus vulgaris*) bacterial pathogens. Euphytica 23:651-656.

Coyne, D.P. & M.L. Schuster. 1983. Genetics of and breeding for resistance to bacterial pathogens in vegetable crops. HortScience 18:30-36.

Coyne, D.P., M.L. Schuster & R. Fast. 1967. Sources of tolerance and reaction of beans to races and strains of halo blight bacterial. Plant Dis. Rptr. 51:20-24.

Coyne, D.P., M.L. Schuster & C.C. Gallegos. 1971. Inheritance and linkage of the halo blight systemic chlorosis and leaf watersoaked reaction in *Phaseolus vulgaris* variety crosses. Plant Dis. Rptr. 565:203-207.

Coyne, D.P., M.L. Schuster & L. Harris. 1965. Inheritance, heritability, and response to selection for common blight (*Xanthomonas phaseoli*) tolerance in *Phaseolus vulgaris* field bean crosses. Proc. Amer. Soc. Hort. Sci. 86:373-379.

Daub, M.E. & D.J. Hagedorn. 1979. Resistance of *Phaseolus* line WBR 133 to *Pseudomonas syringae*. Phytopathology 69:946-951.

Dean, L.L. 1994. Seventy years of bean research. Annu. Rpt. Bean Improv. Coop. 37:9-11.

de Jensen, C.E., R. Meronuck & J.A. Percich. 1998. Etiology and control of kidney bean root rot in Minnesota. Annu. Rpt. Bean Improv. Coop. 41:55-56.

Dickson, M.H., J.E. Hunter, M.A. Boettger & J.A. Cigna. 1982. Selection for resistance in *Phaseolus vulgaris* L. to white mold disease caused by *Sclerotinia sclerotiorum* (Lib.) de Bary. J. Amer. Soc. Hort. Sci. 107:231-234.

Drijfhout, E. 1978. Genetic interaction between *Phaseolus vulgaris* and bean common mosaic virus, with implications for strain identification and breeding for resistance. Agric. Res. Rpt. 872. Centre for Agricultural Publishing and Documentation, Wageningen, Netherlands.

Dursun, A., D.P. Coyne, M.F. Mohamed & G. Jung. 1995. Inheritance of resistance to common bacterial blight in tepary beans. Annu. Rpt. Bean Improv. Coop. 38:162-63.

Fisher, A.G., M.A. Brick, D.R. Wood, M. Stack, H.F. Schwartz, J.B. Ogg, C.H. Pearson, J.F. Shanahan & M. Ballarin. 1995. Registration of Fisher pinto bean. Crop Sci. 35:1511.

Forster, R.L., C.A. Strausbaugh, K. Stewart-Williams & J.R. Myers. 1994. Determination of resistance to BCMV in dry edible bean cultivars and breeding lines. Annu. Rpt. Bean Improv. Coop. 37:1-5.

Fuller, P.A., D.P. Coyne & J.R. Steadman. 1984a. Inheritance of resistance to white mold disease in diallel cross of dry beans. Crop Sci. 24:929-933.

Fuller, P.A., D.P. Coyne, J.R. Steadman & R.F. Mumm. 1984b. Inter- and intra-row inter-genotypic competition influences selection for avoidance of white mold disease in dry edible beans. J. Amer. Soc. Hort. Sci. 109:567-572.

Fuller, P.A., J.R. Steadman & D.P. Coyne. 1984c. Enhancement of white mold avoidance and yield in dry beans by canopy elevation. HortScience 19:78-79.

Gálvez, G.E. & F.J. Morales. 1989. Aphid-transmitted viruses. pp. 333-361. *In* H.F. Schwartz & M.A. Pastor-Corrales (eds.), Bean Production Problems in the Tropics. 2nd ed. CIAT, Cali, Colombia.

Gonzáles-García, V.F. & K.F. Grafton. 1996. Inheritance of rust resistance in common bean of diverse origin. Annu. Rpt. Bean Improv. Coop. 39:174-75.

Goytia-Jiménez, M.A., J.A. Acosta Gallegos, S. Padilla-Ramírez, J. Ortíz-Cereceres & J.D. Kelly. 1998. Heritability of seed yield in common bean under irrigated and drought stressed conditions. Annu. Rpt. Bean Improv. Coop. 41:149-150.

Grafton, K.F., J.R. Venette & D.C. Chang. 1997. Registration of `Maverick' pinto bean. Crop Sci. 37:1672.

Grafton, K.F., G.C. Weiser, L.J. Littlefield & J.R. Stavely. 1985. Inheritance of resistance to two races of leaf rust in dry edible bean. Crop Sci. 25:537-539.

Hagedorn, D.J. & R.E. Rand. 1977. Wisconsin (BBSR) 130 bean breeding line. HortScience 12:356.

Hagedorn, D.J. & R.E. Rand. 1980. Wisconsin (BBSR) 17 and 28 snap bean breeding lines resistant to *Psudomonas syringae*. HortScience 15:208-209.

Hagedorn, D.J., R.E. Rand & S.M. Saad. 1972. *Phaseolus vulgaris* reaction to *Pseudomonas syringae*. Plant Dis. Rptr. 56:325-327.

Haley, S.D., L. Afanador & J.D. Kelly. 1994. Identification and amplification of a random amplified polymorphic DNA marker for the *I* gene in common bean. Phytopathology 84:157-60.

Haley, S.D., P.N. Miklas, J.R. Stavely, J. Byrum & J.D. Kelly. 1993. Identification of RAPD markers linked to major rust resistance gene block in common bean. Theor. Appl. Genet. 86:505-12.

Hall, R. & L.G. Phillips. 1997. Resistance of white bean to white mold: field evaluation of straw test. Annu. Rpt. Bean Improv. Coop. 40:138-139.

Hall, R. & L.G. Phillips. 1998. Field evaluation of the straw test for assessing resistance of dry bean to white mold. Annu. Rpt. Bean Improv. Coop. 41:171-172.

Hassan, A.A., D.H. Wallace & R.E. Wilkinson. 1971. Genetics and heritability of resistance to *Fusarium solani* f *phaseoli* in beans. J. Amer. Soc. Hort. Sci. 96:623-627.

Hill, K., D.P. Coyne & M.L. Schuster. 1972. Leaf, pod, and systemic chlorosis reactions in *Phaseolus vulgaris* to halo blight controlled by different genes. J. Amer. Soc. Hort Sci. 97:494-498.

Honma, S. 1956. A bean interspecifiec hybrid. J. Hered. 47:217-220.

Hunter, J.E., M.H. Dickson, M.A. Boettger & J.A. Cigna. 1982. Evaluation of plant introductions in *Phaseolus spp*. for resistance to white mold. Plant Dis. 66:320-322.

Johnson, W.C., P. Guzmán, D. Mandala, A.B.C. Mkandawire, S. Temple, R. L. Gilbertson & P. Gepts.1997. Molecular tagging of the *bc-3* gene for introgression into Andean common bean. Crop Sci. 37:248-254.

Jung, G., D.P. Coyne, P.W. Skroch, J. Nienhuis, E. Arnaud-Santana, J. Bokoski, H.M. Ariyarathne, J.R. Steadman, J.S. Beaver & S.M. Kaeppler. 1996. Molecular markers associated with plant architecture and resistance to common blight, web blight, and rust in common beans. J. Amer. Soc. Hort. Sci. 121:798-803.

Jung, G., P.W. Skroch, D.P. Coyne, J. Nienhuis, E. Arnaud-Santana, H.M. Ariyarathne, S.M. Kaeppler & M.J. Bassett. 1997. Molecular marker-based genetic analysis of tepary bean-derived common bacterial blight resistance in different developmental stages of common bean. J. Amer. Soc. Hort. Sci. 122:329-337.

Kaiser, W.J., D. Danesh, M. Okhovat & H. Mossahebi. 1968. Diseases of pulse crops in Iran. Plant Dis. Rptr. 52:687-91.

Kaplan, L. 1965. Archeology and domestication in American *Phaseolus* (beans). Econ. Bot. 19:358-368.

Keenan, J. 1988. Varietal differences to infection by halo blight. Annu. Rpt. Bean Improv. Coop. 31: 142-143.

Keenan, J.G., H.D. Moore, N. Oshima & L.E. Jenkins. 1974. Effect of bean root rot on dryland pinto bean production in southwestern Colorado. Plant Dis. Rptr. 58:890-892.

Kelly, J.D. 1995. Use of RAPD markers in breeding for major gene resistance to plant pathogens. HortScience 30:461-65.

Kelly, J.D. 1997. A review of varietal response to bean common mosaic potyvirus in *Phaseolus vulgaris*. Plant Varieties and Seeds 10:1-6.

Kelly, J.D. & M.W. Adams. 1987. Phenotypic recurrent selection in ideotype breeding of pinto beans. Euphytica 36:69-80.

Kelly, J.D., M.W. Adams, A.W. Saettler, G.L. Hosfield, G.V. Varner, M.A. Uebersax & J. Taylor. 1990. Registration of 'Sierra' pinto bean. Crop Sci. 30:745-46.

Kelly, J.D., L. Afanador & S.D. Haley. 1995. Pyramiding genes for resistance to bean common mosaic virus. Euphytica 82:207-12.

Kelly, J.D., S. Haley, L. Afanador, P.N. Miklas & J.R. Stavely. 1994. Application of RAPD markers for disease resistance breeding in beans. Annu. Rpt. Bean Improv. Coop. 37:15-16.

Kelly, J.D., G.L. Hosfield, G.V. Varner, M.A. Uebersax, P.N. Miklas & J. Taylor. 1992a. Registration of 'Alpine' great northern bean. Crop Sci. 32:1509-10.

Kelly, J.D., G.L. Hosfield, G.V. Varner, M.A. Uebersax, N. Wassimi & J. Taylor. 1992b. Registration of 'Aztec' pinto bean. Crop Sci. 32:1509.

Kelly, J.D., J.R. Stavely & P.N. Miklas. 1996. Proposed symbols for rust resistance genes. Annu. Rpt. Bean Improv. Coop. 39:25-31.

Kelly, J.D., J. Taylor, G.L. Hosfield, M.A. Uebersax & G.V. Varner. 1998a. Notice of naming and release of 'Glacier', a new early to mid-season upright, disease resistant great northern bean cultivar for Michigan and the Great Lakes region. Annu. Rpt. Bean Improv. Coop. 41:235-236.

Kelly, J.D., J. Taylor, G.L. Hosfield, M.A. Uebersax & G.V. Varner. 1998b. Notice of naming and release of `Kodiak', a new mid-season disease resistant pinto bean cultivar for Michigan and the Great Lakes region. Annu. Rpt. Bean Improv. Coop. 41:237-238.

Kendrick, J.B. & W.C. Snyder. 1942. Fusarium yellows of beans. Phytopathology 32:1010-1014.

Kmiecik, K.A. & J. Neinhuis. 1998. The use of line G122 as a source of white mold resistance in breeding improved processing snap beans for the Midwest. Annu. Rpt. Bean Improv. Coop. 41:21-22.

Lamey, H.A., M.P. McMullen, P.A. Glogoza, R.K. Zollinger, J.L. Leucke, D.R. Berglund, J.R. Venette & K.F. Grafton. 1998. 1996 Dry bean grower survey of pest problems and pesticide use in Minnesota and North Dakota. North Dakota State Univ. Ext. Serv. Rpt. 44. Fargo, U.S.A.

Legard, D.E. & J.E. Hunter. 1990. Pathogenicity of bean of *Pseudomonas syringae* pv. *syringae* recovered from the phylloplane of weeds and from bean crop residue. Phytopathology 80:938-942.

Legard, D.E. & H.F. Schwartz. 1987. Sources and management of *Pseudomonas syringae* pv. *phaseolicola* and *Psuedomonas syringae* pv. *syringae* epiphytes on dry beans in Colorado. Phytopathology 77:1503-1509.

Lindgren, D., D. Nuland, J. Steadman & D.P. Coyne. 1993. Field Crops: dry bean reaction to common blight. Biological and Cultural Tests 8:98-103.

Lindgren, D., D. Nuland, J. Steadman & D.P. Coyne. 1994. Field Crops: dry bean reaction to common blight. Biological and Cultural Tests 9:135.

Lindgren, D., D. Nuland, J. Steadman & D.P. Coyne. 1995. Field Crops: dry bean reaction to common blight. Biological and Cultural Tests 10:16.

Lindgren, D., D. Nuland, J. Steadman & D.P. Coyne. 1996. Field Crops: dry bean reaction to common blight. Biological and Cultural Tests 11:13.

Lindgren, D., D. Nuland, J. Steadman & D.P. Coyne. 1997. Field Crops: dry bean reaction to common blight. Biological and Cultural Tests 12:21.

Lyons, M.E., M.H. Dickson & J.E. Hunter. 1987. Recurrent selection for resistance to white mold in *Phaseolus species*. J. Amer. Soc. Hort. Sci. 112:149-152.

McClean, P.E., J.R. Myers & J.H. Hammond. 1993. Coefficient of parentage and cluster analysis of north American dry bean cultivars. Crop Sci. 33:190-197.

McElroy, J.B. 1985. Breeding dry beans, *Phaseolus vulgaris* L., for common bacterial blight resistance derived from *Phaseolus acutifolius* A. Gray. Ph.D. dissertation. Cornell Univ. Ithaca, U.S.A.

Michaels, T.E., P.H. Goodwin, J.E. Mayer & M.A. Pastor-Corrales. 1996. Evaluation of a DNA probe for the quantitative detection of common bacterial blight in common bean and its application in a breeding program. Euphytica 90:129-135.

Miklas, P.N. 1998b. Release of small red, pinto, and pink dry bean germplasn lines ICB-6, ICB-12, and ICB-42 with resistance to common bacterial blight. Annu. Rpt. Bean Improv. Coop. 41-225-226.

Miklas, P.N., J.S. Beaver, K.F. Grafton & G.F. Freytag. 1994a. Registration of TARS VC1-4B multiple disease resistant dry bean germplasm. Crop Sci. 34:1415.

Miklas, P.N., R. Delorme, R. Hannan & M.H. Dickson. 1999. Using a subsample of the core collection to identify new sources of resistance to white mold in common bean. Crop Sci. 39:569-573.

Miklas, P.N. & K.F. Grafton. 1992. Inheritance of partial resistance to white mold in inbred populations of bean. Crop Sci. 32:943-948.

Miklas, P.N., K.F. Grafton, J.D. Kelly, H.F. Schwartz & J.R. Steadman. 1998a. Registration of four white mold resistant dry bean germplasm lines: I9365-3, I9365-5, I9365-31, and 92BG-7. Crop Sci. 38:1728.

Miklas, P.N., K.F. Grafton & B.D. Nelson. 1992a. Screening for partial physiological resistance to white mold in dry bean using excised stems. J. Amer. Soc. Hort. Sci. 117:321-327.

Miklas, P.N., K.F. Grafton, G.A. Secor & P.E. McClean. 1992b. Use of pathogen filtrate to differentiate physiological resistance of dry bean to white mold disease. Crop Sci. 32:310-312.

Miklas, P.N., E. Johnson, V. Stone, J.S. Beaver, C. Montoya & M. Zapata. 1996. Selective mapping of QTL conditioning disease resistance in common bean. Crop Sci. 36:1344-1351.

Miklas, P., S. Lambert, G. Mink & M. Silbernagel. 1998b. Many beans with *bc-3* resistance to BCMNV are susceptible to BCMV. Annu. Rpt. Bean Improv. Coop. 41:33-34.
Miklas, P.N., J.C. Rosas, J.S. Beaver, L.H. Telek & G.F. Freytag. 1994b. Field performance of select tepary bean germplasm for the tropics. Crop Sci. 34:1639-44.
Miklas, P.N., J.R. Stavely & J.D. Kelly. 1993. Identification and potential use of a molecular marker for rust resistance in common bean. Theor. Appl. Genet. 85:745-49.
Mink, G.I., J. Vetten, C.W. Ward, P. Berger, F. Morales, J.R. Myers, M.J. Silbernagel & O.W. Barnett. 1994. Taxonomy and classification of legume infecting potyviruses. A proposal from the Potyviridae study group of the plant virus subcommittee of ICTV. Arch. Virology 139:231-235.
Navarrete-Maya, R. & J.A. Acosta Gallegos. 1997. Common blight resistance in wild *Phaseolus vulgaris* L. Annu. Rpt. Bean Improv. Coop. 40:102-103.
Nelson, R. & E.E. Down. 1933. Influence of pollen and ovule infextion in seed transmission of bean mosaic. Phytopathology 23:25.
Nienhuis, J. & S.P. Singh. 1985. Effects of location and plant density on yield and architectural traits in dry beans. Crop Sci. 25:579-584.
Nuland, D.S. & G. Taylor. 1996. Technological progress in dry bean yields. Annu. Rpt. Bean Improv. Coop. 39:50-51.
Park, S.J. 1993. Response of bush and upright plant type selections to white mold and seed yield of common beans grown in various row widths in southern Ontario. Can. J. Plant Sci. 73:265-272.
Park, S.J. & B.N. Dhanvantari. 1987. Transfer of common blight resistance from *Phaseolus coccineus* Lam. to *P. vulgaris* L. through interspecific hybridization. Can. J. Plant Sci. 67:685-695.
Park, S.O., D.P. Coyne, A. Dursun & G. Jung. 1998a. Identifying randomly amplified polymorphic DNA (RAPD) markers linked to major genes for common bacterial blight resistance in tepary bean. J. Amer. Soc. Hort. Sci. 123:278-282.
Park, S.O., D.P. Coyne & G. Jung. 1998b. Gene estimation, associations of traits, and confirmation of QTL for common bacterial blight resistance in common bean. Annu. Rpt. Bean Improv. Coop. 41:145-146.
Park, S.O., A. Dursun, D.P. Coyne & G. Jung. 1996. Identification of RAPD markers linked to major genes for common bacterial blight resistance using bulked segregate analysis in a tepary bean cross. Annu. Rpt. Bean Improv. Coop. 39:128-29.
Patel, P.N. & J.C. Walker. 1965. Resistance in *Phaseolus* to halo blight. Phytopathology 55:889-894.
Petzoldt, R. & M.H. Dickson. 1996. Straw test for resistance to white mold. Annu. Rpt. Bean Improv. Coop. 39:142-143.
Reddick, D. & V.B Stewart. 1919. Transmission of the virus of bean mosaic in seed and observations of thermal death point of seed and virus. Phytopathology 9:445-50.
Ribeiro, R.L. & D.J. Hagedorn. 1979a. Screening for resistance to and pathogenic specialization of *Fusarium oxysporum* f. sp. *phaseoli*, the causal agent of bean yellows. Phytopathology 69:272-276.
Ribeiro, R.L. & D.J. Hagedorn. 1979b. Inheritance and nature of resistance to *Fusarium oxysporum* f. sp. *phaseoli*. Phytopathology 69:859-861.
Richards, B.L. & W.H. Burkholder. 1943. A new mosaic disease of beans. Phytopathology 33:1215-16.
Saad, S.M. & D.J. Hagedorn. 1971. Improved techniques for initiation of bacterial brown spot of bean in the greenhouse. Phytopathology 61:1313-1311.

Saettler, A.W. 1989. Common bacterial blight. pp. 261-283. *In* H.F. Schwartz & M.A Pastor-Corrales (eds.), Bean Production Problems in the Tropics. 2nd ed. CIAT, Cali, Colombia.

Salgado, M.O. & H.F. Schwartz. 1993. Physiological specialization and effects of inoculum concentration of *Fusarium oxysporum* f. sp. *phaseoli* on common beans. Plant Dis. 77:492-495.

Salgado, M.O., H.F. Schwartz & M.A. Brick.1990. Dry bean genotypes resistant to *Fusarium* wilt. Annu. Rpt. Bean Improv. Coop. 34:54-55.

Salgado, M.O., H.F. Schwartz & M.A. Brick.1995. Inheritance of resistance to a Colorado race of *Fusarium oxysporum* f. sp. *phaseoli* in common beans. Plant Dis. 79:279-281.

Salgado, M.O., H.F. Schwartz, M.A. Brick & M.A. Pastor-Corrales.1994. Resistance to *Fusarium oxysporum* f.sp. *phaseoli* in tepary beans (*Phaseolus acutifolius*). Plant Dis. 78:357-360.

Schneider, K.A., M.E. Brothers & J.D. Kelly. 1997a. Marker assisted selection to improve drought resistance in common bean. Crop Sci. 37:51-60.

Schneider, K.A., R. Rosales-Serna, F.J. Ibarra-Pérez , B. Cazares-Enríquez, J.A. Acosta Gallegos, P. Ramírez-Vallejo, N. Wassimi & J.D. Kelly. 1997b. Improving common bean performance under drought stress. Crop Sci. 37:43-50.

Schneiter, A.A., D.W. Burke & J.R. Venette. 1982. Registration of `Pindak' pinto bean. Crop Sci. 22:157.

Schneiter, A.A., K.F. Grafton, D.W. Burke & G.C. Weiser. 1983. Registration of `Holberg' pinto bean. Crop Sci. 23:1219.

Schwartz, H.F. 1989. Halo blight. pp. 285-301. *In* H.F. Schwartz & M.A Pastor-Corrales. (eds.), Bean Production Problems in the Tropics. 2nd ed. CIAT, Cali, Colombia.

Schwartz, H.F. 1994. Bacterial disease of beans. Crop Series, Cooperative Extension publication No. 2.913. Colorado State Univ., Ft. Collins, U.S.A.

Schwartz, H.F., D.H. Casciano, J.A. Asenga & D.R. Wood. 1987. Field measurement of white mold effects upon dry beans with genetic resistance or upright plant architecture. Crop Sci. 27:699-702.

Schwartz, H.F., M.S. McMillan & M.F. Silbernagel. 1989. Occurrence of *Fusarium* wilt of beans in Colorado. Pant Dis. 75:518.

Schwartz, H.F. & J.R. Steadman. 1978. Factors affecting sclerotium populations of and apothecium production by *Sclerotinia sclerotiorum*. Phytopathology 68:383-388.

Schwartz, H.F. & J.R. Steadman. 1989. White mold. pp. 211-230. *In* H.F. Schwartz & M.A. Pastor-Corrales (eds.), Bean Production Problems in the Tropics. 2nd ed. CIAT, Cali, Colombia.

Scott, M.E. & T.E. Michaels. 1988. Inheritance of resistance to common bacterial blight in common bean. Annu. Rpt. Bean Improv. Coop. 31:72.

Silbernagel, M.J. 1969. Mexican strain of bean common mosaic virus. Phytopathology 59: 1809-12.

Silbernagel, M.J. 1980. A rapid screening technique for bacterial brown spot (*Pseudomonas syringae*) and halo blight (*Pseudomonas phaseolicola*). Annu. Rpt. Bean Improv. Coop. 23:81-82.

Silbernagel, M.J. 1990. Genetic and cultural control of Fusarium root rot in bush snap beans. Plant Dis. 74:61-66.

Silbernagel, M.J. 1994. Release of pinto breeding line 92US-1006 with *I* $bc2^2$ resistance to bean common mosaic virus. Annu. Rpt. Bean Improv. Coop. 37:246.

Silbernagel, M.J. & A.N. Hang. 1997a. Release of USWA-6, USWA-8, and USWA-9 small red dry bean germplasm lines. Annu. Rpt. Bean Improv. Coop. 40:151-152.

Silbernagel, M.J. & A.N. Hang. 1997b. Release of pinto dry bean germplasm lines USWA-19 and USWA-20 with virus and root rot resistance. Annu. Rpt. Bean Improv. Coop. 40:153-154.

Silbernagel, M.J. & A.N. Hang. 1997c. Release of USWA-12 and USWA-13 great northern dry bean germplasm lines. Annu. Rpt. Bean Improv. Coop. 40:152-153.

Silbernagel, M.J. & A.N. Hang. 1997d. Release of pink dry bean germplasm lines, USWA-61 and USWA-63, the first with dominant *I* resistance to bean common mosaic virus. Annu. Rpt. Bean Improv. Coop. 40:157-158.

Silbernagel, M.J., A.N. Hang & P.N. Miklas. 1998a. Registration of USWA-27 virus resistant dry bean germplasm. Crop Sci. 38:899-900.

Silbernagel, M.J., A.N. Hang & P.N. Miklas. 1998b. Registration of USWA-20 virus and root rot resistant pinto dry bean germplasm. Crop Sci. 38:899.

Silbernagel, M.J. & H.F. Schwartz. 1988. Preliminary report of *Fusarium* wilt of beans in Colorado in 1987. Annu. Rpt. Bean Improv. Coop. 31:60-61.

Silva, L.O., S.P. Singh & M.A. Pastor-Corrales. 1989. Inheritance of resistance to common bacterial blight in common bean. Theor. Appl. Genet. 78:619-624.

Singh, S.P. 1982. A key for identification of different growth habits of *Phaseolus vulgaris* L. Annu. Rpt. Bean Improv. Coop. 25:92-94.

Singh, S.P. 1992. Common bean improvement in the tropics. Plant Breed. Rev. 10:199-269.

Singh, S.P. 1995. Selection for water-stress tolerance in interracial populations of common bean. Crop Sci. 35:118-124.

Singh, S.P., C. Cajiao, J.A. Gutiérrez, J.García, M.A. Pastor-Corrales & F.J. Morales. 1989. Selection for seed yield in inter-gene pool crosses of common bean. Crop Sci. 29:1126-1131.

Singh, S.P., P. Gepts & D.G. Debouck. 1991a. Races of common bean (*Phaseolus vulgaris*, Fabaceae). Econ. Bot. 45:379-96.

Singh, S.P. & J.A. Gutiérrez. 1982. Sources of some architectural traits in dry bush beans, *Phaseolus vulgaris* L. Annu. Rpt. Bean Improv. Coop. 25:6-7.

Singh, S.P. & J.A. Gutiérrez. 1984. Geographical distribution of the DL_1 and DL_2 genes causing hybrid dwarfism in *Phaseolus vulgaris* L., their association with seed size, and their significance to breeding. Euphytica 33:337-335.

Singh, S.P., J.A. Gutiérrez, A. Molina, C. Urrea & P. Gepts. 1991b. Genetic diversity in cultivated common bean: II. Marker-based analysis of morphological and agronomic traits. Crop Sci. 31:23-29.

Singh, S.P. & A. Molina. 1996. Inheritance of crippled trifoliolate leaves occurring in interracial crosses of common bean and its relationship with hybrid dwarfism. J. Hered. 26:665-679.

Singh, S.P., A. Molina, C.A. Urrea & J.A. Gutiérrez. 1993. Use of interracial hybridization in breeding the race Durango common bean. Can. J. Plant Sci. 73:785-793.

Singh, S.P. & C.G. Muñoz. 1999. Resistance to common bacterial blight among *Phaseolus* species and common bean improvement. Crop Sci. 39:80-89.

Singh, S.P., R. Nodari & P. Gepts. 1991c. Genetic diversity in cultivated common bean: I. Allozymes. Crop Sci. 31:19-23.

Singh, S.P. & C.A. Urrea. 1994. Selection for seed yield and other traits among early generations of intra-and interracial populations of the common bean. Rev. Bras. Genet. 3:299-303.

Stavely, J.R. 1984a. Genetics of resistance to *Uromyces phasioli* in a *Phaseolus vulgaris* line resistant to most races of the pathogen. Phytopathology 74:339-344.

Stavely, J.R. 1984b. Pathogenic specialization in *Uromyces phaseoli* in the United States and rust resistance in beans. Plant Dis. 68:95-99.

Stavely, J.R. 1988. Identification of rust resistance in *Phaseolus vulgaris* plant introductions 90758 through 194333. Annu. Rpt. Bean Improv. Coop. 31:128-29.

Stavely, J.R. 1989. Occurrence of rust resistance in *Phaseolus vulgaris* plant introductions 194333 through 28948. Annu. Rpt. Bean Improv. Coop. 31:128-29.

Stavely, J.R. 1998. Development of comprehensively rust resistant bean germplasm. Phytopathology 88:85.

Stavely, J.R. & K.F. Grafton. 1989. Registration of BelDak Rust resistant-1 and -2 pinto dry bean germplasm. Crop Sci. 29:834-35.

Stavely, J.R., J.D. Kelly & K.F. Grafton. 1994. BelMiDak-rust resistant navy dry bean germplasm lines. HortScience 29:709-11.

Stavely, J.R., J.D. Kelly, K.F. Grafton, C.A. Mullens, A. Straw, R.T. McMillan Jr., J.S. Beaver, P.N. Miklas, J. Steinkle, J.R. Steadman, D.P. Coyne, D.T. Lindgren & M.J. Silbernagel. 1997. Rust resistant bean germplasm releases, 1994-1996. Annu. Rpt. Bean Improv. Coop. 40:120-121.

Stavely, J.R., J.D. Kelly, K.F. Grafton, J. Steinkle, J.R. Steadman, D.P. Coyne, D.T. Lindgren & M.J. Silbernagel. 1994. Rust resistant bean germplasm releases. Annu. Rpt. Bean Improv. Coop. 37:247-248.

Stavely, J.R. & M.A. Pastor-Corrales. 1989. Rust. pp.159-194. *In* H.F. Schwartz & M.A. Pastor-Corrales (eds.), Bean Production Problems in the Tropics. 2nd ed. CIAT, Cali, Colombia.

Stavely, J.R., J.R. Steadman & R. McMillan. 1989. New pathogenic variability in *Uromyces appendiculatus* in North America. Plant Dis. 73:428-32.

Stavely, J.R., J. Steinkle, R.T. McMillan, K.F. Grafton, J.R. Steadman, J.D. Kelly, D.P. Coyne, D.T. Lindgren & M.J. Silbernagel. 1992. Rust resistant bean germplasm releases. Annu. Rpt. Bean Improv. Coop. 35:228-29.

Steadman, J.R. 1983. White mold: a serious yield-limiting disease of bean. Plant Dis. 67:346-350.

Steadman, J.R., J. Beaver, M. Boudreau, D.P. Coyne, J. Groth, J.D. Kelly, M. McMillan, R. McMillan, P.N. Miklas, M.A. Pastor-Corrales, H. Schwartz & J. Stavely. 1995. Progress reported at the 2nd international bean rust workshop. Annu. Rpt. Bean Improv. Coop. 28:1-10.

Steadman, J.R., B.L. Blad & H.F. Schwartz. 1976. Feasibility of microclimate modification for control of white mold disease of bean. Annu. Rpt. Bean Improv. Coop. 19:78-80.

Steadman, J.R., K.F. Grafton, K. Kmiecik, J.M. Kolkman, M. Kyle-Jahn & R. Mainz. 1998. Bean white mold nursery, 1997. Annu. Rpt. Bean Improv. Coop. 41:173-174.

Stewart, V.B. & D. Reddick. 1917. Bean mosaic. Phytopathology 7:61.

Taylor, J.H., N.L. Innes, C.L. Dudley & W.A. Griffiths. 1978. Sources and inheritance of resistance to halo-blight of *Phaseolus* beans. Ann. Appl. Biol. 90:101-110.

Taylor, J.D., D.M. Teverson, D.J. Allen & M.A. Pastor-Corrales. 1996. Identification and origin of races of *Pseudomonas syringae* pv. *phaseolicola* from Africa and other bean growing areas. Plant Pathology 45:469-478.

Urrea, C.A. & S.P. Singh. 1995. Comparison of recurrent and congruity backcrossing for interracial hybridization in common bean. Euphytica 81:21-26.

Valladares-Sánchez, N.E., D.P. Coyne & R.F. Mumm. 1983. Inheritance and associations of leaf, external and internal pod reactions to common blight bacterium in *Phaseolus vulgaris* L. J. Amer. Soc. Hort. Sci. 108:272-278.

Valladares-Sánchez, N.E., D.P. Coyne & M.L. Schuster. 1979. Differential reaction of leaves and pods of *Phaseolus* germplasm to strains of *Xanthomonas phaseoli* and transgressive segregation for tolerance from crosses of susceptible germplasm. J. Amer. Soc. Hort. Sci. 104:648-654.

Velásquez-Valle, R., K.J. Lienert, H.F. Schwartz, M.A. Brick & B. Ogg. 1997. Evaluation of a bean nursery to the *Fusarium* wilt isolate from Colorado. Annu. Rpt. Bean Improv. Coop. 40:93-95.

Velásquez-Valle & H.F. Schwartz. 1997. Reaction of selected Mexican and USA bean lines to *Fusarium oxysporum* f. sp. *phaseoli*. Annu. Rpt. Bean Improv. Coop. 40:95-96.

Voysest, O., M.C. Valencia & M.C. Amézquita. 1994. Genetic diversity among Latin American Andean and Mesoamerican common bean cultivars. Crop Sci. 34:1100-1110.

Walker, J.C. & P.N. Patel. 1964. Inheritance of resistance to halo blight of bean. Phytopathology 54:952-954.

Wallace, D.H. & R.E. Wilkinson. 1965. Breeding for *Fusarium* root rot resistance in beans. Phytopathology 55:1227-1231.

Welsh, M.M. & K.F. Grafton. 1998. Identifying genes for resistance to common bacterial blight of bean. Annu. Rpt. Bean Improv. Coop. 40:104-105.

Woo, S.L., A. Zoina, G. Del Sorbo, M. Lorito, B. Nanni, F. Scala & C. Noviello. 1996. Characterization of *Fusarium oxysporum* f. sp. *phaseoli* by pathogenic races, VCGs, RFLPs, and RAPD. Phytopathology 86:966-973.

Wood, D.R. 1982. Registration of 'Ouray' bean. Crop Sci. 22:1260.

Wood, D.R., A.G. Fisher & M. Ballarin. 1983. Registration of 'Cahone' bean. Crop Sci. 23:399.

Wood, D.R. & J.G. Keenan. 1982. Registration of 'Olathe' bean. Crop Sci. 22:1259-60.

Yerkes, W.D. & G.F. Freytag. 1956. *Phaseolus coccineus* as a source of root-rot resistance for the common bean. Phytopathology 46:32.

Zapata, M., R. Rodríquez & S.P. Singh. 1998. Sources of resistance to *Xanthomonas campestris* pv. *phaeoli* from different geographical origins. Annu. Rpt. Bean Improv. Coop. 41:58-59.

Zaumeyer, W.J. & L.L. Harter. 1941. Inheritance of resistance to six physiologic races of bean rusts. J. Agric. Res. 63:599-622.

Zaumeyer, W.J. & H.R. Thomas. 1957. A monographic study of bean diseases and methods for their control. US Dept. of Agr. Tech Bull. 868. US Govt. Print. Off., Washington DC, U.S.A.

Chapter 10

Improvement of Small-Seeded Race Mesoamerica Cultivars[1]

Shree P. Singh
Plant, Soil, and Entomological Sciences, University of Idaho, 3793 North 3600 East, Kimberly, Idaho 83341-5076, U.S.A.

Key words: gamete selection, integrated genetic improvement, market classes, multiple-crosses, production problems, production regions, selection methods

1. INTRODUCTION

Among all *Phaseolus* beans of American origin, the small-seeded common bean (*P. vulgaris* L.) cultivars occupy by far the largest hectarage in the world (>6 million ha) and have the longest history of genetic improvement. A brief description of these beans, their major market classes, production regions, and production problems will be given in this chapter. This will be followed by the history of how beans have been improved, the genetic progress which has been achieved, current difficulties and challenges, breeding objectives and strategies, and future prospects.

The common bean that is characterized by small seeds, often 100 dry seeds weighing less than 25 g (Schoonhoven & Pastor-Corrales, 1987) belongs to the group designated as race Mesoamerica (Singh et al., 1991a). Among all common bean cultivars, these possess by far the smallest seeds. Some of the wild common beans from the Andean region of South America actually have similar seed size to the cultivars from this group. The wild bean populations, from which these small-seeded beans were domesticated, are distributed from northern Mexico to northern Peru (Gepts et al., 1986;

[1] Published as Idaho Agricultural Experiment Station Journal Article no. 99702. University of Idaho, College of Agriculture, Moscow, ID.

Koenig et al., 1990; Singh et al., 1991a; Toro Ch. et al., 1990). While large variations are found in the phaseolin seed protein patterns (an evolutionary marker) (Gepts, 1988; Gepts et al., 1986) of the wild populations, the cultigens largely possess the "S", "B", or "Sb" phaseolin types (Koenig et al., 1990; Singh et al., 1991a). The latter two phaseolin groups are less frequent in cultigens, but are found both in Middle America and South America, especially the east side of South America (e.g., Brazil, Colombia, and Venezuela) (Koenig et al., 1990). These types also were domesticated in the northern Andes. It is not known why these types did not become as popular there as in Middle America (with the exception of Venezuela) despite their relatively higher yield compared with the large-seeded (> 40 g/100 seeds) Andean counterparts (White & González, 1990; White et al., 1992). It is only more recently that small-seeded cultivars have become popular in Argentina, Bolivia, Brazil, Chile, and other countries of South America.

Most popular landraces and improved cultivars of this group have an indeterminate growth habit, either type III (prostrate, semiclimbing) or II (upright, erect) (Singh, 1982b). In their original domestication regions in Latin America, these small-seeded cultivars traditionally have occupied relatively warmer climates with mean growing temperatures of 22-24°C. The climbing bean cultivars of growth habit IV also are grown in intercropping systems in reduced hectarage in the southern highlands of Mexico (e.g., Veracruz, Chiapas) and Central America. Some climbing bean cultivars have a slightly larger seed size and a shiny seedcoat. The small-seeded determinate bush growth habit (type I) cultivars (e.g., Kupal and Brazil 2), which evolved and were domesticated in Latin America, never became popular. This lack of success is probably due to their inherent low yielding ability (Nienhuis & Singh, 1985), reduced yield stability (Ghaderi et al., 1982; Kelly et al., 1987), and difficulty in flourishing in harsh tropical environments. Presently these type I cultivars (e.g., white pea or navy bean) are mostly grown at higher latitudes in Canada, the U.S.A., and Chile, where early maturity is needed because of the short growing season (<100 days).

Small-seeded cultivars possess central trifoliolate leaflets that are medium to large, ovate or cordate shaped. The flower bracteoles also are large and cordate. For other distinguishing features of these small-seeded race Mesoamerica cultivars, readers should refer to Singh (1989) and Singh et al. (1991a).

2. MAJOR MARKET CLASSES

About a dozen market classes of small-seeded cultivars are grown and consumed globally. The maximum variation among market classes is found in Brazil, Central America, and Mexico. In Brazil, black, beige (Jalinho), cream (Mulatinho), and cream striped (Carioca) are most popular. Carioca gained popularity within the last 40 years (Singh et al., 1998) and Jalinho was introduced less than 20 years ago. Other market classes in Brazil include cream with pink hilum ring (Rosinha), red (Vermelho), purple (Roxo), and brown (Pardo).

In Central America and Mexico, black and red beans are the most popular, although large variation is found among landraces collected from this region. Cuba, the Dominican Republic, Haiti, and Venezuela also grow and consume black beans. In Argentina, Canada, Chile, and the U.S.A., small whites also are grown, in addition to black seeded cultivars. Based on hectarage, the most important market classes of small-seeded beans are black, beige, cream, cream striped, red, and white (Table 1). Of these, black beans are grown in the largest hectarage (>2 million ha) and most widely, from Canada and the U.S.A. to Argentina and Chile. The consumption of small whites is restricted to Canada, the U.S.A., Europe, the Middle East, and North Africa, whereas cream, cream striped, and beige types are largely consumed in Brazil. Cream striped, red, white, and other small-seeded beans also are increasingly being grown and consumed in African countries such as Malawi, Zimbabwe, Zambia, Ethiopia, and the Republic of South Africa.

3. PRODUCTION REGIONS

Because most of the common beans are consumed within the production regions in Latin America, the major bean production regions are strongly associated with the consumption habits for the major market classes. Exceptions occur only in the case of common bean exporting countries, namely, Argentina, Bolivia, and Chile. In Brazil, the world's largest producer (2.5 million MT annually) and consumer of common beans, more than 90% of the production area is under small-seeded cultivars. Black beans are more popular in the southern and central states, and cream is preferred in the northeast, while cream striped and beige are grown essentially all over the country. The brown, pink, cream with pink hilum ring, purple, and red types are grown mostly in Goiás, Minas Gerais, and São Paulo. For further details refer to Thung & Oliveira (1998), Vieira (1967, 1978, 1988), Vieira et al. (1998), and Zimmermann et al. (1988).

In Mexico, small black beans predominate in the Pacific coastal regions such as Nayarit, Veracruz, and Chiapas. Black beans grown in the central humid and semiarid highlands belong to races Jalisco and Durango, respectively, and often possess larger and shiny seeds (Singh, 1989; Singh et al., 1991a). Small-seeded black beans also predominate in Venezuela. In Argentina, black beans (and small amounts of cream striped and white beans) are grown in the northeastern states of Tucuman, Santiago de Estero, Salta, and Jujuy. In Chile, most production of small-seeded black and white beans occurs in the Regions V, VI, and VII, south of Santiago and north of Temuco.

Guatemala and Costa Rica predominantly grow black beans and also some red beans. A similar situation occurs in Cuba, where black beans predominate but large-seeded Andean beans also are grown. The situation is reversed in El Salvador, Honduras, and Nicaragua, where small-seeded red beans predominate. In Mexico and Central American countries, indeterminate cultivars of growth habit II and III predominate up to elevations of about 1,000 meters. Above that altitude, climbing or semiclimbing beans, often possessing larger seed size, gradually predominate. Most traditional bean growing areas are characterized by small farms (<10 ha), using both climbing and type II and III growth habit cultivars that are intercropped with a wide variety of other species (e.g., corn, *Zea mays* L.; cassava, *Manihot esculenta* Crantz; coffee, *Coffea arabica* L.; and sugarcane, *Saccharum officinarum* L.).

In Canada, Manitoba is the leading producer of small whites, followed by Ontario. Saskatchewan and Alberta grow only a very small hectarage of small-seeded beans. Similarly, in the U.S.A., Michigan, North Dakota, and Minnesota are the largest producers of small white and black beans, although these beans also are grown in other midwestern and northwestern states. The principal production regions of small-seeded beans in the Americas are summarized in Table 1.

4. PRODUCTION CONSTRAINTS

In the tropics and subtropics of Latin America, low soil fertility is the major abiotic constraint to production of small-seeded beans. Low fertility soils are often characterized by deficiency of nitrogen and phosphorus, with or without deficiency of some minor elements and toxicity of aluminum and manganese (Araya & Beck, 1995; Howeler, 1980; Thung, 1990; Thung & Oliveira, 1998; see also Chapter 13 by Thung & Rao). Water stress or drought is the second most important constraint that may impede production of common bean (Acosta G. et al., 1989; Singh, 1995; White & Singh, 1991;

White et al., 1988). One of the major causes of low soil fertility problems could be that production of common bean in Latin America (e.g., Brazil) is gradually being pushed to more marginal areas. In certain regions (e.g., coastal El Salvador, Honduras, northern and western Brazil, and the northwestern U.S.A.), high temperatures, especially during flowering and fruiting periods, may limit yield severely (due to blossom and pod abortion) and affect seed quality. Small-seeded beans, in general, possess moderate levels of tolerance to these factors. In some bean production regions in the Latin American highlands (above 2,000 m elevation) and in western Canada, southern Chile, and the northern and northwestern U.S.A., cultivation of these beans also may be affected severely by low temperatures.

Table 1. Major production regions for small-seeded common bean and their constraints

Production Region	Market class	Production constraints[1]
Canada, U.S.A.	Black, white	Anthracnose, rust, CBB, root rots, white mold, late maturity
Southern Mexico, Central America	Black, red	ALS, anthracnose, BCM, BGM, CBB, rust, bean pod weevil, low soil fertility
Cuba, Haiti, Dominican Republic, Venezuela	Black	BGM, CBB, rust, low soil fertility
Argentina, Bolivia, Brazil	Black, beige, cream, cream striped	ALS, anthracnose, BGM, CBB, root rots, rust, white mold, leafhoppers, low soil fertility, water stress

[1]ALS= angular leaf spot, BCM= bean common mosaic, BGM= bean golden mosaic, and CBB= common bacterial blight.

The most widespread foliar diseases of small-seeded beans in the Latin American tropics and subtropics are the following: 1) bean common mosaic (BCM, caused by a potyvirus), especially in red-seeded early maturing landraces in Central America and rosinha and roxo types in Brazil, 2) bean golden mosaic (BGM, caused by a geminivirus), 3) common bacterial blight (CBB) [caused by *Xanthomonas campestris* pv. *phaseoli* (Smith) Dye], 4) angular leaf spot [caused by *Phaeoisariopsis griseola* (Sacc.) Ferr.], 5) anthracnose [caused by *Colletotrichum lindemuthianum* (Sacc. & Magn.) Bri. & Cav.], and 6) rust [caused by *Uromyces appendiculatus* (Pers.) Ung.]. In high rainfall areas of the Latin American highlands (e.g., Guatemala, Colombia), ascochyta blight [caused by *Phoma exigua* var. *diversispora* (Bub.) Boerma] may cause severe defoliation and yield losses. A group of fungi that cause root rots, including species of *Aphanomyces, Fusarium, Macrophomina, Pythium, Rhizoctonia, Sclerotium*, and/or *Thielaviopsis* may take heavy tolls, especially when susceptible cultivars are repeatedly grown in the same fields. In warm humid regions (Brazil, Costa Rica, Guatemala, and Panama), web blight [caused by *Thanatephorus cucumeris* (Frank) Donk.] is the most limiting factor to yield.

In wetter and cooler bean growing environments at higher latitudes (Canada, the U.S.A.), and occasionally in Argentina and Brazil (the winter sowing under central pivot irrigation systems), white mold [caused by *Sclerotinia sclerotiorum* (Lib.) de Bary], halo blight [caused by *Pseudomonas syringae* pv. *phaseolicola* (Burkh.)], and bacterial brown spot (caused by *Pseudomonas syringae* pv. *syringae* van Hall) join the list of major production constraints.

Among insects, leafhoppers (*Empoasca kraemeri* Ross & Moore) are a common problem in the tropics and subtropics, especially in drier northeastern Brazil. In cooler regions (e.g., eastern Canada), potato leafhopper (*E. fabae* Harris) can cause considerable yield losses. In Mexico, Guatemala, El Salvador, Honduras, and northern Nicaragua, bean pod weevil (*Apion godmani* Wagner and to a minor extent *A. aurichalceum* Wagner) may cause up to 100% yield losses in susceptible cultivars. Bruchids [*Zabrotes subfasciatus* (Boheman)] are the major storage insect problem in the tropics and subtropics. *Acanthoscelides obtectus* (Say) is the major storage insect in the cooler highlands and temperate environments if proper care is not taken. Within the last few decades the hectarage of small-seeded beans has also increased in African countries. The greatest production constraint in this region is bean fly (*Ophiomyia phaseoli* Tryon) (Wortmann et al., 1998). A summary of major production constraints of the small-seeded beans is given in Table 1.

5. HISTORY OF IMPROVEMENT

In Latin America, the first organized breeding program for small-seeded common bean cultivars was at the Instituto Agronomico de Campinas (IAC), São Paulo, Brazil. Common bean breeding was probably initiated at IAC in the 1920s (Vieira, 1967, 1988; Voysest, 1983). Pure line selections in local landraces also were made in Mexico in the 1940-1970 period, leading to release of black bean cultivars such as Actopan, Antigua, Jamapa, and others (Voysest, 1983). In El Salvador, the black bean cultivars of the Porrillo series were selected in local and introduced germplasm in the 1960s and 1970s. Similarly, a series of selections in landraces was done in Costa Rica. Some of these selections were later introduced and released in Brazil (Voysest, 1983).

ICA Pijao, released in Colombia in the 1960s, appears to have been the first popular cultivar developed through hybridization and selection.

Subsequently, it was released in Argentina, Cuba, and Guatemala in the 1970s and 1980s. It has also been used as a parent in the development of many cultivars because of its high yield, wide adaptation, upright plant habit, and resistance or tolerance to BCM, BGM, low soil fertility, and leafhoppers.

In the past four decades, many more researchers have begun to work in private and public educational, state, federal, regional, and international institutions devoted to genetic improvement of small-seeded beans. For example, at present, Argentina, Bolivia, Brazil, Chile, Colombia (CIAT), Costa Rica, Cuba, Guatemala, the Dominican Republic, Honduras, Mexico, and Puerto Rico have active breeding programs for genetic improvement of small-seeded beans. In Brazil alone, there are more than a half dozen private and public institutions from Pernambuco to Rio Grande do Sul dedicated to genetic improvement of this group of dry beans.

In the U.S.A., California, Idaho, Michigan, Nebraska, and New York have had the longest history of bean improvement. Genetics and breeding of beans in Nebraska was initiated by R.E. Emerson before 1890 (D.P. Coyne, personal communication, 1998). Michigan was the first state to employ a full-time bean breeder in 1906, and remains the leading state for improvement of small-seeded black and white beans. In addition to Idaho, Michigan, and New York, small-seeded cultivars have also been developed and released by public or private breeders at least in California, Colorado, North Dakota, and Washington. In Canada, evaluation and selection of common bean cultivars was initiated in the 1880s and 1890s (Park & Buzzell, 1995). This work followed the development of cultivars using hybridization and selection in the 1920s. Consequently, breeding programs in Alberta, Ontario, and Saskatchewan have used hybridization and selection to develop and release small-seeded cultivars. Manitoba has initiated a breeding program recently.

6. GENETIC PROGRESS ACHIEVED

Over 20,000 cultivated germplasm accessions and over 13,000 improved lines and cultivars of common bean from around the world have been evaluated repeatedly. The evaluation has taken place in contrasting environments in Colombia and elsewhere. Some observations about the useful genes (traits) that are present or deficient in landraces in comparison with improved modern-day cultivars of different gene pools, races, and market classes of cultivated common bean can now be made. For example, in the case of small-seeded beans of race Mesoamerica, landraces are known to possess several traits. These traits include high yield potential, earliness, upright plant type, moderate tolerance to low soil fertility, water stress, high temperatures, BGM, and leafhoppers, insensitivity to photoperiod, and resistance to angular leaf spot, anthracnose, BCM, root rots, and bean pod weevil. These traits evolved during domestication and were present singly or in various combinations in landraces before their organized breeding was initiated nearly a century ago. Of course, all these traits were not present in any single genotype, but it is not uncommon to find accessions that possess three or more desirable attributes. For example, Compuesto Chimaltenango 2 from Guatemala is high yielding, known to be resistant to rust, angular leaf spot, and anthracnose, and also tolerant to low soil fertility and root rots. Similarly, Ecuador 299 is resistant to rust, angular leaf spot, and anthracnose. Brazil 2 is determinate, insensitive to photoperiod, and resistant to BCM and anthracnose. Carioca, an heirloom cultivar which has gained popularity in recent years (Singh et al., 1998), is high yielding, widely adapted, resistant to BCM, and tolerates low soil fertility, water stress, anthracnose, angular leaf spot, common bacterial blight, leafhoppers, and BGM. Also, it occupies the largest hectarage (at least one million ha) currently sown under any landrace or improved common bean cultivar in the world. Another example is San Cristobal 83, which possesses an attractive red mottled seed color and carries the *I* gene for resistance to BCM. This combination is very rare in either improved cultivars or landraces because of undesirable linkages between genes that control these traits (Cáceres & Morales, 1985; Kyle & Dickson, 1988; Temple & Morales, 1986). San Cristobal 83 is high yielding and tolerant to water stress (Singh, 1995; White et al., 1994).

The major achievements realized through breeding in Canada and the U.S.A. include earliness, adaptation to higher latitude, high yield, upright plant type, combination of *bc-3* and *I* genes for resistance to BCM, and rust and anthracnose resistance (Adams, 1982; Grafton et al., 1993; Kelly et al., 1989, 1994a, b, 1995; Myers et al., 1991). In the tropics and subtropics of Latin America, substantial progress has been achieved with efforts to incorporate resistance to several pests, and other traits. These include resistance to BCM, BGM, common bacterial blight, leafhopper, bruchid, and *Apion*, and upright plant type in red and black beans for Central America (Beebe & Pastor-Corrales, 1991; Beebe et al., 1993; Kornegay & Cardona, 1990, 1991). Most of these traits, with the exception of *Apion* resistance (because it was not required), also have been bred into cream, cream striped, and beige types for Brazil. In addition, high levels of tolerance to water stress and resistance to angular leaf spot and anthracnose were bred into Brazilian bean types (Alberini et al., 1983; Miranda et al., 1979; Pompeu, 1980, 1982; Singh, 1995; Singh et al., 1991b, 1992b, 1998).

While desirable genes for individual traits have been incorporated into improved breeding lines and cultivars, resistance to more than two or three production-limiting factors has seldom been combined with high yield and desirable plant, seed, and adaptation characteristics into one cultivar. This result is largely due to breeding efforts that have often focused on a single trait. Furthermore, a team approach and integrated genetic improvements were not practiced. These factors contribute to the reasons why most released cultivars have not become popular and the impact of common bean breeding has not been adequately realized. Nonetheless, in the last few years, at institutions such as CIAT, where a team of researchers from different disciplines has worked together, it has been possible to breed simultaneously for multiple traits in some bean types such as Carioca and Mulatinho (Singh et al., 1992b, 1998; Thung et al., 1993).

7. DIFFICULTIES AND CHALLENGES

Improved cultivars are grown on essentially all hectarage of small-seeded beans in Argentina, Bolivia, Canada, Chile, and the U.S.A. Coincidentally, all these countries are out of the domestication range of small-seeded beans. Landraces in these areas were probably introduced after the arrival of Columbus. In Brazil, Central America, Cuba, Mexico, and Venezuela, while substantial hectarage is grown under improved cultivars released within the last three decades, large areas are sown to landraces. Failure to adopt improved cultivars is more common in El Salvador, Nicaragua, and Honduras, where high yielding upright improved cultivars resistant to BCM and BGM or bean pod weevil are late maturing and do not possess the desired seedcoat color available in local red seeded landraces. Moreover, as noted earlier, whereas many production-limiting factors may occur in the same region, improved cultivars often carry resistance to only two or three factors at most. Thus, the full potential of improved cultivars is often not realized in growers' fields. Substantial price discounts because of poor seed and culinary quality have also impeded full-scale adoption of newer cultivars. On the other hand, there are no improved cultivars of climbing beans of either red, black, or other market classes available in Latin America.

Unfortunately, only a small fraction of the total available genetic diversity in *Phaseolus* beans has been utilized in improvement programs, and some bean breeders continue to breed cultivars for only one or two traits at a time. In addition, emphasis is placed on individual breeding projects instead of integrated genetic improvement so that functional, multi-disciplinary teams are not present in most countries, and public support for common bean improvement is decreasing rapidly. The decrease is particularly evident in the Latin American tropics and sub-tropics.

Breeders currently engaged in improvement of the small-seeded common bean in the tropics and subtropics may benefit by taking advantage of significant advances in common bean research over the past thirty years. First, many desirable traits exist in high yielding, well-adapted, small-seeded landraces that have not been fully utilized in breeding programs. Second, we now have a much better understanding of the extent and organization of genetic diversity in *Phaseolus* species. Third, a vast pool of genetic variability is available among races, gene pools, wild populations, and secondary and tertiary gene pools (see Chapter 2 by Debouck). Much of this variability has yet to be utilized in breeding programs (Singh, 1992). Fourth, researchers breeding small-seeded beans need to exchange more germplasm and information, and collaborate to become more productive. Fifth, morphological, biochemical, and molecular markers for indirect selection for resistance to diseases and insects and other traits are increasingly available. Sixth, more efficient and useful breeding methods (e.g., recurrent and congruity inbred-backcrossings, single seed descent, and marker-assisted recurrent and gamete selections) are available. Last, a multi-disciplinary team approach and integrated genetic improvement are likely to expedite simultaneous improvement of multiple desirable traits and development of broadly adapted, high yielding, high quality cultivars that are less dependent on water, fertilizers, pesticides, and hand labor (Singh, 1997; Singh et al., 1998). Better use of this available knowledge and technology in breeding programs should bring maximum returns to the growers, consumers, processors, traders, and agricultural industries to help make efficient use of limited natural resources.

8. CURRENT BREEDING OBJECTIVES AND STRATEGIES

8.1 Breeding Objectives

Incorporating high yield (>2,500 kg/ha) and high seed quality in early maturing (<100 days from planting to harvest maturity) upright mechanizable cultivars will be the primary breeding objective. Other desirable traits include resistance to BCM, rust, common bacterial blight, halo blight, bacterial brown spot, anthracnose, white mold, and root rots. These are the long-term objectives of a breeding program seeking widely adapted small-seeded black, white, and other market classes of beans in temperate North America. Water stress, low soil fertility, cold, and high temperature tolerance may also be required for at least some of the northern and northwestern states of the U.S.A. and southern and southwestern Canada where beans are grown without irrigation.

For most of tropical and subtropical Latin America, high yield, high seed quality, upright plant type, and resistance to multiple biotic and abiotic stresses are required for successful cultivars. Diseases of strategic importance for which resistance is required for Central America, Cuba, Mexico, and Venezuela include BCM, BGM, common bacterial blight, anthracnose, angular leaf spot, and rust. For Brazil and Argentina, white mold would be added to the above list. In addition, root rot and/or nematode resistance may be required for intensive large-scale monoculture farming systems in Brazil and for coastal Peru. For insect problems, resistance to leafhoppers and bruchids for the entire region and bean pod weevil in Central American and Mexican cultivars are being sought. Some degree of tolerance to water stress and low soil fertility, especially to acidic soils deficient in phosphorus, nitrogen, and minor elements and sometimes possessing toxic levels of aluminum and manganese is also needed for Brazil, Costa Rica, Cuba, and Mexico, among others. The availability of early maturing cultivars (<70 days to maturity) with all of the above mentioned resistance would facilitate increased adoption of improved cultivars in most countries.

To bring research and development benefits to subsistence farmers and the poorest of the poor growing climbing bean landraces in intercropping systems in Latin America and elsewhere, it is imperative to undertake improvement of these dry beans, especially in public breeding programs. Most of the objectives for resistance breeding for biotic and abiotic stresses may be similar to those for growth habit types II and III cultivars for monoculture cropping systems mentioned earlier. In addition, due consideration would need to be given to inter-genotypic and inter-specific competition among major crops prevalent in the region for intercropping (Clark & Francis, 1985; Francis, 1981; Francis et al., 1978; Hamblin & Zimmermann, 1986; Kawano & Thung, 1982; Woolley et al., 1991).

8.2 Parental Selection and Hybridization

The three-tiered approach for integrated genetic improvement suggested by Kelly et al. (1998; also see Chapters 6 and 8) would be useful for many breeders. This approach should maximize the use of genetic diversity from within each of the market classes of race Mesoamerica and from across other races, gene pools, wild populations, and secondary, and tertiary gene pools. Breeders for small-seeded beans would be making crosses largely among elite x elite parents of the same market class for cultivar development. This process would be the main breeding focus, separate from introgression of useful genes from distantly related germplasm and from gene pyramiding for parental development for specific agronomic traits. For cultivar development, while all selected parents must be well adapted, they must represent the maximum diversity available among that group of elite lines, commercial cultivars, and donor parents of useful genes for all traits to be improved. The elite lines and cultivars may originate from different private and public breeding programs in the region and other similar regions provided they are well adapted. Similarly, a majority of donor parents would be taken from the parental development projects or obtained by pyramiding different sources of useful genes for each character provided they are thoroughly tested for performance and adaptation to the region.

It can be difficult to select useful recombinants for cultivar development from crosses involving unadapted and distantly related germplasm. Thus, for improvement of small-seeded black bean cultivars for Argentina, Bolivia, Brazil, Central America, Cuba, Mexico, and Venezuela (i.e., the entire Latin American tropics and subtropics) elite lines and cultivars available from this entire region could potentially be considered as parents, because they were all derived from the same race. Problems of adaptation due to photoperiod and temperature often are minimal. Parental selection should be considered only after thorough evaluations, and those found superior utilized in crosses. The group of bred lines and cultivars from temperate North America could be used in the tropics as a source of useful genetic variability. Gene flow from tropical to temperate regions will require introgression of photoperiod insensitivity and general adaptation to temperate environments before the exotic germplasm can be used in elite x elite crosses for cultivar development.

Breeders with interest in developing high yielding cultivars possessing multiple resistance to biotic and abiotic factors may want to benefit from the knowledge of evolution, interrelationships, combining ability, and affinity that may exist among germplasms from different races of common bean. For example, black seeded cultivars are found in each of the three races, namely, Durango (e.g., Negro Durango, Negro San Luis, A 800), Jalisco (Negro Puebla 152, Negro 150), and Mesoamerica (ICA Pijao, Jamapa, Midnight, Porrillo, Raven, T 39) of Middle American origin. Complementary genes are found and positive general combining ability for high seed yield and tolerance to water stress and low soil fertility exists among these races (Singh, 1995; Singh et al., 1989, 1992b, 1993). The same also may apply for red, white, cream, beige, and cream striped colored beans of these races. However, from bi-parental interracial crosses it may not be possible to directly recover desirable plant, seed, and adaptation traits. Alternative crossing schemes (e.g., three-way, modified-double, multiple-parent, and backcrosses) (Singh, 1982a; Singh et al., 1998; Urrea & Singh, 1995) that permit ≥75% genetic contribution of the adapted or elite parent(s) of the market class under improvement would be required.

8.3 Multiple Crosses and Selection in F_1

When genes for multiple traits are found in a separate parent, bi-parental and backcrosses are not adequate. To reduce the time required for cultivar development, it may be preferable to make multiple-parent crosses initially and combine all desirable alleles in the first step. This process can be facilitated by evaluation and selection during hybridization and population development (i.e., gamete selection) where dominant and codominant morphological, biochemical, and/or molecular markers are available (Singh, 1994, 1998). For many disease and insect resistances controlled by dominant genes, the multiple-parent F_1 hybrids could be inoculated or infested and selected resistant plants could then be used in further hybridization and/or advancing to subsequent generations. Moreover, they could then be screened for dominant and codominant biochemical and molecular markers, thus assuring recombination and accumulation of the maximum number of favorable alleles early in the crossing and population development phase. In these multiple-parent crosses, the maximum genetic contribution of the most desirable parent(s) of the market class under improvement must be assured. That is, they must be used as the last female parents of the final crosses. Comparatively large numbers of hybrid seeds may be required for each multiple-parent cross for gamete selection in the F_1 for simultaneous improvement of multiple traits.

8.4 Selection Methods Used in Early Generations

Different selection methods have been used for development of common bean cultivars. For example, the F_1 and F_2 were advanced in bulk and a single F_3 plant was selected for development of navy bean cultivar Norstar (Grafton et al., 1993). Different types of bulk selections also were used for development of hundreds of lines in the 1970s and 1980s at CIAT (Singh, 1995; Singh et al., 1992b, 1993). The upright type II growth habit black-seeded cultivar Raven, possessing resistance to anthracnose (*A* gene), BCM (*I* and *bc-2* genes), and rust (*Ur-3* gene), and tolerance to halo blight in Michigan (U.S.A.) was developed by Kelly et al. (1994b), using the pedigree method. They also used the pedigree method of selection to develop the navy bean cultivars Huron and Newport (Kelly et al., 1994a, 1995). The pedigree method was used again for development of lines A 247 and A 295, among many others, at CIAT. Line A 247 (Carioca seed type) was released as Rio Doce in 1987 and line A 295 (Jalinho) was released as EMGOPA Ouro in 1984 in Brazil. These lines carry multiple disease resistance and are high yielding. Single seed descent (SSD) was used for development of navy bean cultivar Mayflower (Kelly et al., 1989). On the other hand, Brazilian researchers have successfully used the recurrent backcross method for incorporation of disease resistance in Mulatinho, Carioca, black (Preto), and other colored beans (Alberini et al., 1983; Miranda et al., 1979; Pompeu, 1980, 1982).

Early generation selection for seed yield in common bean was suggested by Singh et al. (1990) and Singh & Urrea (1995). Urrea & Singh (1994) found that the F_2-derived family method of selection was superior to the SSD and bulk methods commonly used for advancing early generations of hybrid populations. From early generation yield tests (F_2 to F_4) Singh & Terán (1998) identified high- and low-yielding populations which eventually produced high- and low-yielding advanced generation (F_7) lines.

Evaluation and selection methods used in early generations of F_1-derived families depend upon the objectives of the program, environments, and resources available. For example, in temperate environments where only one field crop season per year is feasible, breeders often advance material one or two generations in greenhouses or off-season nurseries out of the production region (Singh et al., 1999). Hence, meaningful evaluations for agronomic traits are not feasible during each generation and families could be advanced by single seed descent or by taking single-seed or single-pod bulks from all plants within each family. The F_1-derived F_4 or F_5 families are then grown in the field for evaluation and selection of promising families and for development of advanced generation lines of commercial value. In tropical and subtropical environments of Latin America (e.g., from coastal Mexico to Brazil) where two or more field nurseries are grown during the year, evaluation and selection for biotic and abiotic stresses and other agronomic traits helps eliminate undesirable recombinants. Promising populations and families thus identified are used to develop lines possessing multiple desirable traits (Singh et al., 1998). Use of biochemical and molecular markers to select families and populations that are harvested in bulk in early segregating generations (F_2 to F_5) may not be feasible at the moment because of the prohibitive costs of screening large number of plants in each generation. For biotic and abiotic stresses that cannot be screened simultaneously, different locations and nurseries may be required for different generations for identifying promising populations and families within populations (Singh et al., 1991b, 1992a, 1998).

For traits inherited by recessive genes, such as susceptibility to leaf chlorosis induced by the geminivirus causing BGM (Urrea et al., 1996;

Vélez et al., 1998) and for seed characteristics, intensive selection in early segregating generations should be avoided. The frequency of desirable recombinants is very low and there is a danger of losing potentially useful recombinants that might arise in later generations. For such traits, it is preferable to initiate evaluation and selection in the F_4generation. Moreover, before planting the F_5 seeds of selected promising families, all non-commercial seeds harvested from the F_4 plants are discarded. The F_5 is space planted under heavy and uniform pressure from diseases such as angular leaf spot, anthracnose, rust, and common bacterial blight. Only plants possessing desirable levels of resistance and plant, maturity, and seed characteristics should be selected and harvested individually. These are then progeny-tested in the F_6. Those found uniform (or true-breeding) for all desirable traits are bulk-harvested for seed increase in the F_7 and subsequent evaluations.

Use of a separate, complementary nursery for each of the major biotic and abiotic stresses, and adaptation, seed yield, and seed quality helps eliminate all undesirable entries inferior to the available elite lines and cultivars (Singh, 1992). Only superior lines of potential commercial value are advanced for further intensive evaluations in contrasting environments and growers' fields, and for seed increase. Thus, truly superior lines are selected as new cultivars for eventual use by growers, processors, traders, consumers, and agricultural industry.

9. FUTURE PROSPECTS

In the tropics and subtropics of Latin America, maximum breeding effort has gone to improvement of small-seeded black, beige, cream, cream striped, and red colored beans during the last four decades. This work has resulted in substantial progress in breeding for resistance to specific biotic and abiotic stresses. However, better seed types, early maturity (Masaya, 1989), and high yield combined with resistance to multiple biotic and abiotic factors (Table 1) are needed to assure cultivar adoption and increased production and consumption of small-seeded beans.

Because the genetic base of common bean cultivars is narrow (Adams, 1977; McClean et al., 1993; Voysest et al., 1994) maximizing use of genetic diversity will require a sustained, focussed, and well-integrated genetic improvement program. The program, in addition to cultivar development, must emphasize useful gene introgression from alien germplasm, and parental development for specific characters through pyramiding genes from different sources. Use of exotic germplasm is more important for bean breeders in Canada, the U.S.A., Europe, and southeast Asia, where a large majority of germplasm is not adapted, is not easily available, and may have specific processing quality requirements that hinder the use of broad-based populations. Genes for early maturity and photoperiod insensitivity would need to be incorporated (i.e., a conversion program) as a priority for their adequate evaluation, selection, and use in improvement programs at higher latitudes.

As bean cultivation continues to be pushed into soils with lower fertility (e.g., Brazil) and other marginal environments (e.g., Argentina, Canada, and the U.S.A.), the necessity of maximizing bean production increases. The area sown to small-seeded bean cultivars will likely increase and there will be increasing demand and justification for development of cultivars specific to such environments and needs because population growth is faster than food production.

REFERENCES

Acosta G., J.A., F.J. Ibarra-Pérez P., A. Pajarito R., R. Ochoa M., R. Lépiz I. & M.W. Adams. 1989. Breeding for tolerance to drought in bean. pp. 416-417. *In* S. Beebe (ed.), Current Topics in Breeding of Common Bean. CIAT, Cali, Colombia.

Adams, M.W. 1977. An estimation of homogeneity in crop plants with specific reference to genetic vulnerability in the dry bean *Phaseolus vulgaris* L. Euphytica 26:665-679.

Adams, M.W. 1982. Plant architecture and yield breeding. Iowa State J. Res. 56:225-254.

Alberini, J.L., M.W. Kranz, L. Oliari & A. Bianchini. 1983. 'IAPAR 5 - Rio Piquiri' e 'IAPAR 7 - Rio Vermelho', novas variedades de feijoeiro para o estado do Paraná. Pesq. Agropec. Bras. 18:393-397.

Araya, R. & D. Beck (eds.). 1995. Memoria del Taller Internacional sobre Bajo Fósforo en el Cultivo de Frijol. Univ. de Costa Rica, San Jose, Costa Rica.

Beebe, S., C. Cardona, O. Díaz, F. Rodríguez, E. Mancia & S. Ajquejay. 1993. Development of common bean (*Phaseolus vulgaris* L.) lines resistant to the pod weevil, *Apion godmani* Wagner, in Central America. Euphytica 69:83-88.

Beebe, S.E & M.A. Pastor-Corrales. 1991. Breeding for disease resistance. pp. 561-617. *In* A. van Schoonhoven & O. Voysest (eds.), Common Beans: Research for Crop Improvement. C.A.B. Int., Wallingford, U.K. & CIAT, Cali, Colombia.

Cáceres O., S.R. & F.J. Morales. 1985. Incorporación de genes de resistencia dominante y recesiva a cepas de virus del mosaico común en dos variedades comerciales de frijol (*Phaseolus vulgaris* L.). Acta Agr. 35(4):7-20.

Clark, E.A. & C.A. Francis. 1985. Bean-maize intercrops: a comparison of bush and climbing bean growth habits. Field Crops Res. 10:151-166.

Francis, C.A. 1981. Development of plant genotypes for multiple cropping systems. pp. 179-215. *In* K.J. Frey (ed.), Plant Breeding II. Iowa State Univ. Press, Ames, U.S.A.

Francis, C.A., M. Prager & D.R. Laing. 1978. Genotype x environment interactions in climbing bean cultivars in monoculture and associated with maize. Crop Sci. 18:242-246.

Gepts, P. 1988. Phaseolin as an evolutionary markers. pp. 215-241. *In* P. Gepts (ed.), Genetic Resources of *Phaseolus* Beans. Kluwer, Dordrecht, Netherlands.

Gepts, P., T.C. Osborn, K. Rashka & F.A. Bliss. 1986. Phaseolin-protein variability in wild forms and landraces of the common bean (*Phaseolus vulgaris*): evidence for multiple centers of domestication. Econ. Bot. 40:451-468.

Ghaderi, A., M.W. Adams & A.W. Saettler. 1982. Environmental response patterns in commercial classes of common bean (*Phaseolus vulgaris* L.). Theor. Appl. Genet. 63:17-22.

Grafton, K.F., K.C. Chang, J.R. Venette & A.J. Vander Wal. 1993. Registration of 'Norstar' navy bean. Crop Sci. 33:1405-1406.

Hamblin, J. & M.J.O. Zimmermann. 1986. Breeding common bean for yield in mixtures. Plant Breed. Rev. 4:245-272.

Howeler, R.H. 1980. Nutritional disorders. pp. 341-362. *In* H.F. Schwartz & G.E. Gálvez (eds.), Bean Production Problems: Disease, Insect, Soil, and Climatic Constraints of *Phaseolus vulgaris*. CIAT, Cali, Colombia.

Kawano, K. & M. Thung. 1982. Intergenotypic competition and competition with associated crops in cassava. Crop Sci. 22:59-63.

Kelly, J.D., M.W. Adams, A.W. Saettler, G.L. Hosfield, G.V. Varner, J.S. Beaver, M.A. Uebersax & J. Taylor. 1989. Registration of 'Mayflower' new bean. Crop Sci. 29:1571-1572.

Kelly, J.D., M.W. Adams & G.V. Varner. 1987. Yield stability of determinate and indeterminate dry bean cultivars. Theor. Appl. Genet. 74:516-521.

Kelly, J.D., G.L. Hosfield, G.V. Varner, M.A. Uebersax, L.K. Afanador & J. Taylor. 1995. Registration of 'Newport' navy bean. Crop Sci. 35:1710-1711.

Kelly, J.D., G.L. Hosfield, G.V. Varner, M.A. Uebersax, M.E. Brothers & J. Taylor. 1994a. Registration of 'Huron' navy bean. Crop Sci. 34:1408.

Kelly, J.D., G.L. Hosfield, G.V. Varner, M.A. Uebersax, S.D. Haley & J. Taylor. 1994b. Registration of 'Raven' black bean. Crop Sci. 34:1406-1407.

Kelly, J.D., J.M. Kolkman & K. Schneider. 1998. Breeding for yield in dry bean (*Phaseolus vulgaris* L.). Euphytica 102:343-356.

Koenig, R.L., S.P. Singh & P. Gepts. 1990. Novel phaseolin types in wild and cultivated common bean (*Phaseolus vulgaris*, Fabaceae). Econ. Bot. 44:50-60.

Kornegay, J. & C. Cardona. 1990. Develoment of an appropriate breeding scheme for tolerance to *Empoasca kraemeri* in common bean. Euphytica 47:223-231.

Kornegay, J. & C. Cardona. 1991. Breeding for insect resistance in beans. pp. 619-648. *In* A. van Schoonhoven & O. Voysest (eds.), Common Beans: Research for Crop Improvement. C.A.B. Int., Wallingford, U.K. & CIAT, Cali, Colombia.

Kyle, M.M. & M.H. Dickson. 1988. Linkage of hypersensitivity to five viruses with the *B* locus in *Phaseolus vulgaris* L. J. Hered. 79:308-311.

Masaya, P. 1989. The development of early-maturing varieties for Central American agricultural systems. pp. 182-192. *In* S. Beebe (ed.), Current Topics in Breeding for Common Bean. CIAT, Cali, Colombia.

McClean, P.E., J. Myers & J.J. Hammond. 1993. Coefficient of parentage and cluster analysis of North American dry bean cultivars. Crop Sci. 33:190-197.

Miranda, P., R.C. Mafra, E.B. Correia & M.A. De-Queiroz. 1979. "IPA-79-19" uma nova variedade de feijão "mulatinho" (*Phaseolus vulgaris* L.) para Pernambuco. Pes. Agrop. Pernamb. 3:105-111.

Myers, J.R., R.E. Hayes & J.J. Kolar. 1991. Registration of 'UI 906' black bean. Crop Sci. 31:1710.

Nienhuis, J. & S.P. Singh. 1985. Effects of location and plant density on yield and architectural traits in dry beans. Crop Sci. 25:579-584.

Park, S.P. & R.I. Buzzell. 1995. Common bean. pp. 1-16. *In* A.E. Slinkard & D.R. Knott (eds.), Harvest of Gold: The History of Field Crop Breeding in Canada. Univ. of Saskatchewan, Saskatoon, Canada.

Pompeu, A.S. 1980. Yields of French bean lines (*Phaseolus vulgaris* L.) of the rosinha and roxinho groups resistant to *Colletotrichum lindemuthianum*. Bragantia 39:89-97.

Pompeu, A.S. 1982. Catu, Aete-3, Aroana-80, Moruna 80, Carioca 80 e Ayso: novos cultivares de feijoeiro. Bragantia 41:213-218.

Schoonhoven, A. van & M.A. Pastor-Corrales (Comps). 1987. Standard system for the evaluation of bean germplasm. CIAT, Cali, Colombia.

Singh, S.P. 1982a. Alternative methods to backcross breeding. Annu. Rpt. Bean Improv. Coop. 25:11-12.

Singh, S.P. 1982b. A key for identification of different growth habits of *Phaseolus vulgaris* L. Annu. Rpt. Bean Improv. Coop. 25:92-95.

Singh, S.P. 1989. Patterns of variation in cultivated common bean (*Phaseolus vulgaris*, Fabaceae). Econ. Bot. 43:39-57.

Singh, S.P. 1992. Common bean improvement in the tropics. Plant Breed. Rev. 10:199-269.

Singh, S.P. 1994. Gamete selection for simultaneous improvement of multiple traits in common bean. Crop Sci. 34:352-355.

Singh, S.P. 1995. Selection for water-stress tolerance in interracial populations of common bean. Crop Sci. 35:118-124.

Singh, S.P. 1997. Progreso, problemas y el futuro de mejoramiento de frijoles Mesoamericanos y la mejora genética integrada. pp. 334-350. *In* S.P. Singh & O. Voysest (eds.), Taller de Mejoramiento de Frijol para el Siglo XXI: Bases para Una Estrategia para América Latina. CIAT, Cali, Colombia.

Singh, S.P. 1998. Uso de marcadores y selección de gametos para el mejoramiento simultáneo de caracteres múltiples de frijol (*Phaseolus vulgaris* L.) para Mesoamérica y el Caribe. Agron. Mesoam. 9:1-9.

Singh, S.P., C. Cajiao, J.A. Gutiérrez, J. García, M.A. Pastor-Corrales & F.J. Morales. 1989. Selection for seed yield in inter-gene pool crosses of common bean. Crop Sci. 29:1126-1131.

Singh, S.P., C. Cardona, F.J. Morales, M.A. Pastor-Corrales & O. Voysest. 1998. Gamete selection for upright carioca bean with resistance to five diseases and a leafhopper. Crop Sci. 38:666-672.

Singh, S.P., P. Gepts & D.G. Debouck. 1991a. Races of common bean (*Phaseolus vulgaris*, Fabaceae). Econ. Bot. 45:379-396.

Singh, S.P., J.A. Gutiérrez, C.A. Urrea, A. Molina & C. Cajiao. 1992a. Location-specific and across-location selections for seed yield in populations of common bean, *Phaseolus vulgaris* L. Plant Breed. 109: 320-328.

Singh, S.P., R. Lépiz, J.A. Gutiérrez, C. Urrea, A. Molina & H. Terán. 1990. Yield testing of early generation populations of common bean. Crop Sci. 30:874-878.

Singh, S.P., A. Molina, C.A. Urrea & J.A. Gutiérrez. 1993. Use of interracial hybridization in breeding the race Durango common bean. Can. J. Plant Sci. 73:785-793.

Singh, S.P., M.A. Pastor-Corrales, A. Molina, C. Urrea & C. Cajiao. 1991b. Independent, alternate, and simultaneous selection for resistance to anthracnose and angular leaf spot and effects on seed yield in common bean (*Phaseolus vulgaris* L.). Plant Breed. 106: 312-318.

Singh, S.P. & H. Terán. 1998. Population bulk versus F_1-derived family methods of yield testing in early generations of multiple-parent interracial and inter-gene-pool crosses of common bean. Can. J. Plant Sci. 78:417-421.

Singh, S.P., H. Terán, C.G. Muñoz & J.C. Takegami. 1999. Two cycles of recurrent selection for seed yield in common bean. Crop Sci. 39:391-397.
Singh, S.P. & C.A. Urrea. 1995. Inter- and intrarracial hybridization and selection for seed yield in early generations of common bean, *Phaseolus vulgaris* L. Euphytica 81:131-137.
Singh, S.P., C.A. Urrea, A. Molina & J.A. Gutiérrez. 1992b. Performance of small-seeded common bean from the second selection cycle and multiple-cross intra-and interracial populations. Can. J. Plant Sci. 72:735-741.
Temple, S.R. & F.J. Morales. 1986. Linkage of dominant hypersensitive resistance to bean common mosaic virus to seed color in *Phaseolus vulgaris* L. Euphytica 35:331-333.
Thung, M. 1990. Phosphorus: a limiting nutrient in bean (*Phaseolus vulgaris* L.) production in Latin America and field screening for efficiency and response. pp. 501-521. *In* N. El Bassam, M. Dambrooth & B.G. Loughman (eds.), Genetic Aspects of Plant Mineral Nutrition. Kluwer, Dordrecht, Netherlands.
Thung, M.T., R.M. Ferreira, P. Miranda, V. Moda-Cirino, M.A. Gava Ferrão, L.O. da Silva, V.V. Dourado, S. Hemp, B. Souza, E. Serpa S., M.J.O. Zimmermann & S.P. Singh. 1993. Performance in Brazil and Colombia of common bean lines from the second selection cycle. Rev. Bras. Genet. 16:115-127.
Thung, M.D.T. & I.P. Oliveira. 1998. Problemas Abióticos que Afectam a Producão do Feijoeiro é Seus Métodos de Controle. EMBRAPA-CNPAF, Santo Antonio de Goias, Brazil.
Toro Ch., O., J. Tohme & D.G. Debouck. 1990. Wild Bean (*Phaseolus vulgaris* L.) Description and Distribution. IBGPR & CIAT, Cali, Colombia.
Urrea, C.A., P.N. Miklas, J.S. Beaver & R.H. Riley. 1996. A codominant randomly amplified polymorphic DNA (RAPD) marker useful for indirect selection of bean golden mosaic virus resistance in common bean. J. Amer. Soc. Hort. Sci. 121:1035-1039.
Urrea, C.A. & S.P. Singh. 1994. Comparison of mass, F_2-derived family and single-seed descent selection methods in an interracial population of common bean. Can. J. Plant Sci. 74:461-464.
Urrea, C.A. & S.P. Singh. 1995. Comparison of recurrent and congruity backcrossing for interracial hybridization in common bean. Euphytica 81: 21-26.
Vélez, J.J., M.A. Bassett, J.S. Beaver & A. Molina. 1998. Inheritance of resistance to bean golden mosaic virus in common bean. J. Amer. Soc. Hort. Sci. 123:628-631.
Vieira, C. 1967. O Feijoeiro-Comum: Cultura, Doenças e Melhoramento. Univ. Rural do Estado de Minas Gerais, Viçosa, Brazil.
Vieira, C. 1978. Cultura do Feijão. Univ. Federal, Viçosa, Brazil.
Vieira, C. 1988. *Phaseolus* genetic resources and breeding in Brazil. pp. 467-483. *In* P. Gepts (ed.), Genetic Resources of *Phaseolus* Beans. Kluwer, Dordrecht, Netherlands.
Vieira, C., T.J. de Paula Jr. & A. Borém (eds.). 1998. Feijão: Aspectos Gerais e Cultura no Estado de Minas. UFV, Vicosa, Brazil.
Voysest, O. 1983. Variedades de Frijol en América Latina y su Origen. CIAT, Cali, Colombia.
Voysest, O., M.C. Valencia & M.C. Amézquita. 1994. Genetic diversity among Latin American, Andean and Mesoamerican common bean cultivars. Crop Sci. 34:1100-1110.
White, J.W. & A. González. 1990. Characterization of the negative association between seed yield and seed size among genotypes of common bean. Field Crops Res. 23:159-175.
White, J., G. Hoogenboom, F.J. Ibarra-Pérez & S.P. Singh (eds.). 1988. Research on Drought Tolerance in Common Bean. CIAT, Cali, Colombia.

White, J.W., R. Ochoa M., F.J. Ibarra-Pérez & S.P. Singh. 1994. Inheritance of seed yield, maturity, and seed weight of common bean (*Phaseolus vulgaris*) under semi-arid rainfed conditions. J. Agric. Sci. (Cambridge) 122:265-273.

White, J.W. & S.P. Singh. 1991. Breeding for adaptation to drought. pp. 501-560. *In* A. van Schoonhoven & O. Voysest (eds.), Common Beans: Research for Crop Improvement. C.A.B. Int., Wallingford, U.K. & CIAT, Cali, Colombia.

White, J.W., S.P. Singh, C. Pino, M.J. Ríos & I. Buddenhagen. 1992. Effect of seed size and photoperiod response on crop growth and yield of common bean. Field Crops Res. 28:295-307.

Woolley, J., R. Lépiz I., T. De A. Portese Castro & J. Voss. 1991. Bean cropping systems in the tropics and subtropics and their determinants. pp. 679-706. *In* A. van Schoonhoven & O. Voysest (eds.), Common Beans: Research for Crop Improvement. C.A.B. Int., Wallingford, U.K. & CIAT, Cali, Colombia.

Wortmann, C.S., R.A. Kirkby, C.A. Eledu & D.J. Allen. 1998. Atlas of Common Bean (*Phaseolus vulgaris* L.) Production in Africa. CIAT, Cali, Colombia.

Zimmermann, M.J.O., M. Rocha & T. Yamada (eds.). 1988. Cultura do Feijoeiro: Factores que Afetam a Produtividade. Associacão Brasileira da Potasa e do Fosfato, Piracicaba, Brazil.

Chapter 11

Improvement of Large-Seeded Race Nueva Granada Cultivars

James S. Beaver
Agronomy and Soils, University of Puerto Rico, Mayaguez, Puerto Rico 00681-9030

Key words: abiotic and biotic constraints, crosses within and between gene pools, interspecific crosses, production regions, transgenic bean

1. INTRODUCTION

Most large-seeded dry bean (*Phaseolus vulgaris* L.) of Andean origin are grown in North America, South America, Europe, Asia, and Africa. This wide geographic distribution has generated new gene combinations that favor adaptation to diverse environments (CIAT, 1996). Large-seeded determinate type I and indeterminate type II and III beans include seed types from the Nueva Granada and Chile races (Singh et al., 1991). Popular seed types include dark and light red kidney, red mottled, cranberry, Canario, Azufrado, and large white beans. The seed size of these Andean beans ranges from 30 to 40 g 100^{-1} seed in warmer climates such as the Caribbean to 50 to 60 g 100^{-1} for beans grown in temperate and other cooler environments.

This chapter describes the biotic and abiotic factors that limit the yield of large-seeded beans of race Nueva Granada in different production regions. This is followed by a discussion of breeding strategies that may be used to improve these beans.

2. PRODUCTION CONSTRAINTS

Because large-seeded beans are grown in a wide range of temperatures, rainfall distributions, and soil types, plant breeders in each region face a unique set of constraints to increase production. Local breeding programs should be able to exploit genotype x environment interaction by developing cultivars for specific production regions (Beaver & Molina, 1997).

In the U.S.A., Canada, and Mexico, large-seeded beans tend to be produced using high levels of inputs. White mold [caused by *Sclerotinia sclerotiorum* (Lib.) de Bary] is an important disease, particularly in high yield environments (Table 1). Continuous cropping of beans can cause severe loss in plant population due to root rots. Many cultivars have halo blight [caused by *Pseudomonas syringae* pv. *phaseolicola* (Burkh.)] resistance and common bacterial blight [caused by *Xanthomonas campestris* pv. *phaseoli* (Smith) Dye] tends to be controlled by planting disease-free seed. Angular leaf spot [caused by *Phaeoisariopsis griseola* (Sacc.) Ferr.] has been reported on red kidney beans in Michigan (Saettler & Correa-Victoria, 1983). In California, high temperatures can reduce bean yield and in Mexico, bean golden mosaic (BGM, caused by a geminivirus) is an important constraint to production. In the Caribbean, BGM, web blight [caused by *Thanatephorus cucumeris* (Frank) Donk.], bean rust [caused by *Uromyces appendiculatus* (Pers.) Ung.], and common bacterial blight are the most important diseases.

In the Caribbean, a significant portion of the red mottled beans is grown in irrigated valleys where BGM, rust, and leafhoppers (*Empoasca kraemeri* Ross & Moore) are the most important biotic constraints. In both the Caribbean (Beaver & Molina, 1997) and the Andean region (Valladolid Chiroque, 1997), there is interest in developing lines with greater heat tolerance in order to plant beans during seasons when rainfall patterns are more favorable. This would permit increased bean production during non-traditional growing seasons. Several diseases, including anthracnose [caused by *Colletotrichum lindemuthianum* (Sacc. & Magn.) Bri. & Cav.], white mold, common bacterial blight, web blight, BCM, BGM, and bean dwarf mosaic, are reported to be important constraints for large white bean production in the southern cone of South America (Ricci, 1997).

In Africa, angular leaf spot, anthracnose, bean stem maggot (*Ophiomyia phaseoli* Tryon), bruchids, root rots, and common bacterial blight are considered the most important biotic constraints for bean production (Wortmann & Allen, 1994; Wortmann et al., 1998). Because most beans in the Caribbean, the Andean region (Murillo et al., 1997; Valladolid Chiroque, 1997), and Africa are produced by small-scale producers using few inputs, both drought and low fertility are considered important abiotic constraints.

Table 1. Important biotic and abiotic constraints in the principal production regions of large-seeded beans of race Nueva Granada

Production region	Biotic constraints[1]	Abiotic constraints
North America	ALS, anthracnose, bacterial brown spot, BGM, CBB, halo blight, root rot, white mold	Drought, high and low temperature
The Caribbean	BGM, bean rust, CBB, leafhoppers, web blight	Drought, high temperature, low soil fertility
Andean region	ALS, Anthracnose, bean rust, CBB, halo blight	Drought, low soil fertility
Southern cone of South America	ALS, anthracnose, BCM, BDM, BGM, CBB, web blight, white mold	Drought
Sub-Saharan Africa	ALS, anthracnose, bean stem maggot, bruchids, CBB, root rot	Drought, low soil fertility

[1]ALS= angular leaf spot, BCM= bean common mosaic, BDM= bean dwarf mosaic, BGM= bean golden mosaic, CBB= common bacterial blight.

3. STRATEGIES TO IMPROVE LARGE-SEEDED BEANS

The breeding pyramid proposed by Kelly et al. (1997, 1998; see also Chapter 8 in this book) is a useful conceptual model for developing a strategy to genetically improve large-seeded beans. The apex of the breeding pyramid includes crosses between elite lines within a particular seed type. These crosses should be an ongoing activity in order to identify breeding lines with greater yield potential and adaptation to specific environments. The intermediate level of the breeding pyramid involves crosses among different seed types within the Andean gene pool in order to introgress new specific traits or greater levels of diversity for quantitative traits. The base of the breeding triangle encompasses crosses between the Andean and Middle American gene pools and interspecific hybridizations. Populations generated from these crosses should be used to accomplish long-term breeding objectives.

Most traits sought by large-seeded bean breeders can already be found within the Andean gene pool. However, significant genetic gain for some of the most important traits will require the development of populations from crosses between the Andean and Middle American gene pools. Interspecific populations may need to be developed and screened for a few traits such as

BGM and common bacterial blight resistance. Finally, transgenic beans may permit unique traits such as herbicide resistance to be incorporated into large-seeded beans.

3.1 Crosses within the Andean Gene Pool

Because large-seeded bean cultivars already have many desirable traits, it is important for plant breeders to become familiar with the Andean gene pool. Breeding programs would benefit from evaluation of large-seeded breeding lines from other production regions. For example, the light red kidney cultivar Indeterminate Jamaica Red has been identified in the Caribbean as a source of heat tolerance (Baiges et al., 1996). Breeding programs in the U.S.A., where high temperature is considered to be a production constraint, have used this source of heat tolerance. Some red mottled landraces from the Dominican Republic have useful levels of resistance to common bacterial blight (Beaver et al., 1992). The dark red kidney cultivar Montcalm and the red mottled cultivar ICA Tundama are sources of resistance to halo blight. The red mottled cultivar PC-50 has adult plant resistance to bean rust (Mmbaga & Steadman, 1992) and Bokosi et al. (1995) identified a simple dominant gene in PC-50 which confers resistance to bean rust pathotypes of diverse origin. Shaik (1984) reported that dense leaf pubescence of light red kidney lines from Jamaica was associated with race non-specific resistance to rust. The red mottled lines CAL 143, AND 277, and AND 279 are sources of resistance to angular leaf spot. Balardin & Kelly (1998) reported that PC-50 has a high level of resistance to anthracnose. The light red kidney cultivars IDIAP-R2 and IDIAP-C1 from Panama have moderate levels of resistance to web blight (Rodríguez E. et al., 1995). The red mottled lines MCR 2301 and MCR 2507 have recessive genes for resistance to BCM whereas MCD 2531 and MCD 2532 have the *I* gene for resistance to BCM (Rodríguez M.A. et al., 1995).

The light beige mottled line DOR 303 has the recessive gene *bgm-2*, which provides a non-chlorotic reaction to BGM (Vélez et al., 1998). The red mottled line San Cristobal 83 has resistance to ashy stem blight [caused by *Macrophomina phaseolina* (Tassi) Goid.] and tolerance to drought (CIAT, 1988, 1996). Red mottled landraces from the Dominican Republic also have been reported to have resistance to ashy stem blight (Echavez-Badel & Beaver, 1987). Northern Andean germplasm accessions such as G 19833, G 19839, G 19842, and G 16140 showed adaptation to low P availability (Lynch & Beebe, 1995). In Haiti, red mottled landraces showed moderate levels of resistance to leafhoppers. In Tanzania, landraces have been identified with high levels of resistance to bean stem maggot (CIAT, 1988, 1996). Large-seeded bean lines with resistance to root knot nematode

(*Meloidogyne incognita*) have been developed in Peru (Valladolid Chiroque, 1997). In Mexico, the Andean cultivar Canario 72 had among the shortest cooking times in the evaluation of a group of cultivars of Middle American and Andean origin (Avila-Rodríguez et al., 1996).

Caution should be used when exchanging bean lines from different production regions. Many diseases that are considered important production constraints, such as common bacterial blight, anthracnose, angular leaf spot, and BCM can be seed-borne. In Uganda, significant differences were found between lines in their ability to transmit *X. campestris* pv. *phaseoli* from seed to seedling. A lower level of transmission would help reduce the importance of seed-borne inoculum in disease development (CIAT, 1996).

Photoperiod and temperature sensitivity can impede the exchange of Andean germplasm between production regions. Simmonds (1993) argued that strong genotype x environment interactions make results site specific, and that the breeding value of a line can be evaluated only after the exotic line is improved for local adaptation. Kornegay et al. (1993) suggested that adaptation of Andean lines to different environments could be achieved through the manipulation of photoperiod response genes. The USDA/ARS bean research program in Mayaguez, Puerto Rico, is attempting to address this problem by developing photoperiod insensitive Andean bean lines of tropical origin that can be tested over a wide range of environments.

Research to identify and utilize sources of tolerance to abiotic stress has been limited for large-seeded beans. Unlike disease resistance, which is often simply inherited, tolerance to abiotic stress is usually a complex trait with low heritability (Simmonds, 1993). Because environmental stress constrains yield potential, an appropriate approach to achieve greater tolerance to abiotic stress would be to select for yield in the presence of the specific environmental stress (Carson & Wicks, 1989). However, Rosielle & Hamblin (1981) suggest that selection for increased yield in stress environments might result in decreased productivity in non-stress environments. They note the importance of choosing testing sites that are representative of production conditions for which the breeder wishes to improve yield. Lynch & Beebe (1995) reported that early generation yield testing of bulked populations and families was successful in the identification of more P-efficient beans.

Most conventional breeding methods for self-pollinated crops would be effective to improve populations derived from crosses between lines within the Andean gene pool. Backcrossing could be employed to incorporate simply inherited traits into existing cultivars quickly. Pedigree selection and single seed descent would be appropriate for populations derived from crosses between elite lines. Some North American bean breeding programs use winter nurseries to advance lines to homozygosity as soon as possible.

Simply inherited traits or traits with high heritabilities can be selected in early generations. Seed color, seed size, and days to maturity often are selected in the F_2 and F_3 generations. Molecular markers should permit plant breeders to select for a greater number of traits in early generations. However, populations must be managed efficiently to reduce the cost of using molecular markers. When several simply inherited traits are to be selected, Beaver & Macchiavelli (1998) demonstrated that screening lines in more advanced generations would require a significantly smaller population size than selection for the traits in early generations. Compared to the Middle American gene pool, less is known about genetics of traits of Andean beans. This information is essential to fully exploit molecular marker technology. Bean breeding programs in developing countries should use regional research networks such as the Bean/Cowpea Collaborative Research Support Program to pool resources and gain access to molecular techniques.

Screening techniques should be rapid, inexpensive, and repeatable. For many traits, field evaluations remain an effective means of evaluating breeding lines. The virulence pattern of pathogens should be established before deploying genes for resistance. Causal agents of many diseases considered important production constraints for Andean beans, such as anthracnose, bean rust, angular leaf spot, BCM, and root knot nematodes, possess different races or strains. Andean bean breeders should combine the dominant *I* and recessive genes such as *bc-3* for BCM resistance. This combination of genes would provide resistance to a broad array of BCM virus and BCM necrosis virus strains (Kelly & Miklas, 1997; Kelly et al., 1995). Pyramiding genes for resistance may require bean breeders to work with specific races or strains of the pathogen. Experimental units for preliminary trials for screening for yield, adaptation, and disease reaction can be as small as single, 1-m row when at least five replications are used (Beaver & Kelly, 1989).

3.2 Crosses Between the Andean and Middle American Gene Pools

Beans from the Middle American gene pool have many unique traits that could be used to improve Andean beans. For example, the small red cultivar Don Silvio (DOR 482) has the recessive gene *bgm-1* (Vélez et al., 1998) and the dominant gene *Bgp* for BGM resistance (Molina Castañeda & Beaver, 1998). Andean beans tend to be more susceptible to Andean isolates of pathogens such as *U. appendiculatus* (Araya et al., 1996), *P. griseola* (Pastor-Corrales et al., 1998), and *C. lindemuthianum* (Balardin & Kelly, 1998). Pyramiding genes for disease resistance from the Andean and Middle American gene pools is a breeding strategy to provide resistance to a wide array of pathotypes. Field evaluations conducted in Honduras found BelMiDak RR-1 to have rust resistance when exposed to highly virulent pathotypes. This line combines *Ur-11*, a Middle American resistance gene, with *Ur-4*, a resistance gene of Andean origin (Stavely et al., 1992). Lines in the same nursery having only the *Ur-11* gene (PI181996) or the *Ur-4* gene (Early Gallatin) were susceptible to rust.

Some bean crosses between lines from the Middle American and Andean gene pools produce hybrid dwarfs (Singh & Gutiérrez, 1984). The black-seeded cultivar ICA Pijao or other Middle American lines (e.g., Brasil 2) that do not possess the *Dl* genes can be used as a bridge in making crosses with beans of Andean origin (Kelly, 1989). Beaver (1993) obtained seed from dwarf F_1 hybrids by promoting adventitious root growth above the cotyledonary node.

Dudley (1982) recommended making at least one backcross if one parent has more favorable alleles than the other parent. One backcross using the Andean line as the recurrent parent should improve the likelihood of selecting desirable progeny from crosses between lines from the Andean and Middle American gene pools. After making the backcross, breeders should consider advancing the lines a few generations to increase the frequency of progeny with Andean characteristics and the desired trait introgressed from the Middle American gene pool (Beaver, 1998).

All bean breeding programs have the goal of developing cultivars with increased yield potential (Voysest, 1997). International cooperation in the development of large-seeded bean cultivars is a fairly recent phenomenon. CIAT initiated a breeding program for Andean beans in 1978 and the Bean/Cowpea Collaborative Research Support Program began working in the Caribbean in 1981. Collaborative trials sponsored by PROFRIJOL in Central America and the Caribbean, PROFRIZA in the Andean Region, and SRBRN in Sub-Saharan Africa have helped to identify breeding lines with greater yield potential and adaptation to specific environments. Singh (1992) reported that large-seeded beans belonging to race Nueva Granada have reduced overall growth and harvest index and are physiologically less efficient than beans from Middle America. The yield potential and yield stability of large-seeded Andean beans could be improved by breeding cultivars with an indeterminate growth habit. Recurrent selection was used to develop indeterminate red mottled and red kidney lines that were up to 10 days later in maturity and yielded 30% more than determinate check cultivars. Although the type II growth habit is rare in the CIAT bean germplasm collection (Kornegay et al., 1992), populations from crosses between the Nueva Granada and Mesoamerica races have produced large-seeded lines with erect, indeterminate plant architecture. However, Kornegay et al. (1992) noted that farmers may prefer determinate (type I) beans due to upright architecture, compact canopy, and early, uniform maturity. However, red mottled landrace cultivars from the Dominican Republic occur with both an indeterminate type III growth habit and moderately early maturity. Laing et al. (1984) noted that the main difference between type III and type II growth habits is that type III lines produce a greater number of branches and a more prostrate canopy. Increased branching provides more sites for pod formation and may contribute to greater yield potential. Beaver et al. (1996) found that the seed yield and yield potential of indeterminate type III red mottled lines generally were superior to determinate lines. High-yielding indeterminate lines were identified that would be suited for irrigated production in the valleys of the Dominican Republic whereas other lines were better suited for low-yielding environments. In Colombia, White et al. (1992) identified both determinate and indeterminate large-seeded lines with desirable yield characteristics.

Adams (1982) discussed the use of ideotypes as a strategy to improve the yield potential of beans. Although large-seeded beans are grown in a wide range of environments, ideotypes can be developed for different bean production regions. This approach exploits genotype x environment interaction and reduces the number of traits that have to be selected for a particular production system. An erect type II growth habit is appropriate for environments where white mold and web blight are endemic. Less lodging and a more open canopy help inhibit the spread of these diseases. In addition, an erect, type II growth habit helps maintain seed quality when climatic conditions are unfavorable at harvest. The type III growth with early maturity is well suited for beans produced on hillsides. Early maturity might reduce yield potential during favorable growing seasons but would help to avoid stress during dry years.

Strategies need to be developed that permit bean production using lower levels of inputs. Disease and insect resistance will help reduce pesticide use. Root rot resistance helps plants use water and absorb soil nutrients more efficiently. There is concern that the high levels of N used in North America to produce kidney and cranberry beans may threaten ground water quality. Indeterminate growth habit and later maturity are traits that could be used to increase biological nitrogen fixation (BNF). Cataño (1990) evaluated the BNF capacity of red mottled landraces from the Dominican Republic and identified lines with superior BNF characteristics. The red mottled line San Cristobal 83 was reported to have early nodulation, which may contribute to greater BNF (CIAT, 1988). Coale et al. (1985) reported that selecting soybeans for higher yield potential had produced cultivars with greater BNF capacity and improved ability to take up available soil N. Buttery et al. (1992) recommended that for most purposes the accumulation of N during the growing season is the most valid measure of N fixation capacity. They recommended BNF selection in the field under conditions of low nitrate availability so that good BNF with indigenous strains of *Rhizobium* can be identified. The inbred backcross method was used by McFerson et al. (1982) to select bean lines with superior BNF capacity. Although this selection method requires a large number of pollinations, it does preserve a greater portion of useful genes from the recurrent parent. This feature would be useful when introgressing enhanced BNF characteristics from exotic germplasm.

In the tropics, year-to-year variability occurs in rainfall (Dennett, 1984). Early maturity permits beans to escape terminal drought stress, but yield potential of early lines can be limiting when rainfall patterns are more favorable (Rosas et al., 1991). In the tropics, the dry season can be used to screen beans in the field for drought tolerance. Supplemental irrigation can be used to prevent drought stress from becoming too severe. In Honduras, Rosas et al. (1991) have screened beans for drought tolerance during the dry season by suspending irrigation at the preflowering stage of development (R5). Drought stress plots received approximately 200 mm of water during the growing season. White et al. (1994) suggested that yield improvement under water stress should be no more difficult than under non-stress environments, although target environments need to be clearly defined due to genotype x environment interaction. However, Singh (1995) reported that heritability of seed yield under drought stress is low for crosses between races Mesoamerica and Nueva Granada. Because drought tolerance is considered to be a quantitative trait, advanced generation lines should be screened in replicated trials. Recurrent selection could be used to accumulate alleles that contribute to drought tolerance.

Field screening for heat tolerance has proven to be effective in Puerto Rico (Baiges et al., 1996). Hall (1990) noted that the design of a breeding program for heat tolerance requires identification of the type of hot weather prevailing in the bean production zone. In northern California, high daytime temperatures (near 35°C) during reproductive development can reduce bean yield (Shonnard & Gepts, 1994). In the Caribbean, nighttime temperatures > 20°C during reproductive development can reduce seed yield. Andean sources of heat tolerance such as the light red kidney cultivar Indeterminate Jamaica Red have been crossed with Middle American sources of heat tolerance such as the small red line DOR 364. These populations are expected to produce transgressive segregants for heat tolerance. Shonnard & Gepts (1994) and Baiges et al. (1996) reported significant additive effects for yield and yield components of beans grown under high temperatures.

3.3 Interspecific Crosses

Xu et al. (1998) noted that transgressive variation is often observed in populations derived from interspecific crosses. The development of interspecific populations, however, is not suited for breeding programs with limited resources. The congruity backcross method (Mejía-Jiménez et al., 1994) has been used by plant breeders at CIAT to transfer common bacterial blight resistance from tepary (*P. acutifolius*) into common bean. The cultivar GN # 1 Sel 27 and the breeding lines OAC 88-1, XAN 159, and VAX 1 to VAX 6 derive their resistance from tepary bean. BGM resistance of the *P. coccineus* accessions G035171 and G035172 from CIAT (F.J. Morales & S.P. Singh, unpublished results) also holds in southern Florida (M. Bassett, personal communication, 1998). Rosas et al. (1991) identified lines derived from interspecific crosses between *P. vulgaris* and *P. acutifolius* that were well adapted to drought conditions in Honduras.

3.4 Transgenic Beans

Recombinant DNA technology permits the development of Andean beans with unique traits. The development of soybeans [*Glycine max* (L.) Merr.] with resistance to the nonselective herbicide glyphosate (Roundup) demonstrated that weed control could be included in the domain of plant breeding objectives (Padgette et al., 1995). The *bar* gene for herbicide resistance was used as a marker in the development of transgenic bean lines with AV1, a BGM virus coat protein gene (Azzam et al., 1996). Herbicide resistance and no-till planting of beans on hillsides would help to control soil erosion. A grass mulch covering the surface of the soil also may help reduce the spread of certain diseases such as web blight. Development of a transgenic bean line with high levels of white mold resistance would have a significant impact on the production of large-seeded beans in cooler environments. In the lowland tropics, developing transgenics with web blight resistance would permit more bean production during growing seasons when rainfall is abundant. Because most Andean beans in the tropics are produced by small-scale farmers with limited resources, transgenic beans would need to have significantly higher yield to justify their development and adoption.

4. CONCLUSIONS

Considerable progress has been achieved during the past 20 years in the genetic improvement of large-seeded beans of the race Nueva Granada. Significant gains in yield potential, disease resistance, and other traits of economic value will require most bean breeding programs to broaden their genetic base. Wider crosses will require evaluation of larger populations. Low-cost, rapid, and repeatable screening techniques are needed for the efficient selection of both simply and quantitatively inherited traits. The availability of molecular markers may expedite the selection of progeny in some populations.

Local bean breeding programs can select for adaptation to specific environments and for resistance to endemic diseases and other biotic and abiotic stresses. Successful cultivar development programs do require a long-term commitment both from the plant breeder and the institution supporting the bean research program. Because the accumulation of desirable traits into a new cultivar may require several cycles of selection, familiarity with the crop and continuity of effort are often the keys to success. In addition, plant breeding programs must maintain the agility to respond to challenges such as new pest problems.

REFERENCES

Adams, M.W. 1982. Plant architecture and yield breeding in *Phaseolus vulgaris* L. Iowa State J. Res. 56:225-254.

Araya, C.M., C.M. Sandlin & J.R. Steadman. 1996. Pathogenic variation of isolates of *Uromyces appendiculatus* from common bean in Andean and Middle American regions. Phytopathology 86:567 (abstract).

Avila-Rodríguez, E.H., F.J. Ibarra-Pérez, M.P. Intriago-Ortega & M.L. Valderrama-Chairez. 1996. Genetic variation for grain quality and nutritional characteristics in common bean (*Phaseolus vulgaris* L.) in cultivars of Andean and Mesoamerican origin. Annu. Rpt. Bean Improv. Coop. 39:258-259.

Azzam, O., O. Díaz, J.S. Beaver, R.L. Gilbertson, D.R. Russell & D.P. Maxwell. 1996. Transgenic beans with the bean golden mosaic geminivirus coat protein are susceptible to virus infection. Annu. Rpt. Bean Improv. Coop. 39:258-259.

Baiges, S., J.S. Beaver, P.N. Miklas & J.C. Rosas. 1996. Evaluation and selection of Andean beans for heat tolerance. Annu. Rpt. Bean Improv. Coop. 39:88-89.

Balardin, R.S. & J.D. Kelly. 1998. Effect of *P. vulgaris* gene pool on *C. lindemuthianum* variability. Annu. Rpt. Bean Improv. Coop. 41:19-20.

Beaver, J.S. 1993. A simple method for producing seed from hybrid dwards derived from crosses between Middle American and Andean gene pools. Annu. Rpt. Bean Improv. Coop. 36:28-29.

Beaver, J.S. 1998. Potential contributions of Andean beans to the genetic improvement of black beans. p. 115-120. *In* R. Lépiz (ed.), Memoria del Taller Internacional de Mejoramiento Genético de Frijol Negro Mesoamericano. PROFRIJOL, Guatemala City, Guatemala.

Beaver, J.S., E. Arnaud-Santana & D.P. Coyne. 1996. Yield stability of determinate and indeterminate red mottled beans. J. Agric. Univ. P.R. 80:187-189.

Beaver, J.S. & J.D. Kelly. 1989. Yield compensation of beans grown in hill-plots. HortScience 24:137-138.

Beaver, J.S. & R. Macchiavelli. 1998. Breeding strategies for pyramiding genes for disease resistance. Annu. Rpt. Bean Improv. Coop. 41:141-142.

Beaver, J.S. & A. Molina. 1997. Mejoramiento de frijol para el Caribe. pp. 353-376. *In* S.P Singh & O. Voysest (eds.),Taller de Mejoramiento de Frijol para el Siglo XXI: Bases para Una Estrategia para America Latina. CIAT, Cali, Colombia.

Beaver, J.S., J.R. Steadman & D.P. Coyne. 1992. Field reaction of landrace components of red mottled beans to common bacterial blight. HortScience 27:50-51.

Bokosi, J.M., D.P. Coyne, J.R. Steadman, D. O'Keefe & J. Reiser. 1995. Inheritance and association of specific rust resistance, stem color, leaf pubescence and flower color in common bean. Annu. Rpt. Bean Improv. Coop. 38:141-142.

Buttery, B.R., S.J. Park & D.J. Hume. 1992. Potential for increasing nitrogen fixation in grain legumes. Can. J. Plant Sci. 72:323-349.

Carson, M.L. & Z.W. Wicks III. 1989. Theory for the selection for yield in disease-stress environments. Phytopathology 79:70-76.

Cataño, H. 1990. Selection of *Phaseolus vulgaris* L. genotypes able to nodulate in the presence of nitrogen. M.S. Thesis. Univ. of Puerto Rico, Mayaguez, Puerto Rico.

CIAT (Centro Internacional de Agricultura Tropical). 1996. Bean Program 1994 Annu. Rpt. CIAT, Cali, Colombia.

CIAT (Centro Internacional de Agricultura Tropical). 1988. Bean Program 1988 Annu. Rpt. CIAT, Cali, Colombia.

Coale, F.J., J.J. Meisinger & W.J. Weibold. 1985. Effect of plant breeding on yield and nitrogen fixation in soybeans under two soil N regimes. Plant Soil 86:357-367.

Dennett, M.D. 1984. The tropical environment. pp. 1-38. *In* P.R. Goldsworthy & N.M. Fisher (eds.), The Physiology of Tropical Field Crops. John Wiley & Sons, New York, U.S.A.

Dudley, J.W. 1982. Theory for transfer of alleles. Crop Sci. 22:632-637.

Echavez-Badel, R. & J.S. Beaver. 1987. Resistance and susceptibility of beans to ashy stem blight. J. Agric. Univ. P.R. 72:403-406.

Hall, A.E. 1990. Breeding for heat tolerance - an approach based on whole-plant physiology. HortScience 25:17-19.

Kelly, J.D. 1989. The presence of dwarf lethal *DL* genes in *Phaseolus* germplasm lines and cultivars. Annu. Rpt. Bean Improv. Coop. 32:73-74.

Kelly, J.D., L. Afanador & S.D. Haley. 1995. Pyramiding genes for resistance to bean common mosaic virus. Euphytica 82:207-212.

Kelly, J.D., J.M. Kolkman & K.A. Schneider. 1997. Breeding for high yield in common bean. pp. 187-204. *In* S.P Singh & O. Voysest (eds.), Taller de Mejoramiento de Frijol para el Siglo XXI: Bases para Una Estrategia para America Latina. CIAT, Cali, Colombia.

Kelly, J.D., J.M. Kolkman & K.A. Schneider. 1998. Breeding for yield in dry bean (*Phaseolus vulgaris* L.). Euphytica 102:343-356.

Kelly, J.D. & P.N. Miklas. 1997. The role of molecular markers in breeding for qualitative and quantitative traits in common bean. pp. 283-293. *In* S.P Singh & O. Voysest (eds.), Taller de Mejoramiento de Frijol para el Siglo XXI: Bases para Una Estrategia para America Latina. CIAT, Cali, Colombia.

Kornegay, J., J.W. White, J.R. Domínguez, G. Tejada & C. Cajiao. 1993. Inheritance of photoperiod response in Andean and Mesoamerican common bean. Crop Sci. 33:977-984.

Kornegay, J., J.W. White & O. Ortíz de la Cruz. 1992. Growth habit and gene pool effects on inheritance of yield in common bean. Euphytica 62:171-180.

Laing, D.R., P.G. Jones & J.H.C. Davis. 1984. Common beans (*Phaseolus vulgaris* L.) pp. 305-352. *In* P.R. Goldsworthy & N. Fisher (eds.), The Physiology of Tropical Field Crops. John Wiley & Sons, New York, U.S.A.

Lynch, J.P. & S.E. Beebe. 1995. Adaptation of beans (*Phaseolus vulgaris* L.) to low phosphorous availability. HortScience 30:1165-1171.

McFerson, J., F.A. Bliss & J.C. Rosas. 1982. Selection for enhanced nitrogen fixation in common bean (*Phaseolus vulgaris* L.). pp. 39-44. *In* BNF Technology for Tropical Agriculture. CIAT, Cali, Colombia.

Mejía-Jiménez, A., C. Muñoz, H.J. Jacobson, W.M. Roca & S.P. Singh.1994. Interspecific hyridization between common and tepary bean: increased hybrid embryo growth, fertility, and efficiency of hybridization through recurrent and congruity backcrossing. Theor. Appl. Genet. 88:324-331.

Mmbaga, M.T. & J.R. Steadman. 1992. Effect of leaf age on uredinia diameter and its epidemiological implications for adult plant resistance. Annu. Rpt. Bean Improv. Coop. 35:128-129.

Molina Castañeda, A. & J.S. Beaver. 1998. Inheritance of normal pod development in bean golden mosaic resistant common beans. Annu. Rpt. Bean Improv. Coop. 41:3-4.

Murillo, A., E. Perlata, J. Pinzón & R. Lépiz. 1997. Logros, problemas y perspectivas para el mejoramiento del frijol arbustivo en el Ecuador. pp. 405-414. *In* S.P Singh & O. Voysest (eds.), Taller de Mejoramiento de Frijol para el Siglo XXI: Bases para Una Estrategia para America Latina. CIAT, Cali, Colombia.

Padgette, S.R., K.H Kolacz, X. Delannay, D.B. Re, B.J. LaVallee, C.N. Tinius, W.K. Rhodes, Y.I. Otero, G.F. Barry, D.A. Eichholtz, V.M. Peschke, D.L. Nida, N.B. Taylor & G.M. Kishore. 1995. Development, identification and characterization of glyphosate-tolerant soybean line. Crop Sci. 35:1451-1461.

Pastor-Corrales, M.A., C. Jara & S. P. Singh.1998. Pathogenic variation in, sources of, and breeding for resistance to *Phaeoisariopsis griseola* causing angular leaf spot in common bean. Euphytica 103:161-171.

Ricci, J.R. 1997. Participación de los agricultores en la identificación y adaptación de nuevos cultivares de poroto: agricultura empresarial en Argentina. pp. 434-448. *In* S. P. Singh & O. Voysest (eds.), Taller de Mejoramiento de Frijol para el Siglo XXI: Bases para Una Estrategia para America Latina. CIAT, Cali, Colombia.

Rodríguez, E., R. García & F. González. 1995. Poroto (*Phaseolus vulgaris* L.): Guia tecnica para su cultivo. Instituto de Investigación Agropecuaria de Panamá, Panamá.

Rodríguez, M.A., H.F. Ramírez, M.C. Valencia, O. Voysest & J.W. White (comps.). 1995. Catalog of advanced bean lines from CIAT. 2nd ed. CIAT, Cali, Colombia.

Rosas, J.C., J.D. Erazo & J.R. Moncada. 1991. Tolerancia a la sequia en germoplasma de frijol común y teparí. CEIBA 32(2):91-106.

Rosielle, A.A. & J. Hamblin. 1981. Theoretical aspects of selection for yield in stress and non-stress environments. Crop Sci. 21:943-946.

Saettler, A.W. & F.J. Correa-Victoria. 1983. Angular leaf spot in seed fields of Michigan red kidney beans. Mich. Dry Bean Digest 8(2):2-3.

Shaik, M. 1984. Race-nonspecific resistance in bean cultivars to races of *Uromyces appendiculatus* pv. *appendiculatus* and its correlation with leaf epidermal characteristics. Phytopathology 75:478-481.

Shonnard, G.C. & P.Gepts. 1994. Genetics of heat tolerance during reproductive development in common bean. Crop Sci. 34:1168-1175.

Simmonds, N.W. 1993. Introgression and incorporation strategies for the use of crop genetic resources. Biol. Rev. 68:539-562.

Singh, S.P. 1992. Common bean improvement in the tropics. Plant Breed. Rev. 10:199-269.

Singh, S.P. 1995. Selection for water stress tolerance in interracial populations of common bean. Crop Sci. 35:118-124.

Singh, S.P., P. Gepts & D.G. Debouck. 1991. Races of common bean (*Phaseolus vulgaris*, Fabaceae). Econ. Bot. 43:379-396.

Singh, S.P. & J.A. Gutiérrez. 1984. Geographical distribution of the DL_1 and DL_2 genes causing hybrid dwarfism in *Phaseolus vulgaris* L., their association with seed size and their significance to breeding. Euphytica 33:337-345.

Stavely, J.R., J. Steinke, R.T. McMillan, K.F. Grafton, J.R. Steadman, J.D. Kelly, D.P. Coyne, D.T. Lindgren & M.J. Silbernagel. 1992. Rust resistant bean germplasm releases. Annu. Rpt. Bean Improv. Coop. 35:228-229.

Valladolid Chiroque, A. 1997. Mejoramiento de frijol para la Costa del Peru. pp. 377-396. *In* S.P. Singh & O. Voysest (eds.), Taller de Mejoramiento de Frijol para el Siglo XXI: Bases para Una Estrategia para America Latina. CIAT, Cali, Colombia.

Vélez, J., M.J. Bassett, J.S. Beaver & A. Molina. 1998. Inheritance of resistance to bean golden mosaic virus in common bean. HortScience 123:628-631.

Voysest, O. 1997. Fundamentos de un eficiente sistema de pruebas de germoplasma de frijol. pp. 417-433. *In* S.P Singh & O. Voysest (eds.), Taller de Mejoramiento de Frijol para el Siglo XXI: Bases para Una Estrategia para America Latina. CIAT, Cali, Colombia.

White, J.W, J.A. Castillo, J.R. Ehleringer, C.H. Cajiao & J. Tejeda. 1992. Effect of growth habit on yield of large seeded bush cultivars of common bean. Field Crops Res. 29:151-161.

White, J.W., R. Ochoa M., F.J. Ibarra-Pérez & S.P. Singh. 1994. Inheritance of seed yield, maturity and seed weight of common bean (*Phaseolus vulgaris*) under semi-arid rainfed conditions. J. Agric. Sci. 122:265-273.

Wortmann, C.S. & D.J. Allen. 1994. Constraints on bean production in Africa. Annu. Rpt. Bean Improv. Coop. 37:202-203.

Wortmann, C.S., R.A. Kirby, C.A. Eledu & D.J. Allan. 1998. Atlas of Common Bean (*Phaseolus vulgaris* L.) Production in Africa. CIAT, Cali, Colombia.

Xu, Yunbi, S.R. McCouch & Z. Shen. 1998. Transgressive segregation of tiller angle in rice caused by complementary gene action. Crop Sci. 38:12-19.

Chapter 12

Improvement of Snap Bean

James R. Myers & James R. Baggett
Horticulture, Oregon State University, Corvallis, OR 97331, U.S.A.

Key words: breeding methods, cultivar development, disease resistance, fresh market, green bean, heat stress, plant architecture, pod concentration, processing, seedstock maintenance

1. INTRODUCTION

In this chapter, we review the status of snap bean research and suggest directions for the future. The snap bean group is defined and differences with dry bean are delineated. The origin of edible podded bean, from their possible genesis in the New World, to expansion of genetic diversity in the Old World, to their reintroduction into the U.S.A. is considered. We examine world commercial production with particular emphasis on North American environments and production constraints. Genetic control for whole plant traits, pod traits, seed traits, and disease resistances of importance is discussed. Finally, we suggest areas in need of research and breeding objectives for the future.

Snap beans comprise a group of common bean (*Phaseolus vulgaris* L.) that have been selected for succulent pods with reduced fiber. The immature pods and seeds are consumed as a green vegetable. While dry bean market classes are based primarily on seed characteristics, snap bean market classes are based on pod characteristics and plant type. Fewer and less distinct market classes of snap bean exist compared to dry bean, but given their importance as a vegetable, breeders have developed many cultivars.

"Snap Bean" is the term used in this paper, but other names such as "French", "Garden" or "Haricot" bean are used to describe the group. In the older literature, "Kidney Bean" may refer to an edible podded bean

(Hedrick, 1972; Zaumeyer, 1963). The term "String Bean" refers to older snap bean cultivars that had fiber in the pod suture, which had to be removed manually before cooking. The stringless trait was discovered in 1870 by C.N. Keeney and has since been widely incorporated into improved snap bean cultivars (Zaumeyer, 1963). Nearly all bush beans today are stringless, while some heirloom pole beans are stringed.

Names such as "Green Bean", "Wax Bean", or "Romano" (also known as "Italian" or "Flat Pod" beans) describe subgroups or market classes. Romano bean pods are green in color, with flat cross-sectional shape, and relatively fiber-free. The traditional snap bean type of southern Europe, the Romano may be one of the predecessors of the contemporary snap bean. In contrast to the Romano, green bean has fleshy pods that are generally oval to round in cross-section. Wax bean pods have a shape similar to green bean pods, but are pale yellow to golden instead of green in color.

Edible-podded beans are handled and consumed in several different ways. Traditionally, freshly harvested pods were cooked for consumption, but with the current emphasis on consuming more fresh vegetables, consumption of raw bean pods has been suggested. This practice should be examined to determine if the antinutritional factors present in mature, dry beans are found in significant quantities in green pods. For off-season consumption, snap beans are canned, frozen, and dehydrated. Pods may be packed whole, or subjected to various cuts, which requires sorting of pods into uniform sizes (Table 1), as well as using cultivars that mature at a certain sieve size. In this chapter, we refer to cultivars that are harvested at 50% 1- to 4-sieve as "full-sieve" beans. Small-sieve cultivars are those that maximize quality and yield at a smaller sieve size.

Table 1. Grade, sieve size, and pod cross-section thickness used for sorting snap bean pods

Canner's grade	Sieve size	Pod cross-section thickness (mm)
1	1	<5.8
1	2	5.8 – 7.5
1	3	7.5 – 8.5
2	4	8.5 – 9.7
3	5	9.7 – 10.9
4	6	10.9
cull	7 and over	>10.9

(After Duncan et al., 1960).

Relative to other vegetable crops, snap bean consumption is moderate in the U.S.A. For example, 1995 per capita consumption of snap bean was 3.4 kg compared to 2.3, 5.0, 5.1, 12.7, 1.7, 2.8, 42.8 kg for broccoli, cabbage, carrot, sweet corn, green pea, bell pepper, and tomato, respectively (Maynard & Hochmuth, 1997). Snap bean consumption was nearly identical to dry bean consumption (3.6 kg) in 1995. Consumers eat most snap beans in canned form (1.8 kg) followed by frozen (0.9 kg) and fresh (0.7 kg) products. In general, consumption of canned products is decreasing while fresh snap bean consumption is increasing. With the growing recognition that more vegetables are needed in the American diet, snap bean consumption should increase in the future. Snap beans are consumed in larger quantities in developed countries, where diets are more varied than in developing countries. On a daily utilization basis, developed countries consumed 3.1 g/day compared to 1.1 g/day in developing countries (Rubatzky & Yamaguchi, 1997). For estimates of snap bean consumption in developing countries, see Silbernagel et al. (1991).

During the past 30 years, the number of public snap bean breeding programs has declined. In the U.S.A., only four public programs have partial effort devoted to breeding and genetics of this crop. Most cultivar development is done in the private sector. Compared to dry bean, snap bean seed sales are more profitable, providing an incentive for private companies to invest in development of improved cultivars. Consequently, much research is proprietary. Brief published descriptions for cultivars are available in seed catalogues. Additional information on pedigree and varietal characteristics is available from the Plant Variety Protection Office (http://www.ams.usda.gov/science/pvp.htm) after a patent has been issued. Silbernagel (1986) and Silbernagel et al. (1991) are the most recent, comprehensive reviews on the subject of snap bean breeding. Other general sources of information include Bliss (1980) on reproductive biology and Wade (1937) as a historical source.

While dry bean is known to be a good source of protein, complex carbohydrates, and soluble fiber, few people are aware of the nutritional content of snap beans. Snap bean provides similar amounts of calories, carbohydrates, and protein when compared to dry bean on a similar moisture basis (Shellie-Dessert & Bliss, 1991). Green bean pods consist of about 88% water compared to 70% for cooked dry beans. This moisture difference accounts for most of the nutritional advantage of dry beans over green pods. Green pods are a superior source of calcium, iron, thiamin, and niacin (Anonymous, 1998c).

2. ORIGIN OF SNAP BEAN

Generally accepted archaeological evidence for common bean indicates that beans were domesticated more than 6,900 years before the present (BP) in Middle America and about 7600 years BP in the Andean region. However, recent dating by accelerator mass spectrometry suggests that common bean was domesticated more recently (Smith, 1995). Increase in seed size and decrease in pod fiber (from the explosively dehiscent pods of the wild bean) are regarded as the primary evidence of domestication (Kaplan, 1981). Contemporary snap bean cultivars have long pods with greatly reduced fiber (<20%), and thickened pod walls. Selection for slender, fleshy pods has also resulted in long, cylindrically shaped seed. Seed shape and extremely low pod fiber could be used to distinguish morphologically between dry and snap beans in the archaeological record. To our knowledge, no such analyses have been reported. One reason for the lack of evidence is that the characteristic seed shape and low pod fiber may be a consequence of recent selection and breeding efforts. In addition, low-fiber pods may not persist well in the archaeological record. In a survey of common bean accessions held by CIAT, Singh (1989) found very few examples of possible landraces of snap bean from Middle- and South America. Snap beans, if present at all in pre-Columbian times, were probably rare, and subject to the capriciousness of the preservation process.

Knowledge of bean genetics suggests that snap beans were derived from dry beans because more genetic changes would be required to derive snap beans from the wild bean than from dry beans. However, Singh et al. (1991) suggested that Native American groups might have originally gathered wild beans for their young green pods. Young pods would have required less cooking time than seeds, an important consideration in pre-ceramic cultures. Contemporary groups in the Americas and Africa eat leaves and extremely young pods of dry beans, suggesting another avenue for the derivation of snap beans. Preference for pods that remain edible later into maturity would have selected for genes causing reduced fiber content. Genes that code for fleshy, succulent pods would have been selected for as well. Native Americans may have derived dual-purpose cultivars or cultivars with an accumulation of snap bean traits that the Europeans then further refined. Much of the explosion in snap bean diversity appears to have happened in Europe after the Columbian exchange. From Europe, snap beans were dispersed to the rest of the world (Gepts et al., 1988).

Based on phaseolin type, snap beans are derived from the Andean center of origin (Brown et al., 1982; Gepts et al., 1986). However, in some contemporary snap bean cultivars, the divisions between Andean and Middle American centers of origin have been blurred by crossing between the two groups as shown by molecular marker analysis (Skroch & Nienhuis, 1995; Skroch et al., 1992). The blending of Middle American and Andean genetic variation may have occurred in breeding programs primarily in Europe, in the U.S.A., or both. Molecular tools may provide a systematic analysis of snap bean ancestry that could answer questions about origin and genetic diversity.

3. WORLD PRODUCTION

3.1 Introduction

Data on world production of snap bean is confounded by the fact that immature green pods of *Vigna unguiculata* L. Walp. are also consumed, predominantly in India and Asia. Snap beans are most important in developed countries (predominantly Spain, France, Italy, Germany, U.K., Netherlands, Portugal, and Hungary in Europe; the U.S.A., Mexico, and Canada in North America; and Australia and South Africa in the Southern Hemisphere). France is the largest producer of snap bean for processing in Europe. Substantial production also takes place in developing countries, some for local consumption, but also large quantities for export. Developing countries with significant production include China (which accounted for 27% of world production of *Phaseolus* and *Vigna* green beans in 1997 (Anonymous, 1998b), Chile, Egypt, India, Indonesia, Mexico, Morocco, Kenya, Tanzania, Thailand, and Turkey.

3.2 North American Production

In the U.S.A., snap bean production is found in the Northeast (New York, Pennsylvania, and Ohio), Eastern Seaboard (New Jersey, Virginia, Delaware, Maryland, and North and South Carolina), Southeast (Florida and Georgia), South Central (Missouri, Oklahoma, Tennessee, and Arkansas), northern Midwest (Wisconsin, Minnesota, Michigan, and Illinois), Southwest (Texas, Arizona, and California), and Northwest (Oregon). The southern regions supply fresh market beans during the winter, spring, and early summer to northern markets in the U.S.A. and Canada, while the northern and central states produce beans mostly for processing. Florida and California are the leading states for production of fresh market beans while Wisconsin, Michigan, Oregon, and New York produce the bulk of beans for freezing and canning (Anonymous, 1998a).

Florida produces fresh market beans on about 12,677 ha (Table 2). The south Florida hectarage may be decreasing in size because of development, restoration of the Everglades, and expanding Mexican production. Beans are planted from September to early March although the bulk of the production is targeted to the Thanksgiving and Christmas markets. Fresh market shipping beans are generally light green in color (although a shift to darker green colors is underway) and are bred for shipping ability and long shelf life. Disease problems include rust [caused by *Uromyces appendiculatus* (Pers.) Ung.], common bacterial blight [caused by *Xanthomonas campestris* pv. *phaseoli* (Smith) Dye], ashy stem blight [caused by *Macrophomina*

Table 2. Area (ha) and yield (MT ha^{-1}) of fresh and processing snap beans[1] in the U.S.A.

State	Area	Yield	Production constraints
Processing			
Midwest			Brown spot, *Aphanomyces* root rot, white mold
Wisconsin	25,641	7.6	
Illinois	6,329	7.5	
Indiana	1,184	5.7	
Michigan	8,812	7.2	
East			White and gray mold, *Fusarium* root rot
New York	8,691	7.8	
Pennsylvania	3,522	5.9	
Pacific Northwest			White and gray mold, root rot complex
Oregon	9,366	13.6	
Other States	18,988	7.3	
Total	82,668	8.1	
Fresh			
Southeast			Bean golden mosaic, fungal and bacterial diseases, heat
Florida	12,677	5.6	
Georgia	6,075	4.5	
Eastern Seaboard			Heat, rust, bacterial blights
Maryland	972	4.4	
Virginia	2,187	4.3	
North Carolina	2,673	4.0	
South Carolina	689	2.7	
East			White and gray mold, *Fusarium* root rot, rust
New York	1,661	5.9	
New Jersey	1,499	4.5	
Ohio	689	5.8	
Midwest			Heat, rust, bacterial blights, white mold
Michigan	689	4.9	
Tennessee	3,240	4.5	
West			Viruses
California	2,835	10.8	
Total	35,924	5.4	

[1]Based on an average of 1995-1997 production figures (Anonymous, 1998a).

phaseolina (Tassi) Goid.], alternaria pod spot (caused by *Alternaria* spp.), root rot (predominantly *Rhizoctonia solani* Kühn), and bean golden mosaic (BGM, caused by a geminivirus). Hurricane Andrew is thought to have brought viruliferous whiteflies from the Caribbean in 1992. Since then, BGM has been present in south Florida. The virus persists during the summer fallow by surviving in indigenous weed species that are also host for the insect vector, silverleaf whitefly (*Bemesia argentifolii* Bellows & Pevering).

Georgia produces more than 6,075 ha of beans for fresh market, although they may be processed, depending on the current market value. Harvests occur in November and May or June. Growers generally use a darker-podded bean than do Florida growers. Cultivars used in this market also tend to be intermediate in shipping characteristics, with lower fiber levels and less pubescence than Florida fresh market beans. Alternaria pod spot, rust, root rot, and common bacterial blight are diseases of importance. Heat tolerance is needed in cultivars grown in this area.

The California production areas of about 2,835 ha include the Indio area of the Coachella Valley, and Riverside County. Climatic conditions for these growing areas are quite different with extreme temperature fluctuations and low rainfall in the Coachella Valley compared to a mild maritime climate in Riverside County. Production is timed for a November harvest for the holiday markets and for early spring. Originally, a slender, medium green, flat podded (Greenpod) type was used. However, round podded types are presently preferred. Darker green pods with good shipping traits are now being sought. Viruses, primarily cucumber mosaic, are the major constraints.

Snap beans for processing are grown on approximately 25,641 ha in Wisconsin. About 3,600 ha are grown in the central sands area under irrigation; the remaining hectarage is grown without irrigation. The season is generally moderate with mild day and cool night temperatures. Yields of 7-12 MT ha^{-1} are common. The cultivars grown in this region tend to be light- to medium-green with fairly large sieve size. Cultivars used in early plantings need cold tolerance. Bacterial diseases such as brown spot [caused by *Pseudomonas syringae* pv. *syringae* (van Hall)], halo blight [caused by *Pseudomonas syringae* pv. *phaseolicola* (Burkh.)], and common blight cause chronic yield reductions and quality problems. Of these, brown spot is the most important disease. White mold [caused by *Sclerotinia sclerotiorum* (Lib.) de Bary] and root rots (predominantly caused by *Aphanomyces euteiches* Drechs. f. sp. *phaseoli* W.F. Pfender & D.J. Hagedorn) are important fungal diseases of the region. Significant summer and fall production in Illinois (about 6,329 ha), Indiana (1,184 ha), and Michigan (8,812 ha) parallel that of Wisconsin in terms of types used and production constraints.

In New York, 8,691 ha of snap beans are grown for processing. Yields are similar, and climate, production constraints, and cultivars used are nearly the same as those of Wisconsin. Disease problems are similar except that Fusarium [*Fusarium solani* (Mart.) Sacc. f. sp. *phaseoli* (Burkh.) W.C. Snyder & H.N. Hans.] and *Rhizoctonia* root rots predominate.

Oregon is the second largest producer of snap beans for processing, at 9,366 ha. With an average of 13.6 MT ha^{-1}, Oregon has significantly higher yield than other regions (the U.S.A. average was 8.1 MT ha^{-1} in 1995-1997).

This yield advantage is due in part to favorable growing conditions and the use of irrigation, and in part to the use of Blue Lake cultivars adapted to the region. The Willamette Valley of Western Oregon, where snap beans for processing are grown, has a mild maritime climate. Nights are cool, and days are warm during the growing season. Severe thunderstorms, early or late freezes, and other unusual weather events are rare during the growing season. Summertime rainfall is low with most water supplied by irrigation. Processors prefer a full sieve bean although preference for additional four-sieve product is growing. The Blue Lake types have superior pod quality with dark-green color, low fiber content, a distinctive flavor and texture, and excellent "holding ability" in institutional situations (i.e., the product looks and tastes good after several hours on the steam table). Production constraints include white mold, gray mold (caused by *Botrytis cinerea* Pers.:Fr.), and heat stress during the day.

Snap beans for processing are produced on the DelMarVa Peninsula (Delaware, Eastern Shore of Maryland, and Virginia). A 'Slenderette' type (3- to 4-sieve) bean has traditionally been grown in this area, but full-sieve types are also used. Heat stress, particularly high nighttime temperatures, is a production constraint. Rust is an important disease in this production area, as are the bacterial blights.

Smaller hectarages where snap beans are produced for processing are found in Texas, Arizona, Arkansas, Missouri, Oklahoma, and Tennessee. The southern areas generally count heat among their environmental constraints, and bacterial blights and rust as disease problems. Production from many of these areas may be shipped to processing plants in Missouri and Arkansas.

A major difference between the processing and fresh market industries is that the processing industry has continued to consolidate, whereas there are many small hectarages scattered throughout the fresh market production regions. From a breeding standpoint, it is much easier to determine the needs of food processors and their growers. Fresh market producers' needs are diverse, and amounts of seed needed by individuals are small. Seeds of processing cultivars are generally sold directly by the seed company whereas many independent seed distributors (as well as direct sales) are involved in fresh market production. There may also be a need for greater cultivar diversity in the fresh market compared to the processing market. However, the produce broker does require uniformity, and in certain markets, one cultivar may predominate.

4. SNAP BEAN LINEAGES

As near as can be determined, settlers brought most edible podded beans from Europe to the U.S.A. (Gepts et al., 1988). There are reports of some cultivars being obtained by colonialists from native Americans (e.g., Martens in Hedrick, 1931), but these events occurred 200 to 250 years after initial Columbian contact. This is ample time for beans taken to Europe to be reintroduced into the Americas and given to native Americans.

Rather than being bred in a systematic manner, the cultivars developed in Europe and the U.S.A. were selected from variation generated by mutations and chance outcrosses in older cultivars. These were introduced as a number of distinct cultivars that were often multiple purpose beans, being consumed as immature pods, mature shell beans, and as dry beans. Pole and bush beans with green and yellow pods are found from an early date suggesting that much of the variation found in the early American cultivars was already present in Europe.

Snap beans have undergone extensive recombination so that today, there are few distinct types in the U.S.A. Flat pod (Romano), Midwest processing, fresh market shipping, and Blue Lake are the main types grown commercially. Flat pod beans are an improved Romano type with large flat pods and strong and distinct flavor. They are generally processed, but some are used as a fresh market bean.

The Midwest processing types (with the exception of bean cultivars developed by Del Monte, which are Blue Lake types) are sometimes referred to as "Tendercrop" types. Tendercrop was released by Zaumeyer in 1958 and was the first snap bean with a strong, upright bush that was suitable for mechanical harvest (Zaumeyer & Thomas, 1959). While no longer used commercially, Tendercrop has had an important influence on the breeding of contemporary cultivars.

For fresh market shipping beans, Harvester was an early cultivar that combined fresh market characters with adaptation to mechanical harvest (Silbernagel, 1986). In general, the pods of these types have higher fiber and a greater degree of pubescence in order to withstand the rigors of machine harvest, long distance shipping, and the need for extended shelf life.

Blue Lake beans have been synonymous with dark green pods, a distinctive mild flavor, and the ability to preserve their appearance after processing and being held on a steam table for several hours. The "Blue Lake" name apparently derives from the Blue Lake area near Ukiah, California (Schultz, 1966). Accounts vary, but either Scotia or White Creaseback was originally grown in this area. The cultivar was taken from the Blue Lake area to the Pacific Northwest in the 1920s (Geise, 1955; Schultz, 1966). Scotia was apparently introduced from Germany in 1896 by

Mr. Dutcher, while White Creaseback seems to have had a much earlier origin in America (Hedrick, 1931). Both cultivars have pod characteristics nearly identical to contemporary Blue Lake cultivars. The original Blue Lake cultivars were pole types, and the processing industry in Oregon developed around the culture of these types. Yields of 20 to 30 MT ha^{-1} were obtained (Moore & Allmendiger, 1954). In the 1950s, other bean growing areas were introducing mechanical harvest of bush types, resulting in increased competition for the more expensive hand labor required for the harvest of the pole types. Dr. W.A. Frazier embarked on a 10 year program to backcross the Blue Lake trait complex into a bush background. This was achieved with some difficulty (Davis & Frazier, 1966; Frazier et al., 1958), and the first bush Blue Lake types were introduced in 1965 (Baggett, 1995). The first cultivars were not a success because of a poor growth habit that resulted in pods on the ground and a trashy harvest. The first successful release was Oregon 58 in 1970 (Frazier et al., 1970), followed by Oregon 1604 in 1972 (Frazier et al., 1972), Oregon 91G in 1980 (Baggett et al., 1981), and Oregon 54 in 1992 (Baggett et al., 1992). Since its introduction, Oregon 91G has dominated the processing industry in Oregon. While higher yielding than 91G, hectarage of Oregon 54 may be declining because field men and growers consider the cultivar to have a non-concentrated pod set.

Although Zaumeyer (1972) concluded that the genetic base of the snap bean was narrow, others have suggested that snap beans have a relatively broad genetic base compared to other common bean market classes (Haley et al., 1994d; Skroch & Nienhuis, 1995). Few studies exist where phylogenetic relationships have been published. Maluf & Ferreira (1983) performed a multivariate analysis to determine genetic distance among ten snap bean cultivars. Data produced by Weeden (1984) could be used to examine genetic relationships among snap bean cultivars. In a study of ten snap bean cultivars, Skroch & Nienhuis (1995) found major separations among those with Andean or Middle American background and, within the Andean group, a split between large- and small-sieve size cultivars. At present, the information available on genetic relatedness and ancestry in snap beans is scattered. A general study using many snap bean cultivars is needed to examine a number of questions concerning snap bean origins and genetic diversity.

5. GENETICS OF IMPORTANT TRAITS

5.1 Introduction

Traits of importance in snap beans can be separated into three categories: whole plant traits, pod and seed traits, and resistance to abiotic and biotic stresses. Whole plant traits include yield and all the factors that are components of yield. These include days to flower, duration of flowering (temporal concentration of pod set), rate of pod fill, number of branches, number of flowers and pods per node, number of reproductive nodes, growth habit, plant architecture, and harvest index. Whereas seed characters determine dry bean market classes, pod characters define snap bean cultivars. Important pod characteristics include pod length, shape (cross-sectional and shape of the distal and proximal ends), spur shape and length, color, straightness, smoothness, rate of seed development, fiber content, internal color and texture, presence of interlocular cavitations, and point of detachment (above or below the pedicel).

5.2 Plant Traits

Yield is more complex in a vegetable crop such as snap bean, compared to a seed crop. As an example, dry bean yield is measured at maturity as seed weight at approximately 12% moisture on a per unit harvest area basis. Snap bean yield is a moving target; growers attempt to harvest at the highest yield possible while maintaining quality. Yield and quality have an inverse relationship (Mullins & Coffey, 1990). To a grower or processor, quality is defined in terms of sieve size, percent seed by weight of total pod weight, pod fiber content, pod smoothness and straightness, pod color, and flavor. With a once-over mechanical harvest, fieldmen must select a time for harvest that produces the highest possible yield and the optimum percentages of sieve sizes before quality deteriorates to an unacceptable level.

Numerous studies have shown that yield of grain crops is a quantitative trait with low heritability, is controlled by many genes and is subject to substantial environmental influence. Because of the additional quality factors involved in yield of a vegetable crop, yield is probably even more complex in snap beans. This fact implies that heritabilities will also be lower. Low heritabilities require extensive replicated trials to determine differences. Perhaps more important to growers and processors than higher yield *per se*, is yield stability. Yield stability can be achieved in part by incorporating resistance to biotic and abiotic stresses, and by breeding for general adaptation.

A contemporary trend in the industry is an increased demand for small-sieve beans. Full-sieve cultivars produce substantial quantities of five- and six-sieve beans, which are used in a cut bean pack. Currently, some processors consider that an optimum harvest should produce 60% 1- to 4-sieve beans. Breeders are developing small-sieve cultivars with maximum of four-, three- and even two-sieve sizes. With the shift to fewer large-sieve pods comes lower yields. The challenge for breeders in the near future is to develop smaller-sieve cultivars that accumulate the same amount of reproductive biomass as do full-sieve cultivars. In reality, some mix of small- and full-sieve cultivars will be grown. Oregon processors continue to grow cultivars with standard size pods because they can obtain acceptable percentages of small-sieve beans by selection of harvest maturity, and still obtain good yields. Standard-sieve size cultivars also permit packing of sliced and cut five-sieve pods, which still have a strong place in the market.

Growth habit and plant architecture in snap beans fall into a range similar to that found in dry beans. Most of the original edible podded beans were climbing type IV vines (Fernández et al., 1986) that are known as pole beans today. Pole beans occupy very little of the contemporary commercial hectarage. Most cultivars, especially for fresh market and processing, have determinate bush (type I) growth habit. The type I growth habit is easier to handle in mechanized agricultural systems and lends itself to a once-over mechanical harvest. Characteristics of bush growth habit that are important for processing or fresh market beans include lodging resistance, plant height, number and length of branches, pod placement in the canopy, and length of the peduncles supporting the pods (Eteve & Fouilloux, 1979; Silbernagel, 1986).

Type I growth habit is inherited as a single recessive gene (*fin*). Within this background, number of internodes and internode length can vary considerably (Debouck et al., 1988; Evans, 1980). Snap bean cultivars bred for mechanical harvest have four to seven nodes with short internodes. Internode length has been reported to show quantitative control (Davis & Frazier, 1966; Detongnon & Baggett, 1989), but also has a large environmental component. Inheritance of number of nodes within the determinate background has not been studied. Lodging resistance appears to be related in part to internode length and has been described as controlled by a single dominant gene (Bliss, 1971) or three recessive genes (Frazier et al., 1958). Other factors including root structure and stem thickness affect lodging, and would suggest that lodging resistance and plant architecture should show quantitative inheritance (Acquaah et al., 1991; Brothers & Kelly, 1993; Kelly & Adams, 1987).

A concentrated pod set maximizes yield in snap beans for a single destructive harvest. The determinate bush growth habit conditioned by *fin* naturally has a more concentrated pod set than do indeterminates. Indeterminate beans begin flowering at a lower node, and generally continue to flower up the stem. The first flower to open on plants with *fin* is a flower on the terminal raceme. Flowering proceeds down the stem to the secondary branches. Snap bean cultivars with a concentrated set have been selected for a near-simultaneous bloom of the terminal and branch node racemes, and for pod fill to occur over a condensed period of time. The inheritance of pod concentration is not known but appears to be quantitative. For processors, a balance must be struck between too high a degree of concentration and too extended a period of pod maturation. Fieldmen find it difficult to determine optimum harvest time in a cultivar that lacks sufficient concentration. There is no clearly defined point at which yield and quality are maximized. On the other hand, cultivars with too concentrated a pod set may proceed through a period of maximum yield and quality too quickly. Processors find it difficult to schedule harvest for such "flashy" cultivars, and may have to bypass fields in hot weather.

Useful quantitative measures for pod concentration have not been established. Several factors may be used; none of them is appropriate for breeding procedures where many lines must be evaluated quickly and cheaply. The components of concentration include developmental timing of anthesis, flowering duration, pod set duration, number of racemes, number of flowers per raceme, and rate of pod fill. Near-simultaneous flowering of the terminal and lateral racemes increases concentration. A limited time period of flowering may also increase concentration; however, it appears that flower duration is not significantly different among concentrated and non-concentrated lines (our unpublished data). Snap beans have been selected for a high number of flowers per raceme even though most of these flowers never develop into mature pods. Limited resources require that many buds and young pods abort so that the most developed pods can continue to develop to maturity. Types with very high flower numbers or long racemes have been observed to produce large numbers of partially developed and unusable pods. More likely, sink strength of the first flush of pods determines when later flowers begin to abort. A high sink strength results in fewer of the later flowers developing into pods, and would probably result in increased concentration. Rate of pod fill controls the rate at which pods reach optimum maturity and quality, and would affect concentration. Pod fill rate can be measured by successive harvests during maturation where yield and sieve sizes are determined for each harvest. Sieve size will shift over time from predominantly low numbered sizes, to the eventual accumulation of predominantly high numbered sieves. For a full-sieve cultivar where 50% 1- to 4-sieve size would be considered optimum, and where three harvests every other day were performed, the rate of maturation can be measured by the slope of a regression line fit to a plot of change in sieve size over time. Steeper regression lines would indicate more rapid pod fill rate and thereby higher pod concentration.

The genetic control of pod concentration needs investigation. We do not know the relative importance of the traits discussed above. It is not likely to be an easy subject for study, given a complexity similar to yield and composed of some of the same components as yield. Concentration also shows a strong genotype by environment interaction.

Pod placement is an important characteristic of plant architecture in both type I and type IV snap bean cultivars. For mechanical harvest, pods should be distributed in the upper half of the plant on stiff, medium length peduncles (Eteve & Fouilloux, 1979). If pods are held too close to the main stem or are located too low on the plant, a bean harvester set to take these pods also picks up unacceptable levels of trash. Pod clusters on long peduncles may bend, allowing pods to touch the ground. Pod clusters should also break apart at harvest because unbroken clusters cannot be easily handled by the harvesting and processing equipment. For pole bean and bush cultivars that are hand picked, long pedicels are preferred because they place the pods outside of the plant canopy where they are easy to see and pick.

Another important trait for mechanical harvest is a strong root system to prevent the bean harvester from pulling the whole plant at harvest. Strong root systems are also important from several other standpoints, including resistance to root rots, increased drought tolerance, and better nodulation with less requirement for nitrogen fertilizer.

5.3 Pod Traits

Pod traits are perhaps the most important aspect of snap bean cultivars. Traits of importance include color (relative internal and external color, and uniformity of color), pod shape, length, cross-sectional shape, straightness, smoothness, fiber content, rate of seed development, and point of detachment (Silbernagel, 1986).

Nearly all beans for fresh market and processing have pods with some shade of green. Pod color shows genetic variation for both intensity and hue. The genetic control of these traits is not well defined, however. Colors range from light- to dark-green with hues ranging from yellow-green as found in many Midwestern processing types to a blue-green color that is characteristic of Blue Lake types. Fresh market cultivars have traditionally had lighter green colors compared to the processing types, but the gap in color differences is narrowing. While not true in every case, many processors would prefer a dark green pod for canning and freezing. Lighter pods are much more acceptable for freezing than canning; in other words, color has been less critical in freezing than in canning. However, freezers have come to want darker pod color also, possibly because of the influence of Blue Lake cultivars although some processors perceive the blue-green hue of Blue Lake pods as "looking old". Cultivars for other processing markets have been developed with a yellow-green pod hue that may be associated with the use of the shiny pod character (*ace*) (Yen, 1957). While not the darkest green, some yellow-green types have an attractive bright green "fresh" appearance in a processed product. The darkest green colors are found when these two color types are combined (our unpublished data). Fresh and processed colors are not necessarily the same. Pods in general will darken with blanching, and a pod that has light color may blanch to an acceptable dark green color. Blanched color should be bright and not muddy. Some beans may be too dark green for present processing markets because they would not blend with existing cultivars. However, with the overall trend towards darker green pods, the darkest green colors available may eventually be acceptable.

Wax bean pod color is controlled by a single recessive gene (*y*), but may be affected by a second gene (*arg*) and perhaps other modifiers (Currence, 1931). The addition of *arg* causes near-white pod color. Wax beans should have bright yellow pods because pale yellow is associated with over-mature beans. Very young pods on a wax bean are light green in color and excessive green in usable pod sizes has been a problem. Breeders have selected for early yellowing of the pods, and have developed small sieve wax bean cultivars that develop mature pods with intense golden color at a smaller size. A few purple-podded cultivars exist, but are not used commercially. These cultivars have *P* and *V* genes and have pods that are solid colored or striped depending on the allele at the [*C Prp*] locus (Bassett, 1996). Other novel pod colors could be developed, including red and silver-green (conditioned by *arg* alone). The anthocyanin pigments responsible for red and purple colors are water-soluble; they do not remain in a processed product.

Length, cross-sectional shape, diameter, and length of the spur and pedicel affect pod shape. Pods of beans for processing are about 10 - 16 cm in length. Pod lengths greater than 16 cm are too long for existing processing equipment. In some fresh markets, short podded beans are preferred. Cross-sectional shape varies from flat to oval, to round to crease-back. Most fresh market beans are oval because this trait is associated with greater durability for shipping, and gives the straightness needed for an attractive appearance. Processors prefer a round pod because round pods are fleshier and because there is a close relationship between sieve size, quality, and maturity when the round pods are sorted in a sieve grader. At its most extreme, pod shape becomes crease-back (Chung et al., 1991). Pod cross sectional shape is a function of pod wall thickness and timing of development. Four genes (E_a, E_b, I_a, and I_b) and possibly others control this trait (Leakey, 1988). Because cross sectional shape is a function of developmental time and absolute wall thickness, it shows quantitative variation. In a study by Chung et al. (1991), pod cross-section index was inherited additively with some degree of dominance. Shape also depends on time of harvest. A cultivar harvested young may show oval pods, at maturity have round pods, and when past prime have crease-back pods. Crosses between two round podded cultivars will produce a substantial number of oval progeny (unpubl. data). The spur, or remnant of the style, can vary in length with some cultivars possessing a long tapered spur and others having a short broad-based spur. Processors prefer spurs that are short and straight because they are easier to remove during preparation of pods for packing.

Sieve size is probably the single most important factor in processed beans. After harvest, unsorted pods are brought into the plant where equipment is used to separate the pods into different sizes based mainly on pod diameter (Table 1). Sieve size categories range from one to seven where pod sizes at both extremes are of little value. Full-sieve beans generally have 50% 1- to 4-sieve size at maturity. Smaller-sieve cultivars may range from those with a maximum sieve size of two at maturity to cultivars that attain maturity at sieve size four. In reality, sieve sizes approximate a normal distribution that shifts over time. Therefore, those cultivars that are not full-sieve may be described in terms of a sieve size distribution at optimum quality (Mullins & Coffey, 1990). The sieve distribution may vary from the expected normal when pod set is disrupted by hot weather, resulting in a "split set".

Pod straightness is important in processing green beans because straight beans make a neater cut or whole pack product and because they flow through the processing equipment more rapidly. Pod straightness is affected by plant habit, and upright bush types and pole beans tend to have straighter pods. However, pods may be inherently curved or even "fish hooked", even in pole beans. Straightness and fiber content are interrelated qualities. Fresh market beans, which have greater fiber content than do beans used for processing, are also straighter (Silbernagel, 1986). Small-sieve beans have straighter pods because pods are inherently shorter. The challenge for breeding processed cultivars is to select for straight pods while maintaining low fiber content. Three major genes control the switch from the highly fibrous dry bean type pod to a relatively fiber free pod of the typical snap bean (Leakey, 1988). Within snap beans, fiber content appears to be quantitatively inherited, with reported values from 0.02% to about 3.0% of pod fresh weight for pods with acceptable quality (Silbernagel & Drake, 1978). The lowest percentages are found in the Blue Lake beans, with fresh market and over-mature beans exhibiting the highest percentages. Fiber content also increases with sieve size and maturity, with some cultivars as high as 20% in mature six -sieve (Silbernagel & Drake, 1978).

Several authors have reported different modes of inheritance for pod suture string (see Leakey, 1988 for a review). The most comprehensive studies are Prakken's (1934) and Drijfhout's (1978a). Prakken (1934) described a single dominant gene (*St*) preventing string formation, and Drijfhout confirmed the finding. Drijfhout (1978a) also described a temperature-sensitive dominant gene (*Ts*). Beans that are stringless at cool temperatures form strings at high temperatures. Modifiers may influence the expression of these genes, and other genes affecting string formation may exist. Interestingly, the revertant string trait that occurs commonly in snap beans shows up as a recessive trait. Progeny rows are required in a stock seed program to eliminate the off-types. Definitive studies with molecular markers are needed to understand and perhaps mitigate the occurrence of fibrous and stringed rogues.

Pod smoothness is related to pod wall fiber and rate of seed development. Fresh market beans with relatively high pod fiber also have relatively smooth pods. Seed size and shape affect pod smoothness because large seeds or oval or round seeds will produce bumps on the pod surface. Selection for small, cylindrical seed minimizes wall bumpiness. In addition, selection for slow seed development can increase smoothness.

The pod interior should consist of solid deep green gel surrounding the seeds. As pods develop, the walls become less fleshy and eventually dry in the maturation process. Cultivars can differ in the retention of fleshiness and the tendency for the fleshy pod wall to break down and become "pithy". Flesh can be green and translucent, which is preferred in processing beans, or opaque with a whitish appearance.

Because of relatively long spacing between seeds in a pod, rapid growth may cause cavities (termed interlocular cavitations) to develop (Kuksal & Seth, 1981). Interlocular cavitations are associated with cooler nights at the early pin pod stage of growth, and has a genetic component. Interlocular cavitations generally are not visible in young pods, but develop as pod and seed size increases.

Pod detachment force is a trait of considerable interest to both fresh market and processing industries. During harvest, the grower would like pods to detach easily and individually from the plant. Fresh market producers consider breakage at the pod neck a disadvantage because such breakage increases pod discoloration (oxidation) and decay during shipping (Lien & Baggett, 1998). At the processing plant, one of the first procedures is to snip both ends of the pod to remove the spur and pedicel. The cultivar EZ Pick which has a low pod force detachment (easy picking) trait was introduced in 1986. Plants with this trait detach from the raceme with less force than normal plants. Point of detachment is along an abscission layer in the pedicel. The OSU program developed breeding lines with this trait, but, when scaled up for trial, found that processors rejected the lines. Lines were rejected because more pods reached the processing plant with pedicels attached, which made snipping them off more difficult. Lien & Baggett (1998) reported that for easy picking lines, 69% of the pods had pedicel attached compared to 29% for Oregon 91G. Fifty percent of the Oregon 91G pods had broken necks compared to 13% of the easy picking lines. Processors would prefer cultivars that had no pedicels at all. It may be possible to select lines that have pods that break at the neck, leaving the pedicel with the stem. While processors may not want to use cultivars with the easy picking trait with present equipment, easy picking cultivars have great potential for fresh market.

Flavor varies considerably from cultivar to cultivar and is a primary reason for the continued popularity of some heirloom bean types with some gardeners. Particularly notable flavors have been observed for the Romano, and the Blue Lake types (Stevens & Frazier, 1967; Stevens et al., 1967; Toya et al., 1974, 1976). In these studies, the authors found that 1-octen-3-ol and linalool were the chemical constituents most closely associated with flavor in snap beans. The pole cultivar Romano contained high levels of these compounds while FM-1L Blue Lake contained relatively high levels of 1-octen-3-ol but low levels of linalool. Gallatin 50 (a white seeded selection from Tendercrop) had low levels of 1-octen-3-ol but high levels of linalool. Low levels of 1-octen-3-ol was inherited as a dominant trait, probably under simple control. Little has been done to follow up on these studies in part because of the time and expense of using gas chromatography in the breeding process. Simultaneous human-subject response and chromatographic analysis could allow a better definition of volatile components associated with flavor.

5.4 Seed Traits

Snap bean seeds have greater sensitivity to injury during harvest, conditioning, planting, and germination, and reduced resistance to cold stress and soil-borne pathogens than dry bean seed. Certain steps can be taken to reduce the chances of mechanical injury, such as avoiding drops over 50 cm and stabilizing seed moisture at around 12% (Dickson & Boettger, 1976; Taylor & Dickson, 1987). However, many problems have a genetic basis.

Selection for pod shape, slow seed development, and the use of a white seedcoat have imposed severe constraints on seed quality and seedling vigor. The long, cylindrical seed shape imposed by selection for long, round, fleshy pods results in a seed that is sensitive to mechanical and imbibitional injury (Spaeth, 1986). Some breeding programs use a drop test to select against lines with exceptional susceptibility to mechanical and imbibitional injury. The test involves dropping seed from a specified distance several times, followed by germination tests.

Slow rate of seed development is in itself a trait for selection. One of the major determinants of pod quality is seed size or seediness. A seed index based on seed length has been developed to gauge quality (Silbernagel, 1979a; Silbernagel & Drake, 1983). Thus, selection for slow seed development will prolong pod quality. In some cultivars, selection for slow seed development has resulted in an increase in ovule abortion. While ovule abortion may be desirable to the processor, it may cause lower seed yield, making the cultivar more difficult to produce.

From about the 1960s until present, most bean cultivars have been bred with white seed. The preference for white seed was dictated by processors, who rejected cultivars with colored seed for canning because the anthocyanins present in the developing seed would leach into the liquor, causing off-color of the product. Colored seeds from over-mature pods can also present problems in a frozen product where a cross-section through a pod and seed will sometimes reveal a pigmented testa as a colored ring around the developing ovule. While white seed is desirable to the processor, it may create problems for the grower and the seed producer. Cultivars with white seed show reduced seedling emergence, particularly in cold, wet soils, compared to colored seed isogenic lines (Dickson & Petzoldt, 1988; Doijode, 1990; Powell et al., 1986). The exact cause of reduced vigor for white seeded lines appears to be related to an increased susceptibility to mechanical damage and reduced defenses against soil borne pathogens (Dickson & Petzoldt, 1988). White-seeded lines do lack tannins and lignins as well as anthocyanins (Agbo et al., 1987), compounds involved in resistance to mechanical stress and pathogens. Seedcoat color affects water uptake as well (Wyatt, 1977).

Breeders have relied almost exclusively on the recessive *p* gene for white seed. The *p* gene, evidently near the beginning of the plant pigment biosynthetic pathway, causes absence of pigment formation in the flowers and seed. One of the authors has observed purple striped pods with *p*, suggesting that *p* is not epistatic to [c^u prp^{st}]. Bassett (1994b) revised the gene nomenclature with regard to pod color but did not address the relationship between *p* and [c^u prp^{st}]. At least two other gene combinations can condition white seeds. One combination is the “Early Wax” system described by Bassett & Blom (1991). The *t* gene for partially colored seed, when combined with two additional genes (*z* and l^{ers}) (Bassett, 1997) extinguishes all color from the testa. Secondly, a triple recessive involving the cartridge buff allele (c^u) at the complex *C* locus, *d*, which conditions hilum ring color, and *j* or the "joker" gene can produce white seed color (Prakken, 1970). While it is an unproven conjecture, white seed, whatever its genetic basis, probably lacks tannins and lignins, and would be susceptible to the problems that plague the *p* gene. One distinctive difference among these pathways to white seed is that if *V* is present, the triple recessive and possibly the Early Wax combinations will have pigmented flowers and vegetative plant parts. Flowers are normally white when recessive *t* is present. However, the addition of *Fcr* and *Fcr-2* to *t* will produce partially colored seed and colored flowers (Bassett, 1993). Presumably, these genes combined with the *z* and l^{ers} would impart white seed with colored flowers. The authors are not aware that this combination has been tested. The significance of producing white seed while other plant parts are pigmented is that anthocyanins and tannins could be expressed in the embryo, roots, and hypocotyl, with possible bactericidal and fungicidal effects. A drawback to using genes other than *p* to obtain white seed is the requirement to manipulate two or more genes. Within a closed genetic population, breeding would not be difficult once conversion had been accomplished. Conversion from *p* to a multigenic system would be onerous, as would the introgression of traits from a different background. Tagging the pigment loci with molecular markers would allow more efficient conversion. Isogenic lines with these different gene combinations need to be developed in order to determine if one genetic system for producing white seed is superior to any other.

Another possible approach to obviating the problems associated in the field with white seed while meeting the needs of the processor involves the use of genes that reduce rather than eliminate pigment expression in the seed. At the *P* locus, p^{gri} is an allele that allows partial expression of color in the testa conditioned by other pigment genes present in the line. For example, p^{gri} with *V* will produce seeds with a gray to light purple testa (*V* with *P* would produce black seeds) (Bassett, 1994a). A similar phenotype is observed with *j*, particularly when c^u is present (Prakken, 1970, unpublished observations). If pigment timing and rate of accumulation is proportional to final amounts, then much less pigment would be expressed at harvest maturity in a snap bean cultivar. At the same time, anthocyanins, tannins, and lignin could accumulate with beneficial consequences for field emergence. The allele p^{gri} is particularly attractive because it involves the use of a single gene. One possible disadvantage may be the need for additional pigment genes to obtain the full benefit of p^{gri}. As above, isogenic lines need to be developed to test these hypotheses.

The galactosyl-oligosaccharides, raffinose and stachyose, found in seeds appear to stabilize cell membranes. These compounds prevent membrane damage during desiccation and rehydration by forming a glassy state upon dehydration (Bernal-Lugo & Leopold, 1995; Blackman et al., 1992; Obendorf, 1997). Spaeth (1986) demonstrated that snap bean seed imbibition under these stress conditions causes transverse cracking of cotyledons, cell membrane damage, and excessive solute leakage. Increasing levels of raffinose and stachyose would result in less membrane damage, improving germination and field emergence. From a screen of common bean germplasm (McPhee & Myers, 1996) and a study of an eight parent diallel (McPhee, 1995), genetic variation for raffinose and stachyose was observed. These sugars were inherited quantitatively, although there appeared to be a major factor for stachyose accumulation. Recombinant inbred populations are under development to examine the relationship between imbibitional injury and oligosaccharide levels.

Research on resistance to imbibitional injury has identified cultural causes and sources of resistance (Dickson & Boettger, 1976; Siddique & Goodwin, 1980; Taylor & Dickson, 1987). One particular mechanism of interest is the semi-hard seed phenomena (Taylor & Dickson, 1987). At low (8%) seed moistures, the testa becomes less permeable and water uptake is delayed. At 12% moisture, semi-hard seed behaves like normal seed with normal germination rates. Semi-hard seeded lines show better field emergence than normal-seeded cultivars (Taylor & Dickson, 1987). In western seed fields, seed moistures are usually less than 10% at harvest, and may drop into the 8% range during storage at ambient conditions. Without priming, such seed could emerge over a prolonged period, resulting in uneven maturity at harvest. Such problems would negate the benefits of improved stand establishment. This trait could be useful only to those companies with the ability to adjust seed moisture levels in large seed lots.

Two traits that affect cotyledon and testa color that are important to snap bean processing quality are the immature white vs. green cotyledon color conditioned by *iw* (Baggett & Kean, 1984) and the persistent green color (*pc*) gene (Dean, 1968). The immature white/green seed color shows the highest level of expression during the seed expansion phase that corresponds to the highest quality stages in the snap bean pod. Immature green seed color is preferred by the industry because it most closely matches the interior pod color. Immature green seed color is dominant to immature white (Baggett & Kean, 1984). Because it is a dominant trait, progeny rows are required to obtain breeder's seed that is free from segregation for the trait. Occasional mutations from immature green to immature white are found, requiring extensive cleanup of an existing line. The designation "G" for the Blue Lake cultivar Oregon 91G stands for the immature green cotyledon selections made after the original Oregon 91 was found segregating for immature white seeds.

The *pc* gene was formally described by Dean (1968) although it had been selected many years before as shown by the Flageolet Vert types grown in France during the 19th century (Hedrick, 1931). A single recessive gene, *pc* prevents chlorophyll breakdown in the seed, pods, and vegetative plant parts. At maturity, plants with *pc* remain green even as they dry down. Increasing numbers of cultivars are being developed with this gene, particularly in fresh market types, because pods do not bleach or lighten even as quality deteriorates. The trait was associated with blossom drop, ovule abortion, and poor seed quality (Dean, 1968) but these problems now have been partially overcome.

5.5 Resistance to Pathogens

The need for disease resistance varies with the production region where snap beans are grown. Diseases that can be problems in most production areas include white mold and the complex of root rot pathogens. Bean common mosaic (BCM, caused by a potyvirus) resistance is needed in all cultivars to prevent spread of this disease. Bacterial diseases (halo blight, brown spot, common blight, and bacterial wilt) and some fungal diseases (anthracnose and rust) are found predominantly in the humid production areas of North America. These areas have fewer problems with viral diseases. In the seed production areas of the arid west of North America, the opposite is generally true. Bacterial and fungal diseases are uncommon unless introduced on seed because the low humidity conditions are not conducive to survival of the pathogens. However, insect vector populations of viral diseases are higher causing greater incidence and spread of diseases such as BCM and BCT (sugarbeet curly top caused by a geminivirus). In the sub-tropics and tropics, additional diseases not present in the temperate regions may be present. For region- and disease-specific information, the reader is referred to other chapters in this book, Hall (1994), Schwartz & Pastor-Corrales (1989), and Silbernagel & Hannan (1992). Region-specific disease needs were discussed above in the section on North American production. In this section, we discuss aspects of diseases that relate specifically to snap beans, particularly those pathogens where substantial progress is still needed.

Root rot pathogens are nearly ubiquitous problems in snap bean production areas. Many snap bean cultivars have rather weak root systems, so additional pruning of the root systems can cause substantial yield loss. Breeding programs have relied on horizontal resistance, and its incorporation has been a slow, incremental task. Several cultivars developed by the USDA Prosser, Washington program have had useful levels of resistance (Dickson & Boettger, 1979; Silbernagel, 1979b, 1987a; Silbernagel & Drake, 1983).

Studies on the interaction of root rot resistance with cultural practices have been conducted (Silbernagel & Doyle, 1985).

Snap bean germplasm lines with rust resistance have been released periodically by the USDA-Beltsville program (Stavely & McMillan, 1992; Stavely & Steinke, 1985). The USDA releases are intended for production in the eastern U.S.A. where rust can be a substantial problem. For more specific information on the genetics of rust resistance, the reader is referred to Kelly et al. (1996).

White and gray mold resistant germplasm has been developed by several programs (Dickson & Hunter, 1989; Kmiecik & Nienhuis, 1998; Mattusch et al., 1982; Miklas & Grafton, 1992; Miklas et al., 1992, 1998, 1999). At present, moderate levels of resistance to white mold are available in breeding lines from the New York program, and in G 122. A QTL associated with white mold resistance from G 122 has been identified from a recombinant inbred population derived from the cross A 55 X G 122 (P. Gepts, personal communication, 1999). The QTL is located in linkage group D 7 and is linked to *P*. Because G 122 has dominant *P*, it will be necessary to break this linkage to bring this QTL into white seeded (recessive *p*) snap bean lines. A number of plant introduction accessions, some of which show resistance, have been identified (Miklas et al., 1999). For snap bean breeders, the New York lines (because they are of snap bean type) represent a source of resistance most easily transferred into a commercial background. None of these sources of resistance offer complete resistance; thus, breeders will need to identify sources that are different, and pyramid them into elite cultivars.

BCT is present in the seed production fields of the western U.S.A. The virus overwinters in wild mustards and is transferred to bean plants when the wild hosts mature and the vector, the sugarbeet leafhopper, searches for a new food source (Hall, 1994). Resistance is conditioned by two dominant genes (Silbernagel, 1986). Several cultivars released by the University of Idaho, USDA-Prosser program, and some seed companies have incorporated BCT resistance (Dean, 1965, 1969; Dean & Hungerford, 1954; Miklas & Silbernagel, 1997; Silbernagel, 1987b). Most programs have relied on field evaluations to identify resistant lines. Because BCT incidence may vary from year to year, field screening is not always effective. The virus is transmitted only by the sugarbeet leafhopper, and greenhouse screening is expensive and time consuming. An *Agrobacterium tumefaciens*-mediated infectious clone of BCT and molecular markers to the resistance genes would greatly facilitate germplasm evaluation and inheritance studies.

BCM is a global problem because it is seed transmitted (Drijfhout, 1978b; Drijfhout et al., 1978). Some strains of BCMV have been referred to as "serotype A isolates". These strains have been transferred to the bean common mosaic necrosis virus (BCMNV) subgroup (Mink et al., 1994).

Whereas BCMV probably coevolved with common bean in the New World, BCMNV appears indigenous to equatorial Africa (Spence & Walkey, 1994). In both the new and old worlds, novel isolates may arise, then travel to other nations when bean seed of a susceptible cultivar is shipped. Resistance conferred by the dominant *I* gene has been widely used in snap beans. *I* gene resistance confers immunity to most isolates of BCMV, but may result in a whole-plant systemic necrosis reaction ("black root") when inoculated with certain isolates under specific growing conditions (Drijfhout, 1978b). In general, certain BCMV isolates will cause a black root reaction when infection occurs while the plants are growing under high (>30°C) temperatures, or when infected with any BCMNV isolate at normal growing temperatures. The addition of $bc\text{-}2^2$ or *bc-3* resistance genes will prevent the black root reaction, but so far, few if any U.S.A. snap bean cultivars have *I* gene protected with $bc\text{-}2^2$ or *bc-3*. Protected *I* gene resistance is more likely to be found in European cultivars where BCMNV isolates have been a problem for many years.

BGM has been a particular problem of common bean in the New World tropics for many years. However, it recently became established in southern Florida where it threatens the winter fresh market production. The virus has a broad host range and can bridge the summer months, when beans are not grown, in certain weed species. Symptoms are severe chlorosis, stunted growth, and misshapen pods. Both plant and pod resistance can occur and three resistance factors have been found. Two genes (*bgm-1* and *-2*) alleviate leaf chlorosis (Vélez et al., 1998), while a third gene (*Bgp*) prevents pod distortion (Castañeda & Beaver, 1998). Resistant cultivars are currently being developed at University of Florida, University of Puerto Rico – Mayaguez, and by private seed companies (McMillan et al., 1998).

5.6 Physiological Stresses

Environmental stresses include air pollution, cold and heat at different stages of growth, fertilizer deficiencies in the soil, high or low pH, drought, and flooding. The factors least under the control of the grower and most amenable to change are those associated with the weather, including cold and heat stress, and air pollution. Here, we briefly address breeding efforts for these areas; we refer the reader to Silbernagel (1986) for a more complete discussion of environmental stress tolerance.

In the eastern snap bean production areas of the U.S.A., ozone contributes to leaf-bronzing and lower yields. Genetic variation for ozone resistance is found in beans and can be incorporated into improved cultivars (Mebrahtu et al., 1990; Meiners & Heggestad, 1974). Resistance to ozone damage is quantitatively inherited, and reported to be controlled by recessive genes with high heritability (Knudson-Butler et al., 1979), or as an additive trait of moderately high heritability with possible epistasis (Mebrahtu et al., 1990). Through breeding in selective environments, many snap bean cultivars have high levels of resistance to ozone damage. OSU Blue Lake cultivars generally have low levels of resistance because they have not been subjected to high ozone environments.

Temperature stress can be a problem in all regions where snap beans are produced. Beans grown in cooler climates may have problems with cold temperature stress during germination and emergence of the crop, and transient problems with heat stress during flowering and pod development. We discussed cold-stress tolerance in the seed quality section. Southern environments may have more problems with heat stress, particularly during the summer months. In the northern U.S.A., high daytime temperatures may cause pollen and ovule abortion while in the south, high nighttime temperatures may be more important. Sources of heat tolerance have been identified, and improved lines have been developed (Dickson & Boettger, 1984a, b; Dickson & Petzoldt, 1989a, b, 1991; Dickson & Thode, 1985; Nakano et al., 1993; Shonard & Gepts, 1994; Silbernagel, 1986). Bouwkamp & Summers (1982) found resistance to stress controlled by a single dominant gene in one case and two genes interacting epistatically with a second parent under high daytime temperatures. In snap beans, the critical period for heat stress was two or three days prior to anthesis and heritabilities for tolerance to heat stress were low (Dickson & Petzoldt, 1989b).

6. BREEDING METHODOLOGY

In breeding snap beans, we should distinguish between procedures used for cultivar development and those used for germplasm enhancement or population improvement. Little has been written on what procedures snap bean breeders use, particularly in private programs. Limited information can be gleaned from the published literature and from Plant Variety Protection applications.

The pedigree method has been the most common breeding procedure used for developing new snap bean cultivars. This method is effective for traits with moderate to high heritabilities. Parental choice is completely controlled, and single plants are selected in segregating generations. Perhaps most important, breeders continue to make single plant selections into advanced generations in order to achieve the high degree of uniformity required in a released snap bean cultivar. The disadvantage is that by selecting traits with high heritability in early generations, genetic variation with low heritability is lost (Fouilloux & Bannerot, 1988).

Recurrent backcrossing has also been used as a standard procedure for the introgression of qualitative traits into an elite background. Silbernagel (1986) outlined a hypothetical breeding program to introgress multiple disease resistance into a Midwestern processing type snap bean that used two successive backcrosses.

Other procedures that may be more efficient for cultivar development include the single seed descent and bulk breeding methods (Fouilloux & Bannerot, 1988). Simulation studies have shown that these methods are more effective than the pedigree method in preserving genetic variation with low heritability. Bulk breeding is more efficient than single seed descent when labor costs and land use are considered (Fouilloux & Bannerot, 1988). Single seed descent allows genetic analysis of traits in a population, whereas bulk breeding does not.

For population improvement, where a cultivar may not be the immediate result, other methods are available. Recurrent selection has generally not been used in snap bean because of the autogamous nature of the crop and the labor involved in random mating the recombination generation. Other techniques that would facilitate recombination are the congruency backcross and inbred-backcross methods. Kmiecik & Nienhuis (1998) used the inbred-backcross method to transfer white mold resistance from G 122 into a snap bean background, although the authors felt that a pedigree selection would have worked as well. The application of the congruency and inbred-backcross methods to breeding snap beans is discussed in Section 8. Breeding methods that incorporate the use of molecular markers are also discussed in that section.

7. SEED MULTIPLICATION

7.1 Seedstock Maintenance

In any seed production system maintaining cultivar purity is important. Through outcrossing, mechanical mixtures, and spontaneous mutation, genetic variation creeps into seed lots. Seedsmen incur significant expense in walking fields during the growing season to rogue off-types, and in putting the seed through elaborate milling systems to eliminate off-types at the seed level.

Among vegetable crops, snap beans require great time and effort to maintain purity. One reason for this need is the occurrence of spontaneous off-types. Probably the most important off-types are the flat or oval podded and stringy podded rogues. These create the greatest havoc in the processing plants and generate the greatest number of customer complaints. The genes responsible for pod shape and fiber appear to be less stable than other genes, and have a higher spontaneous reversion rate to high fiber types (Atkin, 1972; Atkin & Robinson, 1972). Some breeders feel that certain cultivars have a higher mutation rate than others, although no reports quantifying this difference have been published (Silbernagel, 1986). Molecular tools are now available to study the problem of flat and stringy pod rogues.

As detailed above, snap bean seed is more difficult to produce because of sensitivity to mechanical damage. Special equipment is required for harvest and conditioning, and special techniques such as priming allow seedsmen to produce high quality seed. Both the harvest and milling equipment and the cultivars can be improved. Breeders can develop tougher seed as detailed above by selecting for resistance to mechanical and imbibitional damage.

7.2 Seed Production Areas

Because of seed-transmitted diseases, snap bean seed is not usually produced on a large scale in areas where the processed crop is grown. Four main regions in the world responsible for seed production are the Pacific Northwest (PNW) of the U.S.A., East Africa (Tanzania and Kenya), New Zealand, and Chile. Seed production may be developing in areas of Mexico.

The specific localities in the PNW include southern Idaho, eastern Oregon, and the Columbia Basin in Washington. All are cool, extremely arid desert areas. Annual precipitation in these areas is 130 to 280 mm. Precipitation accumulates mainly as mountain snow pack during the winter and supplies the water used for irrigation during the growing season. Most of the seed hectarage is watered by gravity irrigation rather than by overhead sprinklers, although the use of sprinkler irrigation is increasing. Keeping free moisture off of the foliage helps reduce infection and spread of seedborne bacterial diseases. The PNW states also operate under regulations on the amount and type of bean seed that may be imported into each state's seed production areas. Seed previously grown east of the Continental Divide or outside of the U.S.A. is restricted in the quantity that may be imported, and may be subject to serology for bacterial blights and seedborne viruses. Seed fields are inspected by state regulatory agencies during the growing season, and are destroyed if bacterial blight is found. California also produces significant quantities of snap bean seed, and faces many of the same production constraints as do PNW producers.

Companies from the U.S.A. and Europe produce seeds in Chile during the off-season for the Northern Hemisphere. The climate is similar to the PNW; it is arid, with little risk from bacterial diseases. Of special concern is the incidence and spread of BCM and BCMN viruses in non *I*-gene cultivars in this region. Little information is available on this industry because private companies produce the seed and consider production information proprietary.

In Tanzania, snap bean seed for European seed companies is produced in the Arusha-Moshi area. Unlike the New World seed production areas, northern Tanzania receives substantial rainfall (about 1200 mm annually), with beans grown using natural precipitation. Seeds are planted at the beginning of the long rains that begin normally in late March or early April. Because of the high rainfall, the risk of infection with bacterial blights is substantial. Most seed is exported to Europe for production, processing, and consumption within Europe. Because of a European quarantine on common bacterial blight, companies serologically test seed lots before allowing import. Testing is currently done in Europe; testing in Tanzania would allow seed farms to determine which seed lots should be cleaned for export, and which should be sold for consumption. Because seed harvest, threshing and cleaning is done mostly by hand, seed quality is generally quite good, and may be better than western grown seed from the U.S.A. Because East Africa appears to be a center of diversity for halo blight (Taylor et al., 1996; Teverson, 1991) and BCMNV (Spence & Walkey, 1994), particular attention must be paid to seed lots imported from this region.

8. BREEDING OBJECTIVES FOR THE TWENTY-FIRST CENTURY

Throughout this chapter, we have attempted to indicate where our knowledge of a particular aspect of snap bean genetics is lacking. In this section we list some of the constraints imposed on snap beans by genetic, geographic, and economic forces. These constraints will dictate the breeding objectives for the next century. We also suggest ways to overcome these constraints and suggest what may be important future breeding activities. Because snap beans are a biological organism, unexpected events will happen, making predictions a risky business.

The important points about snap bean that have relevance to the breeder are as follows:

- Snap beans represent a closed genetic population. While the genetic base may be fairly broad, incorporating traits from a non-snap bean background takes time and persistence.
- Some traits that are desirable to processors are deleterious to the seed propagation and farm production of the crop. As breeders push for higher quality, the deleterious effects on plant biology may be made worse.
- The needs for a given market are often highly specific, limiting the range of acceptable variation for traits.

- Markets differ from one another. Not only are fresh market beans different from processing beans, but different processors have different environments and needs.
- Snap beans are grown around the world, with adaptation needed to each region. It is impossible to breed one bean that works well in all markets.
- While public snap bean breeding programs may focus on one region with one type of bean, private breeders with the larger seed companies now are breeding globally.

From the six points listed above it is obvious that breeders will need to find ways to counteract the effects of deleterious economic traits. For seed companies with global operations, developing cultivars with low genotype by environment interaction should be an objective. However, there will still be a need for adaptation to specific environments. The specific requirements of growers, marketers, and processors must also be met. Snap beans will probably continue to diverge genetically from dry beans although dry beans will be used as a source of genetic variability. Within snap beans, processing and fresh market types will continue to diverge for traits associated with shipping/processing, but may converge for some quality traits (e.g., color, fiber content).

One of the most important challenges facing snap bean breeders is the identification and transfer of traits from other backgrounds into snap beans. Sufficient criteria to identify useful traits and ways to transfer the traits must be established. The identification of some traits, such as disease resistance or resistance to abiotic stresses in unadapted backgrounds, is straightforward, and once genetic control of the trait has been established, the transfer process may be laborious, but is achievable. Some traits, however, can be evaluated only in an adapted background. With much of the world bean germplasm adapted to the tropics, problems with daylength-sensitivity, and adaptation to specific temperature regimes can obscure performance at the higher latitudes. One solution to this problem is to convert daylength-sensitive material to daylength insensitivity. A process has been initiated to convert tropical dry bean lines (P.N. Miklas, personal communication, 1999), and should give North American breeders access to new sources of germplasm. A similar process should be used with tropically adapted snap bean lines. Some traits that may exist in unadapted germplasm would only be identifiable in a snap bean background (e.g., modifiers of pod fiber content). Thus, there is a need for a similar type of conversion of interesting dry bean materials to a snap bean background. This conversion presents considerably greater challenges than a daylength conversion program because many more genes are involved.

The process of trait transfer to snap beans in general, and to Blue Lake breeding lines specifically, is problematic. The genetic architecture of snap beans has diverged significantly from that of dry beans and may comprise its own co-adapted genetic complex. To our knowledge, no economically useful cultivar has been derived from a single cross between a snap bean cultivar and a closely related dry bean cultivar. The essence of the problem is how to incorporate useful economic traits without breaking apart the snap bean complex of traits. The problem is even more acute for Blue Lake beans because crosses between Blue Lake and other snap beans rarely result in material useable as a Blue Lake cultivar. Blue Lake beans have apparently acquired their own co-adapted genetic complex. In the past, several rounds of crossing, evaluation, and selection were required to recover material suitable for release as a cultivar. As an example, the transfer of pole Blue Lake cultivars into an acceptable bush background took about 20 years (Baggett, 1995).

The lack of recombination in snap bean may be part of the bottleneck to trait transfer. Common bean chromosomes are small with an average of less than two chiasmata per bivalent (Maréchal, 1971). The low rate of crossing over limits the amount of recombination that can occur in any one reproductive cycle. This one factor may determine why quite often in a cross between two separate bean types, types that resemble the parents are recovered in high frequency, with very few that show intermediate or recombinant characters. Certain breeding methods, such as recurrent selection, congruity backcrossing, and the inbred-backcross method can give more opportunities for recombination (Anderson et al., 1996; Bliss, 1981; Urrea & Singh, 1995; Wehrhahn & Allard, 1965).

The use of molecular markers may help facilitate the transfer of traits from an unadapted background into snap beans. Two approaches are conceivable. The conventional approach would be to use molecular markers to tag genes or QTL (quantitative trait loci) of interest, as has been done for several traits in bean (Adam-Blondon et al., 1994a; Alzate-Marin et al., 1997a, b; Haley et al., 1993, 1994a, b, c, d; Johnson et al., 1995; Johnson et al., 1997; Jung et al., 1996; Melotto et al., 1996; Miklas et al., 1996a, b; Olaya et al., 1996; Park et al., 1997; Young & Kelly, 1996). Many of these markers for disease resistance may be of interest to some snap bean breeding programs. The tags would facilitate introgression of the trait of interest during recurrent crossing procedures (such as backcrossing). While this approach would work for relatively simply inherited traits, it will not be as effective for complex traits. The advanced backcross QTL analysis proposed by Tanksley & Nelson (1996) may be a useful method for introgressing genes from a dry bean into a snap bean background, particularly where molecular markers for a trait are unknown. The procedure uses the inbred-backcross method as its basis, with marker analysis initiated in the BC_2 to identify QTL-near isogenic lines. Because hand pollination in bean results in few seed per cross, a program would be limited to a few populations. These populations could be developed by either the inbred-backcross method or the advanced backcross QTL analysis, because of the number of crosses (100-200) required in the BC_1 and BC_2 generations.

While markers are useful for gene pyramiding, simply inherited traits may be easier to evaluate conventionally. It does not make sense to use the molecular technology when cheaper, faster, and more accurate methods are available.

A second approach would take advantage of molecular maps that have been developed for bean (Adam-Blondon, et al., 1994b; Nodari et al., 1992, 1993a, b; Skroch et al., 1996; Vallejos et al., 1992). With the number of RFLP, AFLP, and RAPD markers known, it should be possible to develop a set of polymorphic PCR-based markers to tag a snap bean cultivar's genome, a process described as Whole Genome Selection (Tanksley & Rick, 1980). A marker would be placed on each chromosome arm. From segregating cross progeny, selection for individuals with the most complete set of markers and the trait of interest should result in the recovery of progeny with most of the snap bean traits intact and with the novel trait of interest as well. In simulation studies based on the tomato genome, selection of one marker per chromosome in the first backcross produces equivalent gain to three backcrosses without markers (Tanksley & Rick, 1980). In addition, by selecting the best plant from 30, the recurrent genome can be recovered in 3 backcrosses (Tanksley et al., 1989). The DNA chip technology (Service, 1998) seems applicable to this sort of approach. Overall, molecular tools probably will play an increasingly important role in snap bean breeding, and probably in ways unforeseen at present.

Growers and processors will continue to demand greater productivity while maintaining or improving quality. Incremental increases in yield continue, although these may sometimes be associated with later maturity or increase in pod size and decrease in quality. Increases in yield will ultimately rely on increases in biomass and greater efficiency in partitioning to reproductive growth (including both rate and amount). Pod concentration may also increase. Much more research needs to be done to determine the components of concentration and their relative importance.

One of the specific challenges for the breeder is to develop small sieve size beans while retaining or increasing yield. A small pod contains inherently less dry matter than does a full-sized pod. Breeders will need to compensate for smaller pod size by producing more pods per plant. Increasing pod number requires that more flowers open simultaneously. Here again, pod concentration appears to be a key element to increasing yield. Breeders should not abandon development of full sieve cultivars, as these will continue to be an important type for many markets.

Much more can be done to improve quality and pod traits. The trend will probably continue for darker green pods, and particularly for uniform colors. The genetic variation for pod color exists; the greatest hurdle may be gaining acceptance from the processors. Straighter, smoother pods without an increase in pod fiber need to be developed for processing markets. Slow seed development has been a past breeding objective, but when taken to extremes, may interfere with seed propagation. Processors would prefer pods without seeds because increase in seed size is associated with loss of quality.

Breeding for small seed size may be the best solution to developing cultivars that do not become excessively seedy. Seed size and pod size are positively correlated, and snap bean breeders have exploited this correlation of traits to develop small-sieve size cultivars by transferring small seed size from dry beans into a snap bean background. Whether genes that control seed size exert a pleiotropic effect on pod size or whether genes for seed and pod size are tightly linked is not known. If a pleiotropic effect, then it will be very difficult to develop full-sieve size beans with small seeds. However, within dry bean market classes, seed size may vary while pod size remains nearly the same.

The development of F_1 hybrid beans would be desirable to exploit the heterosis that presumably would be present at economically significant levels. F_1 hybrids would also allow greater uniformity, stability and control of the crop. Because beans are highly self-pollinated, an out-crossing mechanism would be needed. In other crops, cytoplasmic male sterility (CMS) has been an economical means of producing hybrid seed. CMS is known in beans (Bassett & Shuh, 1982; Hervieu et al., 1993), but adequate fertility restoration genes have not been found (He et al., 1995; Jia et al., 1997). Fouilloux & Bannerot (1988) estimated that because of the low seed multiplication efficiency of beans, F_1 hybrid seed would be twice the cost of pure line seed, even if an efficient CMS system were available.

Another important thrust for breeders of processed beans is to improve ease of handling by the grower and processor. Growth habit continues to need improvement with development of sturdy, upright plants that hold their pods in the upper half of the canopy. Plants need strong root systems so that the bean harvester does not pull the entire plant during harvest. Pods should break individually, not in clusters. Levels of trash should be kept to a minimum. In the processing plant, low levels of trash are desirable, as is minimum handling of the product. Pods that break at the neck rather than the pedicel are desirable because these would contribute less trash.

Several diseases warrant particular attention. BGM is potentially one of the most dangerous diseases to snap bean (and dry bean) crops grown in the southern U.S.A. California, Arizona, New Mexico, Texas, Georgia, and South Carolina have resident populations of the silverleaf whitefly during the summer and/or winter and are exceedingly vulnerable to an accident waiting to happen. Surveys of ornamental nurseries in Florida have found a number of flowering ornamentals infected with various geminiviruses. These nurseries are suppliers of bedding plants to the large, nation-wide discount stores and garden centers. These provide direct pipelines for the spread of the vector and pathogen to gardeners in cities and rural agricultural areas of the above-mentioned states. Thus, moving resistance genes for BGM into snap bean cultivars used in the southern states' production areas is an important breeding objective.

Resistance to white and gray mold should receive a strong emphasis. Sources of partial resistance are known, and additional lines with resistance may be identified soon in germplasm collections of common and runner bean. Breeders will need to determine allelic relations among these sources of resistance and pyramid nonallelic sources into elite lines. If transformation systems for common bean become available, then genes for enzymes such as oxalic acid oxidase might be used to confer resistance to white and gray mold.

In the next century, snap beans will continue to be a high value food that provides dietary and culinary diversity to a growing world population. Breeders will continue to develop stable, high yielding cultivars with improved pod quality, and better processing characteristics.

ACKNOWLEDGMENTS

The authors thank Drs. Ron Riley, Novartis; George Kotch, Seminis Vegetable Seeds; and Steve Magnuson, Harris-Moran Seed Co., for helpful conversations and advice during the genesis of this chapter. We also thank several anonymous reviewers who provided thoughtful and helpful comments during revision. Ms. D. Kean assisted with editing and proofreading the manuscript.

REFERENCES

Acquaah, G., M.W. Adams & J.D. Kelly. 1991. Identification of effective indicators of erect plant architecture in dry bean. Crop Sci. 31:261-264.

Adam-Blondon, A.F., M. Sévignac, H. Bannerot & M. Dron. 1994a. SCAR, RAPD and RFLP markers linked to a dominant gene (*Are*) conferring resistance to anthracnose in common bean. Theor. Appl. Genet. 88:865-870.

Adam-Blondon, A.F., M. Sévignac, M. Dron & H. Bannerot. 1994b. A genetic map of common bean to localize specific resistance gene against anthracnose. Genome 37:915-924.

Agbo, G.N., G.L. Hosfield, M.A. Uebersax & K. Klomparens. 1987. Seed microstructure and its relationship to water uptake in isogenic lines and a cultivar of dry beans (*Phaseolus vulgaris* L.). Food Microstructure 6:91-102.

Alzate-Marin, A.L., G. Assis de Carvalho, H. Menarim, G. Soares Baia, T.J. de Paula Junior, E. Goncalves de Barros & M. Alves Moreira. 1997a. Identification of RAPD markers associated with resistance to anthracnose in common bean. Annu. Rpt. Bean Improv. Coop. 40:130-131.

Alzate-Marin, A.L., T.J. de Paula Junior, H. Menarim, T.F.N. de Queiroz, E. Goncalves de Barros & M. Alves Moreira. 1997b. Use of RAPD markers to understand the dominant nature of anthracnose resistance genes present in common bean cultivar AB 136. Annu. Rpt. Bean Improv. Coop. 40:132-133.

Anderson, N.O., P.D. Ascher & K. Haghighi. 1996. Congruity backcrossing as a means of creating genetic variability in self pollinated crops: Seed morphology of *Phaseolus vulgaris* L. and *P. acutifolius* A. Gray hybrids. Euphytica 87:211-224.

Anonymous. 1998a. USDA Agricultural Statistics. Washington DC, U.S.A.

Anonymous. 1998b. FAO Agricultural Production Data. http://apps.fao.org/cgi-bin/nph-db.pl?subset=agriculture.

Anonymous. 1998c. USDA-ARS Nutrient Data Laboratory. http://www.nal.usda.gov/fnic/foodcomp/.

Atkin, J.D. 1972. Nature of the stringy pod rogue of snap beans, *Phaseolus vulgaris*. Search Agric. 2:1-3.

Atkin, J.D. & W.B. Robinson. 1972. Nature of the flat pod rogue of snap beans, *Phaseolus vulgaris*. Search Agric. 2:4-9.

Baggett, J.R. 1995. A historical summary of vegetable breeding at Oregon State Univ. Proc. Oregon Hort. Soc. 86:89-99.

Baggett, J.R. & D. Kean. 1984. Inheritance of immature white seedcoat color in common bean. J. Amer. Soc. Hort. Sci. 109:601-604.

Baggett, J.R., D. Kean, G.W. Varseveld, D.M. Barrett & B.M. Yorgey. 1992. 'Oregon 54' green bean released. Proc. Oregon Hort. Soc. 83:154-159.

Baggett, J.R., W.A. Frazier & G.W. Varsveld. 1981. 'Oregon 91' green bean. HortScience 16:230.

Bassett, M.J. 1993. Interaction of two genes, *Fcr* and *Fcr-2*, with the *t* allele in common bean restores color to flowers. J. Amer. Soc. Hort. Sci. 118:881-884.

Bassett, M.J. 1994a. The *griseoalbus* (gray-white) seedcoat color is controlled by an allele (p^{gri}) at the *P* locus in common bean. HortScience 29:1178-1179.

Bassett, M.J. 1994b. Tight linkage of purple pod character and the complex *c* locus in common bean. J. Hered. 85:288-290.

Bassett, M.J. 1996. List of genes – *Phaseolus vulgaris* L. Annu. Rpt. Bean Improv. Coop. 39:1-19.

Bassett, M.J. 1997. Allelism tests involving two genes, *ers* and *ers-2*, that restrict partly colored seedcoat expression in common bean. J. Amer. Soc. Hort. Sci. 122:802-807.

Bassett, M.J. & A. Blom. 1991. A new genotype for white seed coat discovered in 'Early Wax' snap bean. J. Amer. Soc. Hort. Sci. 116:131-136.

Bassett, M.J. & D.M. Shuh. 1982. Cytoplasmic male sterility in common bean *Phaseolus vulgaris*. J. Amer. Soc. Hort. Sci. 107:791-793.

Bernal-Lugo, I. & A.C. Leopold. 1995. Seed stability during storage: Raffinose content and seed glassy state. Seed Sci. Res. 5:75-80.

Blackman, S.A., R.L. Obendorf & A.C. Leopold. 1992. Maturation proteins and sugars in desiccation tolerance of developing soybean seeds. Plant Physiol. 100:225-230.

Bliss, F.A. 1971. Inheritance of growth habit and time of flowering in beans, *Phaseolus vulgaris* L. J. Amer. Soc. Hort. Sci. 96:715-717.
Bliss, F.A. 1980. Common bean. pp. 273-284. *In* W.R. Fehr & H.H. Hadley (eds.), Hybridization of Crop Plants. Amer. Soc. Agron. & Crop Sci. Soc. Amer., Madison, U.S.A.
Bliss, F.A. 1981. Utilization of vegetable germplasm. HortScience 16: 129-132.
Bouwkamp, J.C. & W.L. Summers. 1982. Inheritance of resistance to temperature-drought stress in the snap bean *Phaseolus vulgaris*. J. Hered. 73:385-386.
Brothers, M.E. & J.D. Kelly. 1993. Interrelationship of plant architecture and yield components in the pinto bean ideotype. Crop Sci. 33:1234-1238.
Brown, J.W.S., J.R. McFerson, F.A. Bliss & T.C. Hall. 1982. Genetic divergence among commercial classes of *Phaseolus vulgaris* in relation to phaseolin pattern. HortScience 17:752-754.
Castañeda, A.M. & J.S. Beaver. 1998. Inheritance of normal pod development in bean golden mosaic resistant common beans. Annu. Rpt. Bean Improv. Coop. 41:3-4.
Chung, W-J., J.R. Baggett & K.E. Rowe. 1991. Inheritance of pod cross-section in beans (*Phaseolus vulgaris* L.). Euphytica 53:159-164.
Currence, T.M. 1931. A new pod color in snap beans. J. Hered. 22:21-23.
Davis, D.W. & W.A. Frazier. 1966. Inheritance of some growth habit components in certain types of bush lines of *Phaseolus vulgaris* L. J. Amer. Soc. Hort. Sci. 88:384-392.
Dean, L.L. 1965. 'Idelight' a curly top and mosaic resistant snap bean. Idaho Agric. Exp. Sta. Bull. 442. Kimberly, U.S.A.
Dean, L.L. 1968. Progress with the persistent-green color and green seed coat in snap beans (*P. vulgaris* L.) for commercial processing. HortScience 3:177-178.
Dean, L.L. 1969. 'Idahchief' and 'Idagem': Two new snap beans resistant to curly top and mosaic. Idaho Agric. Exp. Sta. Bull. 499. Kimberly, U.S.A.
Dean, L.L. & C.W. Hungerford. 1954. 'Idaho Bountiful' and 'Golden Gem' snap beans resistant to mosaic and curly top. Idaho Agric. Exp. Sta. Bull. 217. Kimberly, U.S.A.
Debouck, D.G., H. Rubiano & S.M. del Carmen Menéndez. 1988. Determinate climbers among Argentinean materials of *Phaseolus vulgaris* L. Annu. Rpt. Bean Improv. Coop. 31:189-190.
Detongnon, J. & J.R. Baggett. 1989. Inheritance of stem elongation tendency in determinate forms of common bean. J. Amer. Soc. Hort. Sci. 114:115-117.
Dickson, M.H. & M.A. Boettger. 1976. Selection for seed quality in white seeded snap bean. Annu. Rpt. Bean Improv. Coop. 19:24-25.
Dickson, M.H. & M.A. Boettger. 1979. Release of 12 root rot *Fusarium*, *Pythium* tolerant snap bean lines. Annu. Rpt. Bean Improv. Coop. 22:102.
Dickson, M.H. & M.A. Boettger. 1984a. Heat tolerance in snap beans [*Phaseolus vulgaris*]. Annu. Rpt. Bean Improv. Coop. 27:125.
Dickson, M.H. & M.A. Boettger. 1984b. Effect of high and low temperatures on pollen germination and seed set in snap beans. J. Amer. Soc. Hort. Sci. 109:372-374.
Dickson, M.H. & J.E. Hunter. 1989. Release of six snap bean lines with resistance to white mold and gray mold. Annu. Rpt. Bean Improv. Coop. 32:139.
Dickson, M.H. & R. Petzoldt. 1988. Deleterious effects of white seed due to *p* gene in beans. J. Amer. Soc. Hort. Soc. 113:111-114.
Dickson, M.H. & R. Petzoldt. 1989a. Heat tolerance during bloom in snap beans derived from cold tolerant selections. Annu. Rpt. Bean Improv. Coop. 32:40.
Dickson, M.H. & R. Petzoldt. 1989b. Heat tolerance and pod set in green beans. J. Amer. Soc. Hort. Sci. 114:833-836.

Dickson, M.H. & R.T. Petzoldt. 1991. Breeding for heat tolerance and pod set in snap beans. Annu. Rpt. Bean Improv. Coop. 34:110-111.

Dickson, M.H. & R. Thode. 1985. Breeding for heat tolerance in beans. Annu. Rpt. Bean Improv. Coop. 28:105-106.

Doijode, S.D. 1990. Influence of temperatures on seed vigor, viability and membrane permeability in different french bean cultivars. Plant Physiol. Biochem. 17:19-22.

Drijfhout, E. 1978a. Inheritance of temperature-dependent string formation in common bean (*Phaseolus vulgaris* L.). Neth. J. Agric. Sci. 26:99-105.

Drijfhout, E. 1978b. Genetic interaction between *Phaseolus vulgaris* and bean common mosaic virus with implications for strain identification and breeding for resistance. Agric. Res. Rpt. 872. Center for Agricultural Publishing and documentation, Wageningen, Netherlands.

Drijfhout, E., M.J. Silbernagel & D.W. Burke. 1978. Differentiation of strains of bean common mosaic virus. Neth. J. Pl. Path. 84:13-26.

Duncan, A.A., R.W. Every & I.C. MacSwan. 1960. Commercial production of bush snap beans in Oregon. Oregon State Coll. Ext. Bull. 787. Corvallis, U.S.A.

Eteve, G. & G. Fouilloux. 1979. Adaptation of snap bean to frontal mechanical harvest: Morphological characters of the cultivars. Annu. Rpt. Bean Improv. Coop. 22:71-73.

Evans, A.M. 1980. Structure, variation, evolution and classification. pp. 337-347. *In* R.J. Summerfield & A.H. Bunting (eds.), Advances in Legume Science. Proc. Inter. Legume Conf. Kew, U.K.

Fernández, F. P., P. Gepts & M. López. 1986. Stages of development of the common bean plant. CIAT, Cali, Colombia.

Fouilloux, G. & H. Bannerot. 1988. Selection methods in the common bean (*Phaseolus vulgaris*). pp. 503-542. *In* P. Gepts (ed.), Genetic Resources of *Phaseolus* beans. Kluwer, Dordrecht, Netherlands.

Frazier, W.A., J.R. Baggett, H.J. Mack, A.A. Duncan, G.W. Varseveld, E.K. Vaughan, J.B. Rogers & D. Booster. 1970. 'Oregon 58' bush green pod bean officially released. Oregon Veg. Dig. 19:3.

Frazier, W.A., J.R. Baggett, H.J. Mack, G.W. Varseveld, A. Sheets, L. Martin & E.K. Vaughan. 1972. 'Oregon 1604' bush green pod bean conditionally released. Oregon Veg. Dig. 21:2-4.

Frazier, W.A., J.R. Baggett & W.A. Sistrunk. 1958. Transfer of certain blue lake pole bean pod characters to bush beans. J. Amer. Soc. Hort. Sci. 71:416-421.

Geise, F.W. 1955. The processing bean industry in relation to the northwest. Proc. Oregon State Hort. Soc. (47th Annu. Rpt.) pp. 141-145.

Gepts, P., K. Kmiecik, P. Pereira & F.A. Bliss. 1988. Dissemination pathways of common bean (*Phaseolus vulgaris*, Fabaceae) deduced from phaseolin electrophoretic variability. I. The Americas. Econ. Bot. 42:73-85.

Gepts, P., T.C. Osborn, K. Rashka & F.A. Bliss. 1986. Phaseolin protein variability in wild forms and landraces of common bean (*Phaseolus vulgaris*): Evidence for multiple centres of domestication. Econ. Bot. 40:451-468.

Haley, S.D., L. Afanador & J.D. Kelly. 1994a. Identification and application of a random amplified polymorhic DNA marker for the *I* gene (Potyvirus resistance) in common bean. Phytopathology 84:157-160.

Haley, S.D., L. Afanador & J.D. Kelly. 1994b. Selection for monogenic pest resistance traits with coupling- and repulsion-phase RAPD markers. Crop Sci. 34:1061-1066.

Haley, S.D., L. Afanador, P.N. Miklas, J.R. Stavely & J.D. Kelly. 1994c. Heterogeneous inbred populations are useful as sources of near-isogenic lines for RAPD marker localization. Theor. Appl. Genet. 88:337-342.

Haley, S.D., P.N. Miklas, L. Afanador & J.D. Kelly. 1994d. Random amplified polymorphic DNA (RAPD) marker variability between and within gene pools of common bean. J. Amer. Soc. Hort. Sci. 119:122-125.

Haley, S.D., P.N. Miklas, J.R. Stavely, J. Byrum & J.D. Kelly. 1993. Identification of RAPD markers linked to a major rust resistance gene block in common bean. Theor. Appl. Genet. 86:505-512.

Hall, R. (ed.). 1994. Compendium of Bean Diseases. APS Press, St. Paul, U.S.A.

He, S., Z.H. Yu, C.E. Vallejos & S.A. Mackenzie. 1995. Pollen fertility restoration by nuclear gene *Fr* in CMS common bean: An *Fr* linkage map and the mode of *Fr* action. Theor. Appl. Genet. 90:1056-1062.

Hedrick, U.P. 1931. Beans of New York. Vol. 1 - Part II, Vegetables of New York J.B. Lyon Co., Albany, U.S.A.

Hedrick, U.P. 1972. Sturtevant's edible plants of the world. Dover Publ., New York, U.S.A.

Hervieu, F., L. Charbonnier, H. Bannerot & G. Pelletier. 1993. The cytoplasmic male-sterility (CMS) determinant of common bean is widespread in *Phaseolus coccineus* L. and *Phaseolus vulgaris* L. Current Genet. 24:149-155.

Jia, M.H., S. He, W. Vanhouten & S. Mackenzie. 1997. Nuclear fertility restorer genes map to the same linkage group in cytoplasmic male-sterile bean. Theor. Appl. Genet. 95:205-210.

Johnson, E., P.N. Miklas, J.R. Stavely & J.C. Martínez-Cruzado. 1995. Coupling- and repulsion-phase RAPDs for marker-assisted selection of PI 181996 rust resistance in common bean. Theor. Appl. Genet. 90:659-664.

Johnson, W.C., P. Guzmán, D. Mandala, A.B.C. Mkandawire, S. Temple, R.L. Gilbertson & P. Gepts. 1997. Molecular tagging of the *bc-3* gene for introgression into Andean common bean. Crop Sci. 37:248-254.

Jung, G., D.P. Coyne, P. Skroch, J. Nienhuis, J.M. Bokosi & J.R. Steadman. 1996. RAPD marker linked to gene for specific rust resistance in common bean. Annu. Rpt. Bean Improv. Coop. 39:59-60.

Kaplan, L. 1981. What is the origin of the common bean? Econ. Bot. 35:240-254.

Kelly, J.D. & M.W. Adams. 1987. Phenotypic recurrent selection in ideotype breeding of pinto beans. Euphytica 36:69-80.

Kelly, J.D., J.R. Stavely & P.N. Miklas. 1996. Proposed symbols for rust resistance genes. Annu. Rpt. Bean Improv. Coop. 39:25-31.

Kmiecik, K.A. & J. Nienhuis. 1998. The use of line G122 as a source of white mold resistance in breeding improved processing snap beans for the midwest. Annu. Rpt. Bean Improv. Coop. 41:21-22.

Knudson-Butler, L., T.W. Tibbits & F.A. Bliss. 1979. Inheritance of resistance to ozone in *Phaseolus vulgaris* L. J. Amer. Soc. Hort. Sci. 104:211-213.

Kuksal, R.P. & J.N. Seth. 1981. A note on interlocular cavitations inheritance studies in dwarf French bean (*Phaseolus vulgaris* L.). Progr. Hort. 13:37-39.

Leakey, C.L.A. 1988. Genotypic and phenotypic markers in common bean. pp. 245-347. *In* Gepts, P. (ed.), Genetic Resources of *Phaseolus* Beans. Kluwer, Dordrecht, Netherlands.

Lien, A. & J.R. Baggett. 1998. Pod detachment characteristics of easy picking and normal green beans. Annu. Rpt. Bean Improv. Coop. 41:223-224.

Maluf, W.R. & P.E. Ferreira. 1983. Análise multivariada da divergência genética em feijão-vagem (*Phaseolus vulgaris* L.). Hort. Bras. 1:31-34.

Maréchal, R. 1971. Observations sur quelques hybrides dans le genre *Phaseolus*: II. Les Phénomènes méiotiques. Bull. Rech. Agron. Gembloux 6:461-489.

Mattusch, P., M. Gerlagh, A. Ester & G. Spikman. 1982. Screening of snap bean cultivars for resistance to *Sclerotinia sclerotiorum*. Annu. Rpt. Bean Improv. Coop. 25:48-50.

Maynard, D.N. & G.J. Hochmuth. 1997. Knott's handbook for vegetable growers. John Wiley & Sons, New York, U.S.A.

McMillian, R.T., M.J. Davis, H.J. McLaughlin & J.R. Stavely. 1998. PCR evaluation of fourteen bean golden mosaic virus (BGMV) resistant snap bean germplasm lines for the presence of the virus. Annu. Rpt. Bean Improv. Coop. 41:31-32.

McPhee, K.E. 1995. Determination of environmental and genetic effects controlling raffinose-family oligosaccharide accumulation in common bean (*Phaseolus vulgaris* L.) seeds. Ph.D. dissertation. Univ. of Idaho, Moscow, U.S.A.

McPhee, K.E. & J.R. Myers. 1996. Screening common bean germplasm for raffinose-family oligosaccharide content. Annu. Rpt. Bean Improv. Coop. 38:255-257.

Mebrahtu, T., W. Mersie & M. Rangappa. 1990. Inheritance of ambient ozone insensitivity in common bean (*Phaseolus vulgaris* L.). Environ. Pollu. 67:79-89.

Meiners, J.P. & H.E. Heggestad. 1974. Reaction of snap bean cultivars to ambient air pollutants and to ozone. Annu. Rpt. Bean Improv. Coop. 17:53-55.

Melotto, M., L. Afanador & J.D. Kelly. 1996. Development of a SCAR marker linked to the *I* gene in common bean. Genome 39:1216-1219.

Miklas, P.N., R. Delorme, R. Hannan & M. Dickson. 1999. Using a subsample of the core collection to identify new sources of white mold resistance in common bean. Crop Sci. 39:569-573.

Miklas, P.N. & K.F. Grafton. 1992. Inheritance of partial resistance to white mold in inbred populations of dry bean. Crop Sci. 32:943-948.

Miklas, P.N., K.F. Grafton, J.D. Kelly, J.R. Steadman & M.J. Silbernagel. 1998. Registration of four white mold resistant dry bean germplasm lines: I9365-3, I9365-5, I9365-31, and 92BG-7. Crop Sci. 38:1728.

Miklas, P.N., K.F. Grafton & B.D. Nelson. 1992. Screening for partial physiological resistance to white mold in dry bean using excised stems. J. Amer. Soc. Hort. Sci. 117:321-327.

Miklas, P.N., E. Johnson, V. Stone, J.S. Beaver, C. Montoya & M. Zapata. 1996b. Selective mapping of QTL conditioning disease resistance in common bean. Crop Sci. 36:1344-1351.

Miklas, P.N. & M.J. Silbernagel. 1997. Release of three virus resistant snap bean germplasm lines, USWA-64, USWA-67, USWA-68. Annu. Rpt. Bean Improv. Coop. 40:158-160.

Miklas, P.N., V. Stone, E. Johnson & J. Beaver. 1996a. Selective mapping RAPD markers associated with QTL. Annu. Rpt. Bean Improv. Coop. 39:140-141.

Mink, G. I., J. Vetten, C.W. Ward, P. Berger, F. Morales, J.R. Myers, M.J. Silbernagel & O. W. Barnett. 1994. Taxonomy and classification of legume infecting potyviruses. A proposal from the Potyviridae Study Group of the Plant Virus Subcommittee of ICTV. Arch. Virol. 139:231-235.

Moore, J.F. & D.F. Allemendinger. 1954. Blue lake pole beans in western Washington. Washington Agric. Exp. Sta. Bull. 548. Pullman, U.S.A.

Mullins, C.A. & D.L. Coffey. 1990. Snap bean pod quality as affected by cultivar and harvest. J. Prod. Agric. 3:131-135.

Nakano, H., T. Momonoki, M. Matsuoka & T. Terauchi. 1993. Germplasm of heat tolerant snap bean. Annu. Rpt. Bean Improv. Coop. 36:62-63.

Nodari, R.O., E.M.K. Koinange, J.D. Kelly & P. Gepts. 1992. Towards an integrated linkage map of common bean. 1. Development of genomic DNA probes and levels of restriction fragment length polymorphism. Theor. Appl. Genet. 84:186-192.
Nodari, R.O., S.M. Tsai, R.L. Gilbertson & P. Gepts. 1993a. Towards an integrated linkage map of common bean. 2. Development of an RFLP-based linkage map. Theor. Appl. Genet. 85:513-520.
Nodari, R.O., S.M. Tsai, P. Guzmán, R.L. Gilbertson & P. Gepts. 1993b. Towards an integrated linkage map of common bean. 3. Mapping genetic factors controlling host-bacteria interactions. Genetics 134:341-350.
Obendorf, R.L. 1997. Oligosaccharides and galactosyl cyclitols in seed desiccation tolerance. Seed Sci. Res. 7:63-74.
Olaya, G., G.S. Abawi & N.F. Weeden. 1996. Inheritance of the resistance to *Macrophomina phaseolina* and identification of RAPD markers linked to the resistance genes in beans. Phytopathology 86:674-679.
Park, S.O., D.P. Coyne, J.M. Bokosi & J.R. Steadman. 1997. Identification of RAPD markers linked to a major gene for rust resistance and indeterminate growth habit using bulked segregant analysis in a common bean cross. Annu. Rpt. Bean Improv. Coop. 40:116-117.
Powell, A.A., M.A. Oliveira & S. Mathews. 1986. Seed vigor in cultivars of dwarf french bean (*Phaseolus vulgaris*) in relation to the colour of the testa. J. Agric. Sci. 106:419-425.
Prakken, R. 1934. Inheritance of colours and pod characters in *Phaseolus vulgaris* L. Genetica 16:177-294.
Prakken, R. 1970. Inheritance of colour in *Phaseolus vulgaris* L. II. A critical review. Meded. Landbouwhogeschool Wageningen 70-23:1-38.
Rubatzky, V.E. & M. Yamaguchi. 1997. World Vegetables: Principles, Production, and Nutritive Values. Int. Thomson Publ., Florence, U.S.A.
Schultz, W. 1966. The Blue Legume. Organic Gardening. Sep.:22-27.
Schwartz, H.F. & M.A. Pastor-Corrales (eds.). 1989. Bean Production Problems in the Tropics. 2nd ed. CIAT, Cali, Colombia.
Service, R.F. 1998. Microchip arrays put DNA on the spot. Science 282:396-399.
Shellie-Dessert, K.C. & F.A. Bliss, 1991. Genetic improvement of food quality factors. pp. 649-677. *In* A. van Schoonhoven & O. Voysest (eds.), Common Beans: Research for Crop Improvement. C.A.B. Int., Wallingford, U.K. & CIAT, Cali, Colombia.
Shonnard, G.C. & P. Gepts. 1994. Genetics of heat tolerance during reproductive development in common bean. Crop Sci. 34: 1168-1175.
Siddique, M.A. & P.B. Goodwin. 1980. Maturation temperature influences on seed quality and resistance to mechanical injury of some snap bean genotypes. J. Amer. Soc. Hort. Sci. 105:235-238.
Silbernagel, M.J. 1979a. Seed quality index as an indicator of crop production potential, and a selection tool for the genetic improvement of snap bean seed quality. Annu. Rpt. Bean Improv. Coop. 20:40-42.
Silbernagel, M.J. 1979b. Release of multiple disease resistant germ plasm from the USDA snap bean breeding program at Prosser, Washington. Annu. Rpt. Bean Improv. Coop. 22:37, 40-41.
Silbernagel, M.J. 1986. Snap Bean Breeding. pp. 243-282. *In* M. Bassett (ed.), Breeding Vegetable Crops. AVI, Westport, U.S.A.
Silbernagel, M.J. 1987a. Fusarium root rot-resistant snap bean breeding line FR-266. HortScience 22:1337-1338.
Silbernagel, M.J. 1987b. Flat-podded snap bean breeding line VR-Romano. HortScience 22:1340.

Silbernagel, M.J. & S.R. Drake. 1978. Seed index, an estimate of snap bean quality. J. Amer. Soc. Hort. Sci. 103:257-260.

Silbernagel, M.J. & S.R. Drake. 1983. 'Blue Mountain' bush snap bean. HortScience 18:111.

Silbernagel, M.J. & T.J. Doyle. 1985. Interaction of genetic resistance to fusarium root rot with cultural practices in a white-seeded bush snap bean. Annu. Rpt. Bean Improv. Coop. 28:1-2.

Silbernagel, M.J. & R.M. Hannan. 1992. Use of plant introductions to develop U.S. bean cultivars. pp.1-8. *In* Use of plant introductions in cultivar development, part 2. CSSA Special Publication no. 20. Madison, U.S.A.

Silbernagel, M.J., W. Janssen, J.H.C. Davis & G. Montes de Oca. 1991. Snap bean production in the tropics: Implications for genetic improvement. pp. 835-862. *In* A. van Schoonhoven & O. Voysest (eds.), Common Beans: Research for Crop Improvement. C.A.B. Int., Wallingford, U.K. & CIAT, Cali, Colombia.

Singh, S.P. 1989. Patterns of variation in cultivated common bean (*Phaseolus vulgaris*, Fabaceae). Econ. Bot. 43:39-57.

Singh, S.P., P. Gepts & D.G. Debouck. 1991. Races of common bean (*Phaseolus vulgaris*, Fabaceae). Econ. Bot. 45:379-396.

Skroch, P., G. Jung, J. Nienhuis & D.P. Coyne. 1996. Integration of RAPD marker linkage maps and comparative mapping of QTL for disease resistance in common bean. Annu. Rpt. Bean Improv. Coop. 39:48-49.

Skroch, P.W. & J. Nienhuis. 1995. Qualitative and quantitative characterization of RAPD variation among snap bean (*Phaseolus vulgaris*) genotypes. Theor. Appl. Genet. 91:1078-1085.

Skroch, P., J. Tivang & J. Nienhuis. 1992. Analysis of genetic relationships using RAPD marker data. pp. 26-30. *In* Applications of RAPD Technology to Plant Breeding. Joint Plant Breeding Symposia Series; Crop Sci. Soc. Amer., Amer. Soc. Hort. Sci., Amer. Genetic Assoc. 1 Nov., 1992, Minneapolis, U.S.A.

Smith, B.D. 1995. The Emergence of Agriculture. W.H. Freeman, New York, U.S.A.

Spaeth, S.C. 1986. Imbibitional stress and transverse cracking of bean, pea, and chickpea cotyledons. HortScience 21:110-111.

Spence, N.J. & D.G.A. Walkey. 1994. Bean common mosaic virus and related viruses in Africa. Natural Resources Institute Bull. 63, Chatham, U.K.

Stavely, J.R. & R.T. McMillan Jr. 1992. BARC-rust resistant -bush, fresh-market green bean germplasm. HortScience 27:1052-1054.

Stavely, J.R. & J. Steinke. 1985. BARC-rust resistant -2, -3, -4, and -5 snap bean germplasm. HortScience 20:779-780.

Stevens, M.A. & W.A. Frazier. 1967. Inheritance of oct-1en-3ol and linalool in canned snap beans (*Phaseolus vulgaris* L.). Proc. Amer. Soc. Hort. Sci. 91:274-285.

Stevens, M.A., R.C. Lindsay, L.M. Libbey & W.A. Frazier. 1967. Volatile components of canned snap beans (*Phaseolus vulgaris* L.) Proc. Amer. Soc. Hort. Sci. 91:833-845.

Tanksley, S.D. & J.C. Nelson. 1996. Advanced backcross QTL analysis: A method for the simultaneous discovery and transfer of valuable QTLs from unadapted germplasm into elite breeding lines. Theor. Appl. Genet. 92:191-203.

Tanksley, S. & C.M. Rick. 1980. Isozymic gene linkage map of the tomato: Applications in genetics and breeding. Theor. Appl. Genet. 57:161-170.

Tanksley, S.D., N.D. Young, A.H. Paterson & M.W. Bonierbale. 1989. RFLP mapping in plant breeding: New tools for an old science. Bio/Technol. 7:257-264.

Taylor, A.G. & M.H. Dickson. 1987. Seed coat permeability in semi-hard snap bean seeds: Its influence on imbibitional chilling injury. J. Hort. Sci. 62:183-189.

Taylor, J.D., D.M. Teverson & D.J. Allen. 1996. Identification and origin of races of *Pseudomonas syringae* pv. *phaseolicola* from Africa and other bean growing areas. Plant Path. 45:469-78.

Teverson, D.M. 1991. Genetics of pathogenicity and resistance in the halo-blight disease of beans in Africa. Ph.D. Dissertation, Univ. of Birmingham, U.K.

Toya, D.K., W.A. Frazier & J.R. Baggett. 1976. Inheritance of 1-octen3-ol concentration in frozen pods of bush snap beans, *Phaseolus vulgaris* L. J. Amer. Soc. Hort. Sci. 101:196-198.

Toya, D.K., W.A. Frazier, M.E. Morgan & J.R. Baggett. 1974. The influence of processing and maturity on volatile components in bush snap beans, *Phaseolus vulgaris* L. J. Amer. Soc. Hort. Sci. 99:493-497.

Urrea, C.A. & S.P. Singh. 1995. Comparison of recurrent and congruity backcrossing for interracial hybridization in common bean. Euphytica 81:21-26.

Vallejos, C.E., N.S. Sakiyama & C.D. Chase. 1992. A molecular marker-based linkage map of *Phaseolus vulgaris* L. Genetics 131:733-740.

Vélez, J.J., M.J. Bassett, J.S.Beaver & A. Molina. 1998. Inheritance of resistance to bean golden mosaic virus in common bean. J. Amer. Soc. Hort. Sci. 123:628-631.

Wade, B.L. 1937. Breeding and improvement of peas and beans. pp.251-282. *In* Yearbook of Agriculture, 1937. US Government Printing Office, Washington DC, U.S.A.

Weeden, N.F. 1984. Distinguishing among white seeded bean cultivars by means of allozyme genotypes. Euphytica 33:199-208.

Wehrhahn, C. & R.W. Allard. 1965. The detection and measurement of the effects of individual genes involved in the inheritance of a quantitative character in wheat. Genetics 51:109-119.

Wyatt, J.E. 1977. Seed coat and water absorption properties of seed of near-isogenic snap bean lines differing in seed coat color. J. Amer. Soc. Hort. Sci. 102:478-480.

Yen, D.E. 1957. A shiny-podded mutant in pole bean (*Phaseolus vulgaris* L.). New Zealand J. Sci. Tech. Sec. A. 38:820-824.

Young, R.A. & J.D. Kelly. 1996. RAPD markers flanking the *Are* gene for anthracnose resistance in common bean. J. Amer. Soc. Hort. Sci. 121:37-41.

Zaumeyer, W.J. 1963. Some new Tendercrop mutants. Seed World March 8:8,10.

Zaumeyer, W.J. 1972. Dry beans and snap beans. pp. 224. *In* Genetic Vulnerability of Major Crops. Nat. Acad. Sci., Washington DC, U.S.A.

Zaumeyer, W.J. & H.R. Thomas. 1959. 'Tendercrop', a new slender-podded snap bean. Seed World March 27:8,10.

Chapter 13

Integrated Management of Abiotic Stresses

Michael Thung & Idupulapati M. Rao
Goiânia, GO, Brazil; Centro Internacional de Agricultura Tropical (CIAT), A. A. 6713, Cali, Colombia

Key words: aluminum and manganese toxicities; biological nitrogen fixation; cover crops; crop rotation; drought; green manure; low soil fertility; mineral deficiencies; precision agriculture; soil organic matter

1. INTRODUCTION

The common bean (*Phaseolus vulgaris* L.) production began thousands of years ago as a small component in complex cropping systems on normally fertile lands with cool climates at medium to high altitudes in the Americas. In modern times, with increased demand, bean production has expanded to areas where soil amendments and chemical inputs are required. Beans are now grown extensively on all continents, spanning the latitudes 52° N to 32° S, and from near sea level to higher than 3,000 m elevation (Schoonhoven & Voysest, 1991; Singh, 1992). Beans are grown in monoculture as well as in various intercropping systems, with or without irrigation, from lowland tropics with heavy rainfall and high temperatures to semiarid regions in temperate environments. The great diversity of bean-growing environments implies different abiotic and biotic production constraints (Graham & Ranalli, 1997; Schwartz & Pastor-Corrales, 1989; Singh, 1992; Thung & Oliveira, 1998; Wortmann et al., 1998; see also Chapter 1 by Singh).

Abiotic stresses can be either edaphic or climatic in origin. Their integrated management is needed to sustain bean production while improving soil and water quality. In this chapter, we review the available information on edaphic and climatic stress factors that limit bean yields, and discuss traditional and integrated management systems that overcome these stresses in bean production systems. The reader may obtain additional information in other reviews related to this topic (Howeler, 1980; Flor & Thung, 1989; Masaya & White, 1991; Thung, 1997; Thung & Oliveira, 1998; White & Izquierdo, 1991).

2. EDAPHIC STRESSES

2.1 Soil Chemical Disorders

In regions where low-technology agriculture is still practiced, producing food and fiber depletes soil nutrients. An extreme example of such a system is the no-input, slash-and-burn agriculture practiced in tropical forest margins. Giller & Cadisch (1995) estimate that an average of 20 to 70 kg/ha of N (nitrogen) and K (potassium) are exported annually from agricultural land as crop produce in sub-Saharan Africa, Asia, and Latin America. Thus, nutrient deficiency is chronic and widespread in small subsistence farming systems in these regions, resulting in typically low bean yields (< 700kg/ha). In some irrigated areas (e.g., winter plantings in Central Brazil, northern Pacific coast of Mexico, and Pacific North West U.S.A.), bean yields may reach 4 MT/ha, which implies, that the soil has to supply more than 250 kg/ha of N (including BNF) and 200 kg/ha of K during the crop cycle of approximately 100 days.

2.1.1 Mineral Deficiencies

2.1.1.1 Nitrogen

In most regions, N deficiency limits common bean and associated crop production. An estimated 40% of the bean area in Latin America and perhaps 60% of the bean area in the Middle East and central, southern, and eastern Africa are affected by N deficiency (CIAT, 1992). Yield losses of 5% to 20% (relative to N-fertilized crops) are common throughout bean-growing regions and losses of 30% to 40% are common in N-deficient soils. Only in snap-bean production systems is N deficiency rare, because the benefit of applying N is realized for green-pod production. Most bean farmers know of the N deficiency problem and can easily detect its symptoms. However, by the time this deficiency is detected bean yields have already been reduced, and supplying additional N at an advanced growth stage normally does not fully restore yield potential.

The bean plant's efficiency in using N fertilizer is rather low, being less than 70% of the applied fertilizer. The remaining fertilizer is lost through volatilization and leaching, losses being greater with broadcast application (Amberger, 1990). This author has proposed incorporating straw or organic matter with a low C (carbon)/N ratio to temporarily immobilize N, prevent leaching, and permit the gradual release of N through mineralization. In intensive bean-production systems, where mulch is produced specifically to cover the soil during the planting season, N is applied to the crop before desiccation. This applied N then becomes available from mulch via mineralization during the next bean crop's early growth.

Biological nitrogen fixation (BNF) seldom supplies adequate N during the bean's whole growth cycle. See Section 5.4 for a further discussion on BNF.

2.1.1.2 Phosphorus (P)

In the tropics, P deficiency is most commonly found in Latin America (more than half the area) and Africa (probably between 35% and 50%) (CIAT, 1992; Lynch & Beebe, 1995; Thung, 1990; Thung et al., 1987; Wortmann, 1994; Wortmann et al., 1995). P deficiency is widespread in all acid soils, which are inherently low in soil P, have a high P-fixing capacity, and receive inadequate lime application (Rao et al., 1999). Yields can be reduced by as much as 60% to 75% in severely deficient soils and 50% to 60% in eroded uplands.

During the early 1970s, banding proved to be an economically feasible method of applying P for soils in the Brazilian Cerrados when they were first cultivated. Eventually, farmers found that several years of treatment were needed to improve the P status of these acid infertile soils.

2.1.1.3 Potassium

About 15% of the bean-growing area in Latin America and 20% in Africa may be affected by soil K deficiency (CIAT, 1992). Yield losses can reach 20% in K-deficient soils, but this deficiency is seldom reported in bean literature. Published reports indicate that beans do not respond to K application (Thung, 1991). The lack of response may be because first, beans were regarded as efficient in absorbing K from the soil solution, and, second, experiments were conducted for only a few seasons so that K depletion was not drastic. Third, K's greater mobility in soil allows it to be replenished quickly in soil solution via soil-water movement.

Complete fertilizers, commonly sold (e.g., in Brazil), carry at least 15% of K_2O, which may be sufficient to maintain soil K status and prevent severe K deficiency in many annual crops.

2.1.1.4 Molybdenum (Mo)

Vieira et al. (1992) reported the importance of Mo as a micronutrient for beans grown in the state of Minas Gerais, Brazil. Foliar application at the rate of 12 to 19 g/ha of Mo produced the same yield as side-dressing with 20 kg/ha of ammonium sulfate. Amane et al. (1994) showed that foliar application of 20 g/ha of Mo at 25 days after emergence could replace 30 kg/ha of N as side-dressing. Berger et al. (1995) made these applications practical by combining ammonium molybdate at 20 g/ha of Mo with pelletization of bean seed with Rhizobia. Berger et al. (1996) reported that the application of 20 g/ha between 21 days after emergence until flowering could increase yield by as much as 180%, again in Minas Gerais.

2.1.1.5 Zinc (Zn) and iron (Fe)

Beans are thought to be sensitive to Zn deficiency, which has seldom been reported for tropical acid soils, except in the savannas of Brazil (Cerrados) and Colombia (Llanos Orientales). Zinc deficiency is common in temperate zones, particularly on calcareous soils with pH values higher than 7.0 (e.g., midwestern and northwestern U.S.A., including the states of Colorado, Idaho, Michigan, Nebraska, and Wyoming). Zinc deficiency can be induced by Fe deficiency. Unlike grasses, beans do not release siderophores to facilitate acquisition of Fe from Fe-deficient soils. Beans use strategy "I" mechanisms, involving acidification of the rhizosphere. Iron deficiency is rare in the tropics, whereas it occurs frequently in temperate soils where the soil pH is above 7.0 in many areas.

2.1.2 Mineral Toxicity

2.1.2.1 Acidity

Tropical soils are normally old, highly weathered, strongly leached, and poor in nutrients. The types and characteristics of tropical soils vary greatly from one site to another. However, they share the common problems of soil acidity and low fertility. Soil acidity is still a problem in many tropical and subtropical bean production regions of Latin America and Africa. In Africa, where most of the population is still living in rural areas, chemical inputs are scarce and expensive. Limestone deposits are rare. Organic input, such as cow dung, is removed from the field and used as fuel. Thus, African crops remove more nutrients from the soil than farmers can replenish. This situation eventually leads to severe soil degradation, making restoration more difficult and expensive.

The situation in Latin America is better because this continent has a higher socioeconomic status. For example, in Brazilian Cerrados, soil acidity has been corrected for the past 30 years and the problem is now confined to maintaining soil pH at optimal levels, given the extensive use of acid-forming chemical fertilizers. Nonetheless, use of lime in Brazil is inadequate to completely correct aluminum (Al) or manganese (Mn) toxicity or to maximize the efficiency in use of P fertilizers.

2.1.2.2 Aluminum

Aluminum toxicity occurs in acid soils and limits bean production. About 40% of the bean-growing area in Latin America and 30% to 50% of the bean-production areas of central, eastern, and southern Africa may be affected by Al toxicity (CIAT, 1992). Yield reductions can be as high as 30% to 60% in Al-toxic soils. In many otherwise suitable areas, beans are not grown because of potential Al toxicity.

Al toxicity in beans affects the primary root, forcing the plant to grow secondary roots. Some cultivars, such as A 774, Carioca, Rio Tibagi, and Dade, are more tolerant of high levels of Al in soil solution than are other cultivars. Beans exposed to Al-toxic soil can cope by exuding organic acids from their roots. For example, when exposed to Al stress, the tolerant cultivar Dade exudes 70 times more citric acid than does the susceptible cultivar Romano (Miyasaka et al., 1991).

2.1.2.3 Manganese

Mn toxicity is typically found in soils with high precipitation or poor drainage, low pH, and medium to high organic matter content. About 12% of the bean-production area in Latin America suffers from Mn toxicity, compared with 10% in Africa (CIAT, 1992). Problem areas include central Brazil; Central America; volcanic regions of Mexico; central Chile; central, eastern, and southern Africa; and, possibly, parts of China.

Many acid soils have low Mn oxide content and therefore Mn toxicity is not as prevalent as Al toxicity. Toxicities of both Mn and Al are sometimes present in the same soil (e.g., soils of volcanic origin) but can also be present independently. Mn toxicity reduces plant growth by accumulating Mn in leaf tissue, causing localized necrosis. It also interferes with the uptake of calcium (Ca) and magnesium (Mg), which are usually found in limited quantities in many acid soils. Liming and improved drainage can reduce Mn toxicity in most soils.

2.1.2.4 Arsenic

Polluted water used for irrigation can carry high levels of arsenic. This element can be taken up by beans in large enough amounts to be toxic to the plants (Carbonell-Barrachina et al., 1996).

2.1.2.5 Salinity

About 20% to 30% of the bean-production area in the Middle East and 5% to 10% in Latin America may be affected by soil salinity (CIAT, 1992). Salinity in Latin America is confined to several, small, irrigated regions in Peru, the Dominican Republic, Cuba, Ecuador, and Mexico. But the problem could increase as the amount of irrigated land increases in Brazil, Mexico, and other countries. In salt-affected areas, yield reductions may be between 5% and 40%.

Salinity in beans occurs mostly in irrigated bean-production areas in the Mediterranean region, West Asia, northwestern and western U.S.A., and along the Nile in Africa, where beans are grown under arid or semiarid climates. Smithson et al. (1992) described the problem of salinity in a nonirrigated bean-growing region in Sudan, observing wide variability in tolerance of soil salinity in local cultivars. Nonetheless, beans are classified as highly sensitive to salinity. Ayers & Westcot (1976) showed that 2.3 dS m^{-1} conductivity in soil solution is sufficient to reduce bean yield by 25%, whereas barley or sugarbeet can tolerate as much as five times more salinity. Maas & Hoffman (1977) consider that the limiting value for beans is around 1.0 dS m^{-1}. Bean growth can be reduced markedly when Cl^- concentration is higher than 5 ppm (Debouck, 1976).

In arid and semiarid regions, several factors cause salinity. Either the soils were originally saline and the water is of poor quality, having been used several times for irrigation, or the soils are not saline, but the irrigation water is. In neither case is the water sufficient to leach the salts to below the arable layer.

2.2 Soil Physical Disorders

Any agricultural activity alters the original physical properties of the soil. Even in low-technology agriculture, such as in slash-and-burn farming systems, the bare soil is exposed to the physical impact of raindrops, and soil crusting and erosion are widespread. Trampling by farmers and livestock compacts the soil, leading to erosion. Especially in marginal areas, such as forest margins, erosion leads to a rapid decline in soil fertility.

2.2.1 Soil Erosion

Soil erosion is caused either by natural factors such as climate, relief of terrain, geology, or lack of vegetation cover. Soil erosion occurs not only on land under torrential rain in the tropics, but also on irrigated land in semiarid and arid regions. Even where land is correctly leveled, furrow irrigation causes soil erosion.

Soil crusting prevents rainwater infiltration and promotes runoff. On slopes, the increase in the speed of runoff creates a high risk of loss of soil and nutrients. Topsoil is removed, together with an amount of nutrients that is greater than the average nutrient content of the soil. As much as 200 kg/ha of soils were lost from an Oxisol with a slope of 4.38% that received 403 mm of precipitation. The organic matter (SOM), P, and K contents in this lost soil were more than double those obtained from routine soil chemical characterization (M. Thung, unpublished results).

Improved planting patterns and furrow spacing can reduce this problem. Sojka et al. (1992) found that in coarse-silty soils, planting double rows of beans on a narrow furrow reduced soil losses and significantly increased bean yield. Recently, Sojka et al. (1998) showed that water-soluble polyacrylamides are potent tools for fighting furrow irrigation-induced soil erosion.

2.2.2 Soil Crusting and Compaction

Soil crusting on agricultural land commonly occurs when the bare soil surface receives the direct impact of raindrops, sealing the soil surface against water infiltration and provoking laminar soil erosion. Soil crusting, which occurs immediately after planting, could restrict emergence, thus reducing seed yield through fewer plant population. If the seedling manages to emerge, it typically suffers broken apices and hypocotyl lesions. The physical injuries facilitate infection by soilborne pathogens and prolong exposure to other soil stresses such as salinity and herbicide application.

Soil crusts formed immediately after planting, when seeds are just starting to swell, can be mitigated by mechanical surface scarification with spring-shank blades or a rotary hoe between the rows. Because the risk of damaging the seedlings on or near the soil surface is great, scarification should be conducted only when the crust is too hard for seedlings to emerge. Sometimes, waiting for the seed to germinate and assessing the need to replant is preferable. Soil crusting can be minimized by using a direct planting or no-tillage system, in which crop residues protect the soil surface from raindrops. The residues also reduce soil temperatures and increase soil humidity. These advantages may be counterbalanced by disadvantages, such as increasing incidence of soilborne diseases.

Soil compaction, a major constraint to bean production, is usually a human-induced problem, and can take place in any soil type and at any depth. In mechanized agriculture, soil compaction is caused by several tillage operations, including the use of heavy machinery on very wet soil, excessive preparation of the seedbed (which pulverizes soil aggregates), or the use of improper implements. Declining SOM destabilizes soil macroaggregates, especially during land preparation with improper implements.

Soil compaction can be easily detected with a penetrometer, which measures mechanical resistance to soil penetration, or by extracting samples with a soil-sampling auger and carefully observing the soil structure. Stoner & Freitas (1989) quantified the impact of changes in bulk density on bean root growth in several soils with different clay contents. In sandy loam soil, with only a 16% clay content, roots grew unrestrictedly. In soil with a 53% clay content, bean root growth was restricted at 1.4 g/cm^3 bulk density level. In soil with 70% clay content, no bean roots developed at 1.1 g/cm^3.

Compaction reduces soil aeration and water retention capacity, and impedes normal root development (Canalli & Roloff, 1997; Carpenedo & Mielniczuk, 1990). For example, when a bean root encounters mechanical impedance such as soil compaction, it becomes deformed, branches more frequently, and consumes more oxygen per unit fresh weight to overcome the compaction (Schumacher & Smucker, 1981). Because soil compaction reduces bean root growth, it increases susceptibility to temporary water stress and this, in turn, prevents the crop from reaching its yield potential. This phenomenon is more difficult to see in irrigated bean-production systems, if chemical nutrients are applied adequately and water stress seldom occurs.

3. CLIMATIC STRESSES

3.1 Drought

In Latin America, at least 60% of the bean-production area is estimated to suffer from moderate to severe drought or water stress. Moisture stress occurs in arid (annual rainfall of <250 mm), semiarid (250-500 mm), or subhumid regions (500-1500 mm) (Stephens, 1994). In these areas, wherever bean production depends on the erratic patterns of rainfall distribution, water deficit may occur more than once during a crop's growth cycle. The intensity and duration of stress determine the degree of yield reduction relative to the crop's potential. Moreover, when a dry period (2 to 3 weeks) coincides with flowering, significant yield reduction occurs. In Northeast Brazil, for example, water deficit can reduce annual bean production by about 32% (Magalhães & Millar, 1978); one good harvest is expected every 5 years; and complete crop losses are not uncommon. Despite this, bean cultivation is expanding to semiarid regions rather than to the wet tropical rainforest, because beans tolerate moderate water stress better than they do excess moisture.

3.2 Waterlogging

Excessive soil moisture, or flooding, causes anoxia and changes the cellular ultrastructure of the bean root tips, that is, the amyloplast disappears and a dense lipid layer forms (Papenhuijzen & Roos, 1979). The bean plant thus suffers anaerobic stress. Flooding can occur in subhumid regions, where beans are planted at the onset of the rainy season. Farmers expect a better rainfall distribution pattern when they plant early, but at the same time they face the risk of excessive rainfall at any time between germination and harvesting.

Excessive soil moisture can also be a problem in regions where water tables are high. Brazil alone has several million hectares of wetlands (known as "Várzeas"). This has potential for subterranean irrigation in which water availability in soil is controlled by manipulating the water table through drainage canals. Beans are planted on ridges to create a favorable or drier rooting zone during the dry and cooler period. Planting beans during the hot season is too risky, not only because of the heat, which makes beans more sensitive to flooding, but also because temporary inundations, with the accompanying anaerobic stress (Schumacher & Smucker, 1984; Thompson & Fick, 1981).

In a greenhouse study using the split-root system, beans suffering from anoxia had a reduced uptake of Ca (by 40%) and of NO_3 (by 60%), compared with plants with aerated roots (Schumacher & Smucker, 1987).

3.3 Temperature

Air temperature strongly affects bean growth and development (Masaya & White, 1991). Temperatures in air, the root zone (arable layer of top 20 cm), and subsoil can be key in successful seed germination, formation of tap or lateral roots, and determining the number of days to flowering. Under controlled conditions, temperature for good germination ranged between 15 and 35°C. Temperatures lower than 10°C and higher than 40°C result in poor germination (Hoogenboom & NeSmith, 1992).

In mountainous regions, soil temperatures may be lower than air temperatures, and, in the soil itself, temperatures below the root zone are normally lower than those in the arable layer. Many regions in the higher latitudes are suitable for bean production in terms of rainfall, but the air and soil are, on the whole, too cold. Normally, in the spring, soil temperatures are 2 to 4°C lower than air temperatures and are often too cold for bean seeds to germinate.

In the tropics, high air temperatures are normally accompanied by high soil temperatures in the rooting zone. These temperatures may discourage roots from growing deeper than the normal rooting zone. High temperatures also speed up the crop cycle, which is completed in fewer than 100 days. The effects of high temperature include flower drop, flower and seed abortion, reduced pollen grain viability, impaired pollen-tube formation in the styles, and reduced seed size (Haterlein et al., 1980; Weaver et al., 1985).

In semiarid regions, high temperatures and soil moisture stress usually interact to produce a significantly reduced bean yield. The problem becomes aggravated when night temperatures are also high. Higher temperatures provoke flower drop, thus affecting pod and seed set in commercial cultivars, even in the absence of water stress (Konsens et al., 1991). When day temperatures are high but night temperatures are lower than 25°C beans grow well (Shonnard, 1991).

The interaction between air temperature and photoperiod plays a major role in bean production (Masaya & White, 1991). Many highland cultivars cannot be planted in lowland areas at the same latitude because of the temperature difference. Under controlled conditions, highland bean cultivars can produce flowers and seeds when exposed to a 6-h daylight regime (White & Izquierdo, 1991).

3.4 Air Pollution

In many areas of the world, beans are produced close to large cities or industrial complexes, where polluted air can decrease yields. Ozone and sulfur dioxide are the principal pollutants. Bender et al. (1990) showed that, when beans are placed in open-topped chambers and exposed to ozone at concentrations ranging between 24 and 109 ppb, leaf chlorophyll concentrations are reduced, although vegetative growth is not impaired. Concentrations below 70 ppb have little effect on seed yield but, at 80 ppb, yields are reduced by 12% to 20%. In another study, Astorini et al. (1995) found significant reduction in plant growth at near-ambient levels of ozone stress. In the tropics, ozone stress usually occurs in the dry season when beans are grown under irrigation. Some sensitive cultivars show typical ozone bronzing symptoms in leaves and pods. Mn deficiency can also increase ozone toxicity symptoms (Mehlhorn & Wenzel, 1995).

3.5 Stress from Ultraviolet-B Radiation

Plants sensitive to ultraviolet-B (UV-B) radiation respond in different ways, depending on species and cultivar and on growth conditions (Bornman & Teramura, 1993; Caldwell et al., 1995; Vass, 1996). Differences in sensitivity may be due in part to the type of adaptive mechanisms used against UV-B radiation, including increased quantities of UV-B absorptive substances (e.g., flavonoids), increased reflection of UV-B radiation from epicuticular surfaces, and increasing thickness of cuticles and leaves with increasing UV-B radiation. Genetic variation in common bean was observed for tolerance of UV-B radiation (M. Pinto, personal communication, 1998). Among five cultivars examined, Pinto and Tórtola showed the highest levels of tolerance.

4. EFFECT OF ABIOTIC ON BIOTIC STRESSES

Edaphic and climatic factors can trigger biotic stresses. For example, nutrient deficiencies or toxicities may induce the incidence of diseases (Table 1). Long dry spells and high temperatures may provoke an increase in insect populations, such as *Empoasca kraemeri* Ross & Moore (Schoonhoven & Cardona, 1980); and acíd soils may harbor diseases such as Fusarium wilt. Cloudy days with mild temperatures and high humidity will induce powdery mildew (caused by *Erysiphe polygoni* DC) on bean (Schwartz & Steadman, 1980). With hot and humid weather (e.g., in some regions of Brazil, Costa Rica, and Panama), web blight [caused by *Thanatephorus cucumeris* (Frank) Donk.] is a devastating disease.

Table 1. Influence of macro- and micronutrient disorders on disease incidence in bean

Element/disease	Reaction	References
Nitrogen		
Common bacterial blight	Increase	Hedges & Fisher, 1946
Anthracnose	None, increase	Drobny et al., 1984, 1985
Fusarium root rot	Increase	Huber & Watson, 1970; Maier, 1968
Rhizoctonia root rot	Increase	Davey & Papavizas, 1960
Potassium		
Rhizoctonia root rot	Decrease	Bateman, 1965; Bateman & Miller, 1966
Calcium		
Anthracnose	Decrease	Drobny et al., 1984, 1985
Damping-off	Decrease	Bateman & Lumsden, 1965
Rhizoctonia root rot	Decrease	Bateman & Lumsden, 1965
Magnesium		
Rhizoctonia root rot	Decrease	Bateman, 1965
Copper		
Halo blight	Decrease	Tamietti & Garibaldi, 1984
Iron		
Fusarium root rot	Decrease	Guerra & Anderson, 1985
Boron		
Fusarium root rot	Decrease	Guerra & Anderson, 1985
Sodium		
Rhizoctonia root rot	Increase	Bateman, 1965

5. TRADITIONAL MANAGEMENT TO OVERCOME ABIOTIC STRESSES

5.1 Land Preparation and Use of Machinery

The main purpose of land preparation, using machinery, is to form a firm, but not compacted, seedbed and to control weeds. Proper land preparation improves water retention capacity and the infiltration rate of rain or irrigation water, and reduces waterlogging and temporary inundation. With good soil aeration, the bean plant grows vigorously and can easily compete with weeds during early growth.

Mechanical land preparation is similar in many countries; differences between countries occur in the last phase of soil preparation, which differentiates between rainfed, overhead irrigation, and furrow irrigation systems. In Latin America, irrigated beans occupy a small fraction of the production area, whereas in temperate zones, irrigated (furrow or overhead) beans dominate. In the Mediterranean region, basin irrigation is common.

Appropriate timing and correct soil moisture content are needed for good land preparation. The traffic of heavy machinery should be kept to a minimum to prevent soil compaction and reduce preparation costs. Tractor speed must be kept constant between 4.5 and 5.0 km/h during land preparation and planting. Speeds of higher than 8 km/h separate the soil aggregates and stratify the soil according to aggregate size, thus reducing the efficiency of land preparation. A good combination of two different primary implements and working width is more efficient than increasing working velocity. Tillage that inverts the soil layer not only improves oxygenation, but also contributes to an increase in oxidation of SOM through enhancing soil microbial activity, which, in turn, leads to diminishing SOM (Haynes, 1982).

5.2 Liming

Liming is done not only to reduce Al and Mn toxicities, but also to neutralize soil acidification from heavy annual applications of chemical fertilizers. Use of only calcitic lime may induce Mg deficiency (Andrew, 1977), and use of only dolomitic lime may induce a Ca-to-Mg imbalance. Lopes & Cox (1977) reported that the native savanna topsoils (0-20 cm depth) in the Brazilian Cerrados have exchangeable Ca and Mg levels of 0.25 and 0.09 meq/100 g soil, respectively. In deeper layers, the ratio of Ca-to-Mg is lower, thus increasing Al toxicity and possibly suppressing rooting ability in the subsoil. Supplying Ca alone to high-input bean production systems may not be sufficient; Mg deficiency occurs frequently in high-input bean production at pod formation or later stages.

5.3 Use of Fertilizers

Fertilizer recommendations in Latin America in the past were modest, limiting N rates to 100 kg/ha and P to 66 kg/ha. Potassium was seldom recommended. Farmers tended to ignore even these modest recommendations because of the unreliable rainfall, both in amount and pattern. Without adequate fertilizer use, the need to open new lands will be aggravated, because soils become rapidly exhausted and physically degraded (Borlaug & Dowswell, 1997).

In the late 1980s, fertilizer use increased to maintain productivity. However, even the increased use of chemical fertilizers failed to arrest the decline in bean productivity in mechanized and irrigated systems, under which soil degraded physically as a result of rapid decline in SOM and of increased soil compaction. Moreover, in the past, a recommendation for a specific nutrient was based on comparing tissue analyses with critical nutrient levels. This approach compares the concentration of a single plant nutrient, suspected to be deficient, with the critical levels of that nutrient, and ignores the other elements involved in plant metabolism. This method may be adequate only when some precautions are taken during diagnosis. Tissue sampling must be done at a prescribed age of the plant and at a specific leaf position. Soil samples must be taken at several depths at a certain soil moisture content and the soil type must be carefully described (Thung & Oliveira, 1998).

Because the concept of critical nutrient levels does not reflect the relationships between essential elements, Beaufils (1971) proposed the Diagnostic and Recommendation Integrated System (DRIS), which compares the obtained ratios with the established norms from a database created especially for this purpose. These norms are compiled from the analyses of data obtained from several samples of plant tissue and soil, together with crop yield. Walworth & Sumner (1987) reviewed the DRIS and compared it with the standard procedure of critical nutrient levels. They found that DRIS is more stable and less affected by climate, soils, and plant age. The difficulties of creating the DRIS norms may prevent wide use of this method.

5.4 Biological Nitrogen Fixation (BNF)

Nitrogen application to beans is often criticized for not using the beans' potential for BNF. However, soil conditions must be significantly improved to optimize N fixation by Rhizobia. Bliss (1993) reported about efficient Rhizobia strains that can fix enough N to support bean production from 1 to 2 MT/ha, provided no other limiting factors exist during the whole crop cycle, such as high temperatures, lack of P supply, and presence of water stress and toxic elements. Thus, under irrigated systems, where bean production can be higher than 2 MT/ha, the total N that can be supplied by N fixation may be inadequate. Furthermore, contribution of N by fixation does not normally coincide with the peak demand for N by the bean plant. Practical ways must be found to inoculate fields with efficient Rhizobia, as is done for soybeans. Creating a favorable environment for a bean cultivar that readily nodulates will also improve overall performance. Tsai et al. (1993b) demonstrated that common bean has genetic variability for BNF that can be expressed even when intercropped with maize. Müller et al. (1993) showed that bean cultivars show variability in BNF when grown with different levels of nutrient supply. Beans can also fix N in the presence of high mineral N supply to soil (Silva et al., 1993), and their efficiency in BNF can be further improved by increasing the supply of P, K, and S in the soil (Tsai et al., 1993a).

5.5 Arbuscular Mycorrhizae (AM)

Beans can benefit from associating with AM in several ways: 1) roots develop better because of improved physical properties of the soil (Sutton & Shepperd, 1976), 2) P fertilizer is used more effectively for plant growth and yield (Lopes & Siqueira, 1981), 3) the plant tolerates abiotic stresses better (e.g., low soil P, drought) (Sieverding, 1991), and 4) the plant tolerates root rots (Linderman, 1992).

Beans are classified as AM-dependent (Gerdemann, 1975; Plenchette et al., 1983). This classification suggests that beans rely on AM colonization for better growth. However, it was observed that AM's ability to colonize varies according to bean genotypes. Some cultivars such as Canadian Wonder are less effectively colonized than other cultivars such as Jamapa (Mosse & Thompson, 1984). AM can grow beyond the rhizosphere by elongating their mycelia, thus increasing the soil volume explored by 5 to 200 times (Sieverding, 1991). The mycelia absorb phosphate ions, which then become available to the plant (Barber, 1995; Tinker, 1986).

Frequently, bean roots are well infected by AM, but yield benefits are not obvious. Further research is needed to improve the benefits of AM association for bean production.

5.6 Crop Rotation

Crop rotation is not limited only to sequential growth of annual crops, but also involves managing fallow and green manure crops within the crop sequence. Fallow is important in semiarid ecosystems to replenish soil water used during crop production. Green manure improves soil fertility and product quality. For the effects of green manure see Section 5.8.

Continuous planting of bean without crop rotation is commonly practiced in intensive, irrigated, bean-production systems (e.g., in Calabral, Brazil). In both nonirrigated and irrigated areas, unknown factors have reduced yield over time. When crop rotation was introduced, yield of beans planted after either rice or soybean was higher than that obtained from beans after beans in soils without lime application (Table 2). In Michigan, Christenson et al. (1991) showed that the oat/alfalfa/navy bean/sugarbeet rotation gave the highest bean yield, followed by the maize/navy bean/sugarbeet rotation.

Table 2. Bean yield (kg/ha) as affected by crop rotation and lime application, EMBRAPA-CNPAF experimental station, Goiania, Brazil

Planted after	Without lime	With lime
Bean	1,461	1,910
Rice	1,627	1,965
Soybean	1,733	1,934

Source: M. Thung (unpublished data).

Crop rotation with perennial grasses, such as *Brachiaria plantaginea, B. decumbens*, or *B. brizantha,* reduces the proliferation of soilborne bean diseases. Moreover, grasses planted in bean cropland give a pasture of excellent quality and increase the pasture's carrying capacity from 1 to 3 animals per hectare over 6 months (Spain et al., 1996). Grasses also stabilize the terrain by extensive root formation and sequester large amounts of carbon (Rao, 1998). Extensive root formation prevents soil erosion and improves soil aeration by root decomposition and mineralization. Robinson et al. (1994) showed that rotations of legumes and grasses could increase soil aggregates and improve soil nutrient cycling to the plant through increasing SOM.

5.7 Cover Crops and Crop Residues

During fallow or during the early growth of a bean crop, cover crops can protect bare soil from the impact of raindrops, which seal the soil surface and form crusts. Cover crops can recycle nutrients from deep soil layers to the arable layer, and protect fields from weed proliferation and soilborne diseases. Bottenberg et al. (1997) found that cover crops reduce insect damage in intensive snap bean production. Nasser & Sutton (1993) showed that rice straw could protect bean plants from white mold. Barros & Hank (1993) reported that evapotranspiration in the bean field is reduced by spreading mulch.

5.8 Green Manure Crops

Green manure crops in many ways work similarly as cover crops and crop residues. In that, they protect soil from rapid evaporation of soil moisture and from the physical force of the raindrops during torrential rains. Green manure crops also reduce soil temperature, conserve soil water, increase soil microbial activity, and maintain a balance between beneficial and pathogenic microorganisms.

The organic matter decomposes and builds the colloidal part of the soil, which acts not only as an active component that is responsible for the cation exchange capacity (CEC), but also reduces nutrient loss during heavy rainfall. The advantage of organic manure over inorganic fertilizer is its slow release of nutrients. Miyazawa et al. (1993) reported that incorporating organic matter can also decrease the effects of soil acidity. Decomposing organic matter releases organic acids and CO_2, lowering soil pH and thus also benefiting soils with high pH. Furthermore, it improves soil physical properties, and increases soil CEC and buffering capacity.

Table 3. Bean yield (kg/ha) as affected by continuous planting and green manure incorporation on a soil (oxisol) with low or high P supply, EMBRAPA- CNPAF experimental station, Goiania, Brazil

	High P		Low P	
Cycle	No green manure	Velvet bean incorporated	No green manure	Velvet bean incorporated
1st cycle	1,728	2,081	1,313	1,740
2nd cycle	588	1,272	969	1,062
3rd cycle	1,463	1,309	770	543
4th cycle	854	1,290	504	859
5th cycle	411	519	52	59
Mean	**1,009**	**1,294**	**722**	**853**

Source: M. Thung (unpublished data).

While organic material decomposes, mineralization accelerates and plant availability of nutrients is improved. Treatments with no green manure had reduced bean yields (Table 3). But not all plants can serve as green manure crops; normally, N-fixing plants are preferable. Furthermore, the species should not transmit diseases to the subsequent crop. Significant differences were observed among 19 green manure species, planted at different altitudes, in terms of dry matter production and days to flowering (Thung & Cabrera, 1994). In cooler climates, several green manure species showed a tendency to produce less biomass and to have an increased number of days to flowering.

To study the beneficial effects of incorporated green manure, six species were compared with rice and maize residues. The dry matter production of the eight species ranged from 5.7 to 9.2 MT/ha with low P supply in soil, and from 5.3 to 9.9 MT/ha with high P supply in soil (M. Thung, unpublished results). Bean yields were severely affected by soil P levels. In soil with high P, the greatest mean bean yield was after velvet bean was incorporated (1.45 MT/ha) and the lowest (0.37 MT/ha) was after *Dolichos lablab* was incorporated (Table 4). In low P soil, incorporation of maize straw resulted in the highest bean yield, whereas the lowest yield was obtained when *Crotalaria spectabilis* was incorporated. This result was unexpected because *C. spectabilis* is a N-fixing legume.

5.9 Weeds as Organic Matter and Source of Nutrients

Uncontrolled weed growth competes with the bean crop for nutrients, light, and water. Beans cannot compete effectively with weeds, especially under adverse growing conditions such as nutrient deficiency or water stress. In the forest margins, weed proliferation leads to field abandonment long before soil fertility declines. The nutrient content of the weed biomass can be recycled by properly incorporating it during land preparation.

Thung (1997) evaluated the role of weeds as a source of organic matter and nutrients by using different weed management strategies. Weeds were allowed to grow with either low or high soil P, during 5 months of fallow. These plots were then treated in four ways before land preparation for bean planting: 1) cut and burned, 2) cut and removed from the field, 3) cut and left as mulch, and 4) weeds left undisturbed and incorporated during land preparation (control). The maximum biomass production was 10.5 MT/ha, which was obtained from plots where weeds were allowed to grow undisturbed in the 5-month period. The lowest biomass production was 2.6 MT/ha, which was obtained from plots where weeds were cut twice during the 5-month period and removed from the field. When weeds were incorporated they decomposed and released nutrients.

The potential of weeds to recycle nutrients is shown in Table 5. The quantity of N and K in the weed biomass from soil with low P supply is high enough to meet N and K requirements of beans during early growth. In soil with high P supply, the quantities of N and K were higher than the normal doses of fertilizer recommended for intensive bean production systems in

Brazil. Incorporating weed biomass to low P or high P soil increased the yield of five bean cultivars (Table 6). The cut-and-burn weed management simulated the slash-and-burn agriculture practiced in forest margins. This produced the lowest yield of the weed manipulation methods, regardless of soil P levels. The highest bean production resulted from plots where weeds were undisturbed and incorporated during land preparation. These results show that, in addition to using legumes as green manure crops, weed complexes (both monocotyledons and dicotyledons), with proper management, could also serve as green manure.

Table 4. Yield of six bean cultivars (kg/ha) as affected by different green manure species grown on soil (oxisol) with high and low P, EMBRAPA-CNPAF experimental station, Goiania, Brazil

	Bean yield (kg/ha)					Mean
Green manure	POT 51	IPA 6	A 774	APORE	A 285	
High soil P						
Velvet bean–black	1,767	1,859	1,051	1,624	968	1,454
Crotalaria mucronata	1,574	1,601	1,236	1,485	788	1,337
Velvet bean–speckled	1,412	1,610	1,417	1,436	452	1,265
Maize	1,536	1,458	1,242	1,235	732	1,241
Sunflower	1,805	1,144	1,216	974	613	1,150
Rice	1,495	1,342	1,202	1,085	560	1,137
Crotalaria spectabilis	1,513	880	1,135	574	281	877
Dolichos lablab	666	259	682	163	62	366
Mean	**1,471**	**1,269**	**1,148**	**1,072**	**557**	**1,103**
LSD (5%) cultivars			= 73.0 kg/ha			
LSD (5%) green manure			= 74.0 kg/ha			
LSD (5%) treatment x cultivars			= 207.0 kg/ha			
CV (%)			= 11.5			
Low soil P						
Velvet bean–black	632	670	559	529	437	565
Crotalaria mucronata	554	524	738	386	152	471
Velvet bean–speckled	579	789	350	640	242	520
Maize	911	987	1,027	970	672	913
Sunflower	1,147	944	1,039	778	514	884
Rice	482	757	487	573	430	546
Crotalaria spectabilis	231	369	315	389	304	322
Dolichos lablab	856	488	903	712	527	697
Mean	**674**	**691**	**677**	**622**	**410**	**615**
LSD (5%) cultivars			= 57.0 kg/ha			
LSD (5%) green manure			= 58.0 kg/ha			
LSD (5%) treatment x cultivars			= 161.0 kg/ha			
CV (%)			= 16.1			

Source: M. Thung (unpublished data).

Table 5. Nutrients (kg/ha) acquired by weeds during 5-month fallow period (rainy season of 1992 to dry season of 1993), EMBRAPA-CNPAF experimental station, Goiania, Brazil

	Soil with low P					Soil with high P				
Weed management	N	P	K	Ca	Mg	N	P	K	Ca	Mg
Undisturbed and incorporated	61.0	7.3	134.0	28.5	39.0	189.0	23.2	225.0	110.0	70.0
Cut and left as mulch	130.0	11.0	219.0	55.0	42.0	132.0	12.0	211.0	55.0	24.0
Cut and removed	49.0	5.3	84.0	22.0	11.0	66.0	4.7	116.0	22.0	19.0
Cut and burned	67.0	6.2	95.0	33.0	13.0	99.0	8.0	160.0	40.0	26.0

Source: Thung & Cabrera (1996).

Table 6. Yield (kg/ha) of five bean cultivars grown in fertile soil as affected by four different cultural practices on soil (oxisol) with high and low P, EMBRAPA-CNPAF experimental station, Goiania, Brazil

	Bean yield (kg/ha)					
Weed management	A 774	APORÉ	POT 51	A 285	IPA 6	Mean
High soil P						
Cut and left as mulch	1,504	1,776	1,912	965	1,693	1,570
Undisturbed and incorporated	1,405	1,651	1,862	514	1,156	1,318
Cut and burned	1,141	1,200	1,764	729	1,414	1,250
Cut and removed	1,111	1,384	1,618	629	1,134	1,175
Mean	**1,290**	**1,503**	**1,789**	**709**	**1,349**	**1,328**
LSD (5%) weed management	= 79.0 kg/ha					
LSD (5%) cultivars	= 91.0 kg/ha					
LSD (5%) weed management x cultivars	= 183.0 kg/ha					
CV (%)	= 9.6%					
Low soil P						
Cut and left as mulch	1,046	1,044	1,474	912	1,554	1,206
Undisturbed and incorporated	1,082	1,202	1,653	824	1,302	1,213
Cut and burned	687	441	668	196	854	569
Cut and removed	754	670	1,339	529	1,118	882
Mean	**892**	**839**	**1,284**	**615**	**1,207**	**968**
LSD (5%) weed management	= 122 kg/ha					
LSD (5%) cultivars	= 95 kg/ha					
LSD (5%) weed management x cultivars	= 189 kg/ha					
CV (%)	= 13%					

Source: Thung & Cabrera (1996).

5.10 Farm Manure

The most common farm manure is chicken manure, which is available in large quantities from big poultry farms. The nutrient content is stable, because the poultry farms use standardized formulas to feed their chickens. Chicken manure does not greatly increase SOM content, but it provides the soil with a better buffering capacity (Flaig et al., 1977). Levels of N (5%) and Ca (8%) in chicken manure are high, whereas P levels are low (1.65%), but mostly available in a slow release form. Thus, it constantly releases some macro- and micronutrients during mineralization, making the application of these nutrients unnecessary. For example, chicken manure used alone on a soil with low P was found to increase bean yield from 2.0 to 3.5 MT/ha (Table 7). When combined with green manure (velvet bean), chicken manure increased yields further, especially in low P soils.

Table 7. Bean yield (kg/ha) as affected by chicken manure and P levels, EMBRAPA- CNPAF experimental station, Goiania, Brazil

Manure	Low P	High P	Mean
First year			
Chicken manure (CM)	3,469	3,314	3,391
Velvet bean (VB)	2,525	2,803	2,664
VB + CM	3,669	3,408	3,539
Control	2,027	2,472	2,249
LSD (5%)	284	201	
CV (%)	11.8	11.8	
Second year			
CM	2,296	2,144	2,220
VB	1,434	1,993	1,713
VB + CM	2,408	2,320	2,364
Control	1,666	1,903	1,784
LSD (5%)	263	86	
CV (%)	15.6	15.6	

Source: M. Thung (unpublished data).

5.11 Developing Cultivars that Tolerate Abiotic Stresses

5.11.1 Tolerance of low soil fertility

Phaseolus species exhibit genotypic variation for BNF (Hadarson et al., 1993; Vargas et al., 1991). Even within the common bean, cultivars show differential responses. For example, Rio Tibagi has a lower BNF capacity than does Carioca (Vargas et al., 1994). The capacity for BNF is heritable and a series of lines have been developed by breeding for improved BNF (Bliss et al., 1989; Graham, 1981; Kipe-Nolt & Giller, 1993; Kipe-Nolt et al., 1993). However, these lines often did not perform any better in field experiments than did existing cultivars.

Park & Buttery (1988a, b) developed mutants that produced abundant nodules, even in the presence of nitrate. Unfortunately, the yields of these "supernods" are markedly lower than those of their parents. In 1992, another mutant, derived from RIZ 30, produced almost no nodules at all. Both mutant beans are more useful to researchers than to farmers. Beans have a naturally low-fixing capacity for BNF, compared with other legumes (Hadarson, 1993).

Genetic control of Al tolerance is complex, and screening for Al tolerance is still done in the field and with nutrient solutions (Braccini et al., 1996). Thung et al. (1987) proposed a field screening method to evaluate Al tolerance in beans by using seed yield as a selection parameter. Progress in selection of genotypes has been slow because of the difficulties caused by field spatial variability. Contrasting lines were difficult to find, because the screened germplasm had a narrow genetic base (Thung, 1990).

"Phosphorus efficiency" means efficiency in acquisition and use of P (Rao et al., 1999). "Efficiency in P acquisition" refers to the plant's ability to acquire greater amounts of P per unit root length. "Efficiency in P use" refers to the plant's ability to produce yield per unit of acquired P from soil. Bean genotypes differ in their ability to acquire P from low P soil and use it efficiently for growth and yield (Araujo et al., 1997; Flor & Thung, 1989; Lindgren et al., 1977; Lynch & Beebe, 1995; Salinas, 1978; Yan et al., 1995a, b).

Selection criteria could be developed based on physiological traits and mechanisms that are significantly affected by P supply in the soil. These include the abilities to: 1) mobilize P within the plant, 2) set pods and mobilize photoassimilates to seeds, 3) minimize P storage in seed (as phytic acid), and 4) modify root architecture to exploit greater soil volume. Unfortunately, these traits are not easy to use in a practical breeding program unless simple and reliable molecular markers are developed.

Oliveira et al. (1987), showed that beans were variable for P efficiency. Genetic control of P efficiency is not well known because it is involved in several aspects of plant metabolism (Rao, 1996).

From narrow-based intraracial populations, Singh et al. (1989b) could not select lines better than the parents for combined tolerance to low soil fertility. However, broad-based interracial populations produced higher-yielding lines for both low- and high-input environments (Singh et al., 1989a).

5.11.2 Drought tolerance

Most bean-growing regions suffer from temporary water stress at various times during the cropping season as a result of uneven rainfall distribution. Whether beans have specific mechanisms for resisting drought, is still not clear. However, the phenology of common bean cultivars is highly diverse, which is a major factor in determining yield potential and adaptation to rainfed environments (White & Izquierdo, 1991). For example, many landraces and cultivars originating in the semiarid Mexican highlands show a plastic response to planting date, so much so that with increasingly later plantings, they become earlier flowering (Acosta Gallegos & White, 1995).

Field research under rainfed conditions indicated that carbon isotope discrimination is not a promising indicator for water-use efficiency or adaptation to drought in the common bean, even though useful in other species (White et al., 1994). Other physiological traits, such as shoot dry weight and leaf N concentration appear to be more promising indicators for water-use efficiency in terms of heritability, high general combining ability effects, and correlations with seed yield across trials (White et al., 1994).

Evaluation of diverse shoot genotypes on selected root genotypes of common bean under soil-water deficits showed variation in shoot genotype, but the effect of shoot genotype on growth and yield was small, compared with the effect of root genotype (White & Castillo, 1992). Drought tolerance screening criteria that have already been developed, such as deep rooting capacity (Sponchiado et al., 1989), root density, and water potential (Guimarães & Zimmermann, 1985) are not easily used for large-scale germplasm evaluation. The identification of shoot trait(s) that reflect rooting ability and adaptation to drought will minimize labor-intensive root measurements in a breeding program.

Even though selection criteria for resistance to drought have been identified (Guimarães, 1988; White & Singh, 1991), breeding common bean for adaptation to drought has progressed slowly. Field screening of bean germplasm for adaptation to drought resulted in the identification of several lines, for example, BAT 477, A 195, and San Cristobal 83 (White, 1988). The superior adaptation of BAT 477 to drought was attributed to dehydration postponement through greater root length density and deeper soil-water extraction (Sponchiado et al., 1989). In drought-stricken northeast Brazil, BAT 477 can persist very well. This line combines drought tolerance with resistance to ashy stem blight [caused by *Macrophomina phaseolina* (Tassi) Goid.], which is common in drought areas in Brazil. A more recent line, SEA 5, proved better adapted to drought than BAT 477, mostly because it can partition more assimilates to grain production (I. M. Rao, unpublished results).

Singh (1995) reported an increase of bean yield under drought after he made interracial and intergene pool crosses involving high-yielding and drought-tolerant progenitors derived from varied origins, such as those found in the Mexican highlands. New sources for drought tolerance were found in the bean races Mesoamerica, Jalisco, and Durango (H. Terán & S. P. Singh, unpublished results). Further progress would be expected from their utilization in breeding programs. Moreover, identification and use of DNA-based markers for seed yield and other traits associated with drought tolerance should improve breeding efficiency (Schneider et al., 1997).

5.11.3 Cold tolerance

Zaiter et al. (1993) reported that sufficient genetic variability exists for breeding cold tolerance in beans. Otubo et al. (1996) found that cold tolerant cultivars could be selected under laboratory testing at 12°C in the F_2 populations involving appropriate tolerant parents. Genchev (1988) selected bean lines for cold tolerance at the seedling stage.

5.11.4 Heat tolerance

Bean cultivar IPA 7 (HF 465-63-2) not only tolerates high temperatures but also has moderate resistance to *Fusarium* (Miranda et al., 1997). Shonnard & Gepts (1994) could not identify a simple gene control of heat tolerance at specific development stages, but they did find a significant additive effect, indicating that heat tolerance in beans can be improved.

5.11.5 Salinity tolerance

Beans exhibit genetic variability for tolerance of soil salinity. Zaiter & Mahfouz (1993) showed that, in sand culture systems, the cultivar Badrieh is tolerant at 4 dS m^{-1} salinity, whereas T No. 1 is susceptible. These authors suggested that foliar injury symptoms could be used as selection criteria for salinity tolerance.

6. INTEGRATED MANAGEMENT OF ABIOTIC STRESSES

The basic principles of integrated management of abiotic stresses for sustainable crop production are to limit stresses of edaphic and climatic origin, optimize crop yield, and improve farmers' profits. Fageria & Baligar (1997) extensively reviewed integrated nutrient management and recommended several strategies which need to be put into practice. Five basic principles of soil management are essential for practicing sustainable agriculture (Greenland, 1975). First, chemical nutrients removed by crops must be replenished; second, the physical condition of the soil must be maintained; third, there must be no build-up of pests; fourth, there must be no increase in soil acidity or toxic elements; and fifth, soil erosion must be controlled to be equal to or less than the rate of soil genesis.

To physically improve degraded soils, several steps may be required. For example, terracing can mitigate soil erosion, then subsoiling will reduce soil compaction, and the use of appropriate cultural practices will improve soil structure. Soil chemical disorders can be corrected by fertilizer application to overcome mineral deficiencies, and by liming to increase soil pH and reduce Al and Mn toxicities. Furthermore, judicious use of organic matter and green manure, BNF, and AM inoculation would improve input-use efficiency. Mounting evidence shows that the combined use of inorganic and organic nutrients not only directly remedies limiting factors but also produces beneficial interactions that increase yields above those obtainable with equal quantities of nutrients from a single source. To alleviate climatic or hydrological stresses, practices such as irrigation for drought, drainage for waterlogging, and regulation of high water tables can be employed.

In intensive bean-production systems, soil monitoring must be carried out from the beginning to study the impact of intensive use of agricultural implements over time on soil sustainability. The efficient use of chemical fertilizers must also be monitored. In many intensive agricultural systems, after a certain time, the dosage of chemical fertilizers used will usually increase to maintain productivity (Thung et al., 1998). This is neither economically nor ecologically sound, because excessive application of nutrients may pollute the environment, especially through water runoff. Soil erosion and water quality should be monitored at regular intervals to protect the environment.

The increase in bean yields needed for the next decades will require increasingly science-based management, which, in turn, will require better interaction among farmers, development officers, researchers, and policy makers. To improve the effectiveness and efficiency of inputs, the bean farmer of the 21st century will need to:

- Cooperate with researchers, and county agents/extension workers/crop consultants to select those bean cultivars and crop rotations that match closely with edaphic and climatic stress factors and yield targets;
- Maximize bean yields per liter of water used by growing less thirsty bean cultivars;
- Optimize applications of inorganic and organic fertilizers; and
- Protect the soil from wind and water erosion, waterlogging, salinization, and landslides.

Precision farming is an innovative system that is increasingly being adopted in North America and Europe to achieve the goals just described. It involves the design of detailed cropland maps based on a combination of grid soil sampling, aerial photography, and "on-the-go" yield determinations. Using global positioning systems (GPS), site-specific maps can be drawn to correlate soil characteristics (e.g., fertility, topography, drainage, and wild populations) to crop yield.

Although detailed sampling of soil, generally known as grid sampling, is required, grid plant sampling has not been required in precision farming. Yet, plant analyses can contribute to the objectives of precision farming. For example, infrared photography can be used to observe plant canopy characteristics that are related to particular abiotic stress conditions, although further research is needed to make infrared images meaningful in terms of nutrient element status of the photographed crop. When changes in canopy characteristics are noted, assaying plants in specific areas can help diagnose mineral disorders. Plant analyses therefore can help identify and develop correction measures for specific sites. Thus, biotic and abiotic constraints in the field are pinpointed, and corrective inputs used in precisely those areas where they are needed.

To increase biophysical sustainability, that is, to reduce soil erosion and improve the physical, chemical, and biological properties of soil, farmers should adopt minimal tillage, no-tillage, and direct seeding systems. Reduced tillage minimizes soil compaction through reduced traffic of farm machinery used for plowing and disking. It also reduces fuel costs. Agribusiness is now using precision farming to increase the input-use efficiency.

Gentil (1995) estimated that if all grain production in Brazil would adopt the no-tillage system, Brazil would save on 58% of farm equipment and 78% of combustion fuel costs. Total farm operating costs would be only 47% of conventional farming. Many factors still have to be addressed, especially in tropical areas with high temperatures and humidity. Irrigated, untilled bean crops tend to be shallow rooted, seldom penetrating more than the top 20 cm of soil (Oliveira & Silva, 1990; Pires et al., 1991; Stone & Pereira, 1994). This suggests that roots remain where soil water is available, that is, near the surface. Moreover, because of high soil water content, direct planting increases the incidence of soilborne diseases (Costa & Silveira, 1997; Tu & Tan, 1991). In rainfed conditions, bean roots often penetrate to as deep as 100 cm (Guimarães et al., 1996).

Balbino et al. (1996) suggested that the minimum requirements for a successful direct planting production system in the tropics should include an SOM content at >2%; soil pH above 5.5; and the availability of P, K, Ca, and Mg in soil at more than 20 ppm, 50 ppm, 2.0 meq/100 g soil, and 0.8 meq/100 g soil, respectively. No tillage or direct seeding is now an important component of intensive bean production systems.

Figure 1 provides an overview of the proposed strategies for integrated abiotic stress management for obtaining high yields of beans while sustaining and improving natural resources. Some suggestions on integrated crop management practices are discussed below.

6.1 Land Preparation

During the last 15 years, direct planting and zero or minimum tillage have become common practices in North American and European agriculture to reduce soil erosion and conserve water resources. In the U.S.A. alone, the area under conservation tillage was about 17.3 million hectares, while in Argentina it was 3 million hectares (Steiner et al., 1997). These strategies are now being introduced to other intensive bean production systems in Latin America. Special farm equipment has had to be designed to replace commonly used farm equipment that cannot drill seed through the thick layer of organic matter that develops on the surface of tropical soils. This conservation method depends heavily on herbicides to kill the vegetation and leave it on the soil surface as mulch. In the subtropics, direct planting and minimum tillage increase SOM content (Kern & Johnson, 1993; Sombroek et al., 1993).

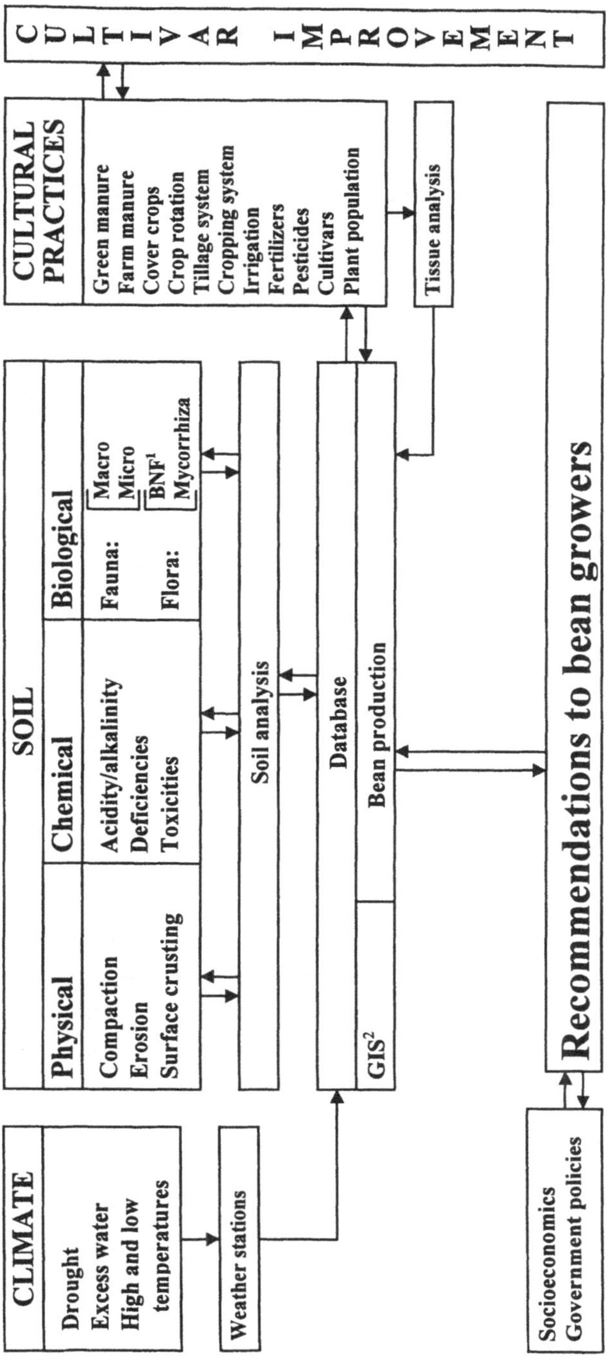

Figure 1. Integrated management of abiotic stresses for obtaining high bean yields while conserving natural resources. [1]BNF = Biological nitrogen fixation. [2]GIS = Geographical information system.

Direct planting or no-tillage practices can produce a bean yield of 2.3 MT/ha with 382 mm of water across the whole growth cycle. In contrast, if the land was prepared conventionally with a moldboard plow, about 535 mm was needed to obtain the same yield (Stone & Moreira, 1995).

To respond to recent worldwide demand for high quality beans, bean farmers in temperate zones are drilling seed directly into residues of the previous crop. Beans must be kept as clean as possible, and should not come into contact with soil particles.

6.2 Choice of Cultivars and Seed Quality

Cultivars are selected according to at least three major factors: first, they must be well adapted to the soil and climatic conditions of the region; second, they must have a potential for high yields and resistance to prevalent diseases, insects, and abiotic stresses; and third, they must respond to strategic application of inputs and be efficient in their use.

High seed quality is necessary to guarantee optimum germination, early vigor, and ensure adequate plant population. Poor quality seed produces defective seedlings (e.g., baldheaded seedlings), with broken cotyledons and poor vigor. Poor seeds are also more vulnerable to soilborne diseases and other stresses. High quality pathogen-free certified seed of recommended cultivars from reputable sources, which often represents less than 15% of total production costs, must be used in the integrated bean production systems.

6.3 Plant Population

An optimal plant population is essential for maximizing production. Planting skips and uneven stands in rows are the most common cause of low yield. Inconsistent tractor speed during sowing creates bouncing and shaking, thus disturbing the seed flow between seed bin and opening disks. The depth of the opening disk is also disturbed and seed is planted at uneven depths, causing poor germination and irregular emergence of seedlings.

Farmers usually adopt a spacing of 0.4 to 0.5 m between rows for most bean types, whether cultivars of indeterminate or determinate growth habit. Smith et al. (1992) tested the influence of spatial arrangement on crop growth and yield, using different row-to-row spacings (16, 36, 56, and 76 cm), and plant densities (100,000; 200,000; 300,000; and 400,000 plants/ha) with beans of three different growth habits (types I, II, and III). No significant differences in yield, seed size, height of first pod insertion, number of branches, and diameter of stem were found among the different spacing and plant density treatments. Only growth habit and year caused significant differences in yield. The tendency for yield reduction at higher plant populations was observed with indeterminate types II and III, while the opposite was true for determinate type I (Nienhuis & Singh, 1985). But the low population density was neither good for selection nor for testing yield potential of bean cultivars (Singh & Gutiérrez, 1990). Fronza et al. (1994) obtained the highest bean yields with 33 cm between rows and at 250,000 plants/ha. Stewart-Williams & Myers (1992), in contrast, reported that seed yield decreased when population density was increased within the 56-cm row-to-row spacing.

6.4 Cultural Practices

High seed yields can be achieved by implementing several integrated cultural practices such as use of adequate fertilization, appropriate disease-and-pest control, careful water management, and weed control. Good economic returns can be expected from mechanization, which is key to bean production expansion in Brazil (Aidar, 1998). Crop rotation with legumes, or with winter cover crops in southern latitudes, is preferable. In drought-stricken areas, beans should be intercropped with maize, because maize creates a favorable microclimate that shades the bean and protects the soil from direct exposure to sunlight (Vieira, 1998). Small farmers in Latin America frequently intercrop beans with maize, a system which is believed to have many beneficial effects such as control of soil erosion and water conservation, among others.

6.5 Soil Fertility Management

Realistic yield goals must be set for each bean production system. Band and side-dress applications of fertilizers for maximum efficiency are already common practices. Still, available manures and wastes are not being exploited in the tropics because of the difficulties in obtaining and transporting large quantities at certain times of the year. Neither is the organization or infrastructure adequate for the use of these amendments. Nitrogen enrichment through planting legumes must be encouraged. Farmers in developing countries should be warned of the risk of polluting ground water through excessive use of chemical N fertilizers, especially in irrigated systems, as is already occurring in developed countries.

6.6 Irrigation and Water Conservation

Efforts are under way to develop cultivar coefficients (kc) of beans in different growth stages (Bergamaschi et al., 1988a, b, c). The bean plant consumes water at different rates, according to plant age, soil type, water availability, and kc. The value of kc is calculated from actual evapotranspiration (Eta), divided by potential evapotranspiration. The kc is about 0.6 between germination and preflowering, increases to 1.1 to 1.4 during the peak demand for water at flowering, and slowly decreases to the initial value of 0.6 during pod filling and at maturity.

The kc is closely correlated with leaf area index (LAI) for bean cultivar Aroana 80 (Bergamaschi et al., 1988b). In the field, when water supply was kept constant at field capacity (>0.05 MPa) from planting to maturity, the average rate of Eta was 3.8 mm/day, out of a total Eta of 360 mm for the cropping season. Bergamaschi et al. (1989) measured the values of maximum evapotranspiration (Etm) at different growth stages, finding only small variations, which were probably explained by the different cultivars planted in different seasons and the final yields obtained. Overall, the Etm value was around 3.5 from germination to flowering (about 45 days), 5.5 during flowering (21 days), and 4.8 from pod formation to maturity (35 days). When water deficits were imposed during the end of pod filling, crop duration was not affected. If water stress was imposed immediately after pod setting, then early maturity resulted. Water stress imposed between initial flowering and full bloom caused a significant delay in maturity (Bergamaschi et al., 1988a). The authors also showed that LAI, soil-water availability, atmospheric evaporative demand, and Eta of beans depended largely on atmospheric conditions (Bergamaschi et al., 1988b).

For the bean cultivar IPA 74-19, Eta is at an average of 3.9 mm/day and at peak demand for water—between late flowering and pod formation—it is 5.3 mm/day (Oliveira & Silva, 1990). The authors also found that 83% of roots concentrated in the top 40 cm of the soil.

Canopy temperatures were higher with lower irrigation frequency and correlated closely with yield components and total yield. This suggests that infrared thermometry may be useful for scheduling irrigation (Pazzetti et al., 1993). But Bergamaschi et al. (1989) observed that on days with high atmospheric evaporative demand, the temperature of the canopy surface measured by infrared remote sensors was higher than the air temperature close to the plant, particularly in plants that were less developed and suffering from water stress.

7. CONCLUSIONS

The modern concepts of soil management were first developed by trial and error; then, to reduce production costs, specific soil treatments such as amendment with lime were developed. Initially, lime applications to acid soils were based on soil pH values. Later, the concept of Al saturation was developed to improve the economic efficiency of lime applications. Management of abiotic stress factors is most efficient when all factors, for example, liming, correct doses of fertilization, adequate land preparation, selection of appropriate cultivars, and suitable cultural practices, are implemented simultaneously during the whole growing season, and when no temporary water deficits occur. In addition, stable government policies and economically viable bean prices for the growers are needed to promote integrated soil and crop management strategies for specific agroecosystems.

At least two major problems need to be solved if integrated management of abiotic stresses is to result in sustainable bean production. One is the management of water resources. Global problems such as deforestation and climatic change have important implications for water quality, quantity, and distribution, salinization, and the life or death of river sources or springs. The other major problem is the adequate management of native and applied nutrients. While fertilizers are applied primarily to correct nutrient deficiency, the socioeconomic implications of using them are seldom recognized.

Farmers' dependency on external inputs can be controlled through on-farm decisions and changes in soil and crop management strategies. Efficient use of chemicals with proper tillage and residue management is key to maintaining the productivity of agricultural land, and to reducing the pressure to look for new lands (e.g., through deforestation). Developing bean genotypes that are less dependent on major inputs such as water, nutrients, and pesticides is another essential strategy for the integrated management of abiotic stresses.

Improving productivity through sustainable agriculture is difficult, because investments and efforts must be over the long term and results are not seen immediately. Low-income farmers, who can scarcely make a living, do not have the capital to invest in sustainable agriculture. They need government support, in terms of long-term commitment with technical and financial backing. Even so, benefits may be seen only after several years, and only at national, regional, or global levels, and are therefore not economically attractive to the individual farmer.

The current trend of conducting only short-term research projects hampers the development of long-term research. This long-term view is needed to establish an in-depth understanding of abiotic stresses and the causes of soil degradation and to develop cultural practices that effectively use modern cultivars and purchased inputs to lead to sustainable bean production. The main challenge to policy makers, research managers, researchers, bean growers, and county agents is to obtain higher bean yields while reducing production costs and conserving natural resources. Integrated management of abiotic and biotic stresses, combined with improved cultivars, can help achieve this goal.

REFERENCES

Acosta Gallegos, J.A. & J.W. White. 1995. Phenological plasticity as an adaptation by common bean to rainfed environments. Crop Sci. 35:199-204.

Aidar, H. 1998. Produção de feijão com alta tecnologia. pp. 84-97. *In* O. Voysest (ed.), An Ecoregional Framework for Bean Germplasm Development and Natural Resources Research. CIAT, Cali, Colombia.

Amane, M.I.V., C. Vieira & G.A.A. Araujo. 1994. Resposta de cultivares de feijão (*Phaseolus vulgaris* L.) às adubações nitrogenada e molíbdica. Rev. Ceres 41:202-216.

Amberger, A. 1990. Harnstoffumsetzung im Boden und NH3-Verluste. VDLUFA-Schriftreihe 32:243-248.

Andrew, C.S. 1977. Nutrition restraints on legumes symbiosis. pp. 261-263. *In* Proc. Workshop on Exploiting the Legume-Rhizobium Symbiosis in Tropical Agriculture. Univ. of Hawaii, Honolulu, U.S.A.

Araujo, A.P., M.G. Texeira & D.L. De Almeida. 1997. Phosphorus efficiency of wild and cultivated genotypes of common bean (*Phaseolus vulgaris* L.) under biological nitrogen fixation. Soil Biol. Biochem. 29:951-957.

Astorini, G., I. Margani, P. Tripodo & P. Manes. 1995. The response of *Phaseolus vulgaris* L. cv. Lit to different dosages of the anti-ozonant ethylenediurea (Edu) in relation to chronic treatment with ozone. Plant Sci. 111:237-248.

Ayers, R.S. & D.W. Westcot. 1976. Water quality for agriculture. Irrigation and Drainage Paper no. 29. FAO, Rome, Italy.

Balbino, L.C., J.A.A. Moreira, J.G. da Silva, E.F. de Oliveira & I.P. de Oliveira. 1996. Plantio direto. pp. 301-352. *In* R.S. Araujo, C.A. Rava, L.F. Stone & M.J. de O. Zimmermann (eds.), Cultura de feijoeiro comum no Brasil. Associação Brasileira para Pesquisa de Potássio e do Fosfato, Piracicaba, Brazil.

Barber, S.A. 1995. Rhizosphere microorganisms, mycorrhizae and root hairs. pp. 157-179. *In* Soil Nutrient Bioavailability: A Mechanistic Approach. John Wiley & Sons, New York, U.S.A.

Barros, L.C.G. & R.J. Hank. 1993. Evapotranspiration and yield of bean as affected by mulch and irrigation. Agron. J. 85:692-697.

Bateman, D.F. 1965. Discussion of the soil environment. pp. 139. *In* K.F. Baker and W.C. Snyder (eds.), Ecology of Soil-Borne Plant Pathogens, Prelude to Biological Control. Univ. of California Press, Berkeley, U.S.A.

Bateman, D.F. & R.D. Lumsden. 1965. Relation of calcium content and pectic substances in bean hypocotyls of different ages to susceptibility of *Rhizoctonia solani.* Phytopathology 55:734-738.

Bateman, D.F. & R.L. Miller. 1966. Pectic enzymes in tissue degradation. Annu. Rev. Phytopathol. 4: 119.

Beaufils, E.R. 1971. Phyisological diagnosis. A guide for improving maize production based on principles developed for rubber trees. Fert. Soil S. Afr. J. 1:1-28.

Bender, J., H.J. Weigel & H.J. Jager. 1990. Regression analysis to describe yield and metabolic responses of beans (*Phaseolus vulgaris*) to chronic ozone stress. Angew. Bot. 64:329-343.

Bergamaschi, H., J.C. Ometto, H.J. Vieira, L.R. Angelocci & P.L. Libardi. 1988b. Deficiência hídrica em feijoeiro. II. Balanço de energia. Pesqui. Agropecu. Bras. 23:745–757.

Bergamaschi, H., J.C. Ometto, H.J. Vieira, L.R. Angelocci & P.L. Libardi. 1988c. Deficiência hídrica em feijoeiro. III. Evapotranspiração máxima e relações com evapotranspiração calculada pelo método de Penman e com a evaporação do tanque "classe A". Pesqui. Agropecu. Bras. 24:387–392.

Bergamaschi, H., H.J. Vieira, J.C. Ometto, L.R. Angelocci & P.L. Libardi. 1988a. Deficiência hídrica em feijoeiro. I. Análise de crescimento e fenologia. Pesqui. Agropecu. Bras. 23:733–743.

Bergamaschi, H., H.J. Vieira, J.C. Ometto, L.R. Angelocci & P.L. Libardi. 1989. Deficiência hídrica em feijoeiro. IV. Alterações micrometeorológicas. Pesqui. Agropecu. Bras. 24:745–757.

Berger, P.G., C. Vieira & G.A.A. Araujo. 1996. Efeitos de doses e épocas de aplicação do molibdênio sobre a cultura do feijoeiro. Pesqui. Agropecu. Bras. 31:473-480.

Berger, P.G., C. Vieira, G.A.A. Araujo & S.T.V. Cassini. 1995. Peletização de sementes de feijão (*Phaseolus vulgaris* L.) com carbonato de calcio, Rhizobio e molibdênio. Rev. Ceres 42:562-574.

Bliss, F.A. 1993. Breeding common bean for improved biological nitrogen fixation. Plant Soil 152:71-79.

Bliss, F.A., P.A.A. Pereira, R.S. Araujo, R.A. Henson, K.A. Kmiecik, J.R. McFerson, M.G. Teixeira & C.C. Silva. 1989. Registration of five high nitrogen-fixing common bean germplasm lines. Crop Sci. 29:240-241.

Borlaug, N.E. & C.R. Dowswell. 1997. The acid lands: one of agriculture's last frontiers. pp. 5-15. *In* A.C. Moniz, A.M.C. Furlani, R.E. Schaffert, N.K. Fageria, C.A. Roselam & H. Canterella (eds.), Plant-Soil Interactions at Low pH: Sustainable Agriculture and Forestry Production. Braz. Soil Sci. Soc., Campinas, Brazil.

Bornman, J.F. & A.H. Teramura. 1993. Effects of ultraviolet-B radiation on terrestrial plants. pp. 427-471. *In* A.R.A. Young, L.O. Bjorn, J. Moan & W. Nultsch (eds.), Environmental UV Photobiology. Plenum Press, New York, U.S.A.

Bottenberg, H., J. Masiunas, C. Eastman & D.M. Eastburn. 1997. The impact of rye cover crops on weeds, insects and diseases in snap bean cropping system. J. Sustainable Agric. 9:131-135.

Braccini, M. do Cl., A. de Lucca e Braccini, H.E.P. Martínez, P.R.G. Pereira & P.C.R. Fontes. 1996. Tecnicas de avaliação da toxidez do alumínio em plântulas de feijoeiro (*Phaseolus vulgaris* L.) cultivadas em solução nutritiva. Rev. Ceres 42:3-16.

Caldwell, M.M., A.H. Teramura, M. Tevini, J.F. Bornman, L.O. Bjorn & G. Kulandaivelu. 1995. Effects of increased solar ultraviolet radiation on terrestrial plants. Ambio 24:166-173.

Canalli, L.B. & R. Roloff. 1997. Influência do preparo e correção do solo na condição hídrica de um latossolo vermelho escuro sob plantio direto. Rev. Bras. Ciênc. Solo 21:99-104.

Carbonell-Barrachina, A.A., F.M. Burlo-Carbonell & J.J. Mataix-Beneyto. 1996. Arsenic uptake, distribution and accumulation in bean plants: human health risk. Fresenius Environ. Bull. 5:1-2, 21-28.

Carpenedo, V. & J. Mielniczuk. 1990. Estudo de agregação e qualidade de agregados de latossolos roxos, submetidos a diferentes sistemas de manejo. Rev. Bras. Ciênc. Solo 4:99-105.

Christenson, D.R, C.E. Briker & R.S. Gallagher. 1991. Crop yield as affected by cropping system and rotation. Mich. Agric. Exp. Stn. Res. Rpt. 516:1-10.

CIAT (Centro Internacional de Agricultura Tropical). 1992. Constraints to and opportunities for improving bean production. A planning document 1993-98 and an achieving document 1987-92. CIAT, Cali, Colombia.

Costa, J.L. da S. & P.M. Silveira. 1997. Influência dos métodos de preparo do solo e rotação de cultura no ocorrência de prodridões radiculares do feijoeiro. Fitopatol. Bras. 22(Supp.):258 (abstract).

Davey, C.B. & G.C. Papavizas. 1960. Effect of dry mature plant materials and nitrogen on *Rhizoctonia solani* in soil. Phytopathology 50:522 (abstract).

Debouck, D.G. 1976. Etude d'une relation entre nutrition minerale equilibrée et morphogenese chez (*Phaseolus vulgaris* L.). Ph.D. dissertation. Faculté des Sciences Agronomiques de l'Stat, Gembloux, Belgium.

Drobny, H.G., G.M. Hoffmann & A. Amberger. 1984. Effect of mineral nutrition on the susceptibility of *Phaseolus vulgaris* to *Colletotrichum lindemuthianum* (Sacc. et Magn.) Br. et Cav. I. Effects of different mineral nutrients and leaf-age on the disease severity. Z. Pflanzenernaehr. Bodenkd. 147:242 (abstract).

Drobny, H.G., G.M. Hoffmann & A. Amberger. 1985. Effect of mineral nutrition on the susceptibility of *Phaseolus vulgaris* to *Colletotrichum lindemuthianum* (Sacc. et Magn.) Br. et Cav. II. Organic and inorganic components in relation to susceptibility with varying N-, K-, and Ca-supply. Z. Pflanzenernaehr. Bodenkd. 148:33 (abstract).

Fageria, N.K. & V.C. Baligar. 1997. Integrated plant nutrient management for sustainable crop production-an overview. Int. J. Trop. Agric. 15(4):1-19.

Flaig, W., B. Nagar, H. Sochtig & C. Tietjen. 1977. Organic materials and soil productivity. Soil Bulletin, 35. FAO, Rome, Italy.

Flor, C.A. & M.T. Thung. 1989. Nutritional disorders. *In* H.F. Schwartz and M.A. Pastor-Corrales (eds.), Bean Production Problems in the Tropics. CIAT, Cali, Colombia.

Fronza, V., C. Vieira, A.A. Cardoso & C.D. Cruz. 1994. Resposta de cultivares eretos de feijão (*Phaseolus vulgaris* L.) a espaçamento entre linhas e níveis de adubação. Rev. Ceres 41:317-326.

Genchev, D. 1988. Screening for cold tolerance in dry bean (planting-emergence). Annu. Rpt. Bean Improv. Coop. 31:210-212.

Gentil, L.V. 1995. Aspectos economicos ao plantio direto. pp. 9-12. *In* Primer Seminario Internacional do Sistema de Plantio Direto. EMBRAPA-CNPT, Passo Fundo, Brazil.

Gerdemann, J.W. 1975. Vesicular-arbuscular mycorrhizae. pp. 575-591. *In* J.G. Torrey & D.T. Clarkson (eds.), The Development and Function of Roots. Academic Press, New York, U.S.A.

Giller, K.E. & G. Cadisch. 1995. Future benefits from biological nitrogen fixation: an ecological approach to agriculture. Plant Soil 174:255-277.

Graham, P.H. 1981. Some problems of nodulation and symbiotic nitrogen fixation in *Phaseolus vulgaris* L.: a review. Field Crops Res. 4:93-112.

Graham, P.H. & P. Ranalli. 1997. Common bean (*Phaseolus vulgaris* L.). Field Crops Res. 53:131-146.

Greenland, D.J. 1975. Bringing the green revolution to the shifting cultivator. Science (Washington DC) 190:841-844.

Guerra, D. & A.J. Anderson. 1985. The effect of iron and boron amendments on infection of bean. Phytopathology 75:989 (abstract).

Guimarães, C.M. 1988. Efeitos fisiologicos do estresse hídrico. pp. 157-174. *In* M.J.O. Zimmermann & M. Rocha Yamada (eds.), Cultura do Feijoeiro. Associação Brasileira para Pesquisa de Potássio e do Fosfato, Piracicaba, Brazil.

Guimarães, C.M., O. Brunini & L.F. Stone. 1996. Adaptação do feijoeiro (*Phaseolus vulgaris* L.) à seca. 1. Densidade e eficiência radicular. Pesqui. Agropecu. Bras. 31:393-399.

Guimarães, C.M. & M.J. Zimmermann. 1985. Deficiência hídrica em feijão. pp. 15-28. *In* Reunión de trabajo sobre mejoramiento de frijol en Brasil com enfasis en tolerancia a sequia. CIAT, Cali, Colombia.

Hadarson, G. 1993. Methods for enhancing symbiotic nitrogen fixation. Plant Soil 152:1-17.

Hadarson, G., F.A. Bliss, M.R. Cigales-Rivero, R.A. Henson, J.A. Kipe-Nolt, L. Longeri, A. Manrique, J.J. Peña-Cabriales, P.A.A. Pereira, C.A. Sanabria & S.M. Tsai. 1993. Genotypic variation in biological nitrogen fixation by common bean. Plant Soil 152:19-23.

Haterlein, J., C.D. Clayberg & L.D. Teare. 1980. Influence of high temperature on pollen grain viability and pollen tube growth in the styles of *Phaseolus vulgaris* L. J. Amer. Soc. Hort. Sci. 105:12-14.

Haynes, R.J. 1982. Effect of liming on phosphate availability in acid soils: a critical review. Plant Soil 68:289-308.

Hedges, F. & H. Fisher. 1946. Association of *Xanthomonas phaseoli* and the common bean mosaic virus, *Marmor phaseoli*. III. The effect of varying amounts of nitrogen on pathogenicity. Phytopathology 36:613.

Hoogenboom, G. & D.S. NeSmith. 1992. Temperature sensitivity of common bean during germination. Annu. Rpt. Bean Improv. Coop. 35:72-73.

Howeler, R.H. 1980. Nutritional disorders. pp. 341-362. *In* H.F. Schwartz & G.E. Gálvez (eds.), Bean Production Problems: Disease, Insect, Soil and Climatic Constraints of *Phaseolus vulgaris*. CIAT, Cali, Colombia.

Huber, D.M. & R.P. Watson. 1970. Effect of organic amendment on soil-borne plant pathogens. Phytopathology 60:22 (abstract).

Kern, J.S. & M.G. Johnson. 1993. Conservation tillage impacts on national soil and atmospheric carbon levels. Soil Sci. Soc. Amer. J. 57:200-210.

Kipe-Nolt, J.A. & K.E. Giller 1993. A field evaluation using the ^{15}N isotope dilution method of lines of *Phaseolus vulgaris* L. bred for increased nitrogen fixation. Plant Soil 152:107-114.

Kipe-Nolt, J.A., H. Vargas & K.E. Giller. 1993. Nitrogen fixation in breeding lines of *Phaseolus vulgaris* L. Plant Soil 152:103-106.

Konsens, I., M. Ofir & J. Kigel. 1991. The effect of temperature on the production and abscission of flowers and pods in snap bean (*Phaseolus vulgaris* L.). Ann. Bot. 67:391-399.

Linderman, R.G. 1992. Vesicular-arbuscular mycorrhizae and soil microbial interactions. pp. 45-70. *In* G.J. Bethlenfalvay & R.G. Linderman (eds.), Mycorrhizae in Sustainable Agriculture. Special Publication, no. 54. Amer. Soc. Agron., Madison, U.S.A.

Lindgren, D.T., W.H. Gabelman & G.C. Gerloff. 1977. Variability of phosphorus uptake and translocation in *Phaseolus vulgaris* L. under phosphorus stress. J. Amer. Soc. Hort. Sci. 102:674-677.

Lopes, A.S. & F.R. Cox. 1977. A survey of the soil fertility status of surface soils under "Cerrado" vegetation in Brazil. Soil Sci. Soc. Amer. J. 41:742-747.

Lopes, E.S. & J.O. Siqueira. 1981. Vesicular-arbuscular mycorrhizas: their potential in phosphate nutrition in tropical regions. pp. 225-242. *In* R.S. Russel, K. Igue & Y.R. Mehta (eds.), The Soil-Root System in Relation to Brazilian Agriculture. Instituto Agronômico do Paraná, Londrina, Brazil.

Lynch, J.P. & S.E. Beebe. 1995. Adaptation of beans (*Phaseolus vulgaris* L.) to low phosphorus availability. HortScience 30:1165-1171.

Maas, E.V. & G.J. Hoffman. 1977. Crop salt tolerance-current assessment. J. Irrig. Drainage Div. Amer. Soc. Civil Eng. 103:115-134.

Magalhães, A.A. & A.A. Millar. 1978. Efeito de déficit de água no período reprodutivo sobre a produção do feijoeiro. Pesqui. Agropecu. Bras. 13(2):55-60.

Maier, C.R. 1968. Influence of nitrogen nutrition on *Fusarium* root rot of pinto bean and on its suppression by barley straw. Phytopathology 58:620.

Masaya, P. & J.W. White. 1991. Adaptation to photoperiod and temperature. pp. 445-500. *In* A. van Schoonhoven & O. Voysest (eds.), Common Beans: Research for Crop Improvement. C.A.B. Int., Wallingford, U.K. & CIAT, Cali, Colombia.

Mehlhorn, H. & A.A. Wenzel. 1995. Manganese deficiency enhances ozone toxicity in bush beans (*Phaseolus vulgaris* L. cv. Saxa). J. Plant Physiol. 148:155-159.

Miranda, P., C.J. da Anunciação Filho, D.G da Cruz & V.F. dos Santos. 1997. Tolerância de cultivares de feijão comum (*Phaseolus vulgaris* L.) a alta temperatura. Pesqui. Agropecu. Pernambucana 8:73-80.

Miyasaka, S.C., J.G. Buta, R.K. Howell & C.D. Foy. 1991. Mechanism of aluminum tolerance in snapbeans. Root exudation of citric acid. Plant Physiol. 96:737-743.

Miyazawa, M., M.A. Oavan & A. Calegari. 1993. Efeito de material vegetal na acidez do solo. Rev. Bras. Ciênc. Solo 17:411-416.

Mosse, B. & J.P. Thompson. 1984. Vesicular-arbuscular endomycorrhizal inoculum production. I. Exploratory experiments with beans (*Phaseolus vulgaris*) in nutrient flow culture. Can. J. Bot. 62:1523-1530.

Müller, S., P.A.A. Pereira & P. Martin. 1993. Effect of different levels of mineral nitrogen on nodulation and N_2 fixation of two cultivars of common bean (*Phaseolus vulgaris* L.). Plant Soil 152:139-143.

Nasser, L.C.B. & J.C. Sutton. 1993. Palhada de arroz pode controlar importante doença de feijoeiro irrigado. Cerrado Pesqui. Tecnol. 3:6 (abstract).

Nienhuis, J. & S.P. Singh. 1985. Effects of location and plant density on yield and architectural traits in dry beans. Crop Sci. 25:579-584.

Oliveira, F.A. & J.J.S. Silva. 1990. Evapotranspiração, índice área foliar e desenvolvimento radicular do feijão irrigado. Pesqui. Agropecu. Bras. 25:317-322.

Oliveira, I.P. de, M.D.T. Thung, J. Kluthcouski, H. Aidar & J.R.P. Carvalho. 1987. Avaliação de cultivares de feijão quanto à eficiência no uso de fósforo. Pesqui. Agropecu. Bras. 22:39-45.

Otubo, S.T., M.A.P. Ramalho, A.F.B. Abreu, J.B. dos Santos, B. Griffing, J.L. Jinks & H.S. Pooni. 1996. Genetic control of low temperature tolerance in germination of the common bean (*Phaseolus vulgaris* L.). Euphytica 89:313-317.

Papenhuijzen, C. & M.H. Roos. 1979. Some changes in the subcellular structure of root cells of *Phaseolus vulgaris* as a result of cessation of aeration in the root medium. Acta Bot. Neerl. 28:491-495.

Park, S.J. & B.R. Buttery. 1988a. Inheritance of nitrate tolerant nodulation in EMS-induced mutants of common bean (*Phaseolus vulgaris* L.). J. Hered. 80:135-138.

Park, S.J. & B.R. Buttery. 1988b. Nodulation mutants of white bean (*Phaseolus vulgaris* L.) induced by ethyl methane sulphonate. Can. J. Plant Sci. 68:199-202.

Pazzetti, G.A., M.A. Oliva & N.F. Lopes. 1993. Aplicação da termometria ao infravermelho à irrigação do feijoeiro: Crescimento e produtividade. Pesqui. Agropecu. Bras. 28:1371-1377.

Pires, R.C. de M., F.B. Arruda & M. Fujiwara. 1991. Profundidade do sistema radicular dos cultivos de feijão e trigo sob pivô central. Bragantia 50:151-162.

Plenchette, C., J.A. Fortin & V. Furlan. 1983. Growth responses of several plant species to mycorrhizae in a soil of moderate P-fertility. I. Mycorrhizal dependency under field conditions. Plant Soil 70:199-209.

Rao, I.M. 1996. Role of phosphorus in photosynthesis. pp. 173-194. *In* M. Pessarakli (ed.), Handbook of Photosynthesis. Marcel Dekker, New York, U.S.A.

Rao, I.M. 1998. Root distribution and production in native and introduced pastures in the South American savannas. pp. 19-42. *In* J.E. Box Jr. (ed.), Root Demographics and Their Efficiencies in Sustainable Agriculture, Grasslands, and Forest Ecosystems. Kluwer, Dordrecht, Netherlands.

Rao, I.M., D.K. Friesen & M. Osaki. 1999. Plant adaptation to phosphorus-limited tropical soils. pp. 61-96. *In* M. Pessarakli (ed.), Handbook of Plant and Crop Stress. Marcel Dekker, New York, U.S.A.

Robinson, C.A., R.M. Cruse & K.A. Kohler. 1994. Soil management. pp. 109-134. *In* J.L. Hatfield & D.L. Karlen (ed.), Sustainable Agriculture Systems. Lewis Publishers and CRC Press, Boca Raton, U.S.A.

Salinas, J.G. 1978. Differential response of some cereal and bean cultivars to aluminum and P stress in an Oxisol of central Brazil. Ph.D. dissertation. North Carolina State Univ., Raleigh, U.S.A.

Schneider, K.A., M.E. Brothers & J.D. Kelly. 1997. Marker assisted selection to improve drought resistance in common bean. Crop Sci. 37:5160.

Schoonhoven, A. van & C. Cardona. 1980. Insects and other bean pests in Latin America. pp. 363-412. *In* H.F. Schwartz & G.E. Gálvez (eds.), Bean Production Problems: Disease, Insect, Soil and Climatic Constraints of *Phaseolus vulgaris*.CIAT, Cali, Colombia.

Schoonhoven, A. van & O. Voysest (eds.). 1991. Common Beans: Research for Crop Improvement. C.A.B. Int., Wallingford, U.K. & CIAT, Cali, Colombia.

Schumacher, T.E. & A.J.M. Smucker. 1981. Mechanical impedance effects on oxygen uptake and porosity of drybean roots. Agron. J. 73:51-56.

Schumacher, T.E. & A.J.M. Smucker. 1984. Effect of localized anoxia on *Phaseolus vulgaris* L. root growth. J. Exp. Bot. 35:1039-1047.

Schumacher, T.E. & A.J.M.Smucker. 1987. Ion uptake and respiration of dry bean roots subjected to localized anoxia. Plant Soil 99:411-422.

Schwartz, H.F. & M.A. Pastor-Corrales (eds.). 1989. Bean Production Problems in the Tropics. 2nd ed. CIAT, Cali, Colombia.

Schwartz, H.F. & J.R. Steadman. 1980. White mold. pp. 111-121. *In* H.F. Schwartz & G.E. Gálvez (eds.), Bean Production Problems: Disease, Insect, Soil and Climatic Constraints of *Phaseolus vulgaris*. CIAT, Cali, Colombia.

Shonnard, G.C. 1991. Genetics of and selection for heat tolerance during reproductive development in common bean. Ph.D. dissertation. Univ. of California, Davis, U.S.A.

Shonnard, G.C. & P. Gepts. 1994. Genetics of heat tolerance during reproductive development in common bean. Crop Sci. 34:1168-1175.

Sieverding, E. 1991. Vesicular-Arbuscular Mycorrhiza Management in Tropical Agroecosystems. Deutsche Gessellschaft für Technische Zusammenarbeit (GTZ), Eschborn, Germany.

Silva, P.M., S.M.Tsai & R. Bonetti. 1993. Response to inoculation and N fertilization for increased yield and biological nitrogen fixation of common bean (*Phaseolus vulgaris* L.). Plant Soil 152:123-130.

Singh, S.P. 1992. Common bean improvement in the tropics. Plant Breed. Rev. 10:199-269.

Singh, S.P. 1995. Selection for water-stress tolerance in interracial populations of common bean. Crop Sci. 35:118-124.

Singh, S.P., C. Cajiao, J.A. Gutiérrez, J. García, M.A. Pastor-Corrales & F.J. Morales. 1989a. Selection for seed yield in inter-gene pool crosses of common bean. Crop Sci. 29:1126-1131.

Singh, S.P. & J.A. Gutiérrez. 1990. Effects of plant density on selection for seed yield in two population types of *Phaseolus vulgaris* L. Euphytica 51:173-178.

Singh, S.P., C.A. Urrea, J.A. Gutiérrez & J. García. 1989b. Selection for yield at two fertility levels in small-seeded common bean. Can. J. Plant Sci. 69:1011-1017.

Smith, J.A., C.D. Yonts, E.D. Kerr & R.G. Wilson. 1992. Yield response of great northern beans to row spacing and plant population. Annu. Rpt. Bean Improv. Coop. 35:35-36.

Smithson, J.B., A.G. Buchara & M.E.K. Ali. 1992. Common bean (*Phaseolus vulgaris* L.) production and research in the Sudan. pp. 130-136. *In* J.B. Smithson (ed.), Proc. CIAT Regional Bean Programme in Eastern Africa, 1990, Debre Zeit, Ethiopia.

Sojka, R.E., M.J. Brown & E.C. Kennedy-Ketcheson. 1992. Reducing erosion from surface irrigation by furrow spacing and plant position. Agron. J. 84:668-675.

Sojka, R.E., R.D. Lentz & D.T. Westermann. 1998. Water and erosion management with multiple applications of polyacrylamide in furrow irrigation. Soil Sci. Soc. Amer. J. 62:1672-1680.

Sombroek, W.G., F.O. Nachtergaele & A. Hebel. 1993. Amounts, dynamics and sequestering of carbon in tropical and subtropical soils. Ambio 22:417-426.

Spain, J.M., M.A. Ayarza & L. Vilela. 1996. Crop-pasture rotations in the Brazilian Cerrados. p. 39-45. *In* R.C. Pereira & L.C. Bhering Nasser (eds.), Simpósio Sobre o Cerrado, 8. First International Symposium on Tropical Savannas. EMBRAPA-CPAC, Planaltina, Brazil.

Sponchiado, B.N., J.W. White, J.A. Castillo & P.G. Jones. 1989. Root growth of four common bean cultivars in relation to drought tolerance in environments with contrasting soil types. Exp. Agric. 25:249-257.

Steiner, K.G., R. Derpsch & K-H. Köller. 1997. Nachhaltiges Management de Bodenresources durch Verzicht auf Bodenbearbeitung. Etwicklung und ländlicher Raum 4:22-25.

Stephens, D.B. 1994. A perspective on diffuse natural recharge mechanism in areas of low precipitation. Soil Sci. Soc. Amer. J. 58:40-48.

Stewart-Williams, K. & J.R Myers. 1992. Effect of population density on seed size in dry bean. Annu. Rpt. Bean Improv. Coop. 35:76-77.

Stone, L.F. & J.A.A. Moreira. 1995. Resposta de duas cultivares de feijão a diferentes lâminas de irrigação, sob diferentes preparos do solo. pp. 1743-1745. *In* Congresso Brasileiro de Ciência do Solo, 25, Viçosa. Resumos expandidos. SBCS and UFV, Viçosa, Brazil.

Stone, L.F. & A.C. Pereira. 1994. Successvo arros-feijão irrigados por aspersão: efeitos de espacamento entre linhas, adubação e cultivares no crescimento, desenvolvimento radicular e consumo d'água de feijoeiro. Pesq. Agropec. Bras. 29:939-954.

Stoner, E.R. & E. Freitas. 1989. Characterization of root-restricting zones in Cerrado soils. pp. 320-321. *In* N. Caudle (ed.), Tropsoils Technical Report, 1986-1987. North Carolina State Univ., Raleigh, U.S.A.

Sutton, J.C. & B.R. Shepperd. 1976. Agregation of sand-dune soil by endomycorrhizal fungi. Can. J. Bot. 54:326-333.

Tamietti, G. & A. Garibaldi. 1984. Eradication of *Pseudomonas syringae* pv. *phaseolicola* in bean seeds in Italy. Istituto Patologia Vegetale, Univ. of Torino, Italy.

Thompson, T.E. & G.W. Fick. 1981. Growth response of alfalfa to duration of soil flooding and to temperature. Agron. J. 73:329-332.

Thung, M. 1990. Phosphorus: a limiting nutrient in bean (*Phaseolus vulgaris* L.) production in Latin America and field screening for efficiency and response. pp. 501-521. *In* N. El-Basam, M. Dambroth & B.C. Loughman (eds.), Genetic Aspect of Plant Mineral Nutrition. Kluwer, Dordrecht, Netherlands.

Thung, M. 1991. Bean agronomy in monoculture. pp. 737-816. *In* A. van Schoonhoven & O.Voysest. (eds.), Common Beans: Research for Crop Improvement. C.A.B. Int., Wallingford, U.K. & CIAT, Cali, Colombia.

Thung, M. 1997. Soil fertility conservation and enhancement in bean production systems in Brazil. pp. 470-486. *In* S.P. Singh & O. Voysest (eds.), Taller de Mejoramiento de Frijol para el Siglo XXI: Bases para Una Estrategia para América Latina. CIAT, Cali, Colombia.

Thung, M. & J.L. Cabrera. 1994. Avaliação de 22 espécies de plantas para fins de adubação verde. *In* Relatorio técnico do CNPAF 1990–1992. Documento no. 51. EMBRAPA-CNPAF, Goiânia, Brazil.

Thung, M. & J.L. Cabrera. 1996. Manejo de plantas daninhas para melhoria da productividade do feijão no ecosistema do Cerrado. pp. 489-491. *In* Reunião Nacional de Pesquisa de Feijão, 5. EMBRAPA-CNPAF, Goiânia, Brazil.

Thung, M., R.M. Ferreira & J.L. Cabrera. 1998. Maintaining the fertility of the Oxisols under intensive bean production systems. pp. 336-371. *In* O. Voysest (ed.), An Ecoregional Framework for Bean Germplasm Development and Natural Resources Research. CIAT, Cali, Colombia.

Thung, M. & I.P. de Oliveira. 1998. Problemas Abióticos que Afetam a Produção do Feijoeiro e Seus Métodos de Controle. EMBRAPA-CNPAF, Santo Antônio de Goiás, Brazil.

Thung, M., J. Ortega & O. Erazo. 1987. Breeding methodolgy for phosphorus efficiency and tolerance to aluminum and manganese toxicities for beans (*Phaseolus vulgaris* L.). pp. 207-222. *In* J.G. Salinas & L.M. Gourley (eds.), Sorghum for Acid Soils. Proc. Workshop on Evaluating Sorghum for Tolerance to Al-Toxic Tropical Soils in Latin America.INTSORMIL-ICRISAT-CIAT, Cali, Colombia.

Tinker, P.B. 1986. Role of rhizosphere microorganisms in phosphorus uptake by plants. pp. 617-653. *In* F.E. Khasawneh, E.C. Sample & E.J. Kamprath (eds.), The Role of Phosphorus in Agriculture. 2nd ed. Amer. Soc. Agron. & Soil Sci. Soc. Amer., Madison, U.S.A.

Tsai, S.M., R. Bonetti, S.M. Agbola & R. Rossetto. 1993a. Minimizing the effect of mineral nitrogen on biological nitrogen fixation in common bean by increasing nutrient levels. Plant Soil 152:131-138.

Tsai, S.M., P.M. da Silva, W.L. Cabezas & R. Bonetti. 1993b. Variability in nitrogen fixation of common bean (*Phaseolus vulgaris* L.) intercropped with maize. Plant Soil 152:93-101.

Tu, J.C. & C.S. Tan. 1991. Effect of soil compaction on growth, yield and root rots of white beans in clay loam and sandy loam soil. Soil Biol. Biochem. 23:233-238.

Vargas, A.A.T., J.S.M. Silveira, J.T. Athayde, A. Athayde & B.E.V. Pacova. 1991. Comparação entre genótipos de feijão quanto à capacidade nodulante e à produtividade com inoculação com rizóbios e/ou adubação de N-mineral. Rev. Bras. Ciênc. Solo 15:267-272.

Vargas, M.A.T., A.R. Suhet, I.C. Mendes & J.R.R. Peres. 1994. Fixação biológica de nitrogênio em solos de cerrados. EMBRAPA-CPAC, Planaltina, Brazil.

Vass, I. 1996. Adverse effects of UV-B light on the structure and function of the photosynthetic apparatus. pp. 931-949. *In* M. Pessarakli (ed.), Handbook of Photosynthesis. Marcel Dekker, New York, U.S.A.

Vieira, C. 1998. Growing beans in favorable environments and in association with crops of high social or economic value. pp. 128-154. *In* O. Voysest (ed.), An Ecoregional Framework for Bean Germplasm Development and Natural Resources Research. CIAT, Cali, Colombia.

Vieira, C., A.O. Nogueira & G.A.A. Araujo. 1992. Adubação nitrogenada e molíbdica na cultura de feijão. Rev. Agric. 67:117-124.

Walworth, J.J. & M.E. Sumner. 1987. The diagnosis and recommendation integrated systems (DRIS). Adv. Agron. 6:149-188.

Weaver, M.L., R. Timm, M.J. Silbernagel & D.W. Burke. 1985. Pollen staining and high temperature tolerance of bean. J. Amer. Soc. Sci. 110:797-799.

White, J.W. 1988. Preliminary results of the Bean International Drought Yield Trial (BIDYT). pp. 126-145. *In* J.W. White, G. Hoogenboom, F.J. Ibarra-Pérez & S.P. Singh (eds.), Research on Drought Tolerance in Common bean. CIAT, Cali, Colombia.

White, J.W. & J.A. Castillo. 1992. Evaluation of diverse shoot genotypes on selected root genotypes of common bean under soil water deficits. Crop Sci. 32:762-765.

White, J.W., J.A. Castillo, J.R. Ehleringer, J.A. García & S.P. Singh. 1994. Relations of carbon isotope descrimination and other physiological traits to yield in common bean (*Phaseolus vulgaris*) under rainfed conditions. J. Agric. Sci. 122:275-284.

White, J.W. & J. Izquierdo. 1991. Physiology of yield potential and stress tolerance. pp. 287-382. *In* A. van Schoonhoven & O. Voysest (eds.), Common Beans: Research for Crop Improvement. C.A.B. Int., Wallingford, U.K. & CIAT, Cali, Colombia.

White, J.W. & S.P. Singh. 1991. Breeding for adaptation to drought. pp. 501-560. *In* A. van Schoonhoven & O. Voysest (eds.), Common Beans: Rresearch for Crop Improvement. C.A.B. Int., Wallingford, U.K. & CIAT, Cali, Colombia.

Wortmann, C.S. 1994. The African network for screening beans for tolerance to edaphic stress - an overview. *In* C.S. Wortmann (ed.), Bean Improvement for Low Fertility Soils in Africa. CIAT, Cali, Colombia.

Wortmann, C.S., R.A. Kirkby, C.A. Eledu & D.J. Allen. 1998. Atlas of Common Bean (*Phaseolus vulgaris* L.) Production in Africa. CIAT, Cali, Colombia.

Wortmann, C.S., L. Lunze, V.A. Ochwoh & J. Lynch. 1995. Bean improvement for low fertility soils in Africa. Afr. Crop Sci. J. 3:469-477.

Yan, X., J.P. Lynch & S. Beebe. 1995a. Genetic variation for phosphorus efficiency of common bean in contrasting soil types. I. Vegetative response. Crop Sci. 35:1086-1093.

Yan, X., J.P. Lynch & S. Beebe. 1995b. Genetic variation for phosphorus efficiency of common bean in contrasting soil types. II. Yield response. Crop Sci. 35:1094-1099.

Zaiter, H.Z., E. Baydoun & M. Sayed-Hallak. 1993. Genotypic variation in common bean in response to cold temperature stress. Annu. Rpt. Bean Improv. Coop. 36:66-67.

Zaiter, H.Z. & B. Mahfouz. 1993. Salinity effect on root and shoot characteristics of common and tepary beans evaluated under hydroponic solution and sand culture. Annu. Rpt. Bean Improv. Coop. 36:80.

Chapter 14

Integrated Pest Management

Howard F. Schwartz & Frank B. Peairs
Bioagricultural Sciences & Pest Management, Colorado State University, Fort Collins, CO 80523-1177, U.S.A.

Key words: diseases, host resistance, insects, pest forecasting, pest management components, primary and secondary elements, weeds

1. INTRODUCTION

The concept of integrated pest management (IPM) has been defined and applied in a variety of contexts by producers and pest management specialists around the world. Jacobsen (1997) recently discussed the following definition of ecologically-based pest management (EBPM) that has been adopted by the National Coalition on IPM: "Integrated pest management is a sustainable approach to managing pests by combining biological, cultural, physical, and chemical tools in a way that minimizes economic, health, and environmental risks."

The goal of IPM is to achieve relief from pests in a manner that ensures safety, profitability, and durability while shifting the perceived focus on pesticide-intensive strategies to a systems approach that emphasizes biological knowledge of pests and their interactions with crops and environments. The shift to a systems approach with primary and secondary tactics requires that we move towards management of all pests rather than target just key pests such as a single insect, pathogen or weed species. Primary tactics rely upon knowledge of the managed ecosystem and its natural processes that suppress pest populations; and secondary tactics rely upon intervention with pesticides, physical, or biological supplementation.

A simplistic categorization of common bean (*Phaseolus vulgaris* L.) IPM strategies during the last 25 years in temperate cropping systems would show that the bean industry has emphasized pest management via pesticides as a primary tool, followed closely by plant resistance, and finally by cultural practices. On the other hand, tropical cropping systems have emphasized traditional cultural practices, followed by plant resistance, with pesticide management often assuming a distant third. There are numerous political, social, and economic reasons for these differences between the cropping systems, but their in-depth discussion is beyond the scope of this review. Suffice it to say that this dichotomy will hold well into the next century and should be recognized by anyone involved in IPM development for bean.

It is also important to note the important common role that host resistance plays across bean production systems. Host resistance is reviewed in great detail in numerous chapters of this book. Host resistance is important, but cannot replace an IPM strategy designed for the dynamic complexes of bean pests found worldwide in varied cropping systems. Reliance on a single control strategy, e.g., host resistance, runs counter to the very core of the IPM philosophy.

The goal of this chapter is to review the principles and practices of IPM elements that have been adopted or proposed for management of biotic stresses, with emphasis upon bean insects and pathogens. This should facilitate their application to diverse cropping systems around the world in the twenty-first century. In this chapter, the term "pest" refers to any pathogen, insect, and/or weed which exerts an adverse effect in terms of yield, quality, or economical viability upon the bean crop.

More in-depth sources of information on specific biotic stress management approaches are available from numerous research papers and reviews including those by Allen et al. (1998), Beebe & Pastor-Corrales (1991), Cardona (1989), Hall (1991, 1995), Hall & Nasser (1996), Hein et al. (1996), Karel & Autrique (1989), Schoonhoven & Voysest (1991), Schwartz & Pastor-Corrales (1989), and Wortmann et al. (1998).

2. TRADITIONAL IPM COMPONENTS

The major components of pest control that provide the foundation for IPM are reviewed for plant pathology in depth by Maloy (1993), and consist of exclusion, eradication, protection, resistance, and biological control. Parallel concepts for insect pests are provided by Pedigo (1999). These are defined below and are related to similar entomological terms. We will then use the plant pathology terminology for the remainder of the chapter.

2.1 Exclusion

This prevents the entrance and establishment of a plant pathogen in an uninfested region by prohibition, interception, and elimination. This is primarily accomplished by implementation of quarantine, embargo, inspection, and certification of plant materials, such as bean seed sources, that are transported between production regions. While clean seed is similarly emphasized in weed management, the parallel concept in entomology, often referred to as legislative control, emphasizes quarantine because of the relatively greater mobility of insects and the lesser importance of seedborne problems.

2.2 Eradication

Emphasis is placed on the removal, elimination, or destruction of a pest from an area in which it is already established. This is primarily accomplished for beans by detection and removal of diseased plants or infested plant debris; elimination of weed hosts and alternate crops, sanitation, and crop rotation; and destruction of pests by disinfestation or disinfection by heat (burning), tillage or pesticides. Such activities would be categorized by entomologists as sanitation, while the term eradication is generally reserved for the removal of a given pest species over a wide area.

2.3 Protection

A protective barrier between the susceptible host and the pest is created. Commonly employed strategies for beans include manipulation of environmental factors, usually by cultural practices such as planting time, row orientation, row spacing/plant density; application of protectant or systemic chemicals such as fungicides, bactericides, insecticides, nematicides, herbicides; and barriers such as distance between plants or fields and multiple cropping. This is a combination of the parallel entomological concepts of pesticide control and environmental modification. The latter is defined as changes in the crop environment to make it either more favorable for natural enemies or less favorable for the pest.

2.4 Resistance

This helps plants escape (via canopy architecture, flowering, maturity), become tolerant (similar disease but higher yield and/or quality versus a susceptible line), or no longer respond to infecting or feeding activities of plant pathogens or insects. True resistance is genetically inherited and is usually expressed as a result of an interaction between a pest and the host at the cellular level that limits pest development and/or elicits plant defense mechanisms. Strategies used with bean include nutrition to increase plant vigor and harden plant tissues to infection or feeding. Plants are selected that are less affected by a pest and there is hybridization and selection of desirable progeny from crosses between susceptible and resistant parents. Pest resistance factors can be characterized, and such factors can be incorporated through traditional and modern breeding strategies, including genetic engineering as reviewed elsewhere in this book.

2.5 Biological Control

It uses natural enemies and competitors to manage insect pests, pathogens, and weeds. This may be accomplished through conservation of existing biological control agents, augmentation of biological control agent population densities through release of additional individuals, or inundation through the release of sufficient biological control agents to reduce the pest population densities to a subeconomic level. An additional approach, often termed classical biological control, is to collect natural enemies of an exotic pest within its native range and release them into areas where the pest has been introduced.

3. HIGHLIGHTS OF WEED MANAGEMENT STRATEGIES

A weed has been defined as a plant which does more harm than good, or as a plant out of place (Hurle, 1998). Weeds are an integral part of any cropping system and are viewed as beneficial components by some. Benefits include: 1) contributing to soil and plant nutrition, 2) sources of animal fodder, 3) sources of human food, 4) helping control soil erosion, 5) contributing to biodiversity in agricultural ecosystems, 6) acting as host plants for natural enemies of pests and plant pathogens, and 7) serving as a mulch to control other weed species. However, weeds are most often viewed as negative components, as they: 1) serve as host plants for pests and vectors of plant pathogens, 2) reduce crop yields by competing for nutrients, moisture, and light, and 3) create harvesting difficulties of crops such as beans (Hurle, 1998; Wooley & Davis, 1991).

Weeds are usually less of a problem for beans grown in association than in sole crops, as crops often fill more than one ecological niche and thus compete effectively with a greater number of weed species. Where weeds are often more of a problem in rotation or relay crops, reliance upon cropping systems, plant density and growth habit, cultivation practices, and herbicide use have been essential. Biological control of weeds remains elusive, yet there is potential for bean production. Herbicide management depends upon many factors, including: 1) chemical selectivity for broadleaf or grassy weeds, 2) effective chemical dose and residues, 3) chemical timing as preplant, preemergence, or postemergence treatment, 4) mode of chemical action via contact or uptake, and 5) the presence of herbicide-resistant microorganisms (Hurle, 1998; Wooley & Davis, 1991).

Hurle (1998) discusses three weed control strategies which are in various stages of evolution and adoption within different agricultural ecosystems: 1) "control at any price" - the objective has been to minimize yield reduction and prevent greater weed infestation (seed bank) in future years, 2) "economic threshold" - the objective has been to implement control only if the cost does not exceed the monetary value of the yield reduction, and 3) "ecological"/ "economic threshold" - the objective is to determine what level of weed presence is needed or desired before control is considered from an economic perspective. This last strategy has arisen largely because during the last 40 to 50 years, agriculture has lost 20 to 40% of its agricultural weed flora due to adoption of more intensive production systems, tighter crop rotations, greater use of nitrogen fertilizer, and higher crop densities.

Hurle (1998) and others see advances in weed science focusing upon the problem of non-uniformity of weed species and their densities within a field which continues to be treated with great precision in a uniform manner. Technology is becoming available to support Global Positioning Satellite (GPS) and Geographical Information System (GIS) site-specific control and sensor-driven control with selective herbicides that reduce environmental exposure and costs. Genetically engineered crops like cotton, soybean, and corn are being developed with resistance to specific herbicides like glyphosate and glufosinate, and someday beans may also benefit from this technological application.

4. HIGHLIGHTS OF INSECT MANAGEMENT STRATEGIES

Most emphasis in insect management has been in the areas of multiple cropping, host resistance, and chemical control. In many cases, crop diversity and host resistance have documented potential but have not been exploited to any practical degree. In reality, most arthropod pest management seems to rely either on the protection afforded by a given cropping system and naturally occurring biological control or on chemical control. Barriers to implementation of IPM include limited knowledge of pest ecology, limited availability of resistant cultivars, over-reliance on chemical control, lack of dynamic economic thresholds, and a paucity of tactics for individual pests. Pest arthropods have been reviewed for Africa (Karel & Autrique, 1989), Latin America (Cardona, 1989), and North America (Hein et al., 1996), and are summarized in Table 1.

Table 1. Key insect pests of common bean in different production regions of the world

Insect	Damage example	Management[1]
Africa		
Ophiomyia phaseoli Tryon (and related species) (Diptera:Agromyzidae)	Seedling stem borer (bean fly)	1,2,3,4,5
Aphis fabae Scopoli (Homoptera: Aphididae)	Suck plant sap, virus vector (black bean aphid)	2,5
Ootheca sp. (Coleoptera: Chrysomelidae)	Defoliator, virus vector (foliage beetle)	1,4,5
Megalurothrips sjostedti (Thysanoptera: Thripidae)	Rasp, suck sap from buds, flowers (flower thrips)	4,5
Maruca testulalis (Geyer)(Lepidoptera:Pyralidae)	Chew on buds, flowers, pods (legume borer)	5
Latin America		
Empoasca kraemeri (Ross & Moore) (Homoptera: Cicadellidae)	Suck sap from leaves (common leafhopper)	1, 2, 4, 6, 7, 8
Diabrotica balteata LeConte (and many relatives) (Coleoptera: Chrysomelidae)	Chew on roots, foliage, flowers, pods; virus vector (banded cucumber beetle)	2, 6
Bemisia tabaci (Gennadius) (and several relatives) (Homoptera: Aleyrodidae)	Virus vector (common whitefly)	4, 5
Delia platura (Meigen) (Diptera: Anthomyidae)	Feed on seed and seedlings (seedcorn maggot)	1, 4, 5
Elasmopalpus lignosellus (Zeller) (Lepidoptera: Pyralidae)	Stem borer (lesser cornstalk borer)	4, 5
North America		
Delia platura (Meigen) (Diptera: Anthomyidae)	Feed on seed and seedlings (seedcorn maggot)	1, 4, 5
Empoasca fabae (Harris) (Homoptera: Cicadellidae)	Suck sap from leaves (potato leafhopper)	4, 6, 7
Epilachna varivestis Mulsant (Coleoptera: Coccinellidae)	Defoliator (mexican bean beetle)	2, 4, 6, 7

Insect	Damage example	Management[1]
Richia albicosta (Smith) (Lepidoptera: Noctuidae)	Chew on leaves, pods (pod borer)	6
Cosmopolitan		
Acanthoscelides obtectus (Say) (Coleoptera: Bruchidae)	Chew on seed in field, storage (bean weevil, bruchid)	4, 5, 9
Zabrotes subfasciatus (Boheman) (Coleoptera: Bruchidae)	Chew on seed in storage (mexican bean weevil)	4, 5, 9

[1]Insect management tactics known to be effective or to have potential: 1 = modified planting date, 2 = multiple cropping, 3 = hilling, 4 = plant resistance, 5 = chemical control, 6 = chemical control with economic threshold, 7 = biological control, 8 = mulches, and 9 = indigenous storage protectants (e.g., vegetable oil).

5. HIGHLIGHTS OF DISEASE MANAGEMENT STRATEGIES

Hall & Nasser (1996) published a thorough review of 33 specific disease management elements that are utilized or recommended individually and collectively to manage 50 bean diseases worldwide. From this list, the following 16 diseases from four pathogen classes are used as collective examples of the more widespread, persistent, and economically threatening diseases of beans in many temperate and tropical cropping systems (Table 2).

A summary of the disease management elements (Allen et al., 1998; Hall & Nasser, 1996; Schwartz & Pastor-Corrales, 1989; Schwartz et al., 1996) utilized for these 16 diseases reveals that a limited number of those elements actually constitute the bulk of current IPM strategies that are employed (Table 3). The major emphasis for most diseases has been on: 1) crop rotation of more than two years between bean crops, 2) timely application of specific pesticides that are effective against the target pest, 3) tillage practices that promote host plant development and/or reduce carryover of the pathogen, 4) weed control to reduce host plant stress from competition, microclimatic influences, and sources of pests (refugia), 5) multiple cropping with non-hosts and crop barriers to reduce pathogen spread, and 6) plant resistance to one or more plant pathogens.

For example, in Table 3 crop rotation is cited as a disease control for 75%, while resistant cultivars are cited as a control for 44% of the 16 diseases. The effectiveness of individual IPM elements is quite varied and unique for specific diseases as illustrated in Table 3. This fact underscores the complexity surrounding the design and implementation of disease management strategies that are appropriate for each disease, cropping system, environment, and resources available to producers.

Table 2. Examples of important plant pathogens and diseases that affect bean production in many temperate and tropical cropping systems

Disease	Pathogen
Fungal	
Angular leaf spot	*Phaeoisariopsis griseola* (Sacc.) Ferr.
Anthracnose	*Colletotrichum lindemuthianum* (Sacc. & Magn.) Bri. & Cav.
Rust	*Uromyces appendiculatus* (Pers.) Ung.
White mold	*Sclerotinia sclerotiorum* (Lib.) de Bary
Pythium diseases	*Pythium* species
Rhizoctonia root rot	*Rhizoctonia solani* Kuhn
Web blight	*R. solani* Kuhn, teleomorph = *Thanatephorus cucumeris* (Frank) Donk
Fusarium root rot	*Fusarium solani* f. sp. *phaseoli* (Burk.) Snyder & Hansen
Bacterial	
Halo blight	*Pseudomonas syringae* pv. *phaseolicola* (Burkh.) Young et al.
Bacterial brown spot	*P. syringae* pv. *syringae* van Hall
Common bacterial blight	*Xanthomonas campestris* pv. *phaseoli* (Smith) Dye
Viral	
Bean common mosaic	Bean common mosaic virus (potyvirus, aphid vector)
Sugarbeet curly top	Sugarbeet curly top virus (geminivirus, leafhopper vector)
Bean golden mosaic	Bean golden mosaic virus (geminivirus, whitefly vector)
Nematode	
Lesion nematode	*Pratylenchus* species
Root knot nematode	*Meloidogyne* species

Hall & Nasser (1996) concluded that cultural practices are essential to the control of 80 % of bean diseases and contribute to the control of all bean diseases. Resistance is utilized for 25, while rotation is utilized for 33 of the 50 diseases cited in their review. Resistance to many plant pathogens and their diseases may not be complete enough to protect the plant from significant loss in yield and quality, so other elements, including cultural practices are essential to supplement the partial gain realized by resistance. In addition, disease resistance may be broken down by interaction of host with pathogen and environment, and/or few cultivars are likely to possess long-lasting resistances to all elements of a disease or pest complex, thereby requiring a more complete protection strategy for overall plant health.

This diversity of practices provides effective, stable, flexible, and adaptable disease management. Cultural management of diseases is a key component of sustainable bean productivity and health, and enhances the effectiveness and stability of other IPM approaches such as host resistance and the judicious use of pesticides, if required.

Table 3. Implementation of integrated pest management (IPM) elements currently available for the major diseases of beans in temperate and tropical cropping systems listed in Table 2

% of 16		Examples of IPM element effectiveness against[1]				
diseases	IPM element	Mold	Fungus	Bacterium	Virus	Nematode
75	Crop rotation	3	3	3	2	3
62	Pesticides	2	2	2	1	1
62	Tillage	2	2	1	1	2
50	Weed control	2	2	2	2	2
50	Multiple cropping	2	1	2	2	2
44	Plant resistance	1	3	3	3	1
44	Pathogen-free seed	2	3	3	3	1
44	Bean debris sanitation	1	3	1	1	1
38	Drainage	2	1	1	1	2
32	Plant spacing	3	2	1	1	2
32	Planting date	1	2	2	2	1
25	Volunteer elimination	1	2	2	2	1
19	Crop separation	1	2	2	2	1
19	Movement in field	1	2	2	1	1
19	Soil fertility	2	1	1	I	2
19	Soil temperature	1	1	1	1	2
13	Soil organic matter	1	1	1	1	2
13	Clean fallow	1	1	1	1	1
13	Mulch	1	2	1	1	1
13	Flooding	2	1	1	1	2
13	Cover crop	1	1	1	1	1
6	Harvest date	1	2	1	1	1
6	Soil pH	1	1	1	1	1
6	Plant architecture	2	2	1	1	1
6	Row direction	2	2	2	1	1
6	Irrigation	3	1	1	1	2
6	Seedling depth	1	1	1	1	2
0	Roguing	1	1	1	2	1
0	Soil testing	1	1	1	1	1
0	Storage	1	1	1	1	1
0	Seed moisture	1	1	1	1	1
0	Trap crops	1	1	1	1	1

[1] 1 = low, 2 = moderate, and 3 = high (Hall, 1991; Schwartz & Pastor-Corrales, 1989; Schwartz et al., 1996).

5.1 IPM Elements

The following primary and secondary IPM elements will be available for adoption by bean producers during the next century (Hall & Nasser, 1996).

Continued emphasis upon adoption of the primary elements will provide the most immediate impact upon priority pathogens and diseases. Adoption of the secondary elements can further enhance the effectiveness of IPM strategies in specific cropping systems in temperate and tropical regions.

5.1.1 Primary IPM Elements

1. Crop rotation; with non-host crops, this is highly effective in reducing soil-borne pest populations; combined with the removal or elimination of infested debris, it reduces the population density of other types of bean pests that multiply within the field. This tactic is also useful against specialized bean pests, especially those with limited mobility, and facilitates management of broadleaf weeds.
2. Elimination of volunteer beans (and infested debris); these can serve as a source(s) of many bean pathogens and contribute to early-season outbreaks and greater pathogenic variability for fungi such as those responsible for diseases like rust, angular leaf spot, and anthracnose. Insect pests such as aphids and thrips can also buildup on volunteer plants before and during the growing season.
3. Clean fallow; burning also removes alternative hosts of bean pathogens, infested debris, insect pests, weed seed and seedlings; however, this approach is often at odds with soil and water conservation programs.
4. Pathogen-free seed; produced in environmental conditions less favorable to pathogen development and certified by appropriate seed agencies, this is the major approach to controlling many diseases of bean, especially those caused by bacterial pathogens or seedborne viral pathogens such as that responsible for bean common mosaic. In addition, weed-free seed is an important strategy to restrict weed species contamination between and within bean production regions.
5. Selection of pest resistant cultivars; there should be an emphasis on adaptation to local growing conditions; avoid late-maturing, long-season, or photoperiod-sensitive cultivars that prolong exposure to and infection by pathogens and pests. While insect-resistant cultivars are generally lacking, well adapted, vigorous plants are always a first line of defense in insect management and compete more successfully with weeds.
6. Soil temperatures of 18°C; temperatures at least this high at planting reduce exposure of seedlings to cool temperatures and long periods of wetness that delay emergence and reduce plant vigor. Seedlings planted in cool temperatures are vulnerable to various pests, including the fungi responsible for root rots and wilts, seedcorn maggots and other seedling pests, and early-season weed competition.

7. Seeding dates; growers should attempt to avoid coincidence of susceptible plants and other disease hosts, as well as insect vectors when possible.
8. Seeding depth 5cm; the depth should normally be 5 cm in moist soil, but should be adjusted for drier soils.
9. Weed suppression; weed control is vital, since weeds can alter the canopy microclimate, provide pathogens with sources of nutrients such as pollen and senesced plant parts, or support nematode and insect-vector populations in addition to competition effects.
10. Tillage (pre-plant, post-emergence); a quality seedbed improves plant health by stimulating root growth, increasing drainage, increasing resistance of roots to pathogens, or decreasing soil compaction, anaerobic conditions, and pathogen populations. Care must be exercised to prevent soil erosion during tillage and preserve soil structure. Tillage is the primary strategy for weed management.
11. Timely harvest; careful timing can reduce losses from pathogens that infect late in crop maturation and reduce the time period that plants and seeds are exposed to inoculum and unfavorable environmental conditions. Storage pests, such as bruchids, that can infest the crop prior to harvest are also minimized in this manner.

5.1.2 Secondary IPM Elements

1. Acceptable pH range of 6 to 8 is preferred for normal physiological development of most bean cultivars.
2. High levels of organic matter in the soil improve physical and chemical conditions for root growth, and may facilitate natural biological control.
3. Adequate and balanced soil fertility is important to optimize bean yields and minimize synergistic effects from pathogens favored by excess or deficient levels of specific nutrients or soil salinity.
4. Soil moisture levels near field capacity, achieved through drainage and careful irrigation, are most suitable for bean growth and health.
5. Controlled use of surface or overhead irrigation is recommended to reduce the spread of pathogens and infested debris and to avoid prolonged periods of high moisture within the plant canopy and on the soil surface.
6. Select the growth habit and population (seeding density) appropriate to reduce disease severity and weed competition. Upright plant architecture, wider row spacing, and lower plant populations can shorten periods of surface wetness, lower relative humidity and free moisture within the plant canopy, and restrict contact of above-ground plant parts with inoculum in the soil.
7. Row orientation should be parallel to the direction of prevailing winds to permit plant canopies and soil surfaces to dry more quickly.

8. Bean plant health is promoted by maximum spacing between plants and rows to minimize plant-to-plant spread of pathogens without sacrificing canopy cover and yield potential on a field basis.
9. Separation of bean fields and use of barriers (e.g., intercropped maize) can reduce intra- and inter-field spread of bean pathogens, insect vectors, and other pests.
10. Roguing eliminates major sources of inoculum prior to or early in the development of an epidemic, and is a useful tool in the production of seed free from various pathogens in subsistence cropping systems.
11. Limiting movement of equipment and personnel within and between fields can restrict the mechanical dissemination of bean pathogens, infested debris, and weed seed.

6. IPM - ADDITIONAL ENHANCEMENTS FOR THE TWENTY-FIRST CENTURY

6.1 Diagnosis

Bean pest-diagnostics will continue to be enhanced by molecular biology applications as more and better diagnostic kits become available. These kits will efficiently detect lower levels of plant pathogens before disease incidence and severity become so high and widespread that they overload the system and reduce the effectiveness of control elements such as pesticides and various cultural practices. Similarly, more efficient sampling and monitoring techniques such as the use of pheromone traps, will improve arthropod pest detection.

A better understanding of pathogen biology and disease development will also enable IPM specialists to develop more effective pest scouting and management calendars in relation to cropping cycles and weather patterns, and to develop more specific thresholds and economic damage estimates (Hall, 1995). However, in many developing and developed countries the absence or continued erosion of extension and IPM personnel that are well-trained and/or have access to adequate facilities required for prompt and accurate diagnoses of bean pathogens and their diseases will reduce IPM capabilities and adoption.

During the mid to late 1990s, technology continued to evolve and enhance remote sensing of pest outbreaks, database development, and record keeping via GPS hardware and GIS software that became more integral to and accepted by IPM personnel. More accurate tracking of historical pest outbreaks has improved the timeliness and accuracy of scouting calendars during succeeding cropping seasons. In temperate areas like Colorado, U.S.A., rust and bacterial pathogens consistently overwinter in debris and volunteer plants develop from bean seed lost during previous harvests (Schwartz et al., 1996, 1998). Similar approaches could be used for priority insect pests and weeds, possibly in combination with site-specific application of pesticides.

6.2 Host Resistance

Host resistance programs may benefit from the transgenic crop modification that allows the introduction of genes for fungal, bacterial, viral, nematode, insect, and weed control from related as well as unrelated plant species or organisms (Knight et al., 1997) into a crop such as bean. These types of advances may reduce the reliance upon conventional pesticides and will supplement other IPM strategies which rely upon cultural practices and biological control. Genetic improvement programs are beginning to benefit from what is now known about the co-evolution of pathogens responsible for diseases such as rust, anthracnose, and angular leaf spot with beans in the Americas. There are also benefits from the development of molecular marker techniques including RAPD (random amplified polymorphic DNA), SCAR (sequence characterized amplified region), RFLP (restriction fragment length polymorphisms), and others (Guzmán et al., 1995; Kelly, 1995; Kelly & Miklas, 1998; Maclean et al., 1995; Young & Kelly, 1997). The development of bean cultivars with resistance to multiple pathogens and pests will continue to become more feasible through the application of these modern genetic techniques. However, the final implementation of transgenic technology for multiple pest resistance may be more limited for legal and economic reasons. Utilization of these advances in genetic improvement is reviewed in Chapters 3 and 4 in this book.

6.3 Pesticides

The pesticide industry is responding to the growing demand for crop protection agents. There are low- use- rate compounds with chemistries that possess a benign environmental profile and have lower toxicity to humans and wildlife (Knight et al., 1997). In addition, the industry is continuing to improve application technology and efficiency to maximize coverage, adhesion of pesticides, relationship of efficiency to particle size, enhanced target exposure, and proper timing (Maloy, 1993). The increased costs associated with such treatments will require improved economic decision making tools such as dynamic economic thresholds, better sampling techniques, and improved forecasting.

The introduction of these expensive crop protection pesticides also will dictate proactive resistance management programs. In resource-limited tropical production systems, it is unlikely that modern crop protection pesticides will be widely adopted. There will probably be a move towards effective research-based home remedies such as Neem or other locally-available botanical pesticides.

There will be an increased emphasis on biological control. Conservation of naturally occurring biological control agents will receive greater interest in all production systems. Applied biological controls also may be implemented for certain difficult problems including soil-borne pathogens and seedling insects. More effective biological control agents, such as those resistant to certain pesticides, may become more available through genetic engineering.

A variety of biological control agents (e.g., species of *Bacillus* and *Trichoderma*) and naturally occurring substances from fungi, bacteria, and higher plants have been isolated and demonstrated to possess antifungal, antimicrobial, and insecticidal properties (biopesticides). Examples include pyrrolnitrin and strobilurin, which affect various fungal processes such as glucose metabolism and mitochondrial respiration, and avermectin, with broad insecticidal activity. Other novel fungicides include famoxadone, pyrimethanil, cyprodinil, quinoxyfen, and spiroxamine that affect these and other metabolic and pathogenic processes of fungi including inhibition of ergosterol biosynthesis (Knight et al., 1997).

Other compounds are being investigated that exert indirect effects against fungal pathogens, such as probenazole, which stimulates the production of antifungal lipids and peroxidases in rice. Systemic acquired resistance (SAR) provides another exciting area for the future, as plants treated with compounds such as analogues of salicylic acid build up their defense response and can be protected from infection by fungal, bacterial or viral pathogens, as well as some insect pests. This diversity of chemistry will help to minimize the development of chemical resistance in plant pathogens by supporting strategies that alternate or mix different modes of action with other IPM elements.

Bean IPM will benefit in the future as more of these types of products become available to supplement current technologies. Precision agriculture and the use of innovative technologies that allow targeted pesticide applications are also expected to influence future pesticide use as will satellite detection and other remote sensing systems (Knight et al., 1997). However, environmental and food safety concerns, in addition to political and special interest group lobbying, continue to place regulatory pressure on the international availability of many existing pesticides. The time and cost involved to label new pesticides for use on minor crops such as bean also hinder availability.

6.4 Pest Forecasting

Pest forecasting efforts will continue to focus upon pests of bean such as rust, which meets the necessary criteria for successful forecasting. These pests must be: 1) widespread in many cropping systems, 2) the cause of consistent yet sporadic damage to the crop, 3) manageable by monitoring the effectiveness of resistant cultivars and/or the timely applications of pesticides to susceptible cultivars, and 4) a rapid communication of forecasts and recommendations can be made to crop consultants and producers (Doesken et al., 1998; Schwartz et al., 1996).

The success of a pest model requires: 1) an understanding of environmental and host influences on the pest development, 2) access to technology that detects, monitors, and analyzes pest development, and 3) knowledge of the pest population dynamics (Doesken et al., 1998; Maloy, 1993). Research is urgently needed to investigate pest biology and identify threshold levels for yield loss potential at different stages of plant development. The influence of responses to microclimatic and long-term (e.g., global warming) effects in varied cropping systems throughout the world also must be determined.

6.5 Technology Transfer

Technology transfer and information networks will be vital for rapid dissemination of information about weather, crop and pest situations, and timely IPM elements that can be implemented by crop consultants and producers. Improvement of dry bean pest management will have to be accomplished in an environment of limited trained personnel and reduced training opportunities, as government resources continue to be diverted from production agriculture and support programs such as IPM. More efficient and effective technology transfer methods will take advantage of the internet and other communication innovations.

Access to archived and current bean literature that has been converted digitally will enhance research and extension projects globally. The Bean Improvement Cooperative and repositories such as the USDA National Agricultural Library and the International Center for Tropical Agriculture (CIAT) library system are invaluable resources. In addition, bean researchers are providing electronic networking via the internet with comprehensive databases such as BeanGenes (P. McClean, email:

mcclean@beangenes.cws.ndsu.nodak.edu) and BeanRef (M. Nenno & W. Nagl, email: nenno@rhrk.uni-kl.de).

7. PROSPECTS AND CHALLENGES

Hall (1995) and Allen et al. (1998) have articulated the most critical needs of and charges to bean research and extension personnel as we enter the next century. We are challenged to reduce the current level of crop losses and to mitigate the impact of pest management actions on human health and environmental quality by releasing improved cultivars with multiple pest resistance that require less and more precise pesticide input. We are challenged to reduce our reliance on one method of pest management. New technologies are important because they increase the number of options that will be available for effective, economical, and flexible pest management programs (Hall, 1995).

As IPM moves into the twenty-first century, several trends are notable (National Research Council, 1996). Whether or not they will be incorporated into modern bean production is a challenge to all associated with the improvement of this important crop. These trends will influence the management of bean pests, although significant differences will remain between temperate and tropical production systems. Expanded and applied ecosystem knowledge will enhance our ability to manage pests in a cost-effective and environmentally sound manner. As our understanding of the structure and function of bean production systems improves, our ability to maintain pest densities below economically significant levels will be enhanced.

There is a need to intensify research on the distribution of physiological races of bean pathogens and inheritance of resistance. If crop losses caused by specific diseases and complexes can be quantified better, priorities can be established. Better information on pathogen variability and distribution, especially considering co-evolution of some pathogens such as those causing rust, angular leaf spot, and anthracnose would be useful. Studies of the biology and epidemiology of bean pathogens have become unfashionable of late and are urgently needed to supplement existing knowledge and provide a sound basis for the development of effective integrated control practices (Allen et al., 1998).

Based upon Jacobsen's (1997) review of IPM elements, bean IPM during the last twenty years would be classified as a "Low IPM" system, characterized by components such as scouting, crop rotation, pathogen-free seed, pest resistant cultivars, cultivation, selective pesticides, and sprayer calibration. During the next 25 years, bean IPM will evolve into a "Medium

IPM" system that will increasingly emphasize pest resistant cultivars, weather-based prediction models, nutrient and water monitoring, use of composts and green manures, precision agriculture-based pesticide treatment, use of induced resistance activators, and pest biotype monitoring. At some point, bean pest management strategies may attain the "High IPM" system, which relies upon pest resistant cultivars, the release of beneficials, use of biocontrol agents and pheromones, trap crops, soil solarization, interactive pest/weather/crop models, and primary reliance upon preventative nonchemical practices.

Regardless of the bean cropping system and its location, IPM will continue its dynamic evolution, as pest management specialists address old and new pest challenges in the twenty-first century.

REFERENCES

Allen, D.J., R.A. Buruchara & J.B. Smithson. 1998. Diseases of common bean. pp. 179-265. *In* D.J. Allen & J.M. Lenne (eds.), The Pathology of Food and Pasture Legumes. C.A.B. Int., Wallingford, U.K.

Beebe, S.E. & M. Pastor-Corrales. 1991. Breeding for disease resistance. pp. 561-617. *In* A. van Schoonhoven & O. Voysest (eds.), Common Beans: Research for Crop Improvement. C.A.B. Int., Wallingford, U.K. & CIAT, Cali, Colombia.

Cardona, C. 1989. Insects and other invertebrate bean pests in Latin America. pp. 505-570. *In* H.F. Schwartz & M.A. Pastor-Corrales (eds.), Bean Production Problems in the Tropics. 2nd ed. CIAT, Cali, Colombia.

Doesken, N.J., H.R. Duke, B.L. Hamblen, J. Kleist, T.B. McKee, M.S. McMillan & H.F. Schwartz. 1998. The Colorado agricultural meteorological network (COAGMET) - A unique collaborative system supporting Colorado Agriculture. 23rd Conf. On Agr. and Forest Meteor., Albuquerque, U.S.A.

Guzmán, P., R.L. Gilbertson, R. Nodari, W.C. Johnson, S.R. Temple, D. Mandala, A.B.C. Mkandawire & P. Gepts. 1995. Characterization of variability in the fungus *Phaeoisariopsis griseola* suggests coevolution with the common bean (*Phaseolus vulgaris*). Phytopathology 85:600-607.

Hall, R. 1991. Compendium of Bean Diseases. APS Press, St. Paul, U.S.A.

Hall, R. 1995. Challenges and prospects of integrated pest management. pp. 1-19. *In* R. Reuveni (ed.), Novel Approaches to Integrated Pest Management. Lewis Pub., Boca Raton, U.S.A.

Hall, R. & L.C.B. Nasser. 1996. Practice and precept in cultural management of bean diseases. Can. J. of Plant Pathol. 18:176-195.

Hein, G.L., F.B. Peairs & S.D. Pilcher. 1996. Insect problems. pp. 57-65. *In* H.F. Schwartz, M.A. Brick, D.S. Nuland & G.D. Franc (eds.), Dry Bean Production and Pest Management. Colorado State Univ., Univ. of Nebraska & Univ. of Wyoming Reg. Bull. No. 562A, Fort Collins, U.S.A.

Hurle, K. 1998. Present and future developments in weed control - A view from weed science. Pflanzenschutz-Nachrichten Bayer 51:109-128.

Jacobsen, B.J. 1997. Role of plant pathology in integrated pest management. Ann. Rev. Phytopathology 35:373-391.
Karel, A.K. & A. Autrique. 1989. Insects and other pests in Africa. pp. 455-504. *In* H.F. Schwartz & M.A. Pastor-Corrales (eds.), Bean Production Problems in the Tropics. 2nd ed. CIAT, Cali, Colombia.
Kelly, J.D. 1995. Use of random amplified polymorphic DNA markers in breeding for major gene resistance to plant pathogens. HortScience 30:461-465.
Kelly, J.D. & P.N. Miklas. 1998. The role of RAPD markers in breeding for disease resistance in common bean. Mol. Breed. 4:1-11.
Knight, S.C., V.M. Anthony, A.M. Brady, A.J. Greenland, S.P. Heaney, D.C. Murray, K.A. Powell, M.A. Schulz, C.A. Spinks, P.A. Worthington & D. Youle. 1997. Rationale and perspectives on the development of fungicides. Ann. Rev. Phytopathology 35:349-372.
Maclean, D.J., K.S. Braithwaite, J.A.G. Irwin, J.M. Manners & J.V. Groth. 1995. Random amplified polymorphic DNA reveals relationships among diverse genotypes in Australian and American collections of *Uromyces appendiculatus*. Phytopathology 85:757-765.
Maloy, O.C. 1993. Plant Disease Control - Principles and Practice. J. Wiley & Sons, New York, U.S.A.
National Research Council. 1996. Ecologically-Based Pest Management: New Solutions for a New Century. Academy Press, Washington DC, U.S.A.
Pedigo, L.P. 1999. Entomology and Pest Management. 3rd ed. Prentice Hall, Upper Saddle River, U.S.A.
Schoonhoven, A. van & O. Voysest (eds.). 1991. Common Beans: Research for Crop Improvement. C. A. B. Int., Wallingford, U.K. & CIAT, Cali, Colombia.
Schwartz, H.F., M.A. Brick, D.S. Nuland & G.D. Franc. (eds). 1996. Dry Bean Production and Pest Management. Reg. Bull. No. 562A, Colorado State Univ., Fort Collins, U.S.A.
Schwartz, H.F., M.S. McMillan & K.J. Otto. 1998. VegNet - Internet and satellite access to bean rust modeling. Annu. Rpt. Bean Improv. Coop. 41:37-38.
Schwartz, H.F. & M.A. Pastor-Corrales (eds.). 1989. Bean Production Problems in the Tropics. 2nd ed. CIAT, Cali, Colombia.
Wooley, J. & J.H.C. Davis. 1991. The agronomy of intercropping with beans. pp. 707-735. *In* A. van Schoonhoven & O. Voysest (eds.), Common Beans: Research for Crop Improvement. C.A.B. Int., Wallingford, U.K. & CIAT, Cali, Colombia.
Wortmann, C.S., R.A. Kirkby, C.A. Eledu & D.J. Allen. 1998. Atlas of Common Bean (*Phaseolus vulgaris* L.) Production in Africa. CIAT, Cali, Colombia.
Young, R.A. & J.D. Kelly. 1997. RAPD markers linked to three major anthracnose resistance genes in common bean. Crop Sci. 37:940-846.

APPENDIX

Correlated linkage maps of common bean. Each linkage group is represented by a separate figure, which includes four linkage groups. From left to right, these originate from the Florida map (Vallejos et al., 1992), the core map (Freyre et al., 1998), Davis map (Nodari et al., 1993a), and the Paris map (Adam-Blondon et al., 1994b). Lines between linkage groups represent shared markers. Boxes between linkage groups B7 and D7 include markers that could only be placed with a log-likelihood difference smaller than 2.00 by Mapmaker. Symbols surrounded by a square box represent genes for phenotypic traits: CBB: QTL for resistance to common bacterial blight, NN: QTL for *Rhizobium* nodulation (Nodari et al., 1993b); SW: QTL for 100-seed-weight (Johnson et al., 1996; Koinange et al., 1996); PL: QTL for pod length (Koinange et al., 1996); *P*: primary pigmentation gene; *Phs*: phaseolin seed protein (Gepts et al., 1993). These figures can also be downloaded from http://agronomy.ucdavis.edu/gepts/map data2.htm.

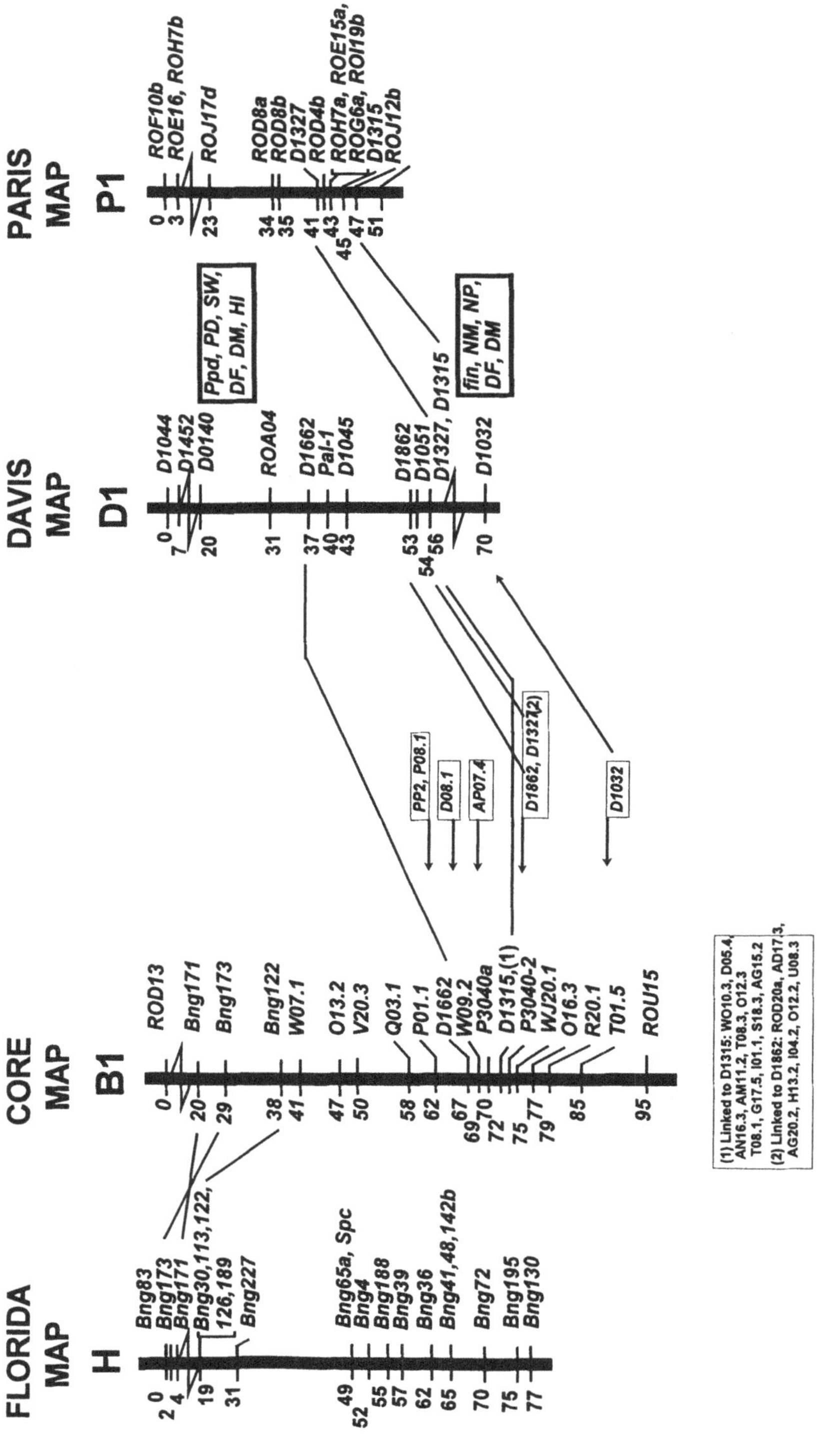
FLORIDA MAP
H
CORE MAP
B1
DAVIS MAP
D1
PARIS MAP
P1
ROF10b
ROE16, ROH7b
ROJ17d
ROD8a
ROD8b
D1327
ROD4b
ROH7a, ROE15a,
ROG6a, ROI19b
D1315
ROJ12b
0
3
23
34
35
41
43
45
47
51
Ppd, PD, SW,
DF, DM, HI
fin, NM, NP,
DF, DM
D1315
D1044
D1452
D0140
ROA04
D1662
Pal-1
D1045
D1862
D1051
D1327,
D1032
0
7
20
31
37
40
43
53
54
56
70
PP2, P08.1
D08.1
AP07.4
D1862, D1327(2)
D1032
ROD13
Bng171
Bng173
Bng122
W07.1
O13.2
V20.3
Q03.1
P01.1
D1662
W09.2
P3040a
D1315,(1)
P3040-2
WJ20.1
O16.3
R20.1
T01.5
ROU15
0
20
29
38
41
47
50
58
62
67
69
70
72
75
77
79
85
95
(1) Linked to D1315: WO10.3, D05.4, AN16.3, AM11.2, T08.3, O12.3
T08.1, G17.5, I01.1, S18.3, AG15.2
(2) Linked to D1862: ROD20a, AD17.3, AG20.2, H13.2, I04.2, O12.2, U08.3
Bng83
Bng173
Bng171
Bng30,113,122,
126,189
Bng227
Bng65a, Spc
Bng4
Bng188
Bng39
Bng36
Bng41,48,142b
Bng72
Bng195
Bng130
0
2
4
19
31
49
52
55
57
62
65
70
75
77

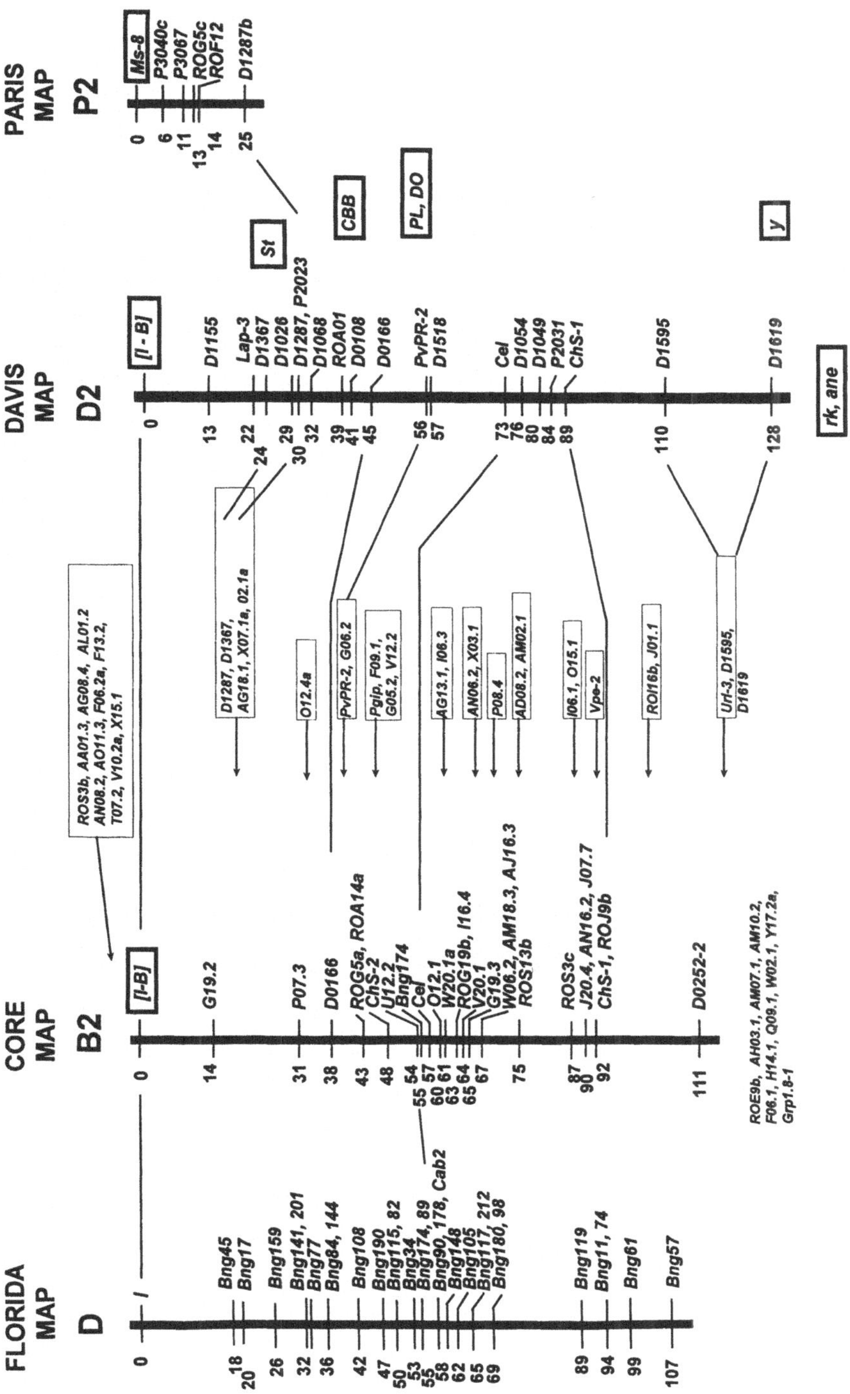

PARIS MAP
P2
Ms-8
P3040c
P3067
ROG5c
ROF12
D1287b
0
6
11
13
14
25
St
CBB
PL, DO
y
DAVIS MAP
D2
[I - B]
D1155
Lap-3
D1367
D1026
D1287, P2023
D1068
ROA01
D0108
D0166
PvPR-2
D1518
Cel
D1054
D1049
P2031
ChS-1
D1595
D1619
0
13
22
24
29
30
32
39
41
45
56
57
73
76
80
84
89
110
128
rk, ane
ROS3b, AA01.3, AG08.4, AL01.2
AN08.2, AO11.3, F06.2a, F13.2,
T07.2, V10.2a, X15.1
D1287, D1367,
AG18.1, X07.1a, 02.1a
O12.4a
PvPR-2, G06.2
Pgip, F09.1,
G05.2, V12.2
AG13.1, I06.3
AN06.2, X03.1
P08.4
AD08.2, AM02.1
I06.1, O15.1
Vpe-2
ROI16b, J01.1
Url-3, D1595,
D1619
CORE MAP
B2
[I-B]
G19.2
P07.3
D0166
ROG5a, ROA14a
ChS-2
U12.2
Bng174
Cel
O12.1
W20.1a
ROG19b, I16.4
V20.1
G19.3
W06.2, AM18.3, AJ16.3
ROS13b
ROS3c
J20.4, AN16.2, J07.7
ChS-1, ROJ9b
D0252-2
0
14
31
38
43
48
54
55
57
60
61
63
64
65
67
75
87
90
92
111
ROE9b, AH03.1, AM07.1, AM10.2,
F06.1, H14.1, Q09.1, W02.1, Y17.2a,
Grp1.8-1
FLORIDA MAP
D
I
Bng45
Bng17
Bng159
Bng141, 201
Bng77
Bng84, 144
Bng108
Bng190
Bng115, 82
Bng34
Bng174, 89
Bng90, 178, Cab2
Bng148
Bng105
Bng117, 212
Bng180, 98
Bng119
Bng11, 74
Bng61
Bng57
0
18
20
26
32
36
42
47
50
53
55
58
62
65
69
89
94
99
107

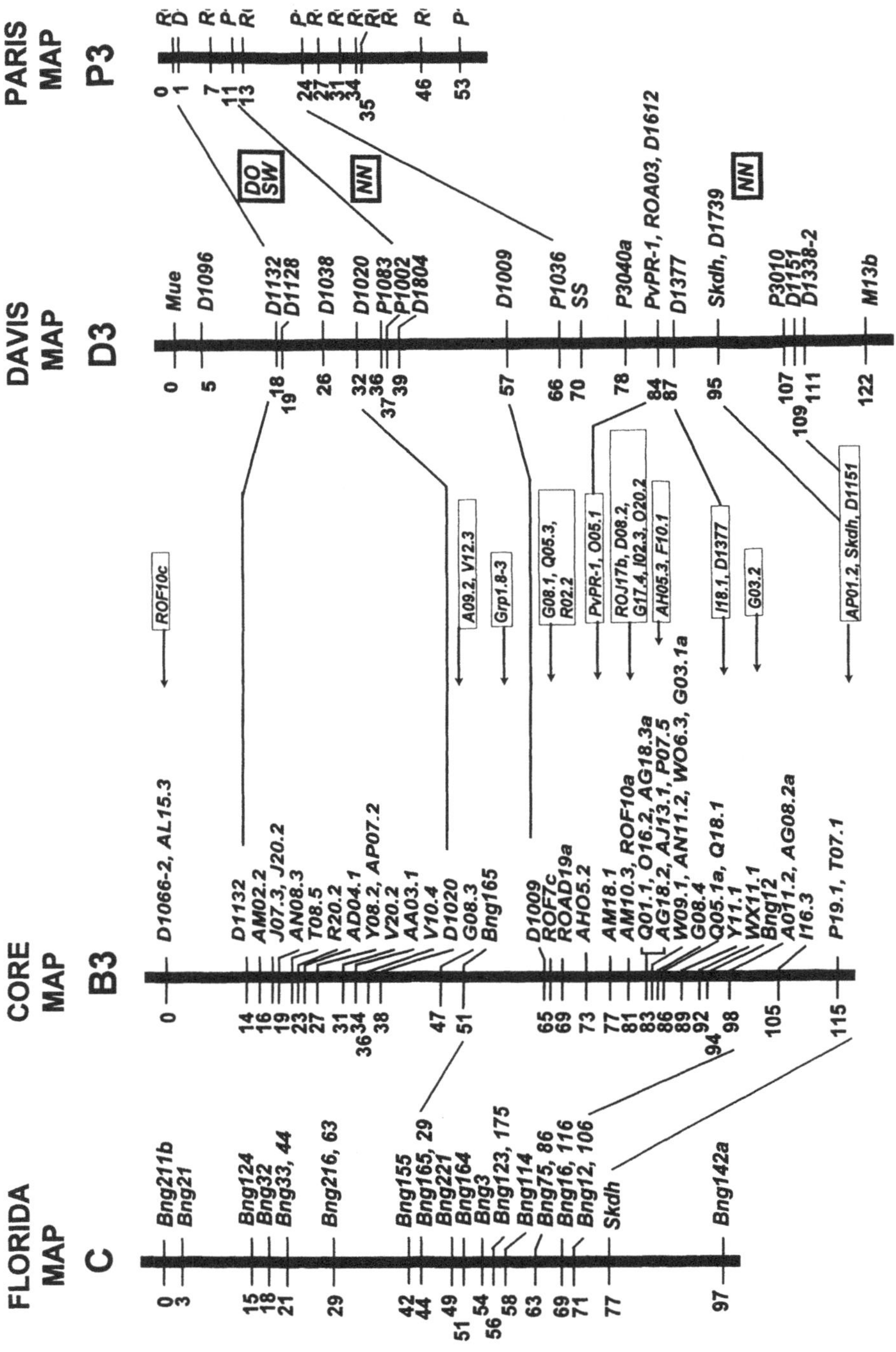
FLORIDA MAP
C
Bng211b
Bng21
Bng124
Bng32
Bng33, 44
Bng216, 63
Bng155
Bng165, 29
Bng221
Bng164
Bng3
Bng123, 175
Bng114
Bng75, 86
Bng16, 116
Bng12, 106
Skdh
Bng142a
CORE MAP
B3
D1066-2, AL15.3
D1132
AM02.2
J07.3, J20.2
AN08.3
T08.5
R20.2
AD04.1
Y08.2, AP07.2
V20.2
AA03.1
V10.4
D1020
G08.3
Bng165
D1009
ROF7c
ROAD19a
AHO5.2
AM18.1
AM10.3, ROF10a
Q01.1, O16.2, AG18.3a
AG18.2, AJ13.1, P07.5
W09.1, AN11.2, WO6.3, G03.1a
G08.4
Q05.1a, Q18.1
Y11.1
WX11.1
Bng12
A011.2, AG08.2a
I16.3
P19.1, T07.1
ROF10c
A09.2, V12.3
Grp1.8-3
G08.1, Q05.3, R02.2
PvPR-1, O05.1
ROJ17b, D08.2, G17.4, I02.3, O20.2
AH05.3, F10.1
I18.1, D1377
G03.2
AP01.2, Skdh, D1151
DAVIS MAP
D3
Mue
D1096
D1132
D1128
D1038
D1020
P1083
P1002
D1804
D1009
P1036
SS
P3040a
PvPR-1, ROA03, D1612
D1377
Skdh, D1739
P3010
D1151
D1338-2
M13b
DO
SW
NN
PARIS MAP
P3

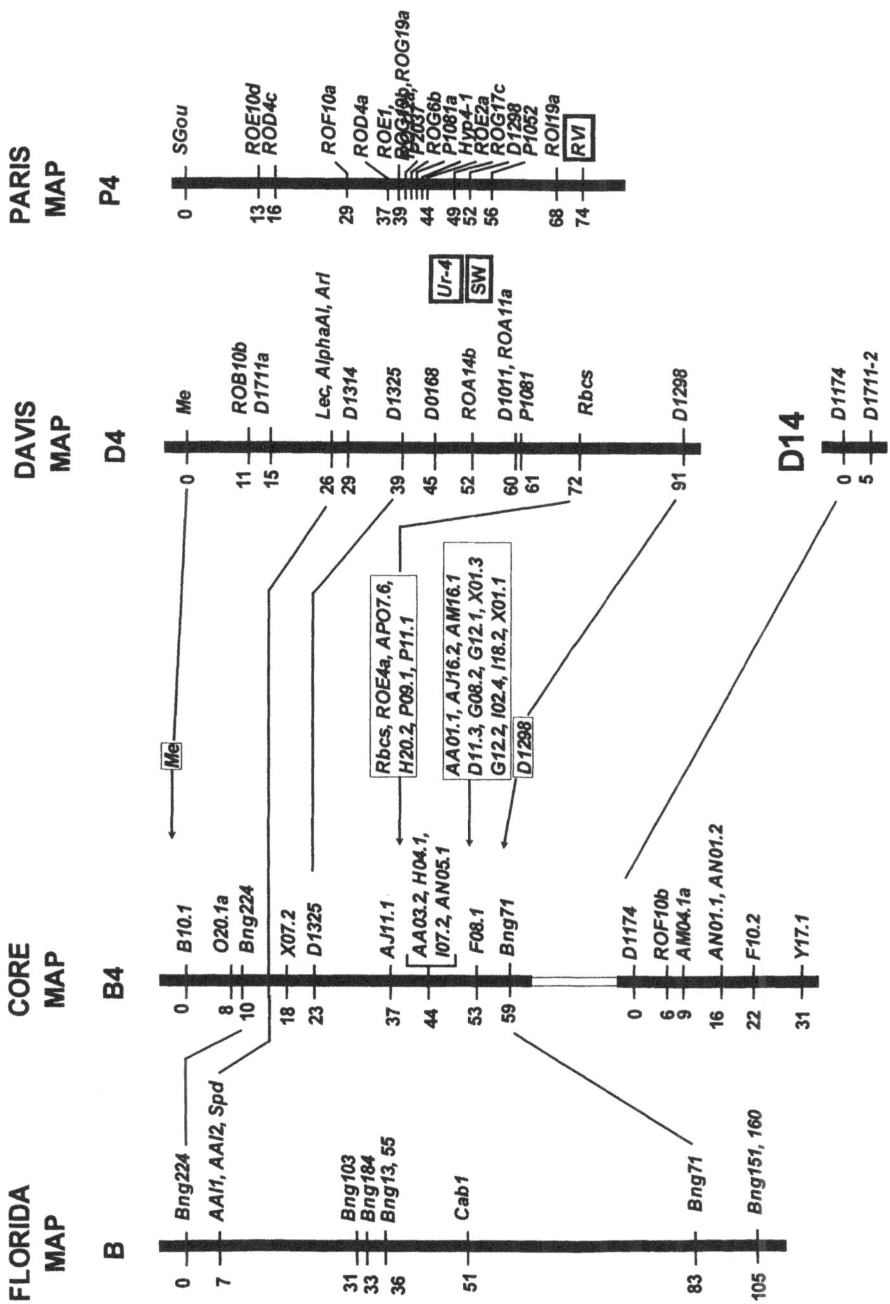
FLORIDA MAP
B
0 Bng224
7 AAI1, AAI2, Spd
31 Bng103
33 Bng184
36 Bng13, 55
51 Cab1
83 Bng71
105 Bng151, 160
CORE MAP
B4
0 B10.1
8 O20.1a
10 Bng224
18 X07.2
23 D1325
37 AJ11.1
44 AA03.2, H04.1, I07.2, AN05.1
53 F08.1
59 Bng71
0 D1174
6 ROF10b
9 AM04.1a
16 AN01.1, AN01.2
22 F10.2
31 Y17.1
Me
Rbcs, ROE4a, APO7.6, H20.2, P09.1, P11.1
AA01.1, AJ16.2, AM16.1, D11.3, G08.2, G12.1, X01.3, G12.2, I02.4, I18.2, X01.1
D1298
DAVIS MAP
D4
0 Me
11 ROB10b
15 D1711a
26 Lec, AlphaAl, Arl
29 D1314
39 D1325
45 D0168
52 ROA14b
60 D1011, ROA11a
61 P1081
72 Rbcs
91 D1298
Ur-4
SW
D14
0 D1174
5 D1711-2
PARIS MAP
P4
0 SGou
13 ROE10d
16 ROD4c
29 ROF10a
37 ROD4a
39 ROE1, ROG19b, ROG19a
44 P2037, ROG6b
49 P1081a
52 Hyp4-1, ROE2a
56 ROG17c, D1298, P1052
68 ROI19a
74 RVI

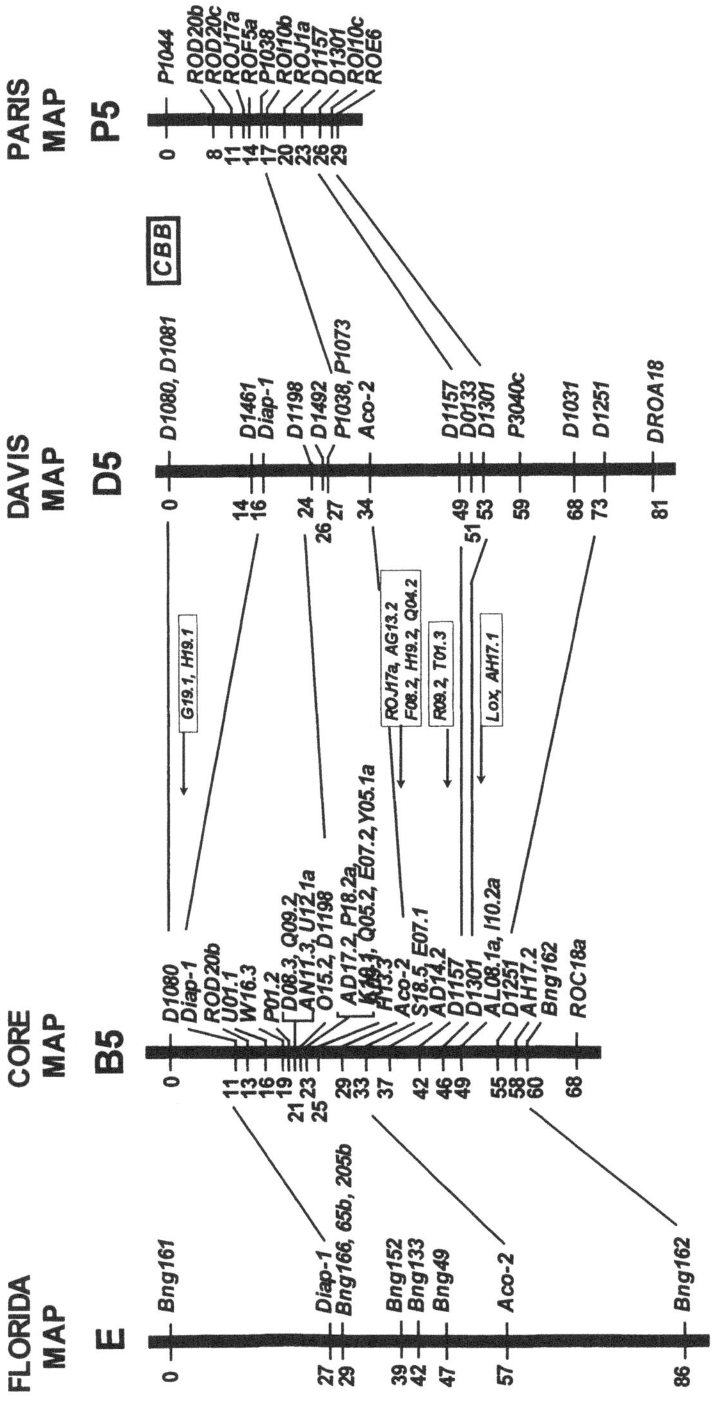
PARIS MAP
P5
DAVIS MAP
D5
CORE MAP
B5
FLORIDA MAP
E
CBB
G19.1, H19.1
ROJ17a, AG13.2
F08.2, H19.2, Q04.2
R09.2, T01.3
Lox, AH17.1

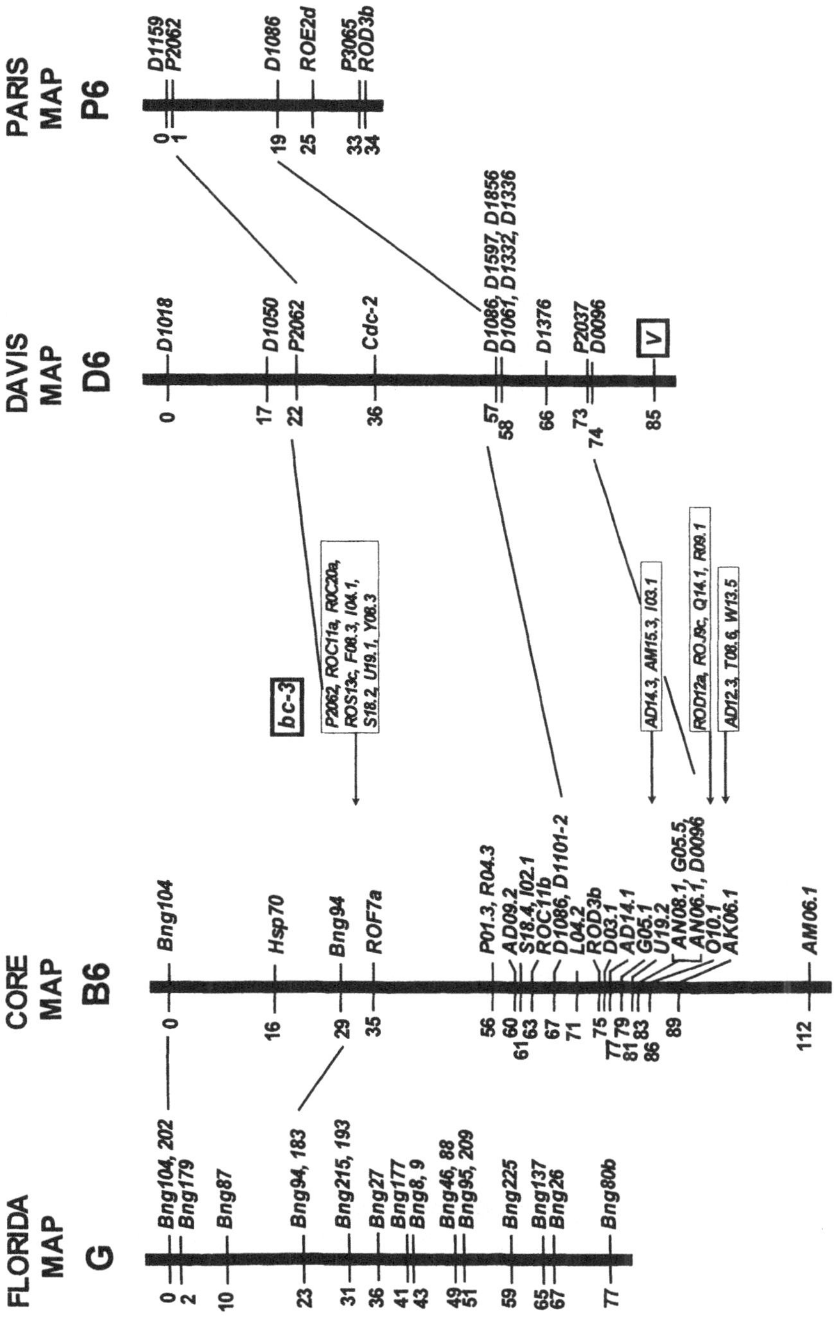
FLORIDA MAP G
0 Bng104, 202
2 Bng179
10 Bng87
23 Bng94, 183
31 Bng215, 193
36 Bng27
41 Bng177
43 Bng8, 9
49 Bng46, 88
51 Bng95, 209
59 Bng225
65 Bng137
67 Bng26
77 Bng80b
CORE MAP B6
0 Bng104
16 Hsp70
29 Bng94
35 ROF7a
56 P01.3, R04.3
60 AD09.2
61 S18.4, I02.1
63 ROC11b
67 D1086, D1101-2
71 L04.2
75 ROD3b
77 D03.1
79 AD14.1
81 G05.1
83 U19.2
86 AN08.1, G05.5,
89 AN06.1, D0096
O10.1
AK06.1
112 AM06.1
bc-3
P2062, ROC11a, ROC20a, ROS13c, F08.3, I04.1, S18.2, U19.1, Y08.3
AD14.3, AM15.3, I03.1
ROD12a, ROJ9c, Q14.1, R09.1
AD12.3, T08.6, W13.5
DAVIS MAP D6
0 D1018
17 D1050
22 P2062
36 Cdc-2
57 D1086, D1597, D1856
58 D1061, D1332, D1336
66 D1376
73 P2037
74 D0096
85 V
PARIS MAP P6
0 D1159
1 P2062
19 D1086
25 ROE2d
33 P3065
34 ROD3b

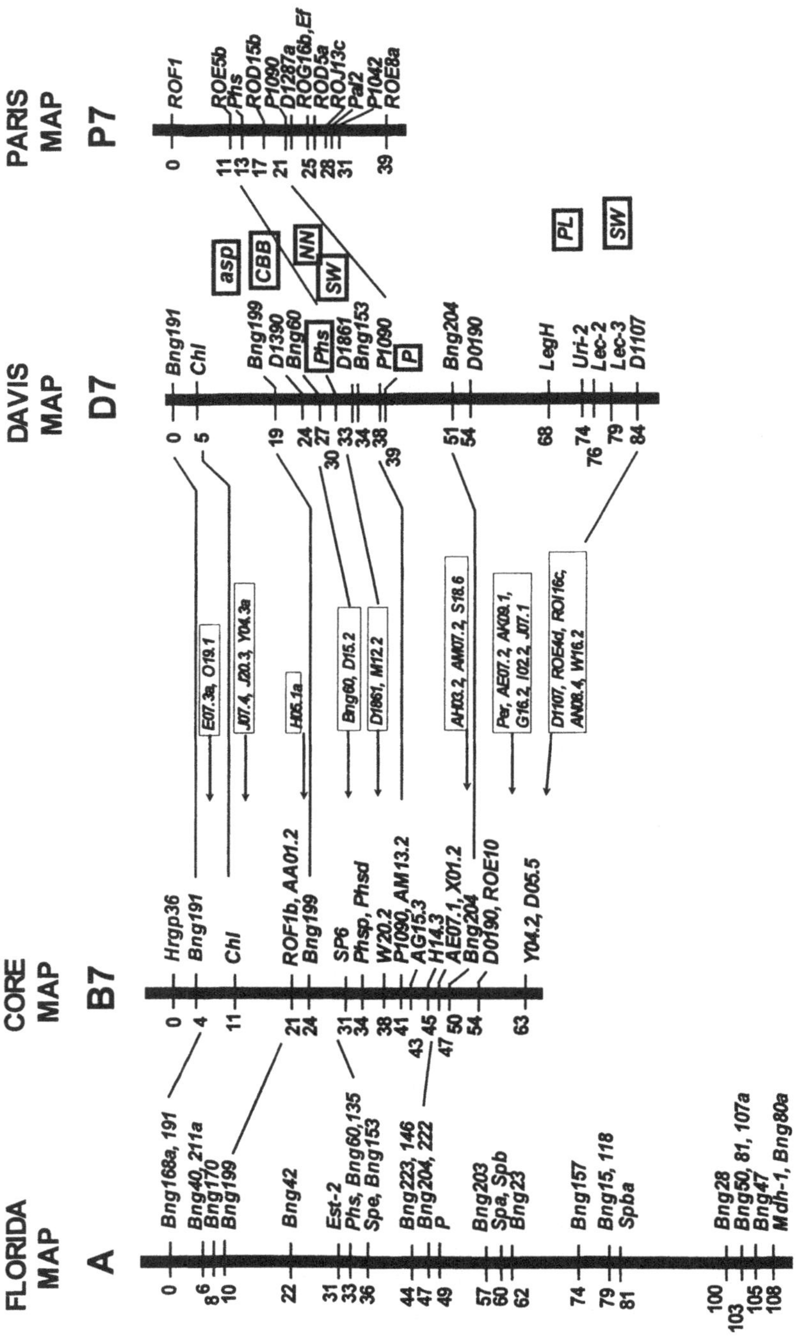
FLORIDA MAP
A
CORE MAP
B7
DAVIS MAP
D7
PARIS MAP
P7

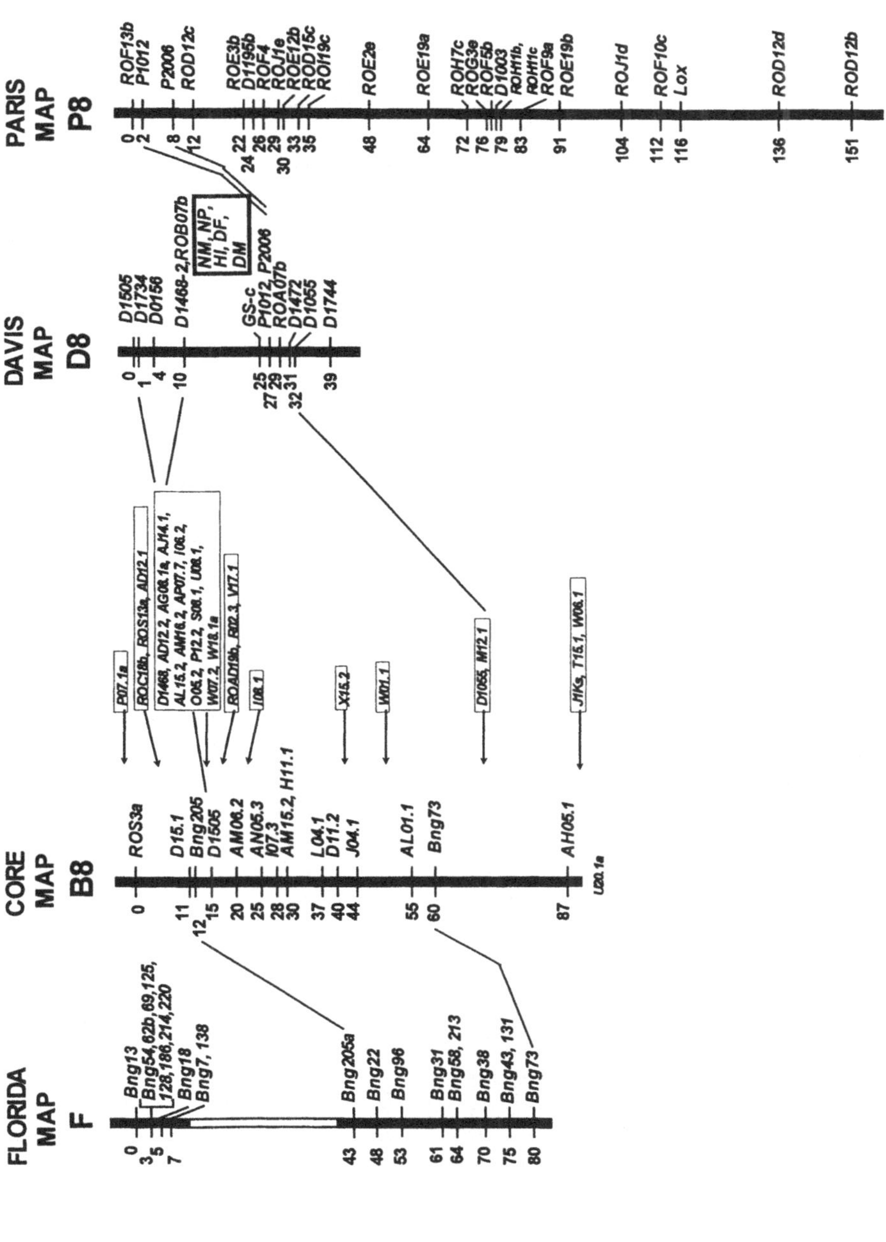
PARIS MAP P8
ROF13b
P1012
P2006
ROD12c
ROE3b
D1195b
ROF4
ROJ1e
ROE12b
ROD15c
ROI19c
ROE2e
ROE19a
ROH7c
ROG3e
ROF5b
D1003
ROF9a
ROE19b
ROJ1d
ROF10c
Lox
ROD12d
ROD12b
DAVIS MAP D8
D1505
D1734
D0156
D1468-2,ROB07b
NM, NP, HI, DF, DM
GS-c
P1012
P2006
ROA07b
D1472
D1055
D1744
CORE MAP B8
ROS3a
D15.1
Bng205
D1505
AM08.2
AN05.3
I07.3
AM15.2, H11.1
L04.1
D11.2
J04.1
AL01.1
Bng73
AH05.1
U20.1a
P07.1a
ROC18b, ROS13a, AD12.1
D1468, AD12.2, AG08.1a, AJ14.1,
AL15.2, AM16.2, AP07.7, I06.2,
O05.2, P12.2, S08.1, U08.1,
W07.2, W18.1a
ROAD19b, R02.3, V17.1
I08.1
X15.2
W01.1
D1055, M12.1
J1Ks, T15.1, W08.1
FLORIDA MAP F
Bng13
Bng54,62b,69,125,
128,186,214,220
Bng18
Bng7, 138
Bng205a
Bng22
Bng96
Bng31
Bng58, 213
Bng38
Bng43, 131
Bng73

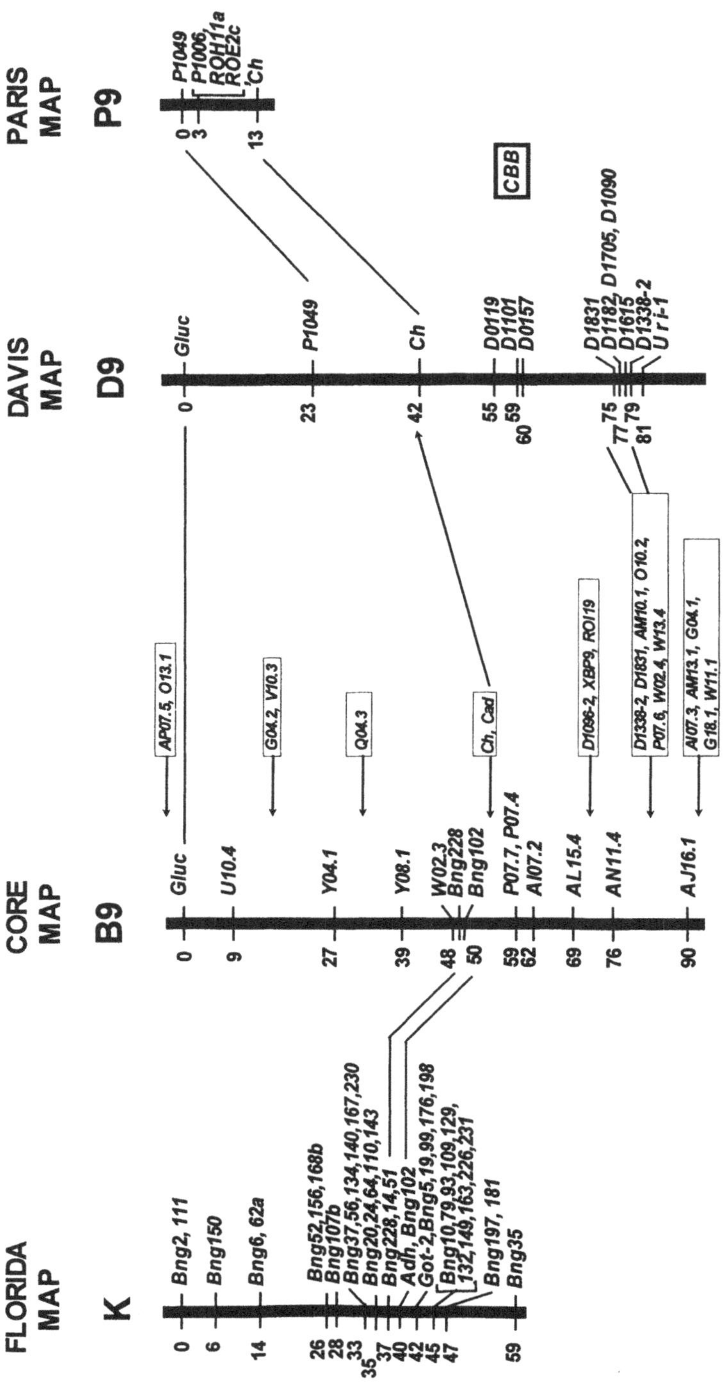
FLORIDA MAP
K
CORE MAP
B9
DAVIS MAP
D9
PARIS MAP
P9
CBB

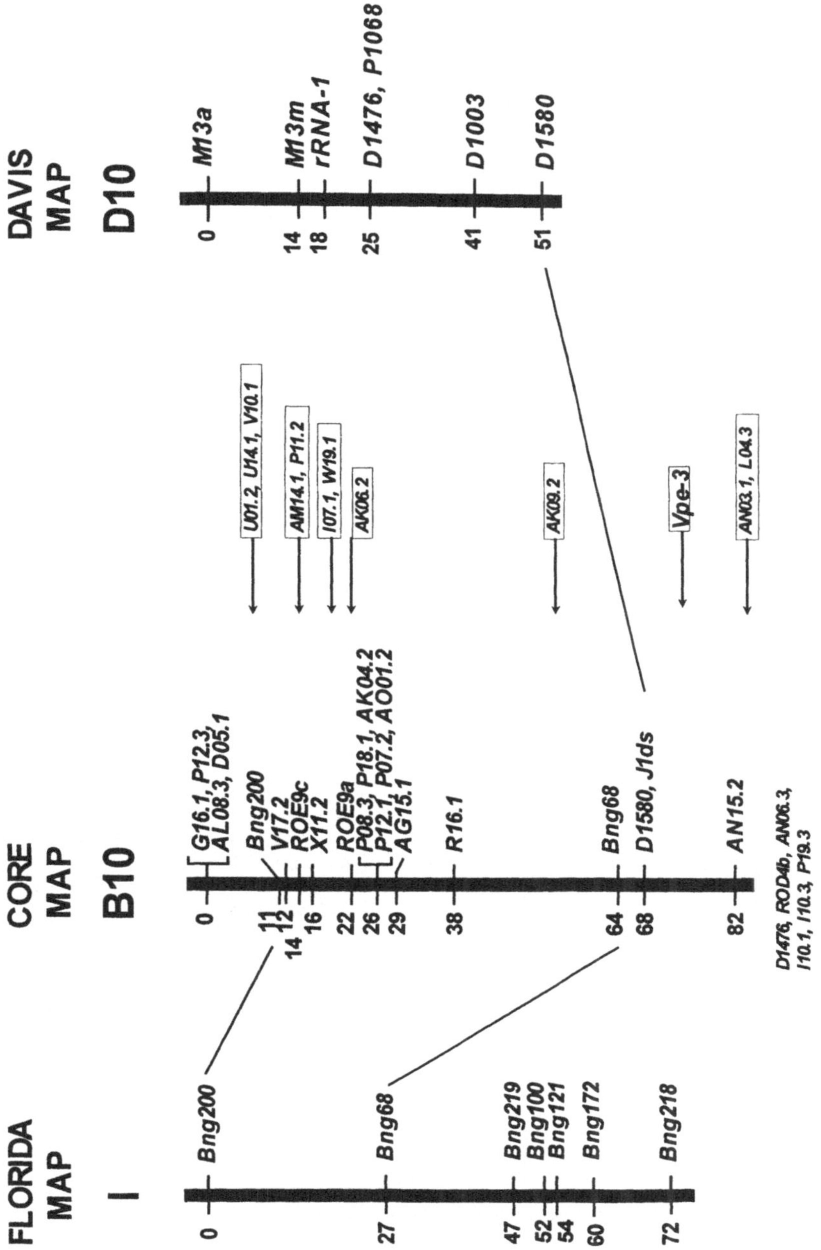
FLORIDA MAP
I
0 Bng200
27 Bng68
47 Bng219
52 Bng100
54 Bng121
60 Bng172
72 Bng218
CORE MAP
B10
0 G16.1, P12.3, AL08.3, D05.1
11 Bng200
12 V17.2
14 ROE9c
16 X11.2
22 ROE9a
26 P08.3, P18.1, AK04.2
P12.1, P07.2, AO01.2
29 AG15.1
38 R16.1
64 Bng68
68 D1580, J1ds
82 AN15.2
D1476, ROD4b, AN06.3, I10.1, I10.3, P19.3
U01.2, U14.1, V10.1
AM14.1, P11.2
I07.1, W19.1
AK06.2
AK09.2
Vpe-3
AN03.1, L04.3
DAVIS MAP
D10
0 M13a
14 M13m
18 rRNA-1
25 D1476, P1068
41 D1003
51 D1580

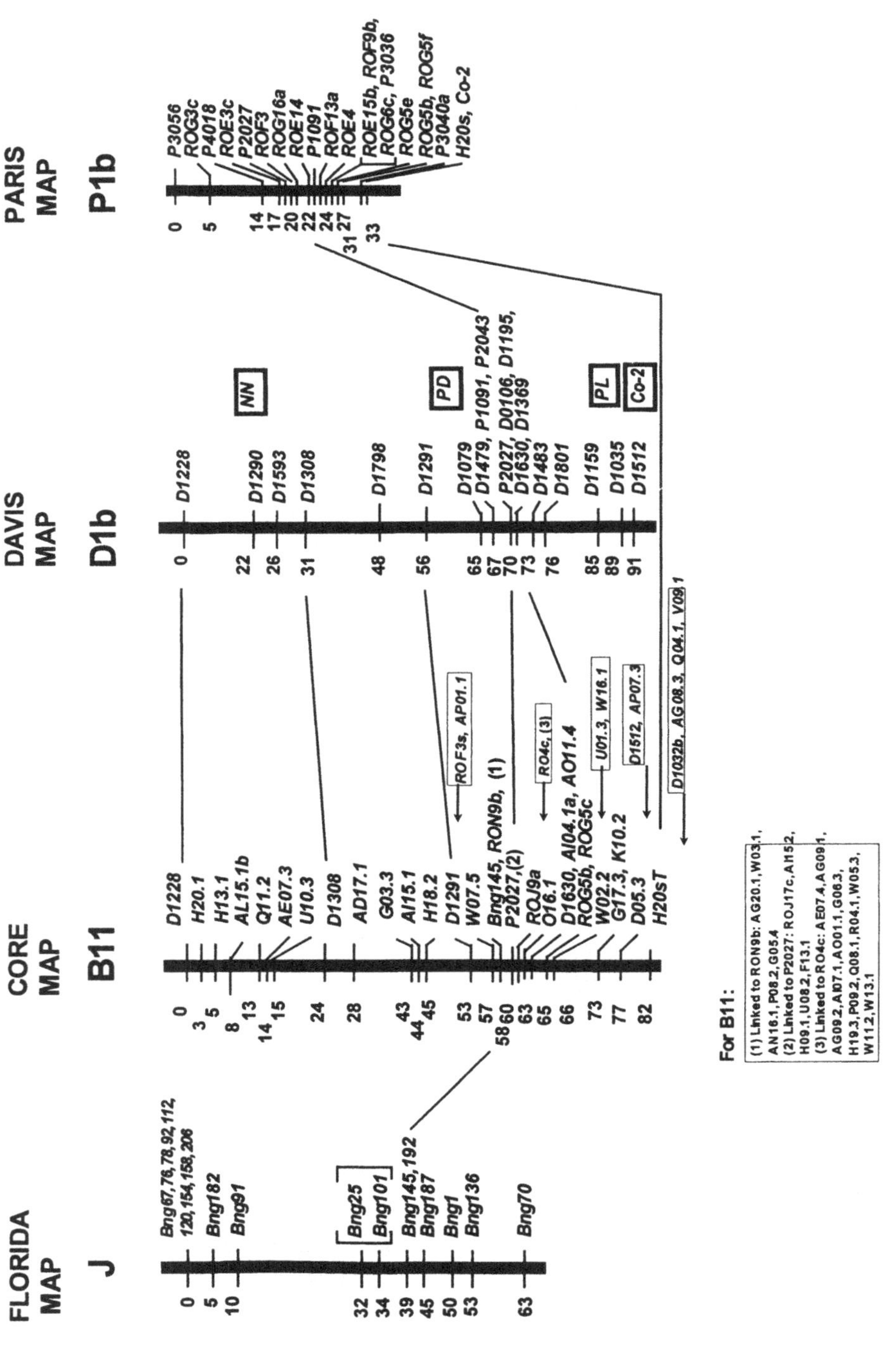
FLORIDA MAP
J
0 5 10 32 34 39 45 50 53 63
Bng67,76,78,92,112, 120,154,158,206
Bng182
Bng91
Bng25
Bng101
Bng145,192
Bng187
Bng1
Bng136
Bng70
CORE MAP
B11
0 3 5 8 13 14 15 24 28 43 44 45 53 57 58 60 63 65 66 73 77 82
D1228
H20.1
H13.1
AL15.1b
Q11.2
AE07.3
U10.3
D1308
AD17.1
G03.3
AI15.1
H18.2
D1291
W07.5
Bng145, RON9b, (1)
P2027,(2)
ROJ9a
O16.1
D1630, AI04.1a, AO11.4
ROG5b, ROG5c
W02.2
G17.3, K10.2
D05.3
H20sT
ROF3s, AP01.1
RO4c, (3)
U01.3, W16.1
D1512, AP07.3
D1032b, AG08.3, Q04.1, V091
DAVIS MAP
D1b
0 22 26 31 48 56 65 67 70 73 76 85 89 91
D1228
D1290
D1593
D1308
D1798
D1291
D1079
D1479, P1091, P2043
P2027, D0106, D1195,
D1630, D1369
D1483
D1801
D1159
D1035
D1512
NN
PD
PL
Co-2
PARIS MAP
P1b
0 5 14 17 20 22 24 27 31 33
P3056
ROG3c
P4018
ROE3c
P2027
ROF3
ROG16a
ROE14
P1091
ROF13a
ROE4
ROE15b, ROF9b,
ROG6c, P3036
ROG5e
ROG5b, ROG5f
P3040a
H20s, Co-2
For B11:
(1) Linked to RON9b: AG20.1, W03.1, AN18.1, P08.2, G05.4
(2) Linked to P2027: ROJ17c, AI152, H09.1, U08.2, F13.1
(3) Linked to RO4c: AE07.4, AG091, AG09.2, AI07.1, AO01.1, G06.3, H19.3, P09.2, Q08.1, R04.1, W05.3, W11.2, W13.1

Index

SPRINGER NATURE

GPSR Compliance

The European Union's (EU) General Product Safety Regulation (GPSR) is a set of rules that requires consumer products to be safe and our obligations to ensure this.

If you have any concerns about our products, you can contact us on ProductSafety@springernature.com

In case Publisher is established outside the EU, the EU authorized representative is:

Springer Nature Customer Service Center GmbH
Europaplatz 3
69115 Heidelberg, Germany

The manufacturer's authorised representative in the EU is Springer Nature Customer Service Centre GmbH, Europaplatz 3, 69115 Heidelberg, Germany. If you have any concerns regarding our products, please contact ProductSafety@springernature.com

Printed and bound by CPI Group (UK) Ltd, Croydon, CR0 4YY
25/09/2026
02222831-0006